Atomic and Molecular Processes with Short Intense Laser Pulses

NATO ASI Series

Advanced Science Institutes Series

A series presenting the results of activities sponsored by the NATO Science Committee, which aims at the dissemination of advanced scientific and technological knowledge, with a view to strengthening links between scientific communities.

The series is published by an international board of publishers in conjunction with the NATO Scientific Affairs Division

A	**Life Sciences**	Plenum Publishing Corporation
B	**Physics**	New York and London
C	**Mathematical and Physical Sciences**	D. Reidel Publishing Company Dordrecht, Boston, and Lancaster
D	**Behavioral and Social Sciences**	Martinus Nijhoff Publishers
E	**Engineering and Materials Sciences**	The Hague, Boston, Dordrecht, and Lancaster
F	**Computer and Systems Sciences**	Springer-Verlag
G	**Ecological Sciences**	Berlin, Heidelberg, New York, London,
H	**Cell Biology**	Paris, and Tokyo

Recent Volumes in this Series

Volume 165—Relativistic Channeling
edited by Richard A. Carrigan, Jr., and James A. Ellison

Volume 166—Incommensurate Crystals, Liquid Crystals, and Quasi-Crystals
edited by J. F. Scott and N. A. Clark

Volume 167—Time-Dependent Effects in Disordered Materials
edited by Roger Pynn and Tormod Riste

Volume 168—Organic and Inorganic Low-Dimensional Crystalline Materials
edited by Pierre Delhaes and Marc Drillon

Volume 169—Atomic Physics with Positrons
edited by J. W. Humberston and E. A. G. Armour

Volume 170—Physics and Applications of Quantum Wells and Superlattices
edited by E. E. Mendez and K. von Klitzing

Volume 171—Atomic and Molecular Processes with Short Intense Laser Pulses
edited by André D. Bandrauk

Volume 172—Chemical Physics of Intercalation
edited by A. P. Legrand and S. Flandrois

Series B: Physics

Atomic and Molecular Processes with Short Intense Laser Pulses

Edited by

André D. Bandrauk

University of Sherbrooke
Sherbrooke, Quebec, Canada

Plenum Press
New York and London
Published in cooperation with NATO Scientific Affairs Division

Proceedings of a NATO Advanced Research Workshop on
Atomic and Molecular Processes with Short Intense Laser Pulses,
held July 20-24, 1987,
in Lennoxville, Quebec, Canada

Library of Congress Cataloging in Publication Data

NATO Advanced Research Workshop on Atomic and Molecular Processes with Short Intense Laser Pulses (1987: Lennoxville, Québec)
Atomic and molecular processes with short intense laser pulses.

(NATO ASI Series. Series B, Physics; vol. 171)
"Proceedings of a NATO Advanced Research Workshop on Atomic and Molecular Processes with Short Intense Laser Pulses held July 20-24, 1987, in Lennoxville, Québec, Canada"—T.p. verso.
"Published in cooperation with NATO Scientific Affairs Division."
Includes bibliographical references and indexes.
1. Multiphoton processes—Congresses. 2. Laser pulses, Ultrashort—Congresses. 3. Coherence (Atomic physics)—Congresses. 4. Lasers in chemistry—Congresses. I. Bandrauk, André D. II. North Atlantic Treaty Organization. Scientific Affairs Division. III. Title. IV. Series: NATO ASI series. Series B, Physics; v. 171.
QC793.5.P427N38 1987 535′.2 88-2509

ISBN-13: 978-1-4612-8269-3 e-ISBN-13: 978-1-4613-0967-3
DOI: 10.1007/978-1-4613-0967-3

Softcover reprint of the hardcover 1st edition 1988
A Division of Plenum Publishing Corporation
233 Spring Street, New York, N.Y. 10013

PREFACE

This volume contains the lectures and communications presented at the NATO Advanced Research Workshop on "Atomic and Molecular Processes with Short Intense Laser Pulses" (NATO ARW 848/86). The workshop was held at Bishop's University, Lennoxville, Que, Canada, July 19-24, 1987, under the directorship of Prof. A.D. Bandrauk, Université de Sherbrooke. A scientific committee made up of Dr. P. Corkum (Laser Physics, National Research Council of Canada), Dr. P. Hackett (Laser Chemistry, National Research Council of Canada), Prof. S.C. Wallace (Dept. of Chemistry and Physics, University of Toronto), and Prof. F.H.M. Faisal (Fakultät für Physik, Universität Bellefeld) was called upon to invite and organize eminent lectures in the fields of

i) Coherence Phenomena in Atomic and Molecular Photoprocesses.
ii) High Intensity Atomic and Molecular Phenomena.
iii) Laser Chemistry

The aim of the workshop was to bring together chemists and physicists in order to discuss and analyze the progress made in the use of short intense laser pulses in understanding coherence phenomena and high intensity, nonlinear radiative effects in atomic and molecular systems.

Nonresonant multiphoton ionization was the predominant theme of the high intensity subsection of the conference. A new frontier of multiphoton ionization was identified - shorter pulses (femtosecond region) seem to imply inhibited ionization. The short pulse behaviour of multiphoton ionization is an important experimental uncertainty at the moment. At higher pressures, nonlinear optics are providing another diagnostic tool to investigate the suggested inhibited ionization with accompanying saturating nonlinearities. Theoretically, the ponderomotive potential and how to include it in a complete theory is a major problem. It is now becoming possible to perform fully time dependent wave function calculations of simple (1&3 dimensions) atomic models. This approach is welcomed since high power laser pulses as short as ten periods are beginning to be produced. It seems appropriate to begin investigating molecules, for which the nuclear degrees of freedom will present a new challenge.

Strong advances are being made in molecular energy transfer processes, where both experiment and theory are being fine-tuned to examine coherence retention during the course of a collision. New ways of altering a collision event with laser pulses shorter than a collision time are now being explored. Contrary to the atomic case discussed above, it is suggested that the rate of molecular transitions can be enhanced due to elimination of destructive interferences.

Substantial growing interest is being manifested by both experimentalists and theorists in the effects of controlled radiation fields for preparing and monitoring molecular states. It has now been demonstrated experimentally that pulse shaping and phase shifting is feasible on timescales of femto to nanoseconds. Applications of this laser pulse control are now in progress for harmonic generation, selective multiphoton excitation, spectral simplification, soliton generation and optical communication. Extensive theoretical work was presented to illustrate the possibility of field tailoring for local mode-selective excitation and laser control of chemical reactions. At the opposite extreme, noisy fields were shown to be also useful for broadband excitation in nonlinear spectroscopy. The discussion of the possibility of control or lack of control of atomic and molecular dynamics in the presence of a laser field brought out the importance of many fundamental concepts of dynamical systems: chaotic and regular spectra, localization and delocalization in phase space, coherence and incoherence, perturbative and nonperturbative regimes.

The above paragraphs essentially resumes the themes of the lectures assembled here into three sections: Coherence Phenomena, High Intensity Multiphoton Processes and Laser Chemistry.

In addition to the generous support of the NATO Scientific Affairs Division, we wish to acknowledge the contributions and cosponsorships of the workshop by the following organizations:

ALCAN International Ltd.
National Research Council of Canada
Natural Sciences and Engineering Research Council of Canada
Université de Sherbrooke
INRS-Energie(Que).

Without these contributions, the workshop on such a timely subject would not have been possible.

André D. Bandrauk
Sherbrooke, October 15, 1987

CONTENTS

COHERENCE PHENOMENA

HIGH INTENSITY MULTIPHOTON PROCESSES

LASER CHEMISTRY

LASER PULSE SHAPING FOR STATE-SELECTIVE EXCITATION

Warren S. Warren and Mark Haner
Department of Chemistry, Princeton University, Princeton, NJ 08544

Our research focuses on the design and application of pulse sequences with *enhanced control over the radiation field* (tailored phase and amplitude modulated pulses or phase shifted sequences) to produce simple and interpretable molecular states in a wide variety of applications (nuclear magnetic resonance, magnetic resonance imaging, and laser spectroscopy). Experimentally generating such "ultracoherent transient" laser pulse sequences with complete control of pulse shapes, phases, and delays previously presented a formidable challenge, but theoretical work has shown that such capabilities would enhance multiphoton pumping[1,2], selectively excite high vibrational levels of the electronic ground state[3], or measure contributions from transition electric dipole-dipole interactions in condensed phases[4]. There have even been recent calculations which show that a well-defined phase relation between multiple lasers[5] or shaped laser pulses[6,7] can promote state-selective chemistry. In this paper we will present the first results from an apparatus which lets us simultaneously tailor pulse shape and control phase shifts on a picosecond timescale, and compare the technological state-of-the-art in various laboratories to the level of sophistication necessary for experimental realization of these concepts.

Effects of Pulse Shaping and Phase Shifting

Obviously the excitation profile from a laser pulse depends critically on the pulse shape, since changing the shape changes the magnitudes of the frequency components. If the pulse is extremely weak linear response holds ($P'(\omega) = \chi E''(\omega)$), and the excited population is proportional to the intensity $I(\omega) = |E(\omega)|^2$. But more intense pulses and pulse sequences generate a much more complex behavior. A detailed discussion of the effects of pulse shaping in laser spectroscopy and magnetic resonance can be found in reference [8]. In addition, extremely sophisticated sequences of rectangular pulses have been technically possible in nuclear magnetic resonance for decades, and are routinely used to dramatically enhance pumping of forbidden transitions, remove the relaxation effects of intermolecular interactions, or increase spectral resolution. These effects cannot be understood by simply Fourier transforming the overall sequence because population in the ground state is strongly depleted. However, the mathematical framework needed to understand these sequences (a density matrix treatment) is well established and easily extended to describe optical multilevel systems.

From the point of view of possible extensions to laser spectroscopy, the most significant complication is that the successful applications in magnetic resonance relied on the inherent simplicity of spin systems. Phrased in the language of optical spectroscopists, protons have the same oscillator strength, all samples are optically thin, and the entire spectrum can be inverted by a single pulse. The reality in laser spectroscopy is very different:

1.Pulses with well-defined flip angles, like the initial $\pi/2$ pulse, rarely exist in molecular spectroscopy. The laser beam will inevitably have some transverse amplitude variation and pulse-to-pulse jitter, and the beam diameter will change over the volume of interest unless the beam is loosely focused. In addition, the exact <u>direction</u> of μ, and hence value of $\mu \cdot E$, depends on molecular orientation. In gas phase molecules such as I_2 (rotational constant = .037 cm^{-1}) many J states are thermally populated at room temperature, and since each state is (2J+1)-fold degenerate the strongest transitions typically have very large values of J' and J". Furthermore, unless there is no nuclear spin, we must consider F = I+J as a good quantum number. The (2F+1) components correspond to differerent values M of angular momentum along the polarization direction of an incident laser beam, and therefore correspond to different values of μ.E (for an R-branch transition with linearly polarized light, $\mu \cdot E$ is proportional to $(F^2-M_F^2)^{1/2}$) Thus even if the individual (v",F")→(v',F') transitions are resolvable (which will only be true in small molecules), the M_F states generate a large number of exactly degenerate transitions with different effective pumping rates. No single rectangular pulse can simultaneously be a $\pi/2$ or π pulse for all of these transitions.[9]

2.Useful sequences often must completely excite one transition, and must not generate excitation elsewhere. This is particularly difficult if relaxation times are comparable to the pulse width, or if the sample is optically dense so that the pulse shape is significantly altered.

Purely amplitude modulated pulses can solve the second problem; we have theoretically derived simple, symmetric, amplitude modulated pulses ("Hermite pulses") which completely excite a fairly rectangular distribution[10], and we have applied such pulses in a wide variety of spectroscopic applications[11-13]. Pulses which are simultaneously amplitude and phase modulated are capable of compensating for such practical complications. As an example, consider the pulse shape

$$E(T) = C\,((\mathrm{sech}(\alpha T))^{1+5i}) \exp(i\omega_o T)$$

where ω_o is the optical carrier frequency and α determines the width of the envelope. Since this envelope is a complex number the area is also complex; in fact the area is meaningless.

The effects of this pulse envelope was first partially understood theoretically in 1975[14] and first demonstrated as a radiofrequency pulse in 1984[15]. It generates an inversion which is localized and uniform, and remains almost completely unchanged as the radiofrequency peak power is varied over two orders of magnitude. In addition, we have recently shown that this pulse attenuates but does not substantially reshape as it propagates through an optically thick sample[16]; thus the inversion profile is nearly independent of optical density, since pulse attenuation does not affect the final state as noted above. Finally, we have shown that the inversion profile from this pulse shape is highly insensitive to T_2 relaxation effects. No pulse shape or sequence has ever before been shown to compensate for a real T_2 relaxation; many sequences refocus static intermolecular interactions of known operator symmetry which would otherwise serve to reduce transverse magnetization in NMR, but this is fundamentally different from compensation for T_2 effects such as relaxation from fluctuating fields of unknown symmetry.

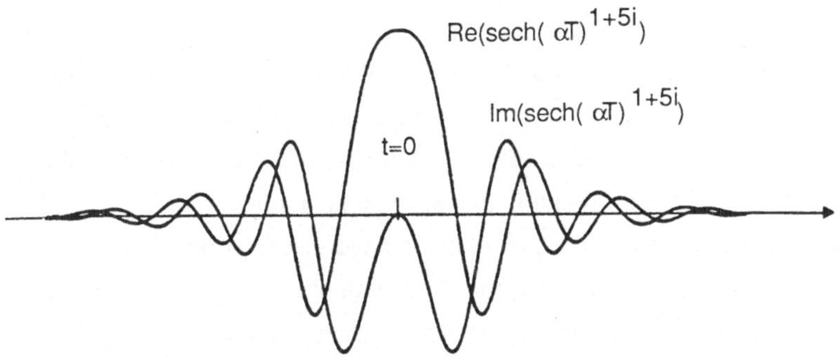

Figure 1. Real and imaginary components of the pulse envelope $\mathrm{sech}(\alpha T)^{1+5i}$. View the real part as modulating one laser beam, and the imaginary part as modulating a second beam at the same frequency but 90° out of phase with the first.

Experimental Generation of Shaped Laser Pulses

Our laser pulse shaping work is conveniently divided into two timescales: nanosecond and picosecond. Nanosecond resolution pulse shaping is accomplished by storing the desired envelope in fast memory chips, clocking out these chips to give voltage pulses, applying the voltage pulses to r.f. modulators such as double balanced mixers to make shaped radiofrequency pulses (phase modulated pulses are created by generating the in-phase and in-quadrature components of the pulse envelope from two different shapers), and then applying the shaped radiofrequency pulse to an acousto-optic modulator along with a single frequency laser[9]. We have used this approach to create amplitude modulated pulses to monitor collisional dynamics[13], and have recently demonstrated laser pulses with the sech$(\alpha T)^{1+\mu i}$ shape on a timescale of roughly 100 nanoseconds[17]. Even though the modulated hyperbolic secant pulse is very complicated, a sixteen-point approximation keeps the important feature of compensation for laser inhomogeneity over roughly a factor of fifteen. We demonstrated that this shape does indeed compensate for laser inhomogeneity by applying it to the (V"=14, J"=60) transition in I_2. This has 119 degenerate M_J states, 61 different values of $\mu \cdot E$, and 15 hyperfine components, so the total fluoresence from a rectangular pulse shows no hint of Rabi oscillations or leveling when the laser power increases. A hyperbolic secant pulse generates much less power dependence to the fluorescence, since most of the degenerate transitions see a large inversion which is nearly power independent. Such pulses invert nearly all the molecules in their path.

Faster pulse shaping requires completely different technologies. In the last year we have demonstrated technological advances which permit *voltage adjustable*, arbitrarily phase and amplitude modulated pulses with megawatt peak powers.[18-21] One ps resolution has been demonstrated experimentally, and our more recent theoretical work predicts 100 fs resolution with our existing devices. In essence, we have designed advanced circuitry which makes an arbitrarily shaped microwave pulse with better than 20 GHz bandwidth, using GaAs FETs or high electron mobility transistors (HEMTs). We have also built an electro-optic amplitude modulator which takes the microwave pulses, plus a continuous laser input, and creates an arbitrarily shaped laser pulse output with roughly 10 picoseconds resolution. We have also demonstrated pulse amplification to megawatt peak powers without significant distortion. Finally, we have experimentally demonstrated that fiber optic pulse compression can reduce the pulse length by at least a factor of ten with predictable envelope changes, and have written programs which predistort the input pulse shape to give any particular desired output.

Picosecond Phase Shifting

Picosecond lasers have frequently been used to make sequences such as photon echoes, but since the repetition rate of such lasers is very low sequences are generated by splitting a single pulse, passing the two pieces over unequal arms of an interferometer, and then recombining. The relative phase as well as the delay between the two combined pulses will then be determined by the exact path length difference: if, for example, two pulses with a $\pi/2$ path length difference and a 1 ns delay are desired, then the path length difference D($\approx$300 mm) must be $(N+1/4)\lambda$, where λ is the carrier frequency of the optical pulses. This implies $N \approx 500{,}000$. Accuracy of ± 0.1 rad for $\lambda = 600$ nm requires knowledge of the pulse carrier frequency to ± 15MHz, which is orders of magnitude less than a typical picosecond pulse bandwidth; thus phase shifted picosecond pulse sequences have only been generated with very short delays[22]. In fact, we recently showed that the concept of phase shifted pulse sequences, in the sense used in NMR, is only meaningful for purely amplitude modulated pulses and probably cannot be generalized to use the output of a normal pulsed laser[23].

We have demonstrated two different laser systems which permit picosecond pulse sequence generation with complete phase control. The first approach used transverse pumping of a dye laser cavity which was previously injected by the output of a ring dye laser[23,24]. Detailed theoretical calculations showed that the laser output would be collinear with the ring laser light and purely amplitude modulated if the injected laser was modematched to the ring laser. This collinearity let us use interference fringes from the cw light to determine the correct phase relation (for example, destructive interference in the interferometer will correspond to a π phase shift, and will generate minimum fluorescence in a two-pulse sequence). This was verified experimentally in iodine. We also showed that frequency mismatch between the injected laser and the ring causes minimum fluorescence to occur when the cw light had a different phase relation[23], as was predicted theoretically[25].

Our more recent work combines pulse shaping and phase shifting into one simple apparatus (Figure 2). The cw laser light partially bypasses the modulator, and is recombined collinearly with the shaped pulses (the required accuracy is about 0.1 mrad). Fringes determine the interpulse phase relation in this approach as well. Experimental demonstration of the effects of phase shifted pulse sequences is shown in Figure 3. We compare two-pulse sequences of Gaussian and Hermite pulses in iodine. The bandwidth of the pulses is about three times larger than the inhomogeneous linewidth. In each case the minimum fluorescence comes at a phase shift of π, indicating that most of the molecules are near enough to exact resonance for the second pulse to undo the effects of the first. The Hermite pulses have a more uniform excitation profile, so molecules which are far from the center of the inhomogeneous

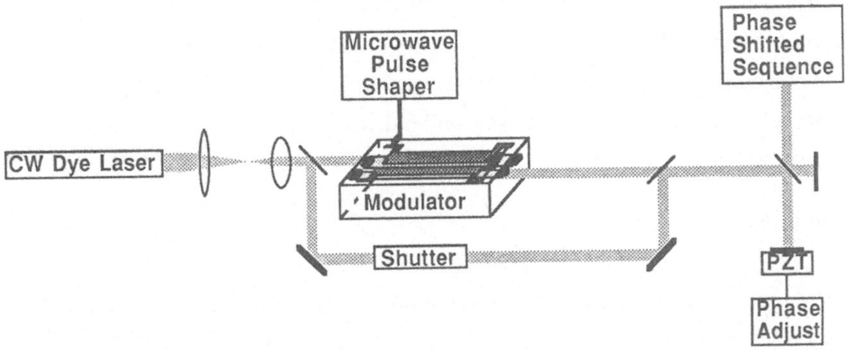

Figure 2.Experimental apparatus for generation of shaped picosecond pulse sequences with interpulse phase control. Fringes on the cw light determine the phase relationship.

line see an excitation which is similar to molecules nearer the center. This gives a more complete null. This approach can be generalized to create an arbitrarily complex pulse sequence.

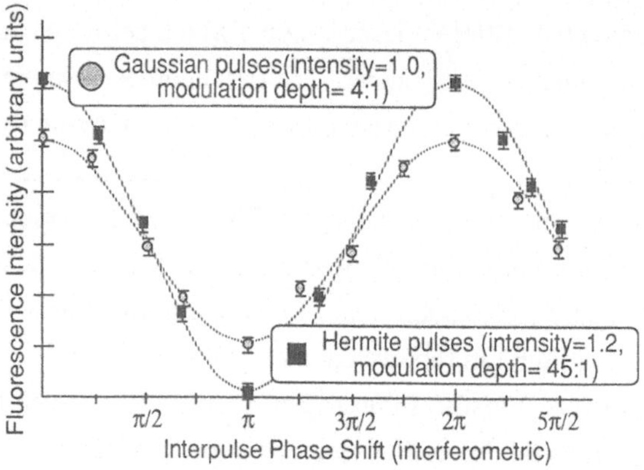

Figure 3. Effects of two-pulse Gaussian or Hermite sequences on fluorescence.

Comparison of Pulse Shaping Techniques

Picosecond and subpicosecond resolution pulse shaping was first accomplished by Heritage's group at Bellcore using a quite different approach[26]. They use spatial amplitude and phase masks inside the grating pair of a fiber-grating pulse compressor to tailor pulse shapes in the frequency domain, as opposed to our work which is fundamentally in the time domain. Their work published to date has required specifically fabricated high resolution amplitude and phase masks, but it seems likely that programmability can be accomplished as well, at least for the amplitude masking. Time domain and frequency domain pulse shaping have complementary advantages and disadvantages, and either might be preferred depending on the specific experimental circumstances:

Single Pulse Excitation

Symmetric, Amplitude Modulated Pulses A symmetric amplitude modulated pulse has a real symmetric Fourier transform. Thus the only values of phase needed for a phase mask are 0 and π. Their phase masking to date has been accomplished with pellicles with holes, which provide such phase shifts in a straightforward way. Extremely simple pulse shapes, such as Gaussian or Hermite pulses, do not have any zero crossings in the frequency domain. In this case frequency domain pulse shaping has the clear advantage of shorter pulse generation, while time domain pulse shaping would be preferred if high frequency selectivity is important.

Nonsymmetric, Amplitude Modulated Pulses The Fourier transform of such pulse shapes is complex, and thus frequency domain pulse shaping requires a phase mask with continuous variations from 0 to 2π. This will require $\approx$50 nm resolution thickness profiling unless it can be programmed as well. Time domain pulse shaping is not fundamentally different for symmetric or asymmetric pulses, and is probably easier in this case.

Dual Color or Multicolor Pulses This can be achieved by masking, and is far easier to do in the frequency domain.

Phase and Amplitude Modulated Pulses Our travelling wave electro-optic modulators are fundamentally amplitude modulators. We can create phase and amplitude modulated pulses by adding a phase modulator, or by modifying the apparatus in Figure 3 slightly to combine two different amplitude modulated pulses with a $\pi/2$ phase shift[27]. We have used the first approach to generate pulses which propagate without shape distortion in optical fibers[28]. The second approach is more appropriate for pulses such as $\mathrm{sech}(\alpha T)^{1+5i}$, which can also be viewed as a frequency swept pulse and would require astronomically large phase shifts.

Creating phase and amplitude modulated pulses in the frequency domain is not fundamentally harder than creating nonsymmetric amplitude modulated pulses. Again there is a question of pulse shape complexity: the phase variations required to create $\mathrm{sech}(\alpha T)^{1+5i}$ are enormous.

Pulse Sequences

Several identical pulses, no phase control This is equally easy with either technology: it requires beamsplitters and mirrors configured into an optical delay line.

Several identical pulses, phase control This is straightforward with the time domain pulse shaping (we start with a cw laser and can use fringes to adjust phase, as in Figures 2 and 3) but requires measurement of the exact carrier frequency of the pulses in the frequency domain approach, which fundamentally starts with a pulsed laser.

Several different pulses In the frequency domain approach this will require multiple masks, because it will be difficult to reprogam a mask in the nanosecond or picosecond delay between pulses. This is possible in the time domain, because we can apply two different shaped microwave pulses to the same modulator.

It should also be recognized that it would be possible to combine these approaches (use an electro-optic modulator plus phase and amplitude masks), which could permit the flexibility to create very sophisticated shapes.

Summary

Laser pulse shaping and phase pulse sequence generation has been experimentally demonstrated with subpicosecond resolution, high powers, and complete phase control out to a nanosecond timescale. The technological developments have in some ways outrun theoretical work: the applications which have been demonstrated to date only hint at the potential. Nonetheless, it is quite clear that this kind of radiation field control, coupled with ongoing theoretical work, can establish a new generation of optical coherent transient techniques as invaluable tools for chemical physics.

This research was supported by the National Science Foundation under grant CHE-8405944. W. S. W. holds a Sloan Foundation Fellowship.

Bibliography

1. J-C. Diels and S. Besnainou, J. Chem. Phys. 85, 6347 (1986)
2. W.S. Warren and A. H. Zewail, J. Chem. Phys. 78, 3583 (1983)
3. J. Bates and W. S. Warren, Advances in Laser Science I (W. Stwalley and M. Lapp, editors, American Institute of Physics, vol. 146, New York, 1986) p.429
4. W.S. Warren and A. H. Zewail, J. Chem. Phys. 78, 2298 (1983)
5. P. Brumer and M. Shapiro, Chem. Phys. Lett. 126, 541(1986)
6. D. J. Tannor and S. A. Rice, J. Chem. Phys. 83, 5013(1985)
7. H. Rabitz, private communication; H. Rabitz, paper in these proceedings
8. W. S. Warren and M. Silver, Adv. Mag. Res. (in press)
9. W.S. Warren, J. Bates, M. McCoy, M. Navratil and L. Mueller, J. Opt. Soc. Am. B3, 488 (1986)
10. W. S. Warren, J. Chem. Phys. 81, 5437(1984)
11. F. Loaiza, K.-T. Lin, W. S. Warren, M. Silver, H. Egloff, G. Laub and B. Kiefer, Health Care Instrum. 1, 188(1986)
12. J. Gutow, M. McCoy, F. Spano and W. S. Warren, Phys. Rev. Lett. 55, 1090(1985)
13. M. A. Banash and W. S. Warren, Laser Chemistry 6, 47(1986)
14. L. Allen and J. H. Eberly, Optical Resonance and Two-Level Atoms (Wiley, New York, 1975)
15. M. S. Silver, R. I. Joseph and D. I. Hoult, Phys. Rev. A31, 2753 (1985)
16. C. P. Lin, J. Bates, J. Mayer and W. S. Warren, J. Chem. Phys. 86, 3750(1987)
17. F. Spano and W. S. Warren, Phys. Rev. A (submitted)
18. M. Haner, F. Spano and W. S. Warren, Ultrafast Phenomena V (G. Fleming and A. Siegman, eds; Springer, Berlin, 1986), p.514
19. M. Haner and W. S. Warren, Applied Optics (in press)
20. "Princeton Team Develops Programmable ps Pulses", Lasers and Applications, April 1987, p.24-26
21. M. Haner and W. S. Warren, Opt. Lett. 12, 398(1987)
22. A. Mukherjee, N. Mukherjee, J.-C. Diels and G. Arzumanyan, Ultrafast Phenomena V(G. Fleming and A. Siegman, editors; Springer, Berlin, 1986) p.266
23. F. Spano, M. Haner and W. S. Warren, Chem. Phys. Lett. 135,97(1987)
24. F. Spano, F. Loaiza, M. Haner and W. S. Warren, Ultrafast Phenomena IV ,99(D. Auston and K. Eisenthal, editors;Springer, New York,1984)

25. F. Spano and W. S. Warren, Laser Applications to Chemical Dynamics(M.A.El-Sayed, ed.)Proc. SPIE 620, 52(1986)

26. A. M. Weiner, J. P. Heritage and R. N. Thurston, Opt. Lett. 11, 153(1986); Opt. Lett. 10, 609(1985); J. P. Heritage, private communication

27. W. S. Warren, Laser Applications to Chemical Dynamics(M. A. El-Sayed,ed.) Proc. SPIE 742, 42(1987).

28. M. Haner and W. S. Warren, Phys. Rev. Lett. (submitted)

QUANTITATIVE MEASUREMENT OF THE ROLE OF PHASE FLUCTUATIONS IN NONLINEAR OPTICAL ABSORPTION PROCESSES

Stephen J. Smith
Joint Institute for Laboratory Astrophysics
University of Colorado and National Bureau of Standards
Boulder, CO 80309-0440

D. S. Elliott
School of Electrical Engineering
Purdue University, W. Lafayette, IN 47907

INTRODUCTION

The role played by fluctuations in optical absorption processes is quite varied and interesting because of the extent to which higher order processes are involved and the dependence of different processes on the intensity of the optical field. At very low intensities it may be sufficient to consider only linear absorption: single photon processes at rates so low that state populations are essentially unperturbed. In this case the role of the optical field is completely representable in terms of its power spectrum (the spectral density as a function of frequency). Higher order statistical properties play no role.

Equivalent to the power spectrum, the lowest order electric field correlation function, $R_E(\tau)$ can be used to represent the radiation field in the linear absorption case:

$$R_E(\tau) = \frac{1}{2} < E(t)\, E^*(t+\tau) > \quad . \tag{1}$$

According to the well-known Wiener-Khintchine theorem the lowest order correlation function is just the Fourier transform of the power spectrum, and vice versa. Linear absorption can be fully described in terms of either of these field representations.

On the other hand, characterization of higher order (nonlinear) absorption processes, which may dominate at high intensities of the radiation field, or for multiphoton processes even at low intensity, requires knowledge of higher order correlation functions. Knowledge of the power spectrum does not suffice. For a theoretical representation of two-photon absorption processes one must know not only the lowest order correlation function (from the power spectrum) but also the next higher order correlation function. For saturated absorption in general, one must know correlation functions to all orders.

That this is true has long been recognized by many theorists involved in these studies. On the experimental front, it has only been recently that techniques have been developed that let us generate a field with known higher order correlations, necessary to carry out a completely characterized measurement of a nonlinear optical process. Typically, the laser fields used to carry out such measurements have not been fully statistically characterized in this sense. The spectral width was usually known, approximately; and the shape of the power spectrum may have been known, approximately; but in the past higher correlation functions have not been controlled or determined.

The Gaussian Random Process

Our program is to develop methods of generating a laser field with a fully characterized spectrum of fluctuations,[1,2] and to apply this field to the study of selected nonlinear processes. Our method is based on the use of very fast optical modulators to impose phase/frequency fluctuations derived from a noise voltage generator on an essentially "monochromatic" laser beam. The laser field generated by this technique can be represented in the form

$$E(t) = E_0 \, e^{-i[\omega_0 t + \phi(t)]} \qquad (2)$$

where E_0 and ω_0 are the constant field amplitude and average field frequency, respectively, and $\phi(t)$ is the fluctuating phase of the field. The fluctuating phase and the fluctuating frequency, $\dot{\phi} = d\phi/dt$, are characterized by the Gaussian Random Process, discussed in the classic works of Rice,[3] and of Wang and Uhlenbeck.[4] Some principal properties are:

1) Correlation functions of all orders depend only on time differences (stationary property). The fluctuations contain no signal.

2) The ensemble average and the time average are interchangeable (ergodic hypothesis).

3) Odd order correlation functions are zero.

4) Higher even correlation functions are constructed from the lowest order correlation function

$$\langle \dot{\phi}(t_1)\, \dot{\phi}(t_2) \cdots \dot{\phi}(t_{2n}) \rangle$$

$$= \sum_{\substack{\text{all} \\ \text{permutations}}} \langle \dot{\phi}(t_i)\, \dot{\phi}(t_j) \rangle \cdots \langle \dot{\phi}(t_k)\, \dot{\phi}(t_\ell) \rangle . \qquad (3)$$

It follows from Eq. (3) that all correlation functions are derivable from the power spectrum (because of the Wiener-Khintchine theorem) if the fluctuations are a Gaussian Random Process.

This is the basis of our technique. The traversal of electrons across the p-n junction of a reversed-biased avalanche diode produces a fluctuating voltage commonly called "shot-noise." This consists of a Poisson distribution of voltage pulses. In the limit of large currents the Poisson distribution becomes a Gaussian Random Process.[3] The next step consists of the conversion of these radio-frequency voltage fluctuations to rf frequency fluctuations using standard linear frequency modulation techniques, which in turn, are imposed on a nearly monochromatic optical frequency laser beam using fast acousto-optic modulators. Random phase fluctuations can be imposed on the laser beam by applying the noise voltage directly to an electro-optic phase modulator.

Noise Spectrum Imposed on a Carrier

The theory required to understand and describe what happens when this rf noise spectrum is imposed on a monochromatic optical frequency carrier was worked out by Middleton.[5] The lowest-order correlation function for the optical field is obtained from equations (1) and (2),

$$R_E(\tau) = \frac{E_0^2}{2} \cos\omega_0\tau \exp[-\frac{1}{2}\int_{-\tau}^{\tau} R_\omega(t) \left(\tau - |t|\right) dt] \tag{4}$$

where $R_\omega(t)$ is the correlation function of the rf noise. The Fourier transform of (4) then gives the power spectrum of the laser field

$$P_E(\omega) = \int_{-\infty}^{\infty} e^{-i\omega\tau} R_E(\tau)\, d\tau \tag{5}$$

in terms of the correlation function $R_\omega(t)$ of the rf noise imposed on it by the fast modulators.

The correlation function of the rf phase/frequency noise has, in our case, a particularly simple and convenient form

$$R_\omega(t) = \langle \dot\phi(t)\, \dot\phi(t+\tau) \rangle = b\beta\, e^{-\beta|\tau|} \, . \tag{6}$$

In the limit $\beta >> b$, this represents the so-called "white noise" spectrum, a Gaussian fluctuation in $\dot\phi$ of constant spectral density 2b to all frequencies (see Fig. 1). The lowest order correlation function equation (6), in this limit becomes

$$R_\omega(t)_{\beta \overrightarrow{>>} b} \langle \dot\phi(t)\, \dot\phi(t+\tau) \rangle = 2b\delta(\tau) \quad . \tag{7}$$

The corresponding power spectrum, from equation (5) is a pure Lorentzian

$$P_E(\omega)_{\beta \overrightarrow{>>} b} \frac{E_0^2}{2} \frac{b}{(\omega-\omega_0)^2 + b^2} \tag{8}$$

with a full-width at half-maximum (FWHM) of 2b. The Lorentzian form for the power spectrum occurs when one allows very fast fluctuations.

This limit is unphysical, because of the relatively large spectral density far from line center. In practice the noise generator has an approximately constant spectral density only to some limit imposed, for example, by stray capacitance, at which it rolls off in some fashion. In fact, we may choose to impose an RC roll-off of characteristic frequency, $\beta/2\pi = 1/RC$. In this case we realize the correlation function given by

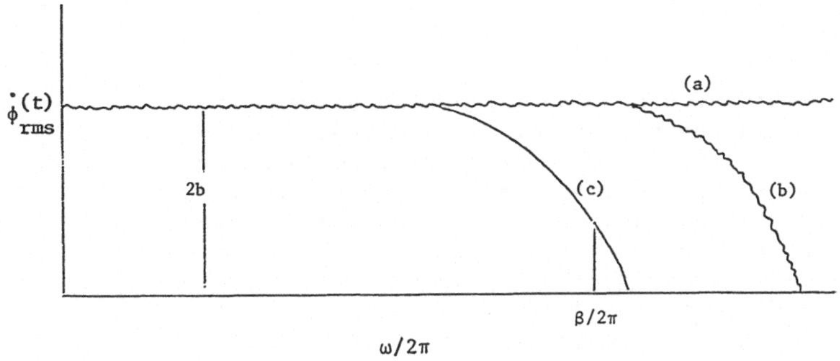

Fig. 1. Representation of the spectral density of fluctuations of $\dot\phi(t)$, (a) in the ideal "white noise" case corresponding to the pure Lorentzian power spectrum, (b) illustrating the occurrence of high frequency roll-off due to such realities as stray capacitance, and (c) illustrating use of a controlled roll-off by imposing an RC ($= 2\pi/\beta$) filter.

equation (6). The technical possibilities are immediately apparent -- by use of selected RC filters (as illustrated in Fig. 2) we can control β. By introducing linear gain or attenuation we can control the factor 2b. We can approach the condition corresponding to equation (7) and the Lorentzian power spectrum, by making β as large as practical relative to b, in order to include fluctuations occurring at high frequency.

The opposite limit $\beta \ll b$ is also quite interesting and useful. In this case

$$P_E(\omega)_{\beta \ll b} \rightarrow \frac{E_0^2}{2} (2\pi/b\beta)^2 \exp\left[-\frac{(\omega-\omega_0)^2}{2b\beta}\right] \tag{9}$$

which is a Gaussian power spectrum. Here the fast fluctuations are excluded, and the radiation field can be thought of as momentarily monochromatic, with slowly varying frequency corresponding to the Gaussian envelope.

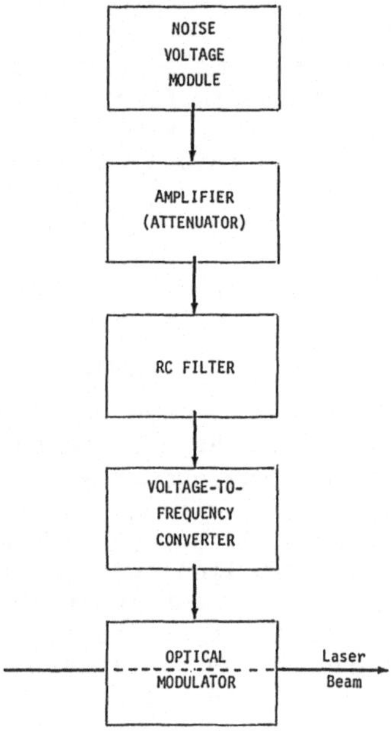

Fig. 2. Fluctations from the noise voltage generator, based on a shot-noise process, are amplified to control spectral density, and filtered to impose an RC roll-off of characteristic frequency $\beta/2\pi$. This fluctuating voltage signal is converted to phase fluctuations using standard FM techniques, and used to drive optical modulation of the laser beam.

IMPLEMENTATION

For any combination of b and β ranging between these two limits the fluctuations retain the properties derived from those of the Gaussian Random Process, so long as all operations imposed in the course of controlling b and β, in the process of converting voltage fluctuations to phase/frequency fluctuations, and in the modulation process itself, may be taken as linear and free of all distortion. The purpose is to maintain the characteristic that the higher order correlation functions of the noise-modulated laser field may be derived from the lowest order correlation function, according to (3) which in turn is determined from measurement of the power spectrum. Thus, it is important to use passive operations as much as possible and to ensure that amplifiers have very large dynamic range to avoid all clipping or distortion.

The actual implementation involves confronting the practical limitations of the various devices and atomic systems involved. Our modulation technique incorporates a number of features, some of which will be mentioned here for illustrative purposes:

● The "monochromatic" laser is a specially designed ring system with an rms linewidth of approximately 200 kHz. In order to consider this contribution negligible we synthesize linewidths of the order of several MHz or larger.

● Some of the principal applications of the system will be to nonlinear absorption in the resonance line from atomic sodium, which has a natural width of about 10 MHz. We want to synthesize linewidths at least comparable to this.

● Acoustooptic modulators work well at (relatively) low frequencies but have bandwidths limited to a few MHz. Electrooptic modulators can operate well out to ~1 GHz but are not efficient at low frequencies. It was necessary to develop a hybrid system of acousto- and electrooptic modulators to produce a power spectrum controlled to ~1 GHz from line center.

● A double pass technique was implemented to eliminate the angular deflection of the laser beam, as a function of frequency of the driving signal to the acousto-optic modulator (AOM).

The schematic arrangement which accomodates these problems reasonably well is represented in Fig. 3. The AOM operates in a 6 MHz bandwidth, effectively as a frequency modulator. The traveling wave electro-optic modulators operate outside the 6 MHz central band, as phase modulators. It is easy to show that the phase modulation corresponds to frequency modulation if a $1/\omega^2$ shaper is introduced.

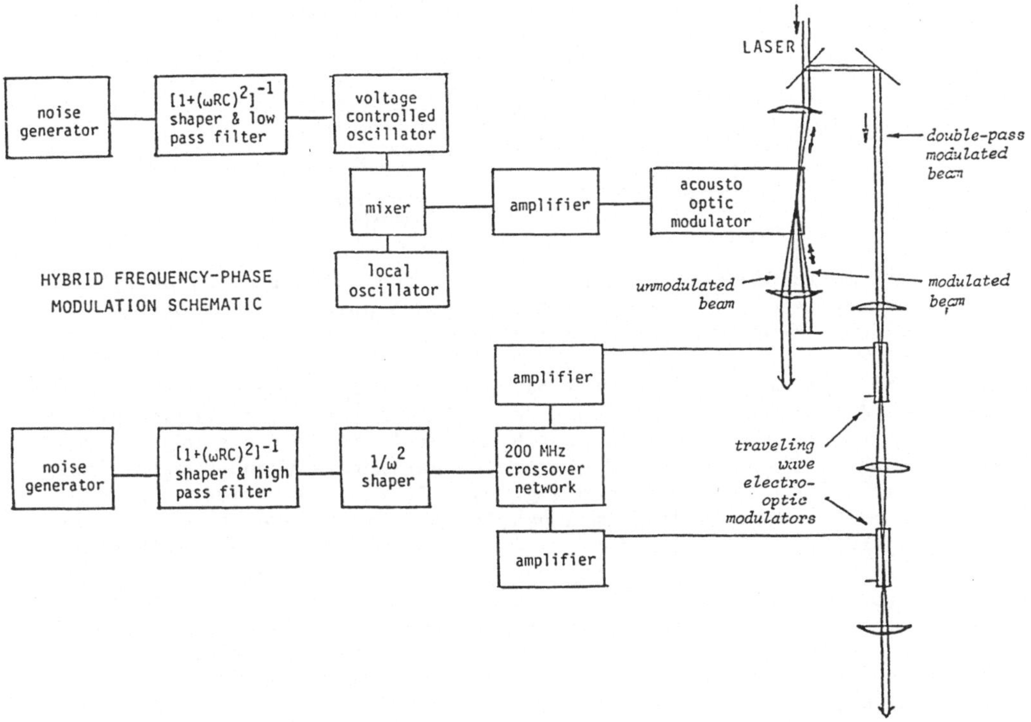

Fig. 3. Schematic diagram of the hybrid noise modulation system discussed in the text.

In order to monitor the random modulation process, a heterodyne beat technique is used. Components of the modulated and unmodulated laser beams are mixed on an avalanche diode, and the difference frequency spectrum is displayed on an rf spectrum analyzer. This spectrum is the power spectrum of the modulated laser beam. In practice, this power spectrum could be adjusted to fit shapes calculated from chosen values of parameters b and β, within about 1 dB, by empirically adjusting the signals driving the several modulators.

INVESTIGATIONS OF NONLINEAR PROCESSES

We now briefly describe some of the principal measurements which have been carried out using the system discussed above for synthesizing specified lineshapes constructed using fluctuations derived from a Gaussian Random Process.

Two-Photon Doppler-Free Absorption in Atomic Sodium

This experiment[6,7] is represented schematically in Fig. 4. The linearly-polarized laser beam is retroreflected through the sodium cell. As is well known, absorption of one photon from the incident beam and one from the retroreflected beam results in cancellation of the Doppler shifts and produces a Doppler-free peak on top of the low-level Doppler-broadened base. The intuitive expectation when using a broad-band laser is that the

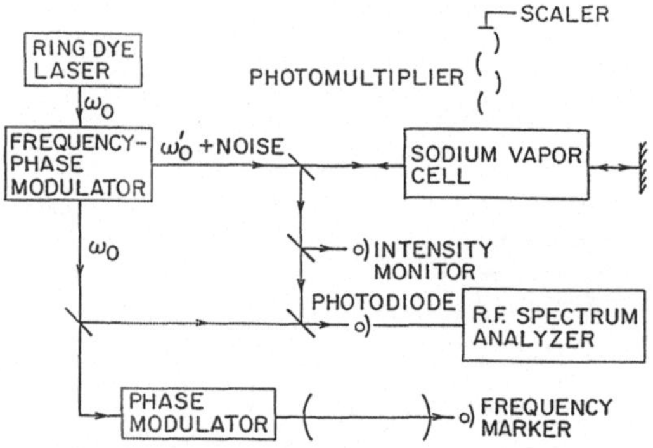

Fig. 4. Schematic diagram showing the use of the noise modulation system as applied to the study of two-photon Doppler-free absorption in the 3S-5S transition in atomic sodium.

width of the Doppler-free peak will be approximately twice the width of the laser line, double the instantaneous deviation of the laser frequency from its center frequency (in addition to the natural width of the upper level). What is found is a marked dependence on the parameters b and β which fully characterize the fluctuations (see Fig. 5). If very fast fluctuations are allowed ($\beta \gg b$, corresponding to a nearly Lorentzian power spectrum) the proportionality factor is found to be 4, illustrating the limitations of intuition in the case of nonlinear processes. For the case $\beta \ll b$ (Gaussian power spectrum) the proportionality factor is found to be 2. These results for the Lorentzian laser line shape were first derived by Mollow[8] on the basis of second-order field correlation functions. Agreement between all of these experimental results and an extension of this theoretical work is excellent.

Another interesting parameter is the time delay τ_D corresponding to round trip travel time of the retroreflected beam. For $\tau_D \sim 1$ ns the

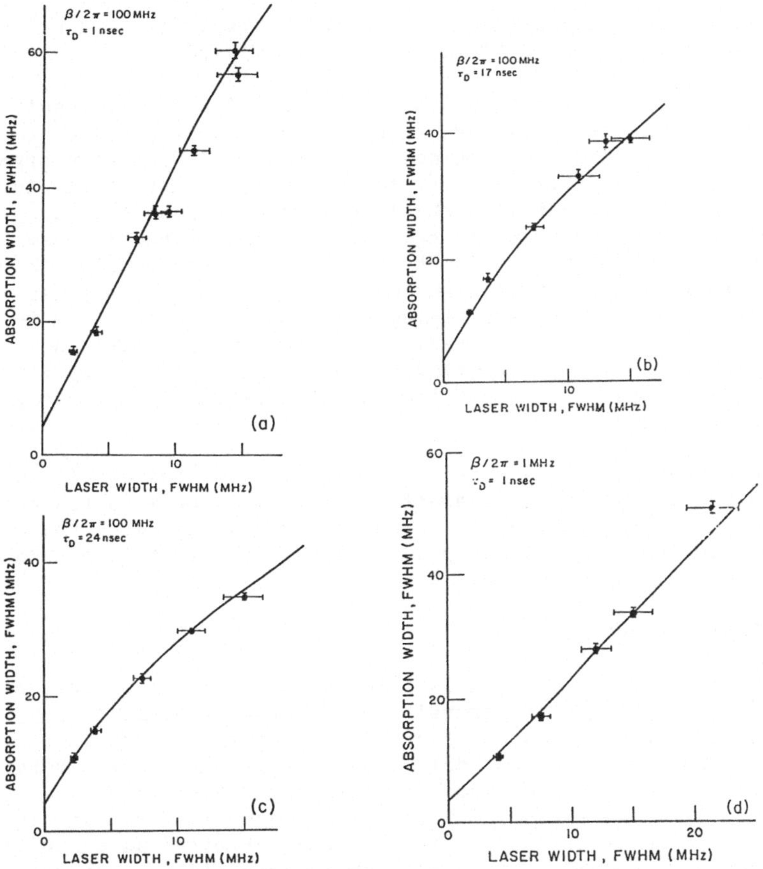

Fig. 5. Width of the two-photon Doppler-free absorption plotted against width of the laser (Ref. 7). Figs. (a) - (c) represent results obtained with nearly Lorentzian power spectra ($\beta/2\pi$ = 100 MHz), with various degrees of decorrelation resulting from different retroreflection path lengths, corresponding to increasing delay times. Fig. (d) represents results obtained with more nearly Gaussian power spectra ($\beta/2\pi$ = 100 MHz). For this case, involving slower fluctuations and longer correlation time ($1/\beta$), the effects of retardation are seen only with optical delay distances larger than are practical in the laboratory.

incident and retroreflected beam remain highly correlated, but for large τ_D there is a pronounced effect due to loss of correlation as the travel time τ_D approaches the correlation time $2\pi/\beta$ of the fluctuations. This is observable in the laboratory, for the case $\beta >> b$, as a decrease in the two-photon absorption peak width, also in agreement with an extension of Mollow's work.

Saturated Absorption in the 3s-3p Transition in Atomic Sodium

The dependence of the optical Autler-Townes effect on laser field phase/frequency fluctuations was also investigated.[9] The intense right-circularly polarized noise-modulated laser is tuned to near resonance with the $3S_{1/2}(F=2,\ M_F=2) \rightarrow 3P_{3/2}(F=3,\ M_F=3)$ transition in a sodium atomic beam which is prepared in the $3S_{1/2}(F=2,\ M_F=2)$ hyperfine component of the ground state. A weak probe laser, also right-circularly polarized, couples to the $4D_{5/2}$ $(F=4, M_F=4)$ state so that the dependencies of the double-peaked optical

Autler-Townes absorption profiles on band-shape, bandwidth intensity, and detuning of the intense laser field can be investigated.

Some of the most interesting and physically significant results come from measurements of the asymmetry of the Autler-Townes doublet as the correlation time $2\pi/\beta$ of the fluctuations is increased. We define asymmetry as $(I_1-I_2)/(I_1+I_2)$ where $I_1(I_2)$ represents the intensity of the Autler-Townes component corresponding to the lower (upper) probe frequency. Figure 6 shows several plots of asymmetry against detuning of the intense laser, for values of $\beta/2\pi$ ranging from 10 MHz to 80 MHz, corresponding to power spectra with increasingly Lorentzian characteristics. The asymmetry reversal evident in Fig. 6d, exhibiting <u>three</u> points on the detuning scale at which peak-height symmetry is realized in the Autler-Townes spectra, is the phenomenon originally observed by Moody and Lambropoulos[10] and by Hogan and Smith[11] which led to the program of quantitative investigation described in this paper.

Those original investigations also gave rise to theoretical investigations. Figure 7 illustrates the excellent agreement between the theory of Dixit, Zoller, and Lambropoulos,[12] and measurements made with the system described here. Thus the role of phase fluctuations in this saturated process may be said to be well understood.

Zero-Field Hanle Effect

Finally, we mention recent measurements of the Hanle effect, carried out in Yb^{174}, on the $^1S_0 \rightarrow {}^3P_1$ transition.[13] Atoms are excited by linear polarization with orientation transverse to an applied magnetic field which may be scanned through zero. At zero field the fluorescence observed in a direction transverse to the direction of alignment (and transverse to the magnetic field axis) is essentially zero. As magnetic field is applied precession occurs and fluorescence is detected. Plotted as a function of magnetic field, the signal shows a pronounced zero field resonance, the shape of which is determined in part by the lifetime of the upper state. Avan and Cohen-Tannoudji[14] showed theoretically that the resonance shapes are also strongly influenced by the statistical properties of the exciting laser field. We are carrying out measurements which confirm this prediction and provide a data base for more detailed theoretical studies. Figures 8a and b illustrate the pronounced change that occurs in Hanle resonances as phase/frequency fluctuations are introduced.

SUMMARY

We have shown that the use of noise derived from a Gaussian Random Process makes it possible to synthesize radiation fields for which, to a good approximation, correlation functions are determined to all orders. These fields are fully characterized by the two parameters b and β which determine the shape of the power spectrum of the laser field (and hence the lowest order correlation function from which all higher order corre lation functions may be derived).

ACKNOWLEDGMENTS

The contributions of M. Hamilton, K. Arnett, and Rajarshi Roy to the experimental programs described above are gratefully acknowledged. Furthermore, the Hanle effect studies were carried out as a collaborative effort with J. Brandenberger of Lawrence University and H. Metcalf of SUNY Stony Brook. The support of the Department of Energy, Office of Basic Energy Sciences is also acknowledged.

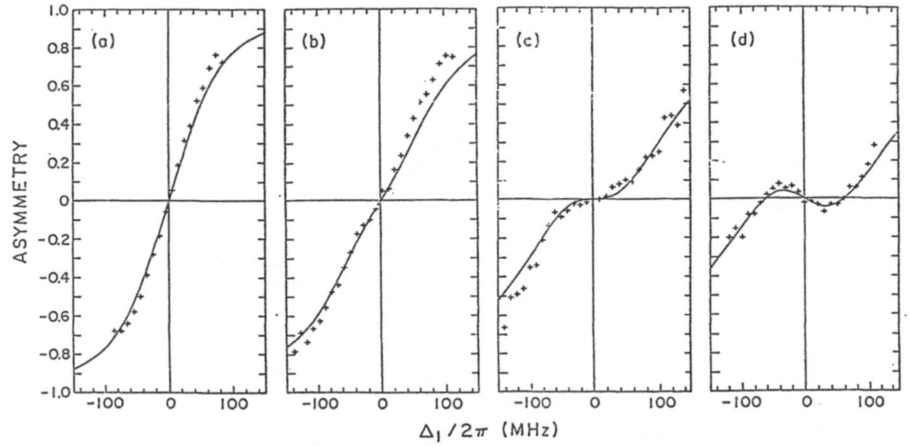

Fig. 6. Autler-Townes asymmetry parameter plotted against detuning of the intense noise-modulated laser (Ref. 9). The laser linewidth is ~14 MHz and the Rabi splitting ~67 MHz throughout. The evolution toward reversed asymmetry with increasing values of $\beta/2\pi$ is illustrated: (a) 10 MHz, (b) 30 MHz, (c) 60 MHz and (d) 80 MHz. The solid curves are calculated according to the theory of Ref. 12.

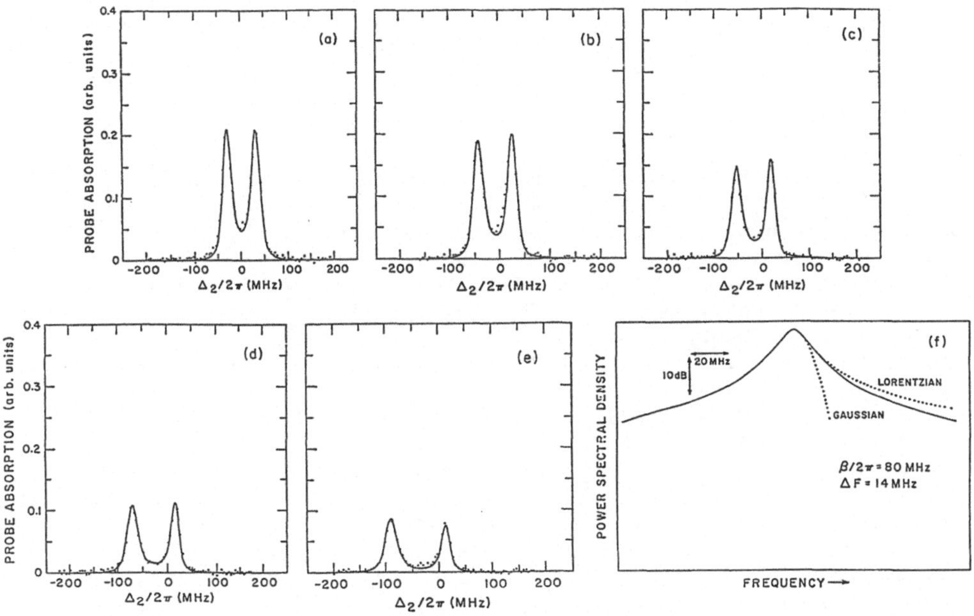

Fig. 7. Comparison of measured Autler-Townes spectra in atomic sodium (dotted curves) with calculations by Dziemballa and Zoller (Ref. 9) based on the theory of Dixit, Zoller, and Lambropoulos (Ref. 12) (solid curves). Rabi splittings are ~64 MHz, and detunings of the intense laser are (a) 0, (b) -20 MHz, (c) -40 MHz, (d) -60 MHz, and (e) -80 MHz. Fig. (f) shows the measured power spectrum of the radiation field used in this measurement, and, for comparison, the pure Lorentzian and Gaussian shapes.

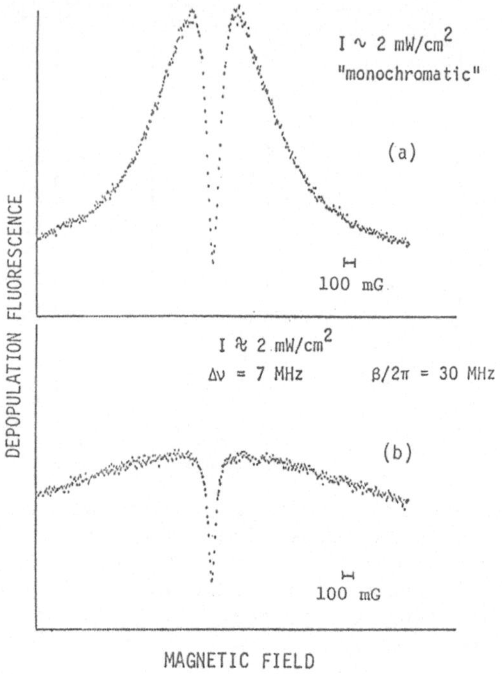

Fig. 8. Hanle resonances as a function of applied magnetic field, for a $^1S_0 \rightarrow {}^3P_1$ transition in Yb^{174} for laser field intensity $I \sim 2$ mW/cm^2 with (a) a "monochromatic" radiation field, and (b) a field with $\Delta\nu = 7$ MHz FWHM and $\beta/2\pi = 30$ MHz.

REFERENCES

1. D. S. Elliott, Rajarshi Roy, and S. J. Smith, Phys. Rev. A 26, 12 (1982).
2. D. S. Elliott, Rajarshi Roy, and S. J. Smith, in Spectral Line Shapes, 2, ed. K. Burnett (deGruyter, Berlin, New York, 1983), pp. 989-998.
3. S. O. Rice, Bell Tell. J. 23, 282 (1944).
4. M. C. Wang and G. E. Uhlenbeck, Rev. Mod. Phys. 17, 323 (1945).
5. D. Middleton, Phil. Mag. 42, 689 (1951); An Introduction to Statistical Communication Theory (McGraw-Hill, New York, 1960).
6. D. S. Elliott, M. W. Hamilton, K. Arnett, and S. J. Smith, Phys. Rev. Lett. 53, 439 (1984).
7. D. S. Elliott, M. W. Hamilton, K. Arnett and S. J. Smith, Phys. Rev. A 32, 887 (1985).
8. B. R. Mollow, Phys. Rev. 175, 1555 (1968).
9. M. W. Hamilton, K. Arnett, S. J. Smith, D. S. Elliott, M. Dziembella, and P. Zoller, Phys. Rev. A 36, 178 (1987).
10. S. E. Moody and M. Lambropoulos, Phys. Rev. A 15, 1497 (1977).
11. P. B. Hogan, S. J. Smith, A. T. Georges, and P. Lambropoulos, Phys. Rev. Lett. 41, 229 (1978).
12. S. N. Dixit, P. Zoller, and P. Lambropoulos, Phys. Rev. A 21, 1289 (1980).
13. K. Arnett, J. Brandenberger, M. Hamilton, and H. Metcalf, private communication.
14. P. Avan and C. Cohen-Tannoudji, J. Phys. B: Atom. Molec. Phys. 10, 171 (1977).

MULTIPHOTON ATOMIC AND MOLECULAR COHERENT EXCITATIONS WITH ULTRASHORT PULSE SEQUENCES

J.-C. Diels

Department of Physics and Astronomy and
The Center for High Technology Materials
The University of New Mexico
Albuquerque, New Mexico 87131

I. COHERENT INTERACTIONS

With the recent progress in short pulse generation, an increasing number of experiments involving energy exchange between light and matter occur in conditions of coherent interactions. The time scale of the experiment becomes shorter than any collision time between particles of the ensemble. The radiation is thus interacting directly with isolated atoms. The condition of coherent interactions can be defined easily in the Bohr atom picture. Let us consider a rotating frame of reference in which the ground state electron appears at rest. A periodic electric field at $\omega = \omega_1/n$ (n integer) can induce transitions (transition rate proportional to E^n) to a higher orbit (of energy $h\omega_1$ above the ground state). In this rotating frame of reference, the periodic motion of the electron is an atomic clock that will remember the phase of the excitation. The electron returns to the original position where it was created at every cycle. Hence, if the atom is truly isolated, it can be returned to the ground state by applying the opposite excitation after an integer number of periods. In conditions of coherent resonant interaction, the transfer of energy between radiation and matter is reversible, but also pulse shape dependent. The reversibility can be exploited to enhance the interaction length of radiation and matter, essentially by using energy preserving pulse sequences. One application is phase matched higher harmonic generation for which long interaction lengths are desirable. We shall see that this problem can be easily solved in the case of two photon transitions. For processes of order $n>2$ however, the presence of level shifts of all order (up to n) results in a very complex dependence on multiphoton absorption and harmonic generation, as will be shown in the example of 4 photon resonance in Hg vapor. New methods to solve the inverse transfer function problem (find the initial condition - pulse shape and phase modulation - that will give optimum result (i.e. optimum harmonic conversion or optical pumping) have to be developed. With the recent progress in pulse synthesizing methods[1,2], one can expect dramatic future developments in the area of multiphoton coherent resonant interactions.

In the case of molecules, there are many possible pathways for the energy transferred to the molecule. Because of the very large number of levels involved in the interaction of an intense picosecond pulse with an anharmonic ladder of levels, one might expect the absorbed energy to be distributed quasi at random among the various levels. We will show instead that it is possible to find a particular shape for the optical excitation, which will result in all the absorbed energy being concentrated in one or two high lying energy levels.

II. TWO PHOTON RESONANT THIRD HARMONIC GENERATION

A two photon resonance has long been known to enhance the third order nonlinear susceptibility for third harmonic generation. However, highest intensity conversion efficiency were reported with non-resonant systems. The reason is that, at the highest fundamental fields which are expected to produce the largest third harmonic field, two photon absorption depletes the fundamental and limits the interaction length. There are two basic questions that we attempted to answer:

1) Can energy preserving pulse shapes be used to reduce the depletion of the fundamental?
2) What is the optimum resonance condition that will lead to the most efficient up-conversion?

The answer to the first question has been given in a theoretical paper by Diels and Georges[3], where it was shown that an optimum energy conversion could be obtained with an energy conserving sequence of two pulses 90° out of phase. An experimental verification of this theory is presented here, as well as the answer to the second question.

In the theory, all other levels than the two-photon resonant levels (labelled 1 - for the ground state - and 2 - for the excited state) are assumed to be off-resonant. The interaction of these levels with the fields $E_n = \mathcal{E}_n e^{ni\omega t}$ is treated by an adiabatic approximation. Therefore, the density matrix of the multilevel system reduces effectively to a 2 × 2 matrix of which the evolution of the off-diagonal element $\sigma_{12}e^{2i\omega t}$ is given by[3]:

$$\left\{\frac{\partial}{\partial t} + i\Delta\omega + \frac{1}{T_2}\right\} \sigma_{12} = i(\rho_{22} - \rho_{11}) \left[\frac{r_{12}}{h^2} \mathcal{E}_1^2 + \frac{\xi_{21}^*}{h^2} \mathcal{E}_1^* \mathcal{E}_3\right] \tag{1}$$

$\Delta\omega$ is the detuning, including the Stark shift, and $1/T_2$ the transverse relaxation, including the phase coherence loss due to photoionization. The two most important parameters in this type of nonlinear interaction are the second order coupling coefficients via two fundamental photons (r_{12}) or via a third harmonic and a fundamental photon (ξ_{21}), defined as:

$$r_{12} = \sum_{\ell} \frac{\mu_{1\ell}\mu_{\ell 2}}{(\omega_{\ell 2} + \omega)} \tag{2}$$

$$\xi_{21} = \sum_{\ell} \left[\frac{\mu_{2\ell}\mu_{\ell 1}}{\omega_{\ell 1} - 3\omega} + \frac{\mu_{2\ell}\mu_{\ell 1}}{\omega_{\ell 2} + 3\omega}\right] \tag{3}$$

μ_{mn} and ω_{mn} are the dipole matrix elements and the frequency of the transition $|m\rangle \rightarrow |n\rangle$.

The resonant contribution of the two level system to Maxwell's equations for the fundamental field is:

$$\frac{\partial \mathcal{E}_1}{\partial z} \alpha - r_{12}\sigma_{12}\mathcal{E}_1^* \tag{4}$$

For strong interactions such that σ_{12} is of the order of unity, one can define a two photon characteristic distance:

$$\ell_2 = \frac{c\ \epsilon_0\ h}{2N\ \omega\ r_{12}} \tag{5}$$

(c = speed of light, ϵ_0 = background dielectric constant, and N = number density of resonant atoms). The resonant contribution to the third harmonic field is instead proportional to ξ_{21} (assuming exact phase matching):

$$\frac{\partial \mathcal{E}_3}{\partial z} \alpha - \xi_{21}\sigma_{12}\mathcal{E}_1 \qquad (6)$$

Efficient harmonic generation can only be maintained over distance for which σ_{12} remains close to unity, i.e. distance of the order of ℓ_2. Therefore, the maximum achievable energy conversion is close to $\eta = (|\xi_{21}|/r_{12})^2$ for a plane wave, as verified by computer calculations. In comparing the weak 2s-4s transition to the stronger 2s-3d, we find that the maximum efficiency ratio is $\eta_{2s-4s}/\eta_{2s-3d} = 10^4$. The benefit of the stronger two-photon transition is that it requires a 20 times smaller field to obtain the peak conversion.

In the experimental set up, the source is a synchronously pumped mode-locked dye oscilator fed into a three stage dye amplifier pumped by a 10 Hz Q=switche Nd-YAG frequency doubled laser. The amplified pulses have an energy of 1 mJ, and a duration of 6 ps. For excitation of the 2s-4s transition, Rh6G is used as a gain medium in both the oscillator and amplifier. DCM dye is used in the case of the 2s-3d transition. The average pulse frequency is tuned by a three plate birefringent filter. Interferometric autocorrelation measurements[4] were made to get a lower limit to the pulse coherence. Each pulse is split into a sequence of two identical pulses of adjustable relative phase and delay by a Mach Zehnder interferometer. The phase and delay are monitored with a He-Ne laser retracing the path of the main beam. As the delay arm of the interferometer is continuously increased, a complete set of data (energy of the frequency doubled input and transmitted signals, third harmonic) is recorded for all values of interpulse phase and delay, and for various input pulse frequencies. The monitoring of the second harmonic of the input signal to the heat pipe (which is the puse sequence transmitted by the Mach Zehnder) is an interferometric second order autocorrelation of the laser pulses. It can be exploited to gain information on the phase modulation of the amplified laser pulses. In the case of our source, the interferometric autocorrelation corresponded to that of a Gaussian pulse[5], with a Gaussian phase modulation given by 1.4 $\exp\{-(t/\tau)\}$. The total phase shift over the pulse is thus less than $\pi/2$.

In the absence of spectral information or single shot measurements, there is an ambiguity in this result. The same fits could be obtained assuming perfectly coherent pulses, with a random distribution of frequencies. Fitting of interferometric autocorrelations averaged over a pulse train provides an upper limit to the pulse chirp, or a lower limit to the coherence time.

Coherent two-photon resonant anomalous propagation of 90° phase shifted pulses was recorded for both the strong 2s-3d and the weak 2s-4s transitions. The maximum third harmonic conversion efficiency was only 10^{-6} for the 2s-3d transition, against 10^{-2} for the 2s-4s transition. The efficiency ratio of 10^{4} is indeed in agreement with theoretical estimates. The maximum energy conversion efficiency of 6% was predicted for a uniform plane wave, which matches reasonably the 1% conversion efficiency measured over the full cross section of a Gaussian beam.

We conclude that while a two photon resonance enhances the third harmonic conversion efficiency at low power levels, it reduces the highest achievable conversion efficiency. For a given fundamental power density and a corresponding interaction length, the ideal two-photon resonant enhancement is that for which the characteristic distance for maximum two photon absorption is of the order of the cell length.

III. FOUR PHOTON RESONANCES

Higher order resonance (n>1) could be used to enhance higher order (2n+1) harmonic generation. As in the two photon case (n=1), one could attempt to minimize the depletion of the fundamental by making use of pulse shapes or pulse sequences that propagate through the resonant medium with minimum energy loss. The higher order (n>1) multiphoton coherent interaction is much more complex than in the case of n=1, because various effects of lower order than 2n have to be taken into consideration. Because of the mixture of phenomena of different orders, new phenomena such as induced resonances and interaction quenching take place, as shown below in the example of the $6^1S_0 - 6^1D_2$ transition in mercury, excited in condition of four photon coherent interaction by pulses near 560 nm.

The basic theoretical approach is the same as for the two photon transition[3,6]. The density matrix of the multilevel atomic system is reduced to a two by two matrix associated with the resonant levels. The temporal evolution of the density matrix elements associated with all the other levels is calculated by an adiabatic expansion to fourth order[7]. The time evolution of the off-diagonal matrix element $\sigma_{12}e^{4i\omega t}$ is given by:

$$\left\{\frac{\partial}{\partial t} + i\Delta\omega + \frac{1}{T_2}\right\} \sigma_{12} =$$

$$i(\rho_{22}-\rho_{11})\left[r_{12}\frac{\mathcal{E}_1^4}{h^4} + b\mathcal{E}_1\mathcal{E}_3 + c\mathcal{E}_1^*\mathcal{E}_5 + d\mathcal{E}_1^2\mathcal{E}_5\mathcal{E}_3^*\right] \tag{7}$$

is the detuning, including a quadratic and quartic Stark shift contribution. The dephasing time $1/T_2$ includes a loss of coherence due to photoionization. The source term is composed of a direct four photon pumping term in $\mathcal{E}_1^4$, and reactions from the third and fifth harmonics, of which the coefficients b and c are expansions up to second order in the fundamental field $\mathcal{E}_1$.

Stark Induced Resonance in Mercury

If we consider the particular level structure of Hg, we note an interesting three photon near resonant enhancement of the four photon 6s-6d transition by the level 6p. As a result, the contribution of the upper level 2 to a "quadratic" Stark shift clearly dominates. The corresponding Stark shift coefficient is the real part α_2' of the susceptibility α_2:

$$\alpha_2(\omega) = \frac{1}{h}\sum\left[\frac{|\mu_{\ell 2}|^2}{\omega_{\ell 2} - \omega} + \frac{|\mu_{\ell 2}|^2}{\omega_{\ell 2} + \omega}\right] \tag{8}$$

As the atom is excited by the electromagnetic wave at ω, the upper level of the four photon resonance shifts by a corresponding amount $\delta\omega_2 = \alpha_2' \mathcal{E}_1^2$. Consequently:

$$\omega_2 = \omega_{20} + \delta\omega_2 \tag{9}$$

However, as a result of the motion of level 2, the value of α_2 is also modified through the resonant denominator of Eq. (8). Hence the Stark shift is further modified, and exhibits an extremely nonlinear dependence on the exciting field.

The Eqs. (8) and (9) can be solved self-consistently to find the polarizability and the Stark shift for each value of the resonant field. This system of equation, however, is not compatible for any value of the field, and ceases to have a solution when the Stark shift brings the 6D - 6P transition nearly into resonance with the radiation field. At any rate, the adiabatic approximation becomes invalid at that point. The fields needed to reach that condition can be reached with focused fs pulses. For instance, the additional 6D - 6P resonance is induced by a Gaussian 200 fs pulse at 560.7 nm (4 photon resonance for the 6s - 6D transition) with an energy density of 4 J/cm^2.

Time resolved measurements of the position of level 2 would be a very sensitive probe of the dipole matrix elements connecting to the level 6p. It would also provide a test of the validity range for the adiabatic approximations.

Interaction Quenching

The set of interaction equations, combined with Maxwell's equations, was integrated numerically using a fifth order Butcher method[8]. The initial fundamental pulse is Gaussian, 3.5 ps FWHM, with a "four photon area" $r_{12}\mathcal{E}_1^4 dt = 1.4$. The incident wavelength adjusted for maximum absorption is at 566.7 nm, or 6 nm above the zero field resonance (560.7 nm). As the pulse propagates through the medium, he third harmonic generation is seen to "saturate" after a very short distance, while the ionization drops from 0.15% to only 0.1 10^{-3}%. This phenomenon is a direct consequence of Eq. (7), in which the factor b has a very large value. The third harmonic being generated with a phase opposite to that of the fundamental, after a relatively short distance the resonant Raman scattering (term $b\mathcal{E}_1\mathcal{E}_3$ in Eq. 7) opposes the four photon resonant pumping (term $r_{12}(\mathcal{E}_1^4/h^4)$ in Eq. 7), and all interaction stops since the right hand side of Eq. 7 vanishes. With this quenching of the interaction by the third harmonic, the ionization also stops. We believe this to be the correct interpretaton for the experimental observation of an apparent "competition" between ionization and third harmonic generation in mercury, reported by Smith[9] and Normand et al[10]. Previous interpretations did not take into account the influence of the four photon resonance.

Since the interaction quenching is related to the relative phase of the fundamental and third harmonic, it can be prevented by designing an exciting signal (modulated pulse or sequence of pulses) that continuously or periodically resets a minimun dephasing (< r/2) between the two fields.

IV. MULTIPHOTON COHERENT MOLECULAR EXCITATION

From the increasing complexity in going from two to four photon resonance, one can extrapolate that the equations for higher order coherent multiphoton propagation and harmonic generation may be excessively tedious to derive. The treatment of infrared multiphoton coherent interactions in molecules appears then nearly inextricable, since in that case, because of the density of vibrational - rotation transitions involved, the resonant levels number by the tens or hundred, rather than being just two. It is precisely this additional complication that leads to very substancial simplification. Indeed, it is always possible to find one or several pairs of levels that are near resonance with a single photon transition. The presence of these near resonance conditions enables us to limit the calculation and modeling to a first order interaction. The shear number of levels and (degenerate) transitions via P, Q and R branches makes it difficult to maintain a clear insight in the physics of the problem. However, the level structure is not random, and therefore there is no reason to treat the interaction as an incoherent one. We show that indeed, it is possible to design a particular form of excitation that results in a population inversion to one or two rotational lines in the v=4 vibrational level. The transverse moment of inertia of the molecule will be assumed negligible. The time dependent interaction equations are simply derived in the usual way by substituting for the wave function in Schrodinger's equation an expansion in unperturbed (by the E field) wave functions $\sum_{vJ} c_{vJ}(r)u_{vJ}(r)$, where v and J refer to the vibrational and rotational quantum numbers, respectively. Using the dipole approximation for the interaction Hamiltonian, we find the following set of equations for the coefficients $c_{v,J}$[11]:

$$\frac{dc_{v,J}}{dt} = \sqrt{\frac{J+1}{2J+3}}\left[E_v c_{v-1,J+1} - E^*_{v+1} c_{v+1,J+1}\right]$$

$$+ \sqrt{\frac{J}{2J-1}}\left[E_v c_{v-1,J-1} - E^*_{v+1} c_{v+1,J-1}\right] + i\Delta_{vJ} c_{vJ} \qquad (10)$$

In Eq. (10), E_v are Rabi frequencies $\mu_{v-1,v}\mathcal{E}/2h$ (where $\mu_{v-1,v}$ is the J-independent part of the dipole matrix element between states v and v-1), and Δ_{vJ} is the multiphoton detuning ω_{v0} + BJ(J-1) - vω (B being the rotational constant).

The general problem is to find a particular shape, modulation and frequency of the exciting field E(t) that will result in selective excitation of a particular level. We have chosen as an example four photon excitation of the CF stretch in the molecule CH_3F, of which the parameters are given in Table I. The excitation consists in a sequence of 4 or 5 identical picosecond pulses. The advantage of this rather restrictive approach to the general "inverse problem" are that it is relatively easy to implement (for instance by pulse splitting and delaying techniques as was done in the case of two photon resonant propagation in lithium vapor), and that the physics of the problem is not completely lost in the numerical computation.

TABLE I

Parameters of CH_3F

<u>Transition frequencies</u>	ω_{01} = 197.7 ps^{-1}
	ω_{12} = 194.7 ps^{-1}
	ω_{23} = 192.1 ps^{-1}
	ω_{34} = 189.2 ps^{-1}
<u>Dipole Moment</u>	μ_{01} = 0.21 Debye
<u>Rotational constant</u>	B ≡ 0.156 ps

Each sublevel J from the Boltzman distribution in the ground state v=0 will couple to a sublevel J+1 and J-1 in the state v=1, which in turn will couple to more levels in v=2 and v=3, to finally couple to all sublevels J-4 to J+4 in the upper level v=4. One can therefore expect that, in general, the interaction will heat the rotational "temperature" in all five vibrational bands considered. This need not always be the case, for particular pulse sequences. To understand how rotational heating can be prevented, let us place ourselves in a frame of reference rotating at the frequency of a particular J-J transition, from a vibrational level v to a level v+1. In that frame of reference, after a short pulse excitation, the off-diagonal matrix elements corresponding to the transitions J→J-1 and J→J+1 will precess with opposite angular velocities B (= 0.16 ps^{-1}). These two quantities will approximately cancel each other after a delay. It is therefore not surprising to find a minimal rotational heating after a sequence

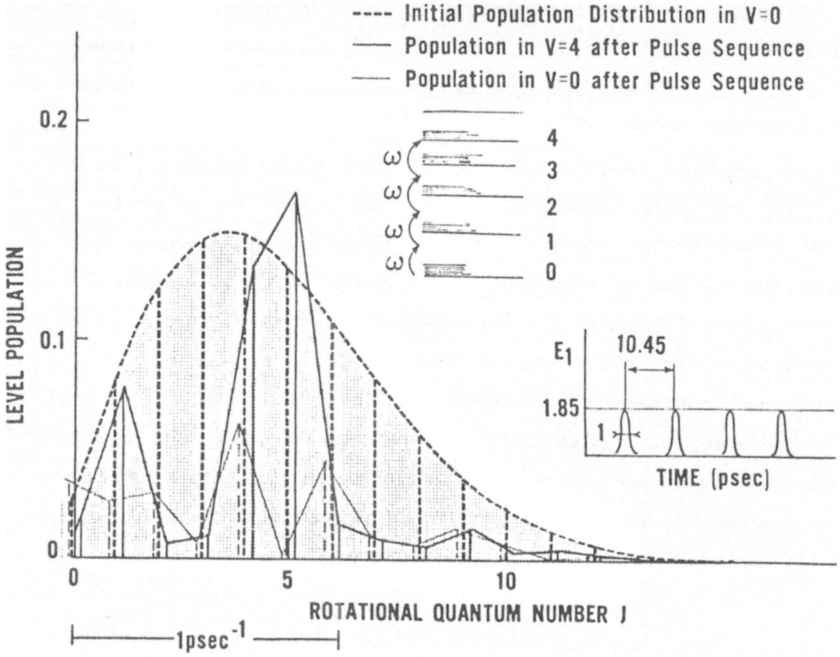

Figure 1. Population versus rotational quantum number after excitation by a four pulse sequence.

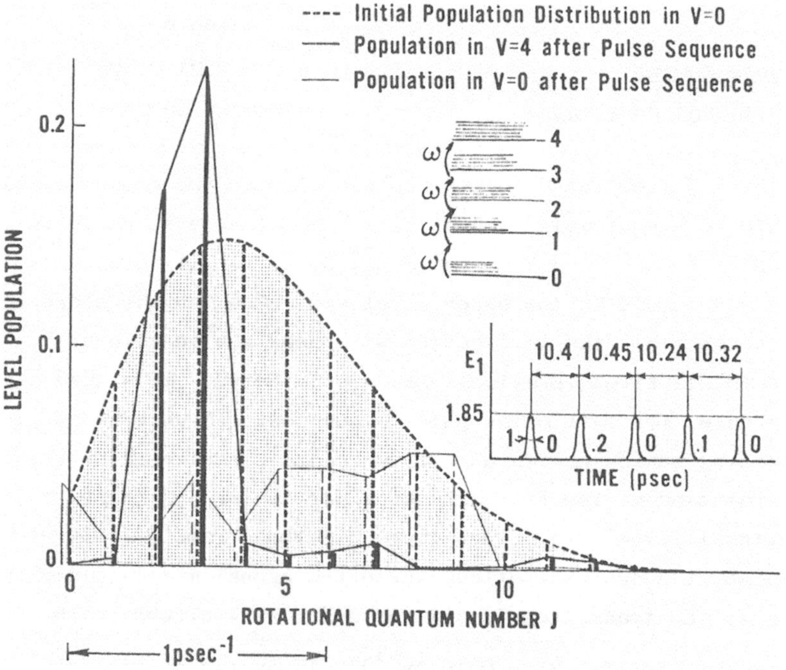

Figure 2. Population versus rotational quantum number after excitation by a five pulse sequence.

of four picosecond pulses equally spaced by $\pi/2B$. Each of the pulse of the sequence has a Gaussian temporal shape, and a duration of 1 ps FWHM. The spectrum of such a pulse centered at 1023.4 cm^{-1}, is sufficiently broad to cover all transition frequencies from ω_{0-1} to ω_{3-4}. With an amplitude corresponding to a peak Rabi rate of 1.85 ps^{-1} for the 0-1 transition, each pulse is capable of inducing some population in all levels up to v=4. After the four pulse sequence however, it can be seen from Fig 1. that most of the population has concentrated itself in level 4. The rotational line distribution in that level is concentrated at the quantum numbers J=3 and J=4.

By successive optimization procedure of the pulse frequency, amplitude, interpulse delay and phase, a five pulse sequence that leads to an even more selective excitation of the v=4 vibrational level has been constructed (Fig. 2). The same pulses are used in Fig. 2 as in Fig. 1. The phases and relative delays of the puses are indicated in the figure. The rotational population distribution in ground state prior to the pulse sequence (Boltzman distribution at $T = 250^{o}k$) is indicated by the shaded area, while the heavy lines show the rotational line distribution in the v=4 state after the pulse sequence. It can also be seen that the ground state population is left uniformly depleted (dashed line) after the pulse sequence. The population in the remaining levels is negligible. We can thus conclude that vibrational and rotational selective coherent multiphoton excitation of molecules is possible with simple pulse sequences. One can expect that more general shape optimization computational techniques will lead to even higher selectivity.

This work was supported by a National Science Foundation Grant (ECS-8406985) and by the Office of Naval Research.

REFERENCES

1. W. S. Warren, NATO advanced research workshop on Molecular Processes with short intense laser pulses, Lennoxville, Quebec, July 1984.
2. J. P. Heritage, NATO advanced research workshop on Molecular Processes with short intense laser pulses, Lennoxville, Quebec, July 1984.
3. J.-C. Diels and A. T. Georges, Phys. Rev. A19, 1589 (1979).
4. J.-C. Diels, J. J. Fontaine, I. C. McMichael, and F. Simoni, Appl. Optics 24, 1270 (1985).
5. H. Vanhergeele, H. J. Mackey, and J.-C. Diels, Appl. Optics 23, 2056-2061 (1984).
6. J.-C. Diels and J. Stone, Phys. Rev. A31, 2397 (1985).
7. N. Mukherjee, A. Mukherjee, and J.-C. Diels, to be publsihed (1987); N. Mukherjee, Ph.D. dissertation, North Texas State University (1987).
8. L. Lapidus and J. Seinfeld, "Numberical Solutions of Ordinary Differential Equations," Academic Press, NY (1971).
9. A. V. Smith, Optics Lett. 10, 341 (1985).
10. D. Normand, J. Morellec, and J. Reif, J. Phys. B16, L227-L232 (1983).
11. J.-C. Diels and S. Besnainou, J. Chem. Phys. 85, 6347 (1986).

TEMPORAL COHERENCE IN ONE-PHOTON EXCITATION OF A MOLECULAR QUASICONTINUUM

Israel Schek* and Joshua Jortner

Department of Chemistry, Sackler Faculty of Exact Sciences, Tel-Aviv University, 69978 Tel-Aviv, Israel
*Dept. of Chemical Physics, Weizmann Institute of Science, Rehovot 76100, Israel

ABSTRACT

The effects of the incoherence of laser phase on Quantum Beats, ringing effects and near-resonance light scattering is investigated.

I. Introduction

The dynamics of the coherent excitation of molecular bound level system in large and medium sized molecules under collisionless conditions, was extensively investigated experimentally as well as theoretically.[1-2] The following manifestations of coherent excitation are encountered and can be interrogated in the photon counting rate (PCR): 1) Ringing oscillations, i.e. coherent exchange of photons between the laser field and the molecular system, i.e. the ground and the excited states; 2) Quantum beats (QB), i.e. pulsations between the coherently excited molecular levels, not necessarily resonant with the laser frequency; 3) Near resonant light scattering (NRLS), such as Rayleigh or Raman effects; 4) Dephasing decay provided by the excitation of a congested, but still well separated (with respect to reactive widths) multilevel system.

In this paper we examine the interplay between the intramolecular and the laser-fluctuations mechanisms in erosion of the coherent features, which are manifested in the PCR detected from a model large molecule with a sparse level manifold and from a model quasicontinuum (QC).

II. Photon Counting Rate of an Incoherent Excitation of a Sparse Level Manifold

The laser field, which excites the molecule, is modeled by a classical field characterized by a stabilized amplitude and fluctuating phase:

$$E(t) = \epsilon(t)\exp\{i[\omega t + \phi(t)]\} + c.c. \tag{2.1}$$

where $\epsilon(t)$ is the field envelope, ω is the mean frequency and $\phi(t)$ is the fluctuating phase. The time development of this fluctuating phase is considered by solving an appropriate stochastic differential equation.

The PCR under realistic conditions, when only a partial knowledge of the field is given[1,2], is expressed in terms of the eigenset of the effective Hamiltonian,[3] $\{|j,vac>\}$, with the corresponding eigenvalues $\{E_j - ir_j/2\}$, where E_j is the energy and r_j - the molecular reactive width. The PCR is

$$I(t) = \sum_{jj'} A_{jj'} \, J_{jj'} \, (t) \tag{2.2}$$

$A_{jj'}$ denotes the time-independent overlap integrals between the eigenstates and the radiative doorway state. The time-dependent dynamical function is given by the double-time convolution integral:

$$J_{jj'}(t) = \int_{-\infty}^{t} d\tau \int_{-\infty}^{t} d\tau' \, \exp[i\omega(\tau-\tau')] \, \epsilon(\tau)\epsilon(\tau') \, S(|\tau-\tau|) \cdot$$

$$\cdot \exp[-(iE_j+r_j/2)(t-\tau) + (iE_{j'} - r_{j'}/2)(t-\tau')] \quad , \tag{2.3}$$

where S is the correlation function of the phase:

$$S(|\cdot\tau|) = \langle \exp[i\phi(t)] \exp[-i\phi(t-\tau)] \rangle \tag{2.4}$$

To provide an explicit expression for the dynamical function $J_{jj'}$ of the PCR we consider the exponentially rising and decaying envelope

$$\epsilon(t) = \epsilon \exp(-\gamma_p |t|/2) \tag{2.5}$$

which corresponds to a Lorentzian wave amplitude. A long time pulse is obtained by setting $\gamma_p \to 0$ (which sharpens the uncertainty limited wave amplitude) and a very short pulse is obtained by setting $\gamma_p \to \infty$ (which smears the wave amplitude). An exponential correlation-function is taken

$$S(\tau) = \exp(-a|\tau|) \tag{2.6}$$

which is obtained when the time-development of the fluctuating phase $\phi(t)$ is described by the phase diffusion model,[4-10] which resembles that of a Brownian particle. The term a is the slow diffusion rate of the laser phase.[9] Using Eqs. (2.5, 2.6), the dynamical function $J_{jj'}(t)$ is a sum of four decaying-oscillating terms: 1) The near resonant fluorescence term, exhibited by the QB between levels $|j\rangle$ and $|j'\rangle$, modulated by the decaying factor $\exp[-(r_j+r_{j'})t/2]$. These QB are manifested by the oscillations at the frequency $\Delta_{jj'} = E_j - E_{j'}$ and are not modified by the stochastic rate a. 2) The pure Raman term, exhibits the decay of the original pulse on the time scale γ_p^{-1}, without any oscillations. 3,4) The ringing effects of the molecular levels $|j\rangle$ and $|j'\rangle$ at the frequencies Δ_j and $\Delta_{j'}$, modulated by the decaying factor $\exp[-(r_j/2+a+\gamma_p/2)t]$. The decay rate is a combination of the intramolecular decay rate r_j, the stochastic rate a and the pulse uncertainty width γ_p.

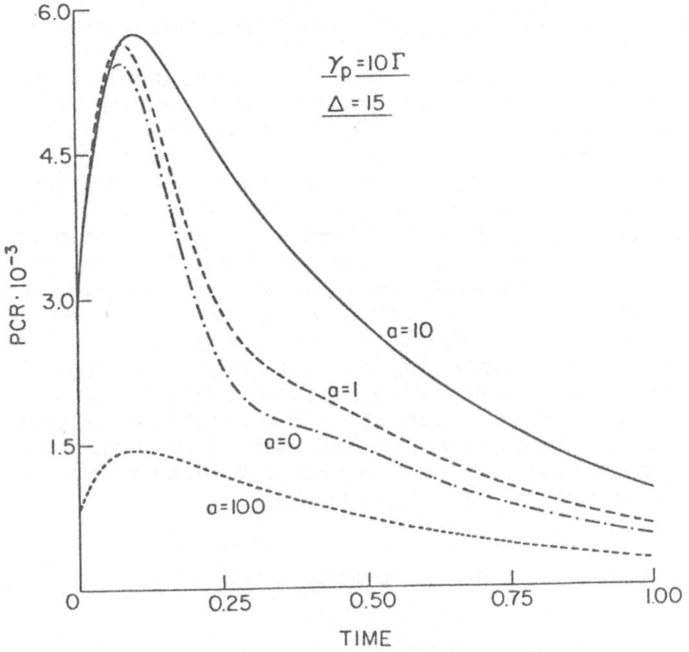

Fig. 1. The photon counting rate detected from a single excited state out of resonance for various values of the stochastic width. The parameters are: r = 1; γ_p = 1.1 (temporally intermediate pulse) and Δ = 15. The values of a are shown on the curves.

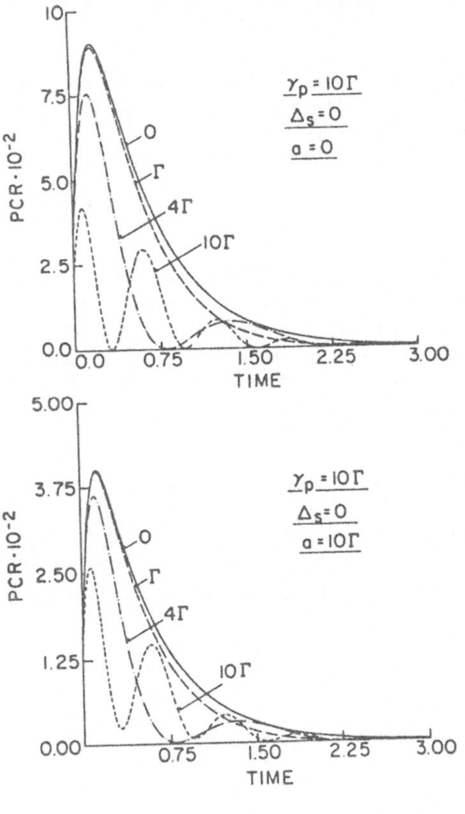

Fig. 2. The quantum beats detected from two excited states in a short pulse excitation for various level spacings Δ_{12} (values shown on the curves). The parameters are: $r_1 = r_2 = 1$; $\gamma_p = 10$; Δ = 0 (the laser frequency is resonant with the center of the excited manifold);
Fig. 3a: a = 0 (coherent pulse).
Fig. 3b: a = 10. Δ_{12} = 0,1,4, and 10.

III. The Erosion of Ringing Effects by the Laser Phase Diffusion

The extremely incoherent case is encountered when the stochastic width a exceeds considerably the temporal width of the pulse γ_p, the molecular reactive widths $\{r_j\}$ and the level detunings $\{E_j-\omega\}$. In this case the two oscillatory decaying terms, which exhibit the ringing effect between the molecular levels and the laser field at the frequency $\Delta_j = E_j-\omega$ are eroded. Fig. 1 shows the time-dependence of the PCR detected from a single state excited by a non-resonant ($\Delta \neq 0$), temporally intermediate ($\gamma_p \approx r$) pulse, for various values of the stochastic width, a. The ringing effect is pronounced for the low values of a (= 0, 1r). In the coherent, temporally long pulse the ringing effect is considerably eroded due to the relatively short depletion-reactive time, which appreciably empties the level before many ringing pulsations do occur. In the coherent, temporally short pulse the ringing effect is practically missing, since the pulse ceases to exist before any ringing pulsation arises. It can be attributed to the broad uncertainty width of the pulse, which effectively smears the ringing.

IV. The Persistence of the Quantum Beats under Incoherent Excitation

The interesting feature exhibited by $J_{jj'}$ is that even under the excitation by extreme stochastic field, the QB terms, i.e. the intramolecular energy pulsation between the sparse levels of the excited manifold, do survive, contrary to the ringing effects. This is evident from the model calculations for a three-level system of Fig. 2. Again, the role of the stochastic rate is only to decrease the overall PCR, but not to change the relative amplitudes of the QB at the various detunings.

The persistence of the QB is also manifested in the long pulses. In the Infinitely long pulse ("truly" energy-resolved experiment) the dynamical function (in the coherent case, a >> r_j):

$$J_{jj'} = [\frac{r_j+r_{j'}}{2} + i\,(E_j-E_{j'})]^{-1}\ \frac{2\epsilon^2}{a} \tag{4.1}$$

This Hanle resonance is not eroded by the stochastic effect, being intramolecular effect. The increase of this stochastic width is not efficient in eroding the QB in the time-resolved experiment. It is apparent that only the amplitude of each of $J_{jj'}$ decreases (proportionally to a), so that the general pattern of the PCR is preserved when the pulse is highly stochastic.

V. Erosion of Dephasing Effects by the Laser Phase Diffusion

Under a coherent pulse excitation of a two-level system the resonant fluorescence and the near resonant light scattering (NRLS) - where the PCR adiabatically follows the exciting pulse are encountered,[11] as monitored, e.g. for I_2.[12] For this system the dynamical function is

$$J_{jj}(t) = \frac{(\gamma_p^2/2)}{\Delta_j^4}\,\epsilon^2\, e^{-r_j t} + \frac{\epsilon^2}{\Delta_j^2}\, e^{-\gamma_p t} \tag{5.1}$$

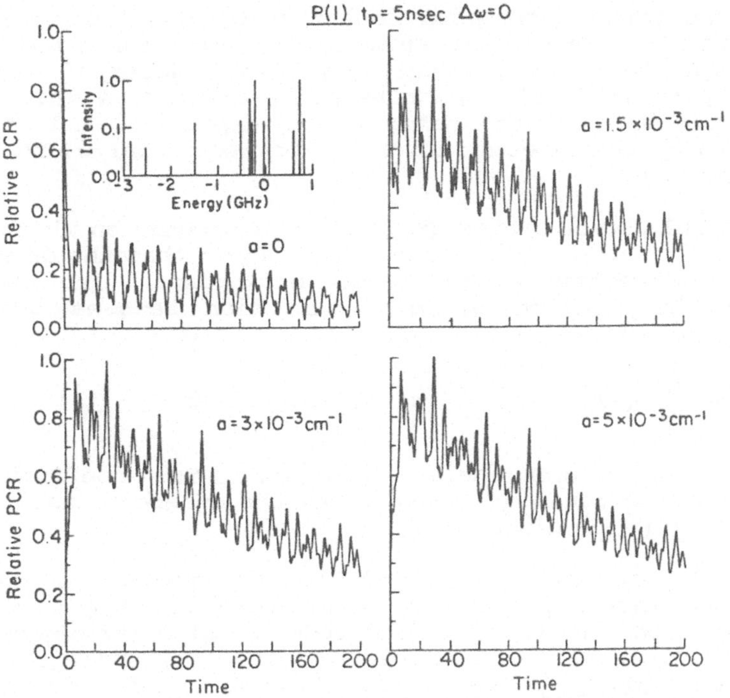

Fig. 3. The PCR for the P(1) line of a pyrazine excited by a short laser pulse (5 nsec), on resonance ($\Delta\omega = 0$) with various slow diffusion rates. The inset shows the spectrum taken from ref. (13).

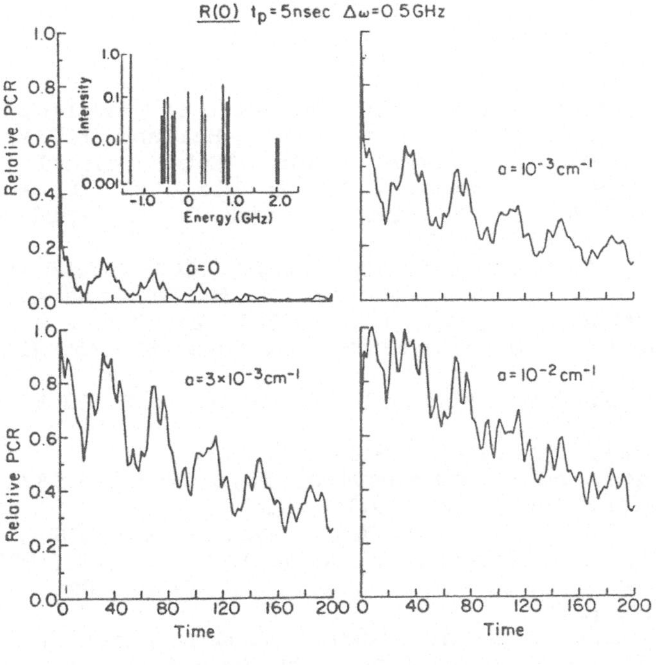

Fig. 4. Like Fig. 3 for R(0), off resonance ($\Delta\omega = 0.5$ GHz).

where the two terms correspond to the resonant fluorescence and NRLS, respectively. A similar situation prevails for a multilevel system, except that when a (moderately) large number of congested but well separated levels are coherently excited, a third decay component will be exhibited being due to dephasing, and which is characterized by the lifetime given by the reciprocal width of the level distribution.

For the multilevel system (MLS) the PCR depends on whether the pulse is short or long. It has been shown above that the QB are not eroded under an extremely incoherent excitation. The NLRS are eroded and the limiting dynamical function becomes under an incoherent time resolved excitation ($a \gg \gamma_p \gg r_j$)

$$J_{jj'}(t) = \frac{2\epsilon^2}{a\gamma_p}\{\exp[-(r_j+r_{j'})t+i(E_j-E_{j'})t] - \exp(-\gamma_p t)\} \qquad (5.2)$$

which resembles a bimolecular reaction. The coefficient is independent of the detuning, contrary to the coherent case and the short component (NLRS) disappears.

As a model system we explored the consequences of the high resolution spectrum of pyrazine-h_4, which was studied by Kommandeur et al.[13] These spectra exhibit a multitude of molecular eigenstates. We simulated excitation of the two members P(1) and R(o) of the $'B_{3u}$(o-o) transition by a 5 nsec pulse. In Figs. 3, 4 for P(1) and R(o) respectively, it is shown that the short component, i.e. the NLRS disappears when the stochastic effects of the laser increase. Since the Groningen group[13] utilized a non-ideal laser, their interpretation of the short decay component under the 5 nsec excitation in terms of NRLS is intriguing and should be carefully considered.

REFERENCES

1. J. Jortner, and S. Mukamel, in: "The World of Quantum Chemistry," R. Daudel, and B. Pullman, eds., Reidel, Holland (1973).
2. J. Jortner, and S. Mukamel, in: "International Review of Science, Physical Chemistry Series Two, Vol. 1, Theoretical Chemical," A. D. Buckingham, and C. A. Coulson, eds., Butterworth, London (1975).
3. S. Mukamel, and J. Jortner, Chem. Phys. Lett. 40:150 (1976).
4. H. Haken, in: "Quantum Optics," S. M. Kay, and A. Maitland, eds., Academic Press, New-York, pp 201-321 (1970).
5. P. Avan, and J. Cohen-Tannoudji, J. Phys. B10:155 (1977); 171 (1977).
6. P. Zoller, J. Phys. B10:L321 (1977); B11:805 (1978); B11:2825 (1978).
7. G. S. Agarwal, Phys. Rev. A18:1490 (1978).
8. A. T. Georges, and P. Lambropoulos, Phys. Rev. A18:587 (1978).
9. K. Wódkiewicz, Phys. Rev. A19:1686 (1979).
10. I. Schek, and J. Jortner, J. Chem. Phys. 81:4858 (1984).
11. S. Mukamel, and J. Jortner, J. Chem. Phys. 62:3609 (1975).
12. P. F. Williams, D. L. Rosseau, and S. H. Dworetsky, Phys. Rev. Lett. 32:196 (1974).
13. B. J. van der Meer, H. Th. Jonkman, J. Kommandeur, W. L. Meerts, and W. A. Majewski, Chem. Phys. Lett. 92:565 (1982).

INTERFEROMETRY IN THE NONLINEAR SPECTROSCOPY OF GASES

P.M. Felker, G.V. Hartland, L.L. Connell,
T.C. Corcoran, and B.F. Henson

Department of Chemistry and Biochemistry
University of California
Los Angeles, CA 90024-1569

INTRODUCTION

It is well known that using a Michelson interferometer to obtain spectroscopic information in a domain Fourier-conjugate to the spectral domain, and then Fourier transforming the data to obtain a spectrum,[1] can have significant advantages over measuring the same spectrum directly in the spectral domain. Fourier transform infrared absorption and emission spectroscopy,[2] Fourier transform UV/visible absorption and emission spectroscopy,[3] and Fourier transform Raman spectroscopy[4] are examples of methods for which such advantages obtain. Recently, there has been increasing interest in applying some of the interferometric techniques used in these linear spectroscopies to *nonlinear* spectroscopic schemes. For example, Hartmann, et al.,[5] Golub and Mossberg,[6] and Morita, et al.[7] have reported results of experiments in which interferometric versions of fully resonant four-wave mixing were used to resolve level splittings in atoms.

In this laboratory we have demonstrated nonlinear interferometric techniques based on coherent Raman scattering[8] and stimulated emission pumping.[9] These techniques, which have particular promise with regard to doing high resolution spectroscopy on sparse gases and molecular beam samples, are the subjects of this paper. In particular, we shall consider the following. First, we shall review some of the rudiments of linear interferometric spectroscopies as a background to nonlinear interferometry. Second, we consider a generic nonlinear interferometric experiment, both to reveal the basics of the experimental scheme and to show how resonances of the sample are manifested in the "interferograms" that are measured in the experiment. Third, we consider the particular case of interferometry applied[8] to coherent Raman scattering[10] (i.e., Fourier transform coherent Raman spectroscopy - *FTCRS)*. Fourth, we present experimental results that demonstrate FTCRS. Fifth, we show that an interferometric version[9] of stimulated emission pumping spectroscopy[11] (i.e., Fourier transform stimulated emission pumping spectroscopy - *FTSEPS)* works in a manner similar to FTCRS. Finally, we conclude by summarizing the advantages of interferometric versions of nonlinear spectroscopies over their spectral domain counterparts.

LINEAR INTERFEROMETRY

As a prelude to a consideration of the use of interferometric techniques in nonlinear spectroscopies, it is useful to review some of the aspects of linear interferometric methods.[1] We consider the experiment depicted schematically in Fig.1. Fig.1a is a level diagram pertinent to the experiment. A species in ground-state $|0\rangle$ is excited in a linear absorption process to excited-state $|1\rangle$. (We take the angular frequency of the transition to be Ω and assume that the line-shape of the resonance is a delta-function $L(\omega) = a\,\delta(\omega-\Omega)$, where a is a constant.) To monitor the absorption process, time- and frequency-integrated fluorescence from $|1\rangle$ is detected, giving rise to signal I. Now, one can envision performing an experiment on this species in one of two ways. First, one might scan the frequency of a narrow band-width excitation source and monitor I. Such a spectral domain experiment would give rise to an I vs. ω spectrum comprised of a delta-function at the resonance frequency Ω.

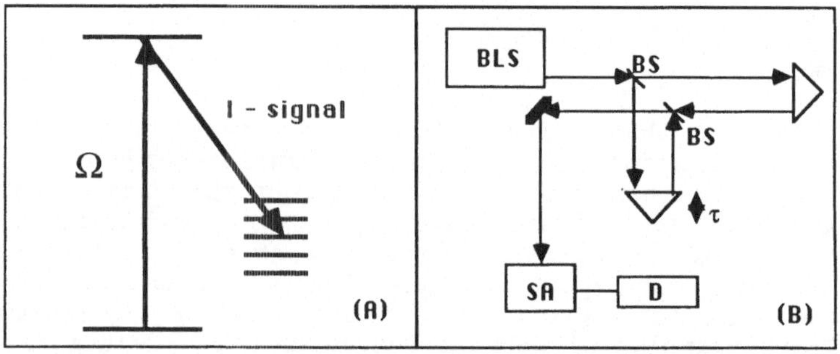

Fig. 1. (A) Level diagram pertaining to the linear spectroscopic process considered in section II of the text. (B) Schematic of a linear interferometric experiment: BLS=broad-band light source, BS=50/50 beam splitter, SA=sample, D=detector, and τ represents the interferometer delay.

Alternatively, one might do the experiment interferometrically. The apparatus for such an experiment is represented schematically in Fig. 1b. A broad band-width light source (spectral density $f(\omega)$) is directed through a Michelson interferometer, one of the arms of which can be varied to give rise to a variable delay τ between the light beams traversing the two arms of the interferometer. The output of the interferometer has spectral density $S(\omega,\tau) = f(\omega)(1 \pm \cos\omega\tau)$. The interferometer output is made to impinge on the sample, and the fluorescence intensity from the sample is measured as a function of τ. The pertinent question with regard to this experiment is how does $I(\tau)$, the interferogram measured in the experiment, relate to the absorption spectrum of the sample?

$I(\tau)$ can be readily found since it is proportional to the integral over ω of the absorption spectrum of the sample times the spectral density of the light source:[1]

$$I(\tau) \sim \int L(\omega)\, S(\omega,\tau)\, d\omega. \qquad (1)$$

Given the preceding expressions for $L(\omega)$ and $S(\omega,\tau)$, eq. (1) gives

$$I(\tau) \sim a[1 \pm \cos\Omega\tau] f(\Omega). \tag{2}$$

More useful, perhaps, the Fourier transform of $I(\tau)$ is

$$\mathcal{I}(|\omega|) \sim a[\delta(|\omega|) \pm \delta(|\omega| - \Omega)]\ f(\Omega) \tag{3}$$

Apart from the $\delta(|\omega|)$ term, $\mathcal{I}(|\omega|)$ just gives the absorption spectrum of the sample. Therefore, by doing an interferometric experiment as described, one obtains the same information as one would if one were to do a spectral domain experiment. Moreover, this statement is true in the more complicated situation involving multiple absorption resonances - i.e., when $L(\omega) = \sum a_i\ \delta(\omega - \Omega_i)$. In this case, it can be shown that the Fourier transform of the interferogram $I(\tau)$ obtained in the experiment will be of the form:

$$\mathcal{I}(|\omega|) \sim \sum a_i\ [\delta(|\omega|) \pm \delta(|\omega| - \Omega_i)]\ f(\Omega_i) \tag{4}$$

Clearly, $\mathcal{I}(|\omega|)$ is the absorption spectrum of the sample times an apparatus function $f(\Omega_i)$.

NONLINEAR INTERFEROMETRY

Now, instead of a linear spectroscopic process, consider a non-linear (multiphoton) one. In particular, consider a process involving the absorption of a photon at frequency ω_1 and the stimulated emission of a photon at ω_2 (Fig.2a). Furthermore, assume that there is a resonance condition such that the spectrum of the sample with respect to this process is given by

$$L(\omega_1, \omega_2) = a\ \delta(\Omega - \omega_1 + \omega_2), \tag{5}$$

where, again, a is a constant and Ω is the resonance frequency of the sample. Finally, assume that there is some means of monitoring whether the process occurs, such a probe scheme giving rise to some signal I proportional to the number of absorption/emission processes that take place.

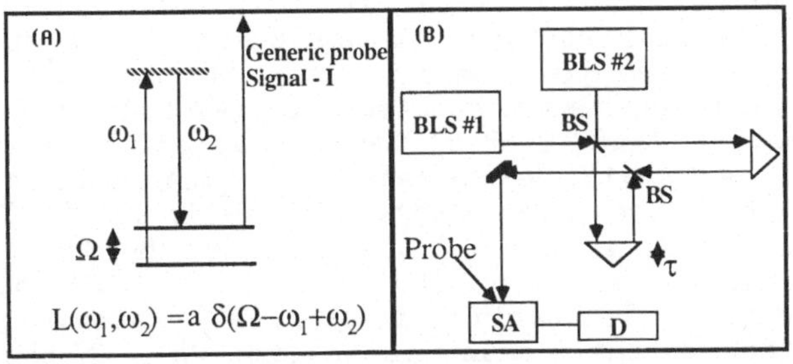

Fig. 2. (A) Level diagram pertaining to the nonlinear process considered in section III of the text. (B) Schematic of a nonlinear interfer-ometric experiment. The abbreviations have the same meanings as in Fig. 1.

If, on the system of Fig.2a, one were to do a spectral domain experiment in which two narrow band-width sources impinged on the sample, and the frequency difference between the two sources $(\omega_1-\omega_2)$ was scanned while I was monitored, one would observe a delta-function at $(\omega_1-\omega_2) = \Omega$. On the other hand, one can also imagine an interferometric experiment analogous to the linear interferometric absorption experiment of Fig.1b. Such an experiment is represented schematically in Fig.2b. In this nonlinear experiment $I(\tau)$ is given by an integral over excitation frequencies of products of spectral densities of the excitation sources times the spectrum of the sample[12]

$$I(\tau) \sim \iint L(\omega_1,\omega_2)\, S_1(\omega_1,\tau)\, S_2(\omega_2,\tau)\, d\omega_1\, d\omega_2 \qquad (6)$$

where $S_1(\omega_1,\tau)$ is the spectral density of broadband source #1 after the interferometer and $S_2(\omega_2,\tau)$ has the same meaning for broadband source #2. As in the linear interferometric experiment $S_1(\omega_1,\tau) \sim f_1(\omega_1)[1 \pm \cos\omega_1\tau]$ and $S_2(\omega_2,\tau) \sim f_2(\omega_2)[1 \pm \cos\omega_2\tau]$. Furthermore, one knows $L(\omega_1,\omega_2)$ - eq. (5). Performing the integration over ω_2 in eq. (6) and doing some algebra, one finds that

$$I(\tau) \sim a \int f_1(\omega_1)\, f_2(\omega_1-\Omega)\, [2 \pm \cos\Omega\tau \pm \cos(2\omega_1-\Omega)\tau \pm 2\cos(\omega_1\tau) \pm 2\cos(\omega_1-\Omega)\tau]\, d\omega_1. \qquad (7)$$

$I(\tau)$ is the sum of a term constant in τ, a resonance modulation term (proportional to $\cos\Omega\tau$), and three high frequency modulation terms. There are several good reasons to neglect the high frequency terms.[8] If one does so, the Fourier transform of $I(\tau)$ is given by

$$\mathcal{I}(|\omega|) \sim a\,[\, 2\delta(|\omega|) \pm \delta(|\omega|-\Omega)\,] \times \int f_1(\omega_1)\, f_2(\omega_1-\Omega)\, d\omega_1. \qquad (8)$$

Just as in the linear interferometric experiment, the Fourier transform of the interferogram (aside from the zero-frequency and high frequency terms) is the spectrum of the sample weighted by an apparatus function. Furthermore, when multiple resonances obtain in the two-photon process:

$$L(\omega_1,\omega_2) \sim \sum a_i\, \delta(\Omega_i - \omega_1 + \omega_2),$$

then $\mathcal{I}(|\omega|)$ takes the form

$$\mathcal{I}(|\omega|) \sim \sum [a_i\, (2\delta(|\omega|) \pm \delta(|\omega|-\Omega_i)\,) \times \int f_1(\omega_1)\, f_2(\omega_1-\Omega_i)\, d\omega_1] \qquad (9)$$

and gives the spectrum of the sample.

FOURIER TRANSFORM COHERENT RAMAN SPECTROSCOPY (FTCRS)

Having considered the general features of nonlinear interferometry in the previous section, we now consider the specific case of interferometry applied to coherent Raman scattering[10] - so-called Fourier transform coherent Raman spectroscopy (FTCRS).[8] A level diagram pertinent to coherent (anti-Stokes) Raman scattering is show in the inset to Fig. 3. We shall be concerned in particular with a CARS experiment in which *1)* the three excitation fields in the CARS process (i.e.,the photons at ω_1, ω_2, and ω_1') are provided by two "broadband" lasers (i.e., ω_1 and ω_1' are from the same laser), *2)* the two laser outputs are directed collinearly (or parallel) through a Michelson interferometer prior to the sample (as in Fig. 2b), *3)* the spectrally and temporally integrated intensity of the coherent Raman scattering is measured as a function of interferometer delay to yield an interferogram $I(\tau)$, and *4)* the Fourier transform of $I(\tau)$ (i.e., $\mathcal{I}(|\omega|)$) is obtained. One can analyze this experiment in a manner similar to the

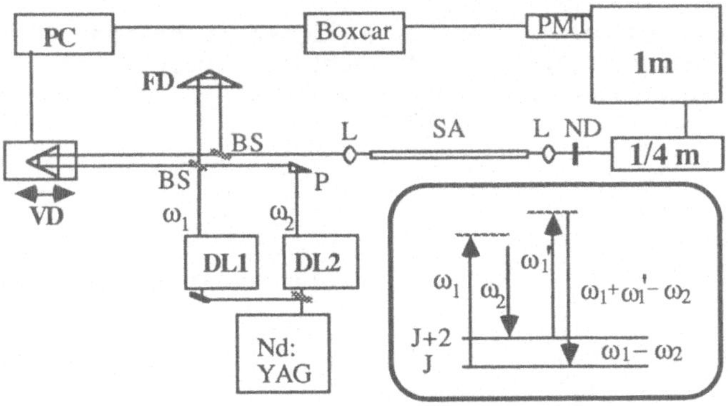

Fig. 3. Schematic of the experimental apparatus used for FTCRS. DL1 and DL2=dye-lasers, BS=beam splitters, P=prism, FD=fixed delay, VD=variable delay, L=lens, SA=sample, ND= filters, 1/4m and 1m = one-quarter and one meter monòchromators, respectively. Inset is a level diagram pertaining to pure rotational CARS in a diatomic.

analysis represented by eqs. (6) to (9). Carrying through such an analysis,[8b] one finds that $\mathbf{J}(|\omega|)$ is the sum of four types of terms: a zero-frequency term, high frequency terms, "broadband" terms, and *a resonance term*. That which interests us here is the resonant part of $\mathbf{J}(|\omega|)$ - i.e., $\mathbf{J}_R(|\omega|)$. Notably, $\mathbf{J}_R(|\omega|)$ can be readily isolated from the other (undesirable) Fourier terms.[8b] $\mathbf{J}_R(|\omega|)$ is just given by the following:

$$\mathbf{J}_R(|\omega|) \sim |L(|\omega|)|^2 \times [\int f_1(\omega_1) f_2(\omega_1-\Omega)\, d\omega_1 \times \int f_1(\omega_1')\, d\omega_1'], \tag{10}$$

where $L(\omega)$ is the Raman susceptibility of the sample, $f_1(\omega)$ and $f_2(\omega)$ are the spectral densities of the laser outputs (before the interferometer), and it has been assumed that every frequency of coherently scattered light is detected with unit efficiency. Clearly, $\mathbf{J}_R(|\omega|)$ is just the product of a function describing the Raman spectrum of the sample times an apparatus spectral window function (the factor in brackets in eq.(10)) determined by the spectral properties of the apparatus. (Note the similarity between eq.(10) and eq.(9).) The upshot is that FTCRS allows one to measure the CRS spectrum of a sample with spectral resolution unaffected by the band-width of the lasers and the detection system used. Spectral resolution is limited instead *1)* by the Raman linewidths of the sample and *2)* by the range over which one scans the interferometer delay (as in linear interferometric methods).

To demonstrate[8] FTCRS experimentally the apparatus depicted schematically in Fig.3 was used. Interferograms on a variety of gaseous species including O_2, N_2, and benzene were measured. Fig.4 shows an example of the results obtained. In a FTCRS experiment on pure rotational transitions in air at atmospheric pressure and room temperature the excitation lasers were set such that the window function of the apparatus (band-width of ca.3 cm^{-1}) overlapped one N_2 resonance and one O_2 resonance near 60 cm^{-1}. The interferogram obtained was Fourier transformed to produce the spectrum of Fig.4a.[8b] The two clearly resolved lines in the Fourier spectrum occur at just the frequencies one would expect them to occur at given knowledge of N_2 and O_2 spectroscopy. Fig.4b is the Fourier transform of a FTCRS interferogram corresponding to pure rotational transitions in benzene. All of the lines observed can be readily assigned to benzene lines.[8b] Thus, Fig.4 clearly demonstrates the theoretical result of eq. (10) - *coherent Raman spectroscopy can be done by using broadband sources and the interferometric technique FTCRS.*[8b]

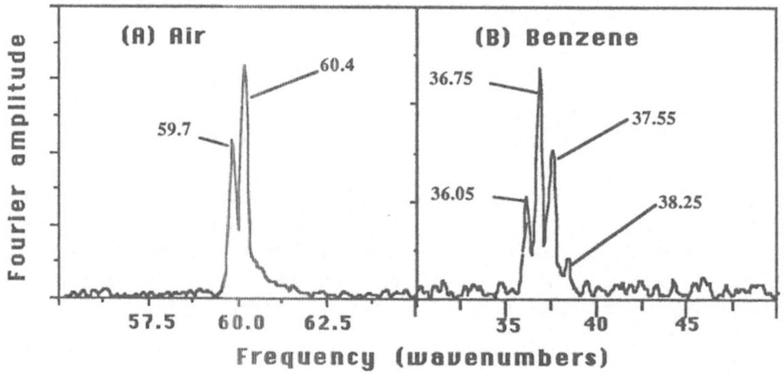

Fig. 4. Fourier spectra of FTCRS interferograms for (A) air and (B) benzene vapor.

FOURIER TRANSFORM STIMULATED EMISSION PUMPING SPECTROSCOPY (FTSEPS)

Coherent Raman scattering is not the only nonlinear process that fits into the general scheme of Fig.2a. Stimulated emission pumping spectroscopy (SEPS)[11] is an example of another process that does. Given this, it should be possible to demonstrate an interferometric, Fourier transform version of SEPS.

Fig.5a shows a level diagram pertinent to SEPS.[11] A species in ground-state level |0> is excited by one laser (the "pump" laser) to excited-state level |1>. Subsequently, the output of a second laser (the "dump" laser) drives the stimulated emission transition from |1> to a final ground-state level |2>. This "dump" process is monitored by measuring the frequency- and time-integrated intensity of spontaneous emission from |1>.

Typically, SEPS is done in the spectral domain by monitoring spontaneous emission intensity from a sample while the frequency of a narrow band-width dump laser is scanned. As stated above, however, by using interferometry one should also be able to do SEPS in a domain Fourier-conjugate to the spectral domain. To demonstrate[9] this the outputs of two broad-band lasers (pump and dump) were directed collinearly through a Michelson interferometer (as in Fig.2b). The output of the interferometer impinged on a supersonic molecular beam of I_2 seeded in He. Time- and frequency-integrated fluorescence was then detected (at right angles to the laser beams and the molecular beam axis) as a function of the delay between the arms of the interferometer.

Fig.5b shows FTSEPS experimental results[9] on I_2. The upper (unmodulated) trace corresponds to a situation in which the dump laser was tuned off of any possible stimulated emission resonance. The lower trace corresponds to a situation in which SEP was possible; the pump

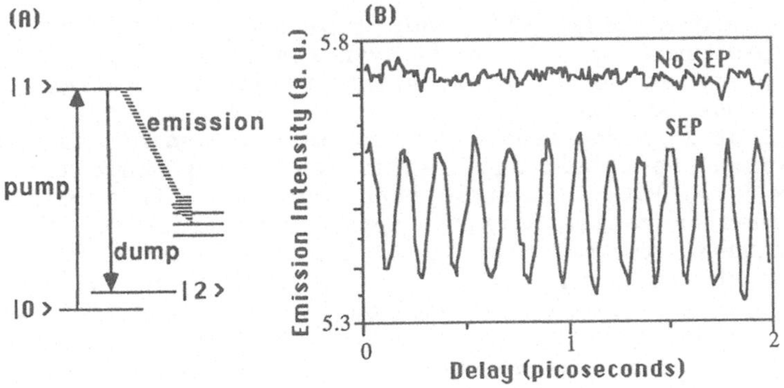

Fig. 5. (A) Level diagram pertaining to stimulated emission pumping spectroscopy. (B) FTSEPS interferograms on jet-cooled I_2.

laser was tuned to the v'=19 - v"=0 vibronic resonance of the molecule and the dump was tuned to the v'=19 - v"=1 resonance. Fourier analysis of a longer version of the modulated interferogram reveals that the modulations correspond to a frequency of 214 ± 1 cm^{-1},[9] a value within experimental error of accepted values[13] for the v"=1, v"=0 splitting of I_2. These results show that SEPS, too, can be done interferometrically as per the general nonlinear interferometric scheme of section III. In this regard it is pertinent to add that we have observed modulations like those in Fig.5b at interferometer delays two orders of magnitude greater than the coherence times of our laser sources. In addition, we have observed manifestations of the rotational level structure in FTSEPS experiments on I_2. The upshot is that spectral resolution in FTSEPS, like that in FTCRS (and linear interferometric methods like FTIR), is unaffected by the band-widths of the excitation sources involved.

CONCLUSIONS

In closing it is pertinent to address two questions.

First: Why use interferometric methods over spectral domain methods in nonlinear spectroscopies? There are a number of reasons to suppose that interferometric versions of nonlinear spectroscopies eventually will become more useful (at least in some applications) than their spectral domain counterparts. These reasons include the following:
1) Generally, it is considerably cheaper and more convenient to use high-power, broad-band lasers that need not be tuned during the course of an experiment, than it is to use high-power, narrow-band lasers whose frequencies must be scanned. Now, the latter type of light source is required if one wishes to do high resolution nonlinear spectroscopy in the spectral domain. On the other hand, one can use the former type of source and still obtain high spectral resolution if one uses interferometric methods. *2)* Linear interferometric spectroscopies are known for the routine accuracy that they can provide in wavenumber measurements.[1] One expects such accuracy to obtain in nonlinear interferometric spectroscopies, too. *3)* Fourier

transform techniques offer the ready capability of obtaining a high resolution spectrum (e.g., 0.01 cm^{-1}) over a wide spectral range (e.g., tens to hundreds of cm^{-1}) by measuring a single interferogram.[1] In contrast, without sophisticated and expensive electro-optical control in the tuning of a narrow-band laser, it is not possible in a spectral domain experiment to obtain such resolution over a range much greater than ca. 1 cm^{-1}. *4)* Interferometric techniques are often characterized by well defined or easily measured apparatus functions.[1] In spectral domain techniques, on the other hand, the apparatus function depends intimately on the spectral profiles of the excitation sources, one of whose wavelength is typically scanned during the experiment. The point is that in situations where knowledge of the apparatus function is crucial (e.g., in the determination of line-shapes), one is often better served by a Fourier transform technique than a spectral domain one.

Second: What other nonlinear spectroscopies can one subject fruitfully to interferometric schemes? Others have previously demonstrated interferometric versions of fully resonant four-wave mixing.[5-7] Besides these, FTCRS, and FTSEPS, one expects that, for example, Raman-induced Kerr effect spectroscopy (RIKES)[10] and spectroscopies based on stimulated Raman scattering (SRS)[10] may also be performed interferometrically in a Fourier-conjugate domain. Regarding the determination of excited electronic state splittings using nonlinear interferometry, we have obtained preliminary results using a method based on the ground-state depletion that occurs in a sample subsequent to its interaction with a vibronically resonant light pulse. Work is currently in progress both to develop new techniques and to demonstrate their efficacy in studies of gaseous and molecular beam samples.

ACKNOWLEDGMENTS

It is a pleasure to acknowledge Prof. M. A. El-Sayed for generous loans of equipment. Work was supported, in part, by the Petroleum Research Fund administered by the ACS, Research Corp., UCLA Academic Senate, Newport Corp., and a NSF Presidential Young Investigator Award to PMF.

REFERENCES

1. For reviews of (linear) Fourier spectroscopies and interferometric methods see, for example, (a) R.J. Bell, *Introductory Fourier Transform Spectroscopy* (Academic, New York, 1972); (b) J.W. Brault, "Fourier transform spectrometry" (pre-print); and references contained in these works.
2. There is a vast literature on FT-IR. Early papers by two of the pioneers of this technique and Fourier transform spectroscopy in general are (a) P.B. Fellgett, J.Phys.Radium **19**, 187 (1958); (b) P. Jacquinot, Rep.Progr.Phys. **23**, 267 (1960).
3. P. Luc and S. Gerstenkorn, Appl.Opt. **17**, 1317 (1978).
4. For example, (a) T. Hirschfeld and B. Chase, Appl.Spectrosc.**40**, 133 (1986); (b) V.M. Hallmark, C.G. Zimba, J.D. Swalen, and J.F. Rabolt, Spectrosc. **2**, 40 (1987); and references in both of these papers.
5. (a) D. Debeer, L.G. van Wagenen, R. Beach, and S.R. Hartmann, Phys.Rev.Lett. **56**, 123 (1986); (b) R. Beach, D. Debeer, and S.R. Hartmann, Phys. Rev. A **32**, 3467 (1985).

6. J.E. Golub and T.W. Mossberg, in *Ultrafast phenomena,* Vol. 5, edited by G.R. Fleming and A.E. Siegman (Springer, Berlin, 1986), p. 164.
7. N. Morita, T. Yajima, and Y. Ishida, in *Ultrafast phenomena,* Vol. 4, edited by D. H. Auston and K. B. Eisenthal (Springer, Berlin, 1984), p.239.
8. (a) P.M. Felker and G.V. Hartland, Chem.Phys.Lett. **134**, 503 (1987); (b) G.V. Hartland and P.M. Felker, J.Phys.Chem. - in press.
9. P.M. Felker, B.F. Henson, T.C. Corcoran, L.L. Connell, and G.V. Hartland - to be submitted.
10. For example, (a) M.D. Levenson, *Introduction to Nonlinear Laser Spectroscopy* (Academic, NY, 1982); (b) Y. R. Shen, *The Principles of Nonlinear Optics* (Wiley, NY, 1984); and references therein.
11. For example, C. Kittrell, E. Abramson, J.L. Kinsey, S.A. McDonald, D.E. Reisner, R.W. Field, and D.H. Katayama, J.Chem.Phys. **75**, 2056 (1981).
12. Strictly, this holds true for independent, stationary, stochastic excitation fields.
13. S. Gerstenkorn and P. Luc, *Atlas du Spectre d'Absorption de la Molecule d'Iode entre 14800 - 20000 cm*$^{-1}$ (CNRS, Paris, 1978).

EFFECTS OF EXCITED STATE ABSORPTION ON OPTICAL PULSE PROPAGATION

Yehuda B. Band

Department of Chemistry, Ben-Gurion University, Beer-Sheva
Israel and Allied-Signal Corporation, Morristown, N.J. 07960
U.S.A.

1. Introduction

In this paper I review the effects of excited state absorption on intense pulse propagation in gas, liquid or solid phase molecular media. I consider the propagation of short (compared with the lifetime of the excited state, τ) intense pulses in media with large excited state absorption, and also discuss briefly propagation of long pulses too. In particular, I consider materials known as reverse saturable absorbers (RSA's). A RSA is a material whose excited state absorption cross section at wavelength λ, $\sigma_{ex}(\lambda)$, is larger than the ground state absorption cross section, $\sigma_{gr}(\lambda)$, and which possesses four other necessary properties detailed below. Recently, the optical properties of RSA's and their applications to a number of areas in optical pulse processing and laser physics have been described.[1-6] RSA's can be used as optical pulse shorteners and optical energy limiters for laser pulses whose temporal widths are small compared with τ.[1] A RSA in conjunction with a saturable absorber can be used to mode-lock lasers whose gain media have high saturation energies (the saturable absorber cuts the leading edge of the pulse and the RSA cuts the trailing edge of the pulse)[2]. Moreover, RSA's stabilize high gain mode-locked or CW lasers against the onset of relaxation oscillations.[2] A combination of RSA and saturable absorber can also be used as an extracavity optical pulse compressor to eliminate the undesirable wings of short optical pulses which may have been produced by the process of superfluorescence.[7] Such a pulse compressor can be inserted between the amplifier stages of a laser amplifier chain or after a single amplifier. RSA's can also be used as power limiters and pulse smoothers for pulses whose rate of change of intensity is slow compared with τ, the decay time of the first excited state of the RSA.[1] There are many examples of RSA's at specific wavelengths, however, a suitable RSA may not be readily available, at least not readily known, for a given wavelength of interest. This is due in part to the paucity of information available regarding excited state absorption cross sections as compared with ground state absorption cross sections. Thus, choosing appropriate candidates for RSA's at a given wavelength can prove difficult. However, as an illustration of how common RSA is let us consider the familiar laser dye rhodamine 6G. Fig. 1 shows the ground state absorption and excited state absorption minus emission cross sections vs wavelength for rhodamine 6G as mesasured by Hammond[8]. It is clear from the figure that in the region around 435 nm, the excited state absorption minus emission cross section is larger than the ground state absorption cross section.

Here we review the properties and applications of RSA's.[1-6] We analyze the propagation of optical pulses through RSA's, describe how to design RSA's for a desired wavelength, describe some of the applications of RSA's mentioned above, and report some novel pulse propagation properties in such materials[9].

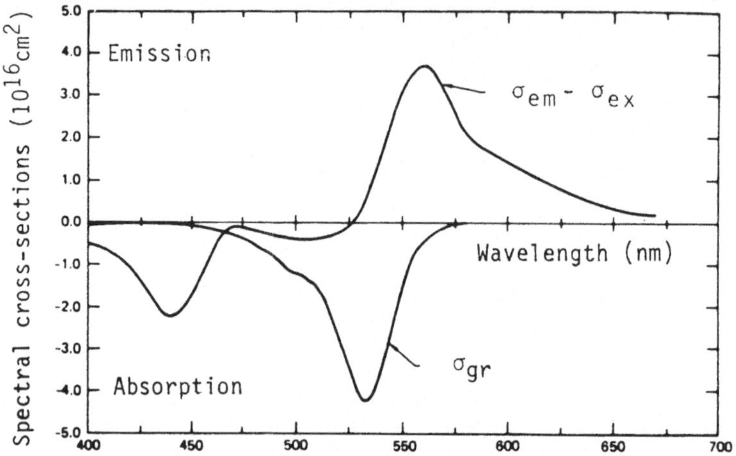

Fig. 1. Induced spectra of the S_1 and S_0 states of rhodamine 6G fluoroborate in ethanol.

2. Criteria for Successful RSA's

The criteria for molecules (in the gas, solid or liquid phase) to be successful reverse saturable absorbers are: (1) $\sigma_{ex}(\lambda)$ should be larger than $\sigma_{gr}(\lambda)$ at the desired wavelength. (2) $\sigma_{gr}(\lambda)$ should be sufficiently large to insure that the incident pulse saturates the ground state transition (at least somewhat). Otherwise excited state absorption can not occur. The pulse may be focused into the RSA sample so this is not too severe a criteria. (3) The second excited state of the RSA (the terminal level of the excited state absorption) should decay quickly (compared with the incident pulse width) back down to the first excited state) and neither the first nor the second excited states should decay to other levels during the pulse. This criterion helps insure the retention of the population in the first excited state and therefore the continued presence of excited state absorption. (4) the emission cross section $\sigma_{em}(\lambda)$ should be small at the desired wavelength. Recall that for large molecules the emission is shifted to the red relative to the absorption so it is generally possible to meet this criteria. (5) the decay time, τ, of the first excited state back down to the ground state should be large compared to the pulse width, τ_p, for pulse-smoothing or power limiting applications, and should be small compared to the pulse width for pulse-shortening or energy limiting applications. This criterion will be explained more fully below.

3. Analysis of Propagation of a Pulse through a RSA

Consider a plane wave pulse with incident time dependent intensity $I_p(t)$ and with central wavelength λ. Let us describe the dynamics of the propagation of the pulse through a RSA of width L. The rate equation for the excited state population density of the RSA at position x and time t, N(x,t), and the wave equation for the intensity of the pulse of light within the RSA at position x at time t, I(x,t) take the form

$$\partial N(x,t)/\partial t = \{(N^0-N)\,\sigma_{gr} - N\sigma_{em}\}\, I/h\nu \; - \tau^{-1}N\,, \qquad (1)$$

$$\partial I(x,t)/\partial t + c\,\partial I(x,t)/\partial x \;=\; -c\{(N^0-N)\sigma_{gr} - N\sigma_{em} + \sigma_{ex}N\}\, I \qquad (2)$$

Here, N^0 is the concentration of RSA molecules, σ_{gr} is the RSA ground state absorption cross sections at wavelength λ, τ is the excited state lifetime of the RSA, σ_{em} is emission cross sections at wavelength λ, σ_{ex} is the excited state absorption cross section at wavelength λ, $h\nu$ is the photon energy, and c is the speed of light in the medium. In Eq. (1) we assumed that the doubly excited state of the molecule decays back down to the excited state on a time scale fast compared to the width of the pulse and therefore no term involving σ_{ex} appears. Initially before the arrival of the input pulse N(x,0)=0 and I(x,0)=0. The boundary condition is $I(0,t)= I_p(t)$. In what follows we shall assume that the emission cross section σ_{em} is negligible according to criterion (4) above.

There are three dimensionless parameters which aid us in analyzing the Eqs. (1-2): (a) τ_p/τ, (b) σ_{ex}/σ_{gr}, (c) I_p/I_{sat}, where the saturation energy is defined as $I_{sat}=h\nu/\tau\sigma_{gr}$ (or E_p/E_{sat} where E_p is the pulse energy per unit area, $E_p = \int dt\, I_p(t)$, and $E_{sat} = \tau I_{sat}$ is the saturation energy per unit area). When $\tau_p/\tau >> 1$ then the steady state approximation can be made in Eqs. (1-2). Eq. (1) is solved to yield

$$N(x) = \sigma_{gr} N^0 I(x)/\{I_{sat}[1 + I(x)/I_{sat}]\}. \tag{3}$$

Substitution of this result into Eq. (2) yields the differential equation

$$dI/dx = -\sigma_{gr} N^0 I(x)\{1 - (1-\sigma_{ex}/\sigma_{gr})I/(I(x) + I_{sat})\}. \tag{4}$$

In the limit $I_p << I_{sat}$ Eq. (4) gives $I(x) = I_p \exp(-\sigma_{gr} N^0 x)$ whereas if $I_p >> I_{sat}$, $I(x) = I_p \exp(-\sigma_{ex} N^0 x)$. Thus, the output $I(x=L)$ varies as a function of I_p. Since $\sigma_{ex} > \sigma_{gr}$ the relative output $I(L)/I_p$ decreases with increasing I_p the RSA behaves as a power limiter. Moreover, if there are slowly varying fluctuations in I_p, the RSA will behave as a pulse smoother, transmitting relatively more (less) output when I_p small (large).

When the pulse width is small, $\tau_p/\tau \leq 1$, and the medium is not optically thin, no analytic solutions to Eqs. (1-2) are known. Numerical solution of the equations show[1] that for $\sigma_{ex}/\sigma_{gr} > 1$ the medium behaves as an energy limiter and a pulse shortener if the pulse energy is sufficient to begin saturation, $E_p/E_{sat} > 1$. The leading edge of the pulse begins to saturate the medium, and upon saturation the transmission of the remaining part of the pulse decreases. The larger the pulse energy, the greater the reduction of the trailing edge of the pulse. Since the trailing edge of the pulse is eaten away, the pulse is shortened. However, to shorten the pulse from both the leading and trailing edges, the combination of a saturable absorber and a RSA is required.[3] Thus, the RSA behaves as an energy limiter and a pulse shortener for short pulses. For very short pulses other factors must be considered, for example, rotational diffusion time of the molecules[4], nonequilibrium distributions of vibrational levels of the electronic state manifolds, heating effects due to decay of the doubly excited state[9], etc, but we will not deal with these issues here.

4. Engineering RSA's for Desired Wavelengths

Since RSA's for arbitrary wavelengths are not readily known, it is important to be able to design molecules which behave as a RSA at a desired wavelength.[5] Briefly, the idea is to use a dimer molecule formed from monomers whose ground state absorption peaks near the wavelength of interest. If synthesis of the dimer with fixed orientation of the monomers at desired distances from one another is possible, and the dimer is sufficiently stable to withstand dissociation upon absorption of two quanta of radiation, and a number of other criteria are satisfied, the dimer may behave as a RSA at the desired wavelength. The dimer ground state absorption is to a state in which the excitation is spread over both monomeric units and the dimer excited state absorption commences from this state to the doubly excited electronic state in which both monomeric units are excited. To zeroth order in the interactions between monomers, the energy separation between the ground and first excited states and between the first excited and doubly excited states are identical and equal to the monomer ground state transition energy. The transition enrgies of these two transition bands will shift due to the mononmer interactions and if the desired wavelength is near the peak of the excited state absorption band the condition $\sigma_{ex} > \sigma_{gr}$ can be met. It is then expected that RSA can occur since the other criteria listed in Sec. 2 can also be met. The details of the interactions which give rise to this phenomenon are contained in Ref. 5.

5. Pulse Compressor

We show that an optical pulse compressor, a device to remove the leading and trailing edges of a subnanosecond pulse, can be made of a first stage composed of a saturable absorber (SA) whose function is to reduce the leading edge of the pulse, and a second stage composed of a reverse saturable absorber (RSA) which functions to reduce the trailing edge of the pulse.

Often, short intense pulses of light are produced in oscillator- amplifier chains wherein the oscillator prepares a short stable pulse which passes through a succession of amplifier stages. However, there are a number of deleterious processes (e.g. superfluorescence) which degrade the quality of the short pulse produced by the oscillator upon traversal through the amplifier chain.[1]

Here we describe an optical compressor which can be used between amplifier stages or after a single amplifier stage to remove the temporally wide noise surrounding a central pulse. If an RSA is used in conjunction with a SA, both the leading and trailing edges of a pulse can be reduced in intensity relative to the central portion of the pulse. Thus, the temporal width of a light pulse with central wavelength λ can be compressed by selective elimination of the tails of the pulse. This compression reduces the noisy tails produced via superfluorescence. We demonstrate that for effective extra-cavity pulse compression and pulse shortening, the RSA and SA should be separated into distinct compartments, with the SA upstream relative to the RSA.

The intensity of the leading edge of the pulse is reduced using a SA with appropriate absorption cross section, concentration, optical thickness, etc. Very little absorption by the SA occurs after the leading edge of the pulse has passed since the SA transition is already saturated and few molecules are left in the ground state to absorb the trailing edge photons. The correct choice of the above mentioned parameters insures that only the desired fraction of the leading edge of the pulse is reduced in intensity. Using a RSA with the correct ground state absorption cross section, excited state cross section, concentration, optical thickness, etc., the intensity of the trailing edge of the pulse is selectively reduced by excited state absorption. The leading edge and central portions of the pulse are reduced somewhat by ground state absorption and consequently the excited state population density of the RSA begins to build up. By the time the desired portion of the trailing edge of the pulse has passed through the RSA, the excited state is sufficiently populated to insure that the absorption is dramatically increased due to excited state absorption.

For the case where the RSA and SA are mixed, the leading edge of the pulse populates the RSA as well as the SA. We separate the SA and RSA into separate compartments with the SA upstream relative to the RSA. This separation is desirable so that the leading edge of the pulse populates the SA excited state and not the RSA excited state, for then the absorption decreases with increasing time and the central portion of the pulse experiences minimum absorption. Only after saturation of the SA and passage of the central portion of the pulse through the RSA will the RSA excited state be populated and cause increased absorption. It is possible to optimize the choice for parameters of the SA and RSA for effective pulse compression.[4] A numerical example of pulse compression of a ps pulse whose central wavelength is 266 nm using a Coumerine dye as the RSA and quaterphenyl as the SA is presented in Ref 3. Fig. 2 shows the incident pulse shape and the pulse shape after emerging from the SA-RSA pulse compressor for this example. Both pulses are normalized relative to their own maximum intensity, i.e. we plot $I_0(t)/I_0(0)$ for the incident pulse (where $I_0(t)$ is the incident pulse intensity whose maximum occurs at t=0) and $I_{out}(t)/I_{out,max}$ for the output pulse. The ratio of the total output energy to total input energy is 43% (most of the reduction is in the noise which accounts for 75% of the input pulse). The intensity reduction at the maximum intensity is about 34% (i.e. $I_{out,max}/I_0(0)$ is 0.66), whereas the reduction of the leading and trailing edges of the pulse are about 86% and 60% respectively. The leading edge of the noise is almost completely eliminated and the trailing edge is drastically reduced. Thus, pulse compression can be successfully performed without significantly reducing the central pulse.

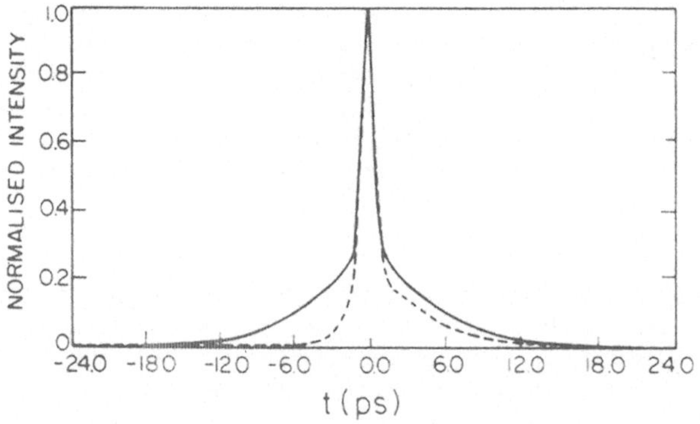

Fig 2. Incident pulse shape versus time (solid line) and exit pulse shape after passage through the medium (dashed line), each normalized to its maximum.

III. Passive Mode-Locking

We consider the effects of introducing SA and RSA within a laser cavity to mode-lock lasers. The introduction of SA and RSA into the laser cavity is shown to be particularly useful for passive mode-locking of lasers with gain media with high gain saturation energies wherein individual pulses cannot saturate the gain. Moreover, introduction of a RSA into the laser cavity stabilizes the laser against the onset of relaxation oscillations. In passively mode-locked lasers containing gain media with saturation energy too high to be saturated by individual mode-locked pulses, the trailing edge of the pulse continues to see gain and therefore the pulse is not terminated.[10] There is no mechanism to shorten the pulse to less than the SA relaxation time. Many examples of such gain media exist, including Nd:YAG, Nd:glass, ruby, and alexandrite. We shall show that inclusion of a RSA as an additional passive element will succeed in turning off the gain of the laser during the trailing edge of the pulse due to the increased absorption by the excited state of the RSA. Thus, for a gain medium with high gain saturation, the pulse width can be limited to the gain bandwidth of the gain medium and not the relaxation time of the SA.[2] Another problem of passively mode-locking high gain medium CW lasers is there instability against relaxation oscillations.[7] The addition of a RSA into the laser cavity suppresses relaxation oscillations. We thereby slay two birds with one stone in the CW passive mode-locking case.

An analytic solution to steady-state passive mode locking with a RSA can be obtained and a closed form solution for the pulse shape is determined. We consider a homogeneously broadened laser medium with small enough cross section σ_L and long enough lifetime τ_L so that a single pulse does not drive down the gain. The bandwidth of the laser is limited by an additional intracavity element which is narrow compared with the laser medium bandwidth. The SA and RSA are taken sufficiently thin so that longitudinal variations of the populations and intensities can be neglected. The effect of each element of the laser is assumed small enough so that the exponentials can be expanded and truncated to the first two or three orders. With these assumptions an integrodifferential equation for the temporal dependence of the amplitude of the electric field, A(t), can be obtained and the solution to this nonlinear operator equation can be obtained. It takes the form[7]

$$A(t) = (E_0/2\tau_p)^{1/2}\,\mathrm{sech}(t/\tau_p) \qquad (5)$$

and the values of the pulse energy E_0 and the pulse width τ_p are determined in terms of the parameters of the laser medium, the SA and the RSA. The details of the solution and a numerical example are presented in Ref. 7.

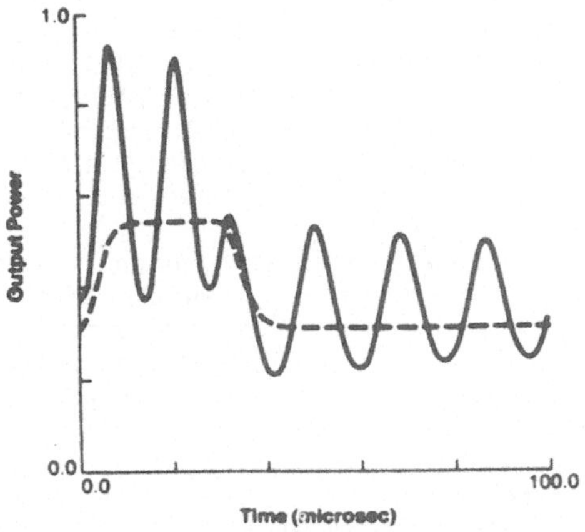

Fig. 3. Temporal response of a CW alexandrite laser to a 20 percent change in the pumping rate for 30 μs, without a reverse saturable absorber (solid line) and with a reverse saturable absorber (dashed line). The saturable absorber lowered the initial output power of the laser 15 percent.

The mode-locked pulses must be able to grow from noise to be self-starting. Therefore a perturbation of the field having a frequency which is a multiple of the round trip cavity time must be unstable to allow the passive mode-locked laser to self start. However, the laser must be stable against relaxation oscillations, otherwise it will not be CW and the output will be spiky. The stabilizing effect of including a RSA into the
laser cavity is illustrated in Fig. 3. The temporal response of a CW alexandrite laser to a 20% change in the pumping rate for 30 μs is shown as a solid curve, whereas the dashed curve shows the behavior with the addition of the RSA above RSA. The inclusion of the RSA lowers the steady-state output by 15%, but as Fig. 3 shows, the laser becomes stable against relaxation oscillations.

6. Propagation Effects Due to Increased Vibrational Temperature

There is a great deal of energy released in the nonradiative decay of the doubly excited state back down to the excited state manifold. This energy goes into the vibrational state manifold of the molecule before eventually dissipating into translational and rotational degrees of freedom of the molecular solution. There is also energy released in the decay of the excited state to the ground state level but only that part of the energy which does not go into spontaneous fluorescence passes through the vibrational manifold of the molecule. The increase in the vibrational energy of the molecule can affect optical pulse propagation. Recently, measurements of the rate of transfer of excitation into the vibrational manifold of states, and from the vibrational manifold of states into translational (and rotational) energy of the molecular solution were reported.[11] It was found that the time for energy transfer into the vibrational manifold is on the order of a ps whereas the time for transfer from vibrations into other bath degrees of freedom is significantly longer (for a solution of anthracene in C_2Cl_4 these times were measured to be 2 ps or less and 20 ps respectively). We shall assume that the time for energy transfer into the vibrational manifold is small compared with the pulse width and can therefore be neglected, and we denote the inverse of the transfer time from vibrations into other bath degrees of freedom, the vibrational energy transfer rate, as γ_{vt}. As discussed in Ref. 11, there is evidence for treating the vibrational excitation in terms of a vibrational temperature. We can write a rate equation for the vibrational temperature (in energy units) in the following form:

$$\partial_t T(x,t)=N_m^{-1}\{E_{dc}\,\sigma_{ex}|A|^2 2\varepsilon c_0/\omega+E_{ca}(1-y)\tau^{-1}\}N(x,t)-\gamma_{vt}(T(x,t)-T_b) \qquad (6)$$

where y is the quantum yield for decay of 2 to 1, $y=\tau/\tau(\text{radiative})$, E_{dc} and E_{ca} are the energy differences between the doubly excited and excited state and the excited state and ground state respectively, N_m is the number of vibrational modes that accept the transferred energy, and T_b is the bath temperature. In deriving Eq. (6) we have taken the vibrational temperature of the ground and excited levels as equal because of the coupling of the vibrational manifolds of these states due to absorption and stimulated emission (and spontaneous decay)and we have also assumed that the thermal distribution within these levels is a Boltzmann distribution whose effective temperature is the vibrational temperature obtained by solving Eq. (6). The first term on the right hand side of Eq. (6) is due to excited state absorption and the almost immediate decay of the excited state. This process adds an energy of E_{dc} to the vibrational manifold. The second term on the right hand side of Eq. (6) is due to spontaneous nonradiative decay of the excited state level. The last term is due to the energy transfer from vibrations into other bath degrees of freedom.

The effective cross sections and refractive indices depend upon the vibrational temperature.[9] Therefore, the propagation of an intense pulse through the medium raises the vibrational temperature and this rise in temperature affects the optical properties of the medium. For short intense optical pulses, the change of the cross sections due to the change of the vibrational temperature can be substantial, as is demonstrated in Ref 9. Fig. 4 shows the absorption coefficient calculated for rhodamine 6G to an intense (20 mJ/cm^2) short (20 ps) pulse of light whose central wavelength is 569 nm with and without taking the vibrational heating of the molecule into effect. Clearly there is a substantial difference between the results. Thus, excited state absorption can drastically effect pulse propagation of short intense light pulses.

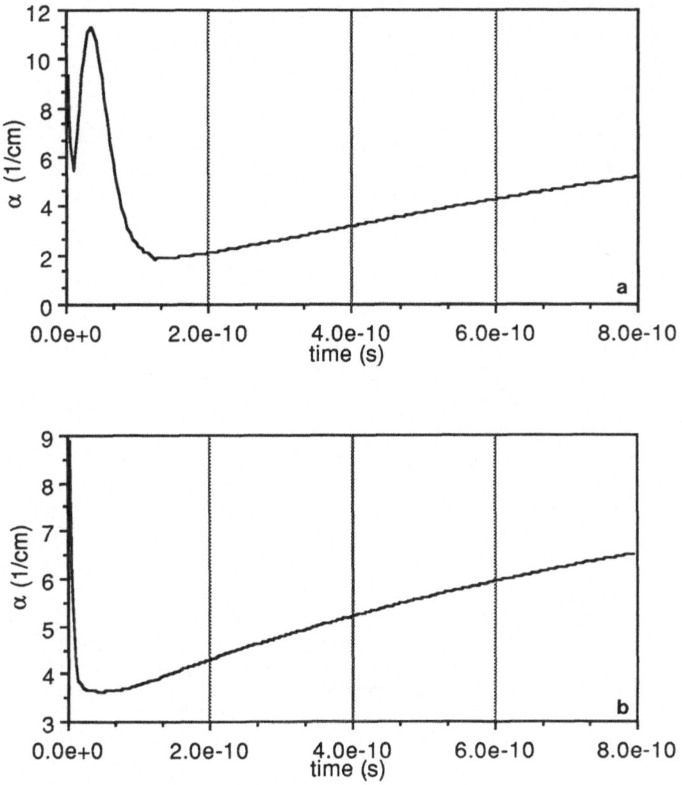

Fig. 4. Absorption coefficient α of the incident pulse vs. time. (a) molecular temperature varies during pulse, (b) frozen molecular temperature.

References

1. D. J. Harter, M. L. Shand, and Y. B. Band, J. Appl. Phys. 56, 865 (1984).
2. D. J. Harter, Y. B. Band, and E. P. Ippen, IEEE J. Quant. Electron. QE-21, 1219 (1985).
3. Y. B. Band, D. J. Harter, and R. Bavli, Chem. Phys. Lett. 127, 280-384 (1986).
4. Y. B. Band, J. Chem. Phys. 83, 5453-5457 (1985); Phys. Rev. 84, 326-332 (1986).
5. Y. B. Band and B. Scharf, Chem. Phys. Lett. 127, 381-386 (1986).
6. Y. B. Band, "Optical Properties of Reverse Saturable Absorbers", Methods of Laser Spectroscopy , (Plenum, N. Y., 1986), pp. 117-123; Y. B. Band and R. Bavli, "Extracavity Pulse Compressors and Intracavity Mode-Locking with Saturable and Reverse Saturable Absorbers", Methods of Laser Spectroscopy , (Plenum, N. Y., 1986), pp. 123-129.
7. W. Koechner, Solid-State Laser Engineering, (Springer-Verlag, N.Y., 1976), pp. 162-168.
8. P. Hammond, IEEE J. Quant. Electron. QE-16, 1157 (1980).
9. Y. B. Band and R. Bavli, "Nonlinear Optical Pulse Propagation in Molecular Media", Phys Rev. A, (in press).
10. H.A. Haus, IEEE J. Quant. Electron. QE-11, 169 (1976).

PROPAGATION OF ULTRASHORT LASER PULSES IN AN N-LEVEL MEDIUM

S. Chelkowski and A.D. Bandrauk

Département de chimie, Faculté des Sciences
Université de Sherbrooke
Sherbrooke, Québec, Canada J1K 2R1

INTRODUCTION

The coherent resonant interaction of ultrashort laser pulses with two-level system - self-induced transparency, soliton formation and pulse amplification - has been exhaustively discussed previously in the literature[1-3]. The main characteristic features of this phenomenon are:

a) conservation of certain pulse areas during pulse propagation along the z-axis expressed by the "area theorem"[3]:

$$\frac{dS}{dz} = \frac{-\alpha}{+2} \sin S \quad , \quad \text{"-" attenuator, "+" amplifier} \quad , \tag{1}$$

where

$$S = \frac{p_{12}}{\hbar} \int_{-\infty}^{\infty} \varepsilon(\tau,z)\, dt \quad , \tag{2}$$

is a dimensionless pulse area, $\varepsilon(t,z)$ is a slowly varying electric field envelope, p_{12} is the dipole moment of the considered transition and α is an absorption coefficient related to the nonhomogeneous broadening of the medium due to the Doppler shift of the moving atoms levels.

b) self-induced transparency phenomenon for areas $S = 2 \cdot n \cdot \pi$ which are related to the formation of n-solitons described by sine-Gordon equation:

$$\frac{\partial^2 \sigma}{\partial z \partial \tau} = - \frac{2\pi n_o \omega p_{12}^2}{\hbar} \sin \sigma \quad , \tag{3}$$

where

$$\sigma = \int_{-\infty}^{\tau} \varepsilon(\tau',z)\, d\tau' \quad , \quad \tau = \tau - z/c \tag{4}$$

and n_o is the number of molecules or atoms per cm^3.

c) maximal absolute pulse amplification for $S = \pi, 3\pi, \ldots$ in an amplifier.

In this note we investigate the possibility of observing similar phenomena when a medium consists of systems having N-equidistant, non-degenerate levels instead of the usual two-levels. We do this by considering the Maxwell-Schroedinger equations instead of the usual Bloch-Maxwell equations.

CALCULATION OF TRANSITION AMPLITUDES

Supposing that the pulse duration is shorter than any relaxation time of the medium and considering the exact resonance case, we can write the state vector of each atom or molecule in the form:

$$|\psi(t)> = \sum_{J=1}^{N} c_j \, e^{-i/\hbar E_j t} \, |\psi_j> \quad , \tag{5}$$

where $|\psi_j>$ are energy eigenvectors of each molecule. Denoting the radiative interaction potential with the system in the dipole approximation by V and using Schroedinger's time dependent equation one gets

$$\begin{aligned} i\hbar \dot{c}_1 &= V_{12} \, e^{-i/\hbar \, (E_2-E_1)t} \, c_2 \\ &\vdots \\ i\hbar \dot{c}_k &= V_{k,k-1} \, e^{i/\hbar \, (E_k-E_{k-1})t} \, c_{k-1} + V_{k,k+1} \, e^{i/\hbar \, (E_k-E_{k+1})t} \, c_{k+1} \\ &\vdots \\ i\hbar \dot{c}_N &= V_{N,N-1} \, e^{i/\hbar \, (E_N-E_{N-1})t} \, c_{N-1} \end{aligned} \tag{6}$$

where

$$V_{ij} = - p_{ij} \, \varepsilon(t,z) \cos (kz-\omega t) \quad ,$$

p_{ij} is the dipole moment of the transition from the i-th level to the j-th level and it is supposed that only transitions between adjacent levels are allowed i.e., that p_{ij} is non-zero only if $j = i + 1$ and $j = i - 1$. We have assumed that our system has equidistant levels, i.e., $E_{j+1} - E_j = \hbar\omega_o$ and that the incident light is resonant with the system, which means that $\omega = \omega_o$. Moreover, pulse duration although short is supposed to be much longer than optical cycle $2\pi/\omega$. Then rotating wave approximation (R.W.A.) is expected to be valid. It consists in neglecting terms containing factors $e^{\pm i2\omega t}$ in (6). Thus one gets the equations

$$\begin{aligned} \dot{c}_1 &= \frac{i\varepsilon(t,z)}{2\hbar} \, p_{12} c_2 \\ &\vdots \\ \dot{c}_j &= \frac{i\varepsilon(t,z)}{2\hbar} \, [p_{j-1,j} c_{j-1} + p_{j,j+1} c_j] \\ &\vdots \\ \dot{c}_N &= \frac{i\varepsilon(t,z)}{2\hbar} \, p_{N-1,N} \, c_{N-1} \end{aligned} \tag{7}$$

These equations were solved in ref. 5 for the plane wave case (E = const). In this note we solve it for any function $\varepsilon(t,z)$ and then we study the envelope modification when the pulse travels along z-axis. Introducing the same variable

$$\sigma = \frac{p_{12}}{\hbar} \int_{-\infty}^{t} \varepsilon(t',z) \, dt' \tag{8}$$

as used in two-level problem, one can transform the equation (7) into

$$\frac{d}{d\sigma}\begin{bmatrix} q_1 \\ \vdots \\ q_N \end{bmatrix} = iA \begin{bmatrix} q_1 \\ \vdots \\ q_N \end{bmatrix} \tag{9}$$

$$A = \begin{bmatrix} 0 & d_1 & 0 & 0 & \cdots & & & \\ d_1 & 0 & d_2 & 0 & \cdots & & & \\ 0 & d_2 & 0 & d_3 & \cdots & & & \\ \cdot & \cdot & \cdot & \cdot & \cdots & & & \\ \cdot & \cdot & \cdot & \cdot & \cdots & d_{N-2} & 0 & d_{N-1} \\ & & & & & 0 & d_{N-1} & 0 \end{bmatrix} \tag{10}$$

where

$$d_j = 1/2\ p_{j,j+1}/p_{12}$$

$$q_1 = e^{ikz}c_1 \quad , \quad q_2 = c_2$$

$$q_3 = e^{-ikz}c_3 \quad , \quad q_4 = c_4$$

$$q_5 = e^{ikz}c_5 \quad , \quad \text{etc}$$

The particular solutions of (9) are obtained by the substitution $q_j = v_j(\lambda)e^{i\lambda\sigma}$ which leads to the eigen equation of matrix A

$$A = \begin{bmatrix} v_1(\lambda) \\ \vdots \\ v_N(\lambda) \end{bmatrix} = \lambda \begin{bmatrix} v_1(\lambda) \\ \vdots \\ v_N(\lambda) \end{bmatrix} \tag{11}$$

which was discussed in ref. 5. If all coefficients d_j are different from zero all eigenvalues are different and thus the general solution of (9) satisfying initial conditions $q_j|_{t=-0} = \delta_{jm}$ has the form

$$q_j = \sum_{k=1}^{N} v_m(\lambda_k)v_j(\lambda_k)e^{i\lambda_k\sigma} \tag{12}$$

The characteristic equation of A determining its eigenvalues λ_k is

$$D_N(\lambda) = 0 \tag{13}$$

where $D_m(\lambda)$ is the polynomal of order m defined by relations

$$D_m(\lambda) = -\lambda D_{m-1}(\lambda) - d_{m-1}^2 D_{m-2}(\lambda) \tag{14}$$

for $m \geq 2$, $D_o(\lambda) = 1$, $D_1(\lambda) = -\lambda$.

The components of normalized eigenvectors are[5]

$$v_m(\lambda_k) = (-1)^{m-1} A(\lambda_k)\ (d_o d_1 \ldots d_{m-1})^{-1} D_{m-1}(\lambda_k) \tag{15}$$

where

$$A(\lambda_k) = [\sum_{m=0}^{N-1} (d_o, d_1 \ldots d_m)^{-2} D_m^2(\lambda_k)]^{-1/2}$$

For the sake of polarization calculations in the next section it will be useful to prove that the eigenvalues are symmetrically distributed around zero. Indeed, from (14) it follows that

$$D_N(-\lambda) = (-1)^N D_N(\lambda) \qquad \text{for} \qquad \lambda \neq 0$$

Therefore, if λ_k is the root of equation (13) $-\lambda_k$ is its root also and one has

$$v_m(-\lambda_k) = (-1)^{m-1} v_m(\lambda_k) \tag{16}$$

For N-odd one of the eigenvalues is zero while for N-pair all eigenvalues are different from zero. Thus the solution of (9) may be rewritten in the following form:

$$q_j = \sum_{k=1}^{M} v_m(\lambda_k) v_j(\lambda_k) [e^{i\lambda_k\sigma} + (-1)^{m+j} e^{-i\lambda_k\sigma}] + \frac{1-(-1)^N}{2} v_m(0) v_j(0) \tag{17}$$

where

$$M = \begin{cases} N/2 & \text{for N-pair} \\ (N-1)/2 & \text{for N-odd} \end{cases}$$

CALCULATION OF MEDIUM POLARIZATION

The electric dipole moment of each atom or molecule is given by

$$p = \langle\psi|\mu_D|\psi\rangle = \sum_{j=1}^{N} [q_j^* q_{j+1} p_{j,j+1} \; e^{i\phi} + c.c.]$$

where μ_D is dipole moment operator and $\phi = kz - \omega t$. It is convenient to decompose the dipole moment into a part in phase with the electric field and a part out of phase by π. Thus

$$p = p_{12} [\delta \sin \phi + C \cos \phi] \tag{18}$$

$$C = \sum_{j=1}^{N-1} q_j^* q_{j+1} p_{j,j+1}/p_{12} + c.c. \tag{19a}$$

$$\delta = i [\sum_{j=1}^{N-1} q_j^* q_{j+1} p_{j,j+1}/p_{12} - c.c.] \tag{19b}$$

Inserting (17) into (19) yields C=0, which is a consequence of the present model where the levels are equidistant and resonant with the incoming pulse and

$$\delta(\sigma) = \sum_{\substack{k=1 \\ n=1}}^{M} [\delta_{km} \sin[(\lambda_k+\lambda_m)\sigma] + \gamma_{km} \sin[(\lambda_k-\lambda_m)\sigma]] + \frac{1-(-1)^N}{2} \sum_{k=1}^{M} [\alpha_k \sin(\lambda_k\sigma)]. \tag{20}$$

where

$$\delta_{k_n} = -4\, v_m(\lambda_k)v_m(\lambda_n) \sum_{j=1}^{N-1} (-1)^{m+j}\, v_j(\lambda_k)v_{j+1}(\lambda_n)\, p_{j,j+1}/p_{12}$$

$$\gamma_{k_n} = -4\, v_m(\lambda_k)v_m(\lambda_n) \sum_{j=1}^{N-1} v_j(\lambda_k)\, v_{j+1}(\lambda_n)\, p_{j,j+1}/p_{12} \qquad (21)$$

$$\alpha_k = -4\, v_m(0)v_m(\lambda_k) \sum_{j=1}^{N-1} [v_j(0)\, v_{j+1}(\lambda_k)-v_{j+1}(0)v_j(\lambda_k)]p_{j,j+1}/p_{12}$$

PULSE MODIFICATION DURING ITS PROPAGATION IN MEDIUM

The wave equation in polarized medium is

$$\frac{\partial^2 E}{\partial z^2} - \frac{1}{c^2}\frac{\partial^2 E}{\partial t^2} = \frac{4\pi n_o}{c^2}\frac{\partial^2 P}{\partial t^2} \qquad (22)$$

where n_o is the number of atoms or molecules per cm^3, p is the dipole moment given by (18) and (20). Supposing that the field envelope and the polarization envelope δ are slowly varying one gets

$$\frac{\partial \varepsilon}{\partial t} + C\frac{\partial \varepsilon}{\partial z} = 2\pi\, n_o\, \omega\, p_{12}\, \delta \qquad (23)$$

This equation determines the envelope modification during the pulse propagation in the medium. While for the two-level system case $\delta = \mp \sin\sigma$ ("−" for attenuator, "+" for amplifier), for N-level system case δ is a linear combination of sine functions given by (20). Recalling that $\varepsilon = \hbar/P_{12}\ \partial\sigma/\partial t$ and introducing the variable $\tau = t - z/c$ one gets

$$\frac{\partial^2\sigma}{\partial\tau\partial z} = \frac{2\pi n_o \omega P_{12}^2}{\hbar}\,\delta(\sigma) \qquad (24)$$

After inserting (20) this equation becomes a multi sine-Gordon equation. Using the same method as for the two-level case with non-homogeneous broadening one derives the area theorem:

$$\frac{dS}{dz} = \alpha\delta(S)/2 \qquad (25)$$

where again $S = p_{12}/\hbar \int_{-\infty}^{\infty} \varepsilon\, dt$ and α is the same absorption coefficient as for two-level case[3].

Multiplying (22) by $c/8\pi\ \varepsilon$ and integrating time variable from $-\infty$ to ∞ yields

$$\frac{dI_p}{dz} = -\frac{n_o\hbar\omega}{8}\,A(S) \qquad (26)$$

where $I_p = c/8\pi \int_{-\infty}^{\infty} E^2\, dt$ is the total energy contained in the pulse per cm^2 and

$$-A(S)=\int_o^S \delta(\sigma)d\sigma = 2\sum_{\substack{k=1\\n=1}}^{M} \frac{\delta_{kn}}{\lambda_k+\lambda_n}\sin^2[\frac{\lambda_k+\lambda_n}{2}]S + 2\sum_{k\neq n}\frac{\gamma_{kn}}{\lambda_k-\lambda_n}\sin^2[\frac{\lambda_k-\lambda_n}{2}]S$$
$$+ (1-(-1)^N)\sum_{k=1}^{M}\frac{\alpha_k}{\lambda_k}\sin^2\frac{\lambda_k S}{2} \qquad (27)$$

The functions $\delta(S)$ and $A(S)$ are displayed for N = 5 in figures 1 and 2 respectively.

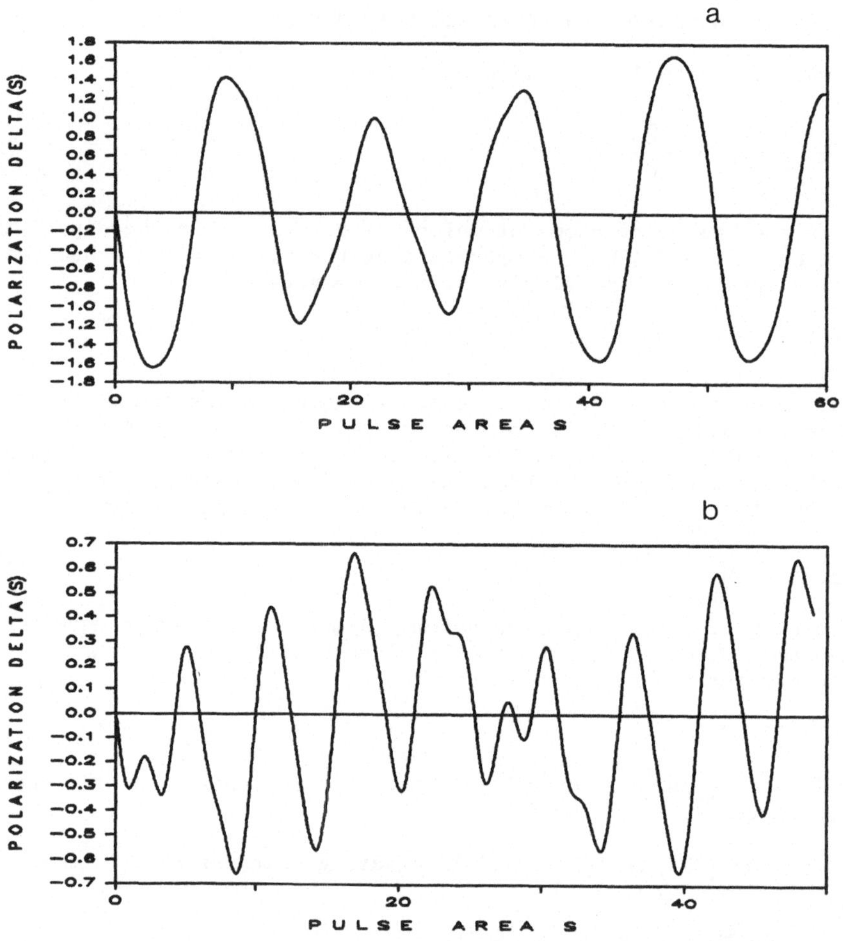

Figure 1. Polarization envelope $\delta(S)$ for the N=5 system which initially was in level 1. The transition dipole moments are: a) equal, b) $p_{j,j+1} = \sqrt{j}\, p_{12}$.

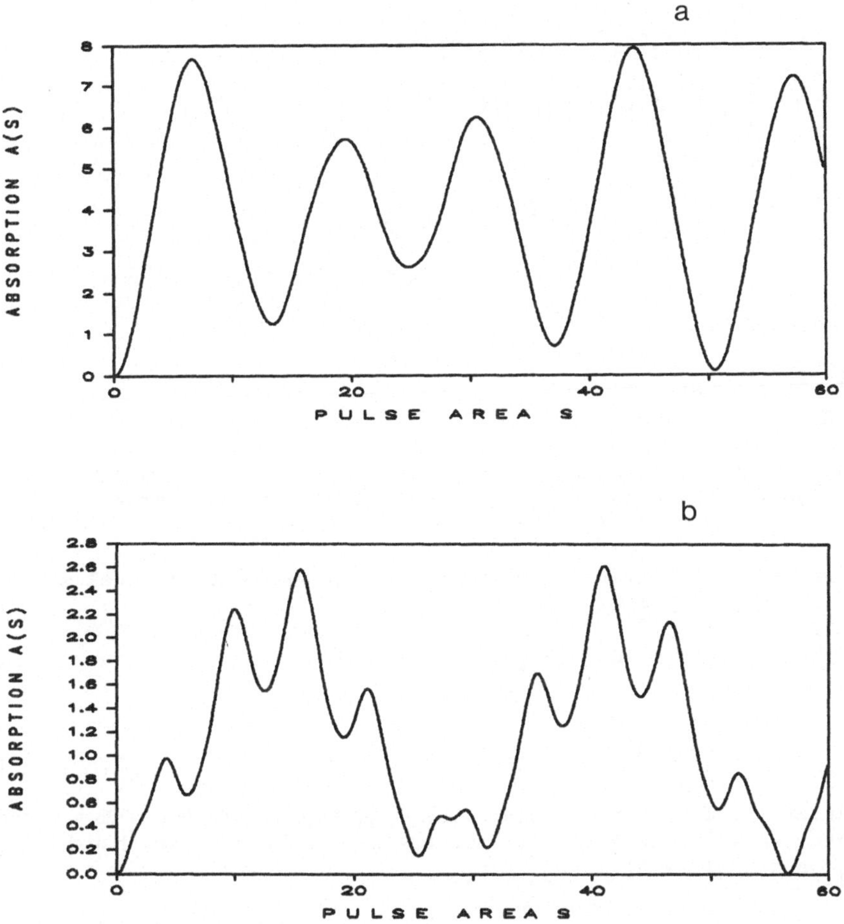

Figure 2. Absorption A(S) for the N=5 system which initially was in level 1. The transition dipole moments are: a) equal, b) $p_{j,j+1} = \sqrt{j}\, p_{12}$.

THREE-LEVEL SYSTEM

Two and three-level systems have particular properties since in these cases the transition amplitudes q_i are periodic functions of σ and only for these cases one is able to prove the existence of solutions of the equation (23) in form of distorsionless travelling localized waves (solitary waves, which in the case of sine-Gordon equation appear to be stable solitons).

Let us consider now two different initial conditions:
a) before pulse arrival the system was in level m = 1. Then

$$\delta(\sigma) = a_1 \sin(\lambda_1\sigma) - a_2 \sin(2\lambda_1\sigma) \tag{28}$$

$$A(\sigma) = \frac{2a_1}{\lambda_1} \sin^2\left(\frac{\lambda_1\sigma}{2}\right) + \frac{a_2}{\lambda_1} \sin^2(\lambda_1\sigma) \tag{29}$$

where

$$\lambda_1 = \frac{1}{2}\left[1 + (p_{23}/p_{12})^2\right]^{1/2}, \quad a_1 = \frac{1}{8}(p_{23}/p_{12})^2 \lambda_1^{-3}$$
$$a_2 = \frac{1}{8}\left[1 - (p_{23}/p_{12})^2\right] \lambda_1^{-3}$$

and thus the areas

$$\frac{p_{12}^2+p_{23}^2}{2\hbar} \int_{-\infty}^{\infty} \varepsilon \, dt = n\pi \tag{30}$$

are conserved and self induced transparency will occur for n = 0,2,4.... The equation (23) becomes a double sine-Gordon equation. Properties of its solitary waves were discussed in ref. 7 and by authors quoted therein.

b) Before the pulse arrival the system was in the central level m = 2. Then

$$\delta = \frac{1 - (p_{23}/p_{12})^2}{2\lambda_1} \sin(2\lambda_1\sigma) \tag{31}$$

and

$$- A = \frac{1 - (p_{23}/p_{12})^2}{2\lambda_1} \sin^2(\lambda_1\sigma) \tag{32}$$

A very interesting situation occurs when $p_{23} = p_{12}$. Then $\delta = A = 0$ and the medium behaves like a vacuum. Otherwise, the medium is absorbing if $p_{23} > p_{12}$ and amplifying if $p_{23} < p_{12}$.

DISCUSSION

The purpose of our investigation was to compare the coherent pulse propagation in two-level systems with its propagation in multilevel systems. Our main results can be summarized as follows:

a) The function $\sigma(\tau,z)$ determining electric field envelope via the equation $\varepsilon = \hbar/p_{12}\, \partial\sigma/\partial\tau$ is a solution of the multisine-Gordon equation (equations (24) and (20)), while for two-level case the corresponding equation was simple sine-Gordon equation. The multisine-Gordon equation does not have distorsionless solutions with $S \neq 0$ when $\delta(\sigma)$ is not periodic, which happens in general for $N > 3$, since then the conditions $A(S) = 0$ and $\delta(S) = 0$ and cannot be simultaneously satisfied except the case of area S = 0.[4]

b) For $N > 3$ one observes the complete vanishing of absorption only for the pulse having the area $S = 0$. Such pulse may be a pulse consisting of two identical pulses whose phase difference is fixed and equal to π. Propagation of such pulses in two-level degenerate systems was experimentally investigated in ref. 8. We think that this effect constitutes some universal phenomenon consisting in inverting coherently by the second pulse the effect produced by the first pulse. This phenomenon certainly deserves some more thorough experimental investigation.

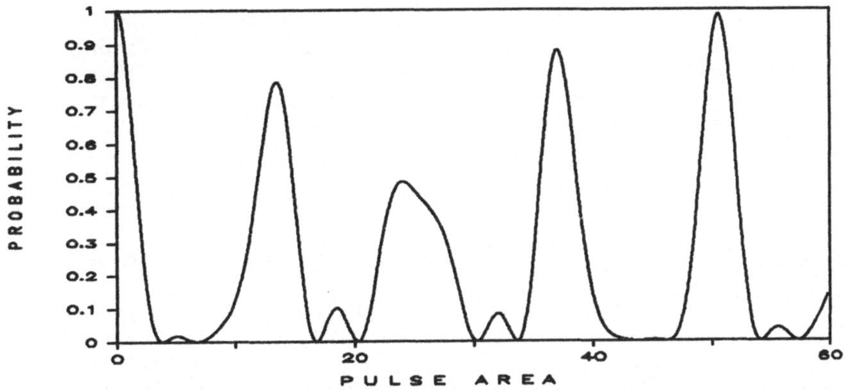

Figure 3. Probability of returning to the initial level m=1 after the passage of the pulse of area S for N=5 and equal transition dipole moments.

c) The equation $dS/dz = \alpha/2\ \delta(S) = 0$ has a series of roots $0, S_1, S_2, ..$ $0 < S_1 < S_2 \ldots$ for which the pulse area does not change. The areas $0, S_2, S_4, \ldots$ appear to be stable ones in attenuating media i.e. any pulse area should tend to one of these values after the pulse has passed the distance.

d) Though for $N > 3$ there are no areas corresponding to total vanishing of absorption, except the pulses of zero area mentioned in b), the absorption as function of pulse area exhibits deep minima corresponding to stable areas. Inspection of the figure 3 suggests that they coincide probably with the maxima of probability for finding the system in fundamental level $m = 1$ after the passage of the pulse.

The formula (25) describes the area changes when a pulse propagates along z-axis in a medium consisting of moving molecules or atoms. It is supposed that Doppler shifted frequencies of atoms are symmetrically distributed around light frequency. This ensures that the phase difference between electric field ε and the polarization p is exactly equal to π, which means that C = 0.

One may conclude that the possibility of existence of stable solitary waves is related to the periodic character of the transition amplitudes for N = 2 and N = 3 while for general N-level case they appear to be quasiperiodic, but not chaotic[9], thus yielding a multisine-Gordon equation for the function $\sigma(\tau,z)$ to satisfy. The existence of soliton-like solutions and, vice versa, maximum amplification will be therefore related to the degree of integrability of such nonlinear equations. This aspect is being pursued at the moment.

We are perfectly aware that in reality one does not deal with equidistant N-level systems but we think that the considered model approximates to some extent the light propagation in a molecular system in which some group of levels are visibly better tuned to the light frequency than the remaining ones and thus the discussed effects, although blurred by the neglected effects, should be observed in nature. The effect of the mentioned detuning should be treated numerically for each given molecular system separatly.

REFERENCES

1. J. Allen, J.H. Eberly, "Optical Resonance and Two-Level Atoms", J. Wiley & Sons, Inc. (1975).
2. G.L. Lamb, Jr., Rev. Mod. Phys. 43, 99 (1971).
3. G.L. Lamb, Jr., "Elements of Soliton Theory", J. Wiley & Sons Inc. (1980).
4. C.K. Rhodes, A. Szöke, A. Javan, Phys. Rev. Lett. 21, 1151 (1968).
5. Z. Bialynicka-Birula, I. Bialynicki-Birula, J.M. Eberly, B.W. Shore, Phys. Rev. A 16, 2048 (1977).
6. S. Chelkowski, A.D. Bandrauk, in preparation.
7. D.K. Campbell, M. Peyrard, P. Sodano, Physica 19D, 165 (1986).
8. H.P. Grieneisen, J. Goldhar, N.A. Kurnit, A. Javan, M.R. Schlossberg, Appl. Phys. Lett. 21, 559 (1972).
9. A.D. Bandrauk, L. Claveau, see this volume.

HIGH SPECTRAL BRIGHTNESS DYE LASERS

Orson L. Bourne

Division of Chemistry
National Research Council of Canada
Ottawa, Canada. K1A 0R6

In this report we give a brief overview of the methods presently available for the direct generation of high peak power (> 1 MW), high spectral brightness ($\Delta\nu < 0.033\ cm^{-1}$), tunable, visible radiation. These criteria removes from consideration such methods as Raman scattering (Stokes and anti-Stokes) and frequency addition of some primary sources. All of the methods to be described operate on the same principle which involves a tunable low power source coupled to a high gain (> 1000) pulse amplifier chain.

We will not attempt to give an analysis of the general properties of dye lasers. The reader is advised to consult publications listed in the references (1,2,3). It is our intention to identify and describe the various methods of obtaining the spectrally pure low power source (LPS), and then a description of the salient features of the high gain amplifier chain will be given. We will identify the strengths and weaknesses of each LPS and then speculate on the potential of a likely successor/-competitor to dye laser technology. Finally we will give examples of areas in which such system are currently employed.

Historically, the first such LPS was developed by Hänsch (4). A schematic of this now very popular approach is given in Fig. 1. The salient features of this design are (a) a grating used in the Littrow configuration, (b) some form of beam expansion optics and (c) one or more Fabry Perot etalons. The grating serves three purposes. It acts as a retro-reflector (in the Littrow configuration the diffracted beam is collinear with the incident beam), it selects the wavelength and it is part of the cavity pass band control. For a grating operating in this configuration the resolution is proportional to ℓ the illuminated length. Since the beam cross section of this oscillator is usually small it is necessary to include some form of beam expanding optics to fill the grating. This is usually accomplished in one of two ways. Firstly by including a telescope (4) within the cavity. This has the advantage of reducing the beam divergence of the radiation incident onto the grating and thus increasing the resolving power of the grating beyond its passive limit. This improvement is usually in the same ratio as the magnification and can be as large as 100. Alternatively one can use a prism ensemble (5,6) to magnify the beam, but unlike the telescope

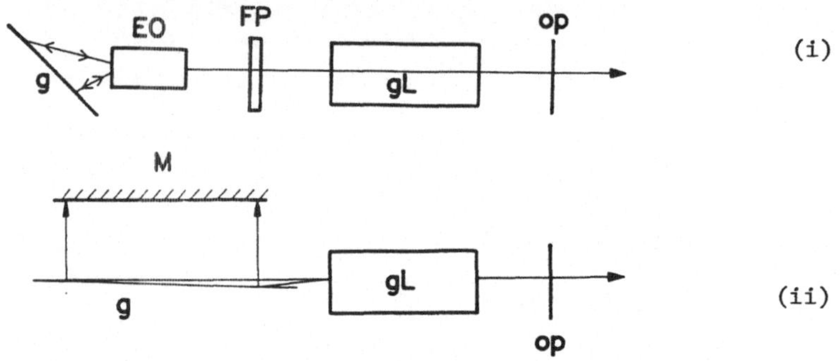

Fig. 1. A schematic of the layout of the optical elements of a Hänsch type (i) and grazing incidence (ii) pulse dye oscillator designs. g = grating, EO = expanding optics, FP = Fabry Perot etalons, gL = gain length, op = output mirrors, M = retro reflecting mirror.

arrangement this does not reduce the beam divergence. However because a prism is a dispersive element this form of expander can and does act as a secondary frequency filter. The reader is referred to an article by Trebino et al.(7) for a critical comparison of the merits of beam expanders used in this type of LPS. In all cases the beam expander has the tertiary effect of reducing the energy density on the surface of the grating and therefore protecting the delicate optical surface from damage. The final optical elements are the etalons which are used to restrict the cavity pass band to the desired limit. These etalons can either be internal or external to the resonator, and depending on the point of view, the use of etalons can be considered to be the greatest strength or largest weakness of this LPS design. From the viewpoint of being advantageous one can always include a sufficient number of etalons to reduce the cavity pass band to its transform limit. However the price to be paid comes in the form of a more complex oscillator. This complexity introduces three additional disadvantages. These are, (a) unless some form of automated technique is used to link all of the tuning elements together the continuous tuning range is governed by the FSR of highest power etalon, (b) the oscillator is more difficult to align and (c) the insertion of such elements into the cavity will reduce the overall efficiency of the oscillator. Tuning of the oscillator is achieved by rotating the grating and/or altering the position of the peak of the pass band of the etalons by means of either angle tuning or pressure tuning.

An alternative to the Hänsch design was introduced indpendently by Shoskan et al. (9) and Littman et al.(10). Figure 2b shows the typical layout of such an oscillator. Unlike the Hänsch design, the grating is not used in the Littrow configuration but at a high angle of incidence (~89°), hence the name grazing incidence oscillator. To complete the cavity an auxiliary reflector is used to return the diffracted beam through the gain volume. This reflector also serves as the wavelength tuning element. It can either be a mirror as in the initial design or a grating as used in subsequent versions (11). The use of a grating at grazing incidence removes the need for beam expanding optics but it is

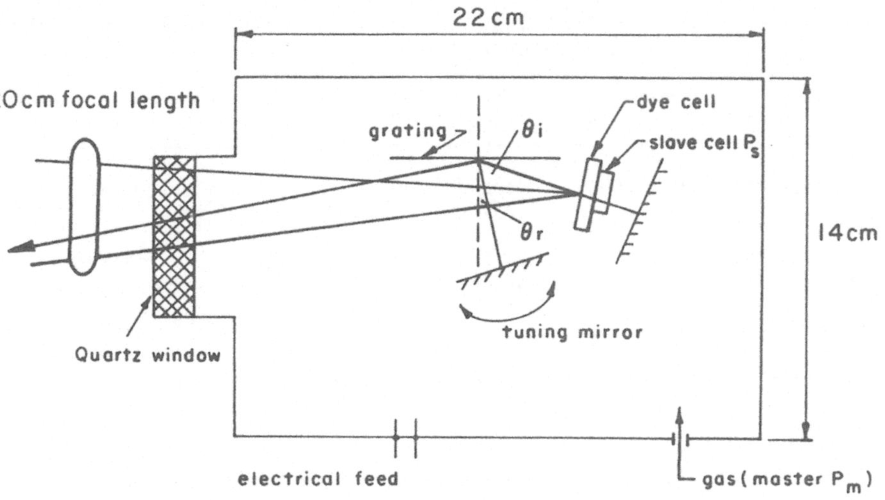

Fig. 2. A schematic of a pressure tune SLM dye laser based on a design first proposed by Littman (13).

not uncommon to include a prism expander, thus allowing the grating to be used at a more efficient angle (12). The advantage of this design is its ability to generate narrow linewidth (< 1 GHz) with at most two cavity filters. This makes the alignment and the continuous tuning of this type of laser relatively simple. This together with the Hänsch design dominates the commercial dye laser market. Although it is possible to operate either of these versions on a single longitudinal mode (SLM), it is not their preferred mode of operation. Recently Littman (13) has introduced a version of the basic grazing incidence design which is capable of stable SLM operation. This design goal was achieved by reducing the cavity length to such a point that in the short time in which the population is inverted 10 cavity round trips are possible, and by spatially restricting the optical path by tightly focussing the pump beam. If some attention is paid to the geometric layout of the optical elements it is possible to tune this laser on a single longitudinal mode over many wavenumbers. This method is suitable for a large scan range (tens of wavenumbers) but not suitable for precision tuning (tens of linewidths) due to the inherent limitation of any mechanical motion. For example, for a cavity of length 5 cm operating at a frequency of 25000 cm^{-1}, and a linewidth of 0.02 cm^{-1}, a frequency change of 0.02 cm^{-1} will correspond to a change in cavity length $d\ell$ of 5 nm. Technically to achieve such a small change in a reliable and consistent manner by mechanical means will be difficult. Pressure tuning is an alternative which may be applied to this laser design as the oscillator is sufficiently compact that it is practical to enclose it entirely in a single pressure vessel. Since $d\ell = kdp$, where k is a constant and dp is the change in pressure in the optical cavity, for a gas such as nitrogen a change of 5 nm represents a pressure change of 0.25 Torr. Such a pressure change can be produced in a reliable and consistent manner. Unfortunately, since not all of the cavity length is sensitive to pressure (the refractive indices of the dye cell and dye solution are independent of pressure), it will be impossible to tune this oscillator over more than 2 mode spacings on a SLM by pressure alone without some form of correction being made. The author (14) has demonstrated that this disadvantage can be eliminated if a slave chamber

Table 1. This table qualitatively compares the major features of three pulse laser oscillator designs.

Design	Δν FWHM	Energy	Tuning range	Advantage	Disadvantage
Hänsch	determined by etalons	~100 μJ	dependent on number of etalons	versatile Δν	complex
Grazing incidence	~750 MHz	~100 μJ	lasing range of dye	simple	limited Δν
SLM (13)	~125 MHz	~30 μJ	< 50 cm^{-1}	SLM	careful alignment procedure

is included into the optical path and its pressure varied according to the ratio $\frac{\Delta P_s}{\Delta P_m} = \frac{L_s + L_d}{L_s}$ where ΔP_s = pressure change in the slave chamber, ΔP_m = pressure change in remainder of the optical path, L_s= optical length of the slave chamber and L_d = optical length of the pressure insensitive regions of the optical path. With the addition of this modification it was possible to tune the laser over at least 3.75 cm^{-1} in a reliable and linear manner. A schematic of this device is shown in Fig. 2 and finally table 1 summarizes the performance characteristics of the pulse oscillator designs just described.

The LPS designs described above are pulse oscillators. An alternative approach is to pulse amplify the output of a CW SLM ring dye laser (15,16,17). The linewidth of such lasers can be < 1 MHz ($3\times10^{-5}cm^{-1}$) and they can be tuned over a number of wavenumbers on a SLM. Their disadvantage lies mainly in their expense and their complexity. For example they are at least 13 optical elements in a narrow line ring resonator. This does not include any external interferrometers which may be needed for added frequency stability.

Up to this point very little has been said about the pump source for these oscillators, we will address this in this paragraph. We will restrict our discussion to the oscillators pumped by other lasers. The choice between the available pump sources is partially a matter of preference and practical needs. For example, if one wants to operate at a variable repetition rate of 0-100 Hz over the widest possible spectral range (380-750) nm, then the choice should be an excimer laser. However if the operating region is to be restricted to those frequencies attainable by the Rhodamine family of dyes then the choice should be a frequency doubled Nd/YAG to take advantage of the excellent overlap between the peak of the absorption curve of these dyes and the output of this laser. Alternatively, if a high repetition rate (KHz) is required then the choice should be a copper vapor pump source (18). Finally for CW operation the only available sources are either a frequency double Nd/YAG or an argon ion laser.

The final choice of visible LPS to be considered are diode lasers. The linewidth of these systems when used with an external resonator can be superior to that of a ring laser. Output linewidths as low as 20 KHz have been published in the literature. Unfortunately, the lower wavelength range of these systems does not extend much below 700 nm but

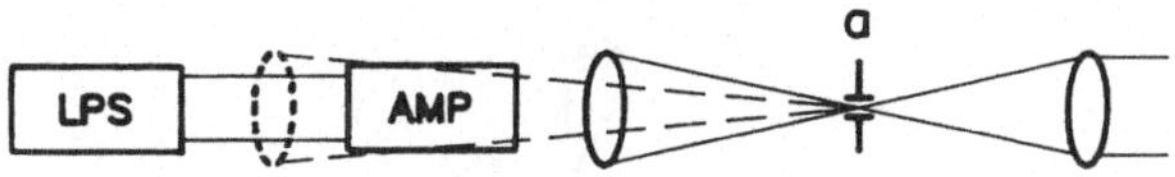

Fig. 3. A schematic of the optical arrangement used in a typical spatial filter. A variation of the standard design is shown in dotted lines. a = pin hole aperature, AMP = amplifier module.

given the rapid growth of this technology this lower limit should be extended in the foreseable future.

As outlined in the introduction, to achieve power levels of > 1 MW all of the various LPS must be pulse amplified. In any laser amplifier it is essential to control the build up of amplified spontaneous emission (ASE). A figure of merit for the background noise of an amplifier is the noise equivalent power (NEP). From the work of Ganiel (2) we can estimate that for a dye such as Rhodamine 6G this will be ~1 $W/cm^2/cm^{-1}$. Therefore it is essential that the injected signal into the amplifier exceed this value. Even if these injected power levels are attainable, this is not sufficient in themself to ensure the complete elimination of ASE. To achieve this the injected intensity into the amplifier must exceed the saturation intensity (I_s) of that amplifier. The reason for this is fairly straightforward. Unless the amplifier is saturated it is operating under small signal gain conditions. i.e. $A_{out}=A_{in}e^{g\ell}$ where g is the gain of the amplifier and ℓ is its length. This is an expotential function and it is obvious that if the injected signal is not exactly coincidental with the peak of the gain curve of the dye, any small supremacy which the injected signal may have had over the NEP of the amplifier can be quickly eliminated. For this reason it is not uncommon to find that it is not only necessary to have different dye concentrations in the separate modules of a dye amplifier chain, but also a completely different dye (14) is sometimes required. The design goal in any amplifier chain is thus to control the growth of ASE until the injected signal is at the saturation level. This is usually accomplished by employing all, one or a combination of three techniques. These are spatial filtering, spectral filtering and temporal filtering. In the case of spatial filtering one relies on the fact that the propagation characteristics and/or focussing properties of the ASE will be inferior to that of the injected signal. Fig. 3 shows the optical layout of such a filter. As an example let us assume that the ASE originates from a 1 mm^2 section (side pump amplifier) and is diffraction limited. For radiation at 25000 cm^{-1} this implies that the divergence angle θ_1 of the beam is ~50 mrad. Typical divergence angles from the previously described oscillators are $\theta_2 \sim 1$ mrad. The discrimination power of a spatial filter is simply $\sim(\theta_1/\theta_2)^2$, and from the numbers quoted above values of 2500 are attainable. In the special situation where only a single amplifier module is to be used or on the first stage of a multi-stage chain the focussing lens in Fig. 3 can be mounted on the LPS side of the amplifier (dotted in Fig. 3). In this manner the ratio (θ_1/θ_2) can be further enhanced since it is now the geometric projection of the ASE onto the aperature which is competing with the focussed amplified signal as opposed to the focussed ASE. The disadvantage with this latter approach is the ASE is bidirectional and therefore the

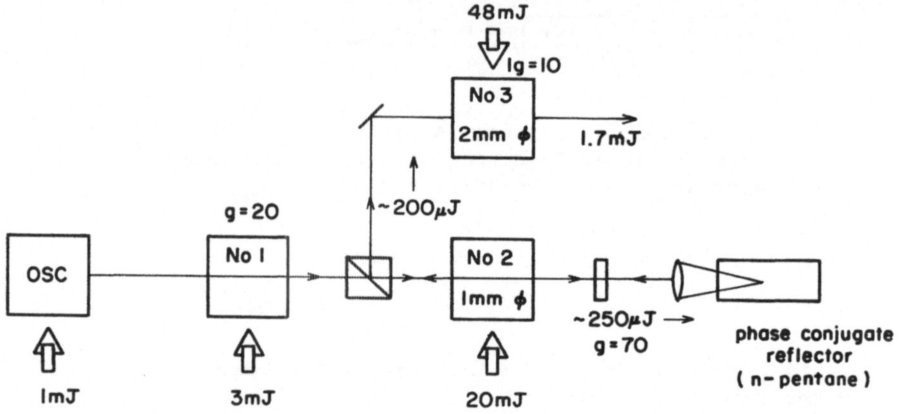

Fig. 4. A schematic of the layout of a pulse amplifier chain incorporating a phase-conjugate mirror to control the build up of ASE.

possibility exists that this lens may produce a focus within the LPS itself and hence degrade its performance. Spatial filtering has the advantage of being wavelength independent. However, if the operating frequency range is limited then spectral filtering with etalons should be considered. The same advantages and disadvantages that were discussed previously for etalons when used in an oscillator is applicable in this situation. Another approach is to take advantage of the fact that a non-linear process such as stimulated Brillouin scattering is intensity sensitive (spatial filtering) and spectrally sensitive (efficiency $\alpha \frac{I}{\Delta\nu}$) i.e. it can be used as a combination spatial and spectral filter. Such a filter has been incorporated into amplifier chains developed by the author (14,19). This method has the additional advantage of being a phase conjugate process, which means that in principle it can correct many of the phase distortions which may be generated by the amplication process. The final filtering technique to be considered is that of temporal filtering. It is known that the ASE can precede the stimulated emission from an oscillator. Therefore if a suitable delay is arranged between the pump of the oscillator and the pump of the amplifier one can ensure that the amplifier is seeded with the stimulated emission prior to the onset of gain in the amplifier (20).

The final choices to be considered in an amplifier design are the choice of pump, the shape of amplifier module and its pumping geometry. Here again personal preference and practical considerations are the guiding factors in the choice of pump. With respect to the amplifier modules one may chose between a quartz cuvette or a Bethune prism cell (21). The pumping geometry can be either longitudinal or transverse for the quartz cuvette but only transverse for the Bethune cell. Transverse pumping of a rectangular quartz cuvette is ideally suited for use with the excimer laser since the geometry of its output is ideally suited for focussing with a cylindrical lens. Although there is no reason why with the aid of suitable optics the circular cross section of the Nd-YAG or CVL output cannot be altered to accommodate this pumping configuration, it is more usual to use these pump sources in the longitudinal pumping configuration. The Bethune cell is a novel design which overcomes the asymmetric nature of the output from transversely excited cells. In this design the judicious positioning of a hole in a prism ensures that this hole is uniformly excited from all sides. If beam quality is of paramount importance and net gain can be sacrified then this cell is the preferred choice. Fig. 4 is a schematic of an amplifier chain which incorporates some of the techniques just described (14).

To conclude we will briefly highlight some of the areas in which such systems are used. They have been used as the injection sources for other high power lasers (16,17). Various techniques are available to up or down convert the radiation from these systems to other parts of the spectrum. The system described by Mahon et al. (12) for VUV generation and that used by Rabinowitz et al. (22) for the generation of radiation in the 1-12 μm region using stimulated Raman scattering are but two examples. Perhaps the most ambitious use of this technology is the system developed at Livermore for use in their (AVLIS) program (23). This laser system consists of a series of high average power dye lasers (oscillator/amplifier) pumped by an array of CVL lasers. The linewidth of this system is < 100 MHz and it is capable of generating 5 KW of output at a repetition rate of > 10 KHz. Equally as important, the capital and running costs of this laser system are consistent with a commercially viable process.

References

1. Principle of Dye Lasers Operations, F. P. Schäfer, Springer-Verlag, Berlin (1973).
2. G. Ganiel, A. Hardy, G. Neumann and D. Treves, IEEE Journal of Quantum Electronics, QE-11 881-892 (1975).
3. F.P. Schäfer: Laser Physics, Systems and Techniques, Ed. W.J. Firth and R.G. Harrison, SUSSP Publications, Edinburgh, 1982, pp. 271.
4. T. W. Hänsch, Appl. Opt., 11 895 (1972).
5. D. C. Hanna, P. A. Karlckainen, R. Wyatt, Opt. Quantum Electron., 7 115 (1975).
6. G. K. Klauminzer, U.S. Patent No. 4016504.
7. R. Trebino, J. P. Roller and A. E. Siegman, IEEE Journal of Quantum Electron., QE-18 1208:1213.
8. R. Wallenstein and T. W. Hänsch, Appl. Optics, 13 1625:1628 (1974).
9. I. Shoshan, N.N. Danon and U. P. Oppenhein, J. of Applied Phys., 48 4495:4497 (1977).
10. M. G. Littman and H. J. Metcalf, Appl. Optics, 17 2224:2227 (1978).
11. M. G. Littman, Optics Letters 3 138-140 (1978).
12. R. Mahon and F. S. Tomkins, IEEE Journal of Quantum Electron., QE-18 913:920 (1982).
13. M. G. Littman, Appl. Optics, 23, 4465:4468 (1984).
14. O. L. Bourne and D. M. Rayner, Submitted to Optics Communications.
15. T.F. Johnson Jr, R. H. Brady and W. Proffitt, Appl. Opt. 21, 2307:2316 (1982).
16. O. L. Bourne and A. J. Alcock, Appl. Phys. Lett., 42, 777:779, (1983).
17. R. T. Hawkins, H. Egger, J. Bokor and C. K. Rhodes, Appl. Phys. Lett., 36, 391:392, (1980).
18. M. Broyer, J. Chevaleyre, G. Delacretaz and L. Wöste, Appl. Phys., B35, 31:36 (1984).
19. O. L. Bourne, A. J. Alcock and Y. S. Huo, Rev. Sci. Instrum., 56, 1736:1739 (1985).
20. T. J. McKee, J. Lobin and W. A. Young, Appl. Opt., 21, 725:728 (1982).
21. D. S. Bethune, Appl. Opt., 20, 1987 (1981).

22. P. Rabinowitz, B. N. Perry and N. Levinos, IEEE Journal of Quantum Electron., QE-22, 797:802 (1986).
23. Lawrence Livermore National Laboratory 1980, Laser Program Annual Report UCRL-50021-80, pg.10-1 - 10-68.

LIGHT-INDUCED COLLISIONAL ENERGY TRANSFER - REVISITED

P. R. Berman

Physics Department
New York University
New York, NY 10003

INTRODUCTION

Light-induced collisional energy transfer (LICET) refers to a process in which excitation energy is transfered from one atom to another via a combined collisional-radiative reaction.[1] Two atoms, A and A', enter the collision in states i and i', respectively. As a result of both the collisional interaction and the absorption or emission of a photon, the atoms undergo a transition taking them to final states f and f', respectively (see Fig. 1). The LICET reaction can be written as

$$A_I + \hbar\Omega \rightarrow A_F , \tag{1}$$

where $|I\rangle = |ii'\rangle$ and $|F\rangle = |ff'\rangle$ represent product states in the atomic basis. Owing to energy considerations, the LICET cross section vanishes unless the collisional and radiative interactions occur *simultaneously*.

In a typical LICET experiment,[2-5] atoms A and A' are contained in a heat-pipe oven. Pulsed laser radiation is used both to produce the initial state $|I\rangle$ and also to provide the photons of frequency Ω which drive the LICET reaction. The LICET cross section is measured as a function of detuning

$$\Delta = \Omega - \omega_{FI} = \Omega - (E_F - E_I)/\hbar$$

of the applied laser field frequency from the over-all transition frequency ω_{FI}. The LICET profile which emerges from such studies, as shown in Fig. 2 as a function of $\Delta\tau_c$ (τ_c = collision duration), exhibits a marked asymmetry about $\Delta\tau_c = 0$. The experimental profiles are in good qualitative agreement with theoretical predictions. One may then ask, "What is there left to discuss?".

It is well known that physicists can *always* find something to discuss about any problem, no matter how much attention it has already received in the literature. Consequently, it is not a question of what there is left to discuss but rather a question of what I choose to discuss at this time. I will concentrate on two aspects of the LICET problem. The first

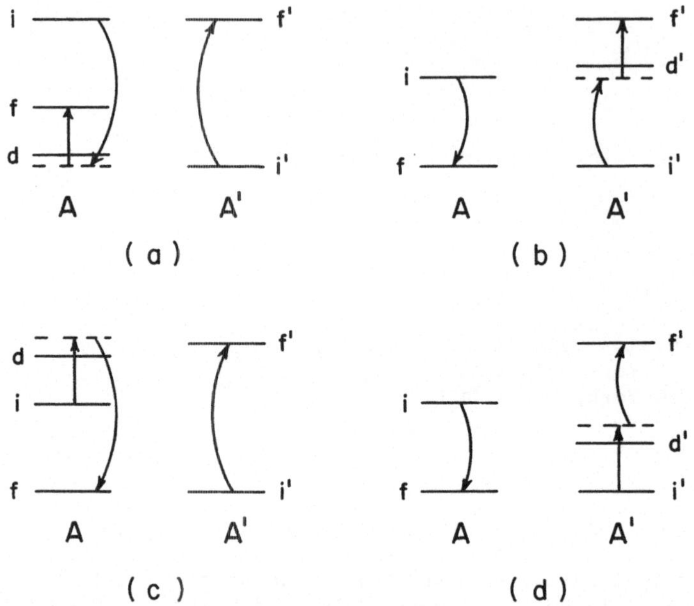

Fig. 1. Various LICET excitation schemes for the reaction $A_i + A'_{i'} + \hbar\Omega \rightarrow A_f + A'_{f'}$. The curved arrows represent the collisional coupling and the straight arrows the radiative coupling. The reactions proceed via an intermediate state which is nearly resonant with either the initial or final state.

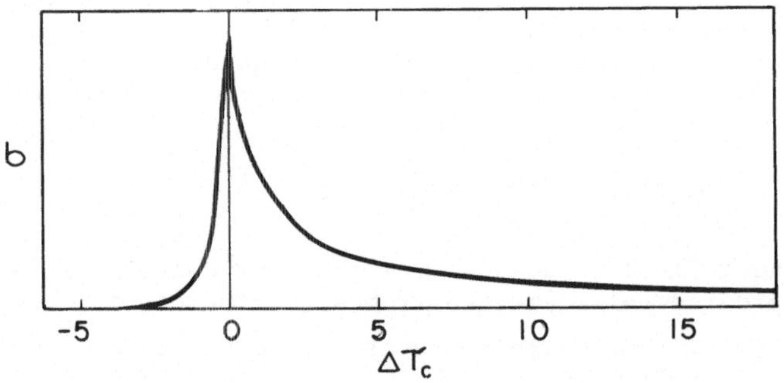

Fig. 2. Typical LICET profile [cross section σ (arbitrary units) as a function of dimensionless detuning $\Delta\tau_c$] showing a long quasistatic tail for $\Delta > 0$.

of these is the detuning dependence of the LICET cross section in the so-called quasistatic wing (that is, the long tail which occurs for positive Δ in Fig. 2). The quasistatic wing dependence on detuning has been a "thorn in the side" of the agreement between theory and experiment. Theory[1] predicts a $|\Delta|^{-0.5}$ fall-off for a dipole-dipole collisional interaction while experiment[2-5] quite often reveals a fall-off which varies more like $|\Delta|^{-0.8}$. Even if one modifies the theory to allow for alternative collisional interactions, there is no way to consistently account for this discrepancy within the context of the conventional theory. I will present a simple generalization[6] of the conventional theory which provides a much better agreement between theory and experiment.

The second aspect of the problem I will discuss is the final-state magnetic sublevel coherence produced in the LICET reaction, again for the quasistatic wing. There has been considerable theoretical and experimental interest in obtaining the final state polarization produced in Collisionally-Aided Radiative Excitation (CARE)[7], a collisional-radiative process closely related to LICET; however, little attention has been devoted to the analogous problem of final-state polarization in LICET. I will give a qualitative picture of the manner in which LICET can lead to final state magnetic coherence, emphasizing the differences between the LICET and CARE reactions.

QUASISTATIC-WING DEPENDENCE OF LICET CROSS SECTIONS

It is not my purpose here to go into a detailed calculation of the LICET cross section. Explicit calculations relating to the effects that are reported herein either have appeared recently or will soon appear.[6] Instead, I shall sketch a picture that, hopefully, illustrates the relevant physical ideas.

In virtually all LICET reactions, the excitation cross sections are sufficiently small to necessitate finding a level scheme with a nearly resonant intermediate state. Such states are indicated by the solid lines labeled d or d' in Figs. 1a - d. The intermediate states in the composite atomic basis shall be labeled by $|D\rangle$. For the diagrams of Fig. 1,

$$|D\rangle = |df'\rangle \quad \text{(Fig. 1a)}; \qquad |D\rangle = |fd'\rangle \quad \text{(Fig. 1b)};$$

$$|D\rangle = |di'\rangle \quad \text{(Fig. 1c)}; \qquad |D\rangle = |id'\rangle \quad \text{(Fig. 1d)}.$$

For Figs. 1a and 1b, states $|I\rangle$ and $|D\rangle$ are nearly resonant, while for Figs. 1c and 1d, states $|D\rangle$ and $|F\rangle$ are nearly resonant.

It is convenient to introduce quasi-molecular states $|I(R)\rangle$, $|D(R)\rangle$, $|F(R)\rangle$, where R is the A-A' internuclear separation. The quasi-molecular energy levels are shown in Fig. 3 as a function of R. As $R \sim \infty$, one recovers the composite atomic state energies. Figure 3a corresponds to the LICET reactions shown in Figs. 1a and 1b, while Fig. 3b corresponds to the reactions of Figs. 1c and 1d. In reality, the energy F(R) in Fig. 3a and I(R) in Fig. 3b would also depend on R owing to non-resonant collisional interactions (i.e., non-resonant van der Waals interactions). It is assumed for illustrative purposes, however, that the nearly-resonant collisional interaction of states $|I\rangle$ and $|D\rangle$ (Fig. 3a) or $|D\rangle$ and $|F\rangle$ (Fig. 3b) is much stronger than any non-resonant collisional interactions; consequently, for the range of internuclear separations shown in Fig. 3, significant variations in energy levels F(R) (Fig. 3a) or I(R) (Fig. 3b) are negligible.

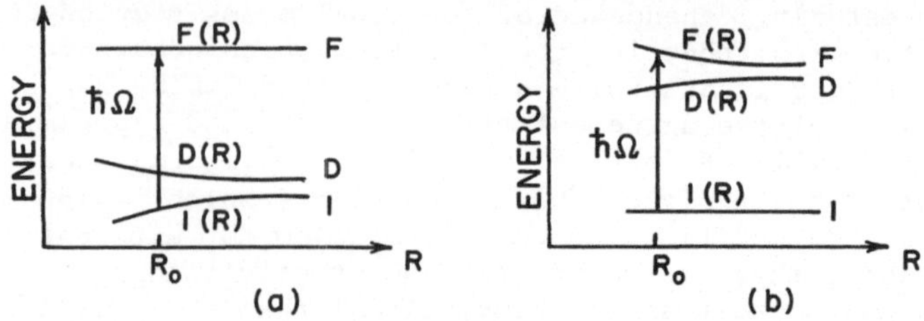

Fig. 3. Quasi-molecular energy levels as a function of A-A' internuclear separation R. The asymptotic values I, D, and F are the energies of the composite atomic states. Figure 3a corresponds to the LICET excitation schemes of Figs. 1a and 1b while 3b corresponds to that of Figs. 1c and 1d. For the frequency shown, the I(R) → F(R) transition is resonant at $R=R_o$.

The nearly-resonant collisional interaction gives rise to the repulsion of energy levels I(R) and D(R) (Fig. 3a) or D(R) and F(R) (Fig. 3b). One can neglect any coupling between the nearly-resonant quasi-molecular states (i.e., rotation of the internuclear axis does not mix these states) provided that

$$\omega_{DI}\tau_c >> 1 \text{ (Fig. 3a)}; \quad \omega_{FD}\tau_c >> 1 \quad \text{(Fig. 3b)} . \tag{2}$$

It is assumed that inequalities (2) are valid, which is a good assumption for typical LICET experiments in which $\omega_{DI}\tau_c$ (or $\omega_{FD}\tau_c$) $\gtrsim$ 10. As a consequence of the adiabaticity of the quasi-molecular states (hereafter referred to simply as molecular states), state $|D(R)>$ is never populated for the level scheme of Fig. 3a and, if an I(R) → F(R) transition is made in Fig. 3b, there is no subsequent transfer to level D(R). In other words, *molecular* state $|D(R)>$ enters this problem as a virtual state only. It does *not* follow that composite *atomic* state $|D>$ acts only as a virtual level, since molecular state $|I(R)>$ (Fig. 3a) or $|F(R)>$ (Fig. 3b) contains an admixture of state $|D>$.

The LICET reaction is now simply understood as a transition from molecular state $|I(R)>$ to $|F(R)>$ resulting from the absorption of a photon of frequency Ω. The asymmetry of the LICET profiles is easily explained in terms of this molecular picture. For detunings Δ > 0, there are internuclear separations for which the field is resonant with the molecular transition; for detunings Δ < 0, no such resonances exist. Consequently, the probability for excitation in the quasistatic wing Δ > 0 is significantly larger than that for excitation in the so-called antistatic wing Δ < 0.*

An estimate for the LICET cross section can be obtained once the coupling between initial and final states is established. Assume that the coupling strength (in frequency units) between initial and final *molecular* states is given by χ(R). To *lowest order* in the coupling parameter, the final state amplitude a_F is given by

* If $\omega_{DI} < 0$ (Fig. 3a) or $\omega_{FD} < 0$ (Fig. 3b), the quasistatic wing occurs for Δ < 0 and antistatic wing for Δ > 0.

$$|a_F| = \left| \int_{-\infty}^{\infty} \chi[R(t)]\, e^{i\{\Omega t - \int_0^t \omega_{FI}[R(t')]\, dt'\}}\, dt \right| , \tag{3}$$

where R(t) is now an implicit function of t evaluated along the classical collision trajectory and $\omega_{FI}(R)$ is the frequency separation between molecular levels F(R) and I(R). If $\Delta = (\Omega - \omega_{FI}) >> \tau_c^{-1}$, the major contribution to the integral comes from those values of $t = t_o$ (or internuclear separations $R = R_o$) for which $\Omega = \omega_{FI}(R_o)$. For the energy levels shown in Fig. 3, there are two times, symmetric about t = 0, for which $\Omega = \omega_{FI}(R_o)$, provided that $\Delta > 0$ and $\Omega < \omega_{FI}(b_o)$, where b_o is the impact parameter at closest approach. Neglecting any interference effects between the contributions from the two crossing points, it is possible to integrate Eq. (3) by the method of stationary phase (assuming $\Delta\tau_c >> 1$) take its square modulus, and average over impact parameter. In that manner, assuming straight-line collision trajectories, one obtains the LICET cross section

$$\sigma = \frac{8|\chi(R_o)|^2 \pi^2 R_o^2}{v\, |d\omega_{FI}(R_o)/dR_o|} , \tag{4}$$

where v is the A-A' relative speed, and R_o is defined by $\Omega = \omega_{FI}(R_o)$.

Equation (4) could have been predicted using dimensional analysis. The cross section depends on the square of the radius of excitation R_o^2 and the square of the product of coupling strength $\chi(R_o)$ times the time spent in the crossing region. The time spent in the crossing region varies as $|d\omega_{FI}(R_o)/dt_o|^{-\frac{1}{2}} \simeq [v\, |d\omega_{FI}(R_o)/dR_o|]^{-\frac{1}{2}}$. It remains to specify $\chi(R)$ and $\omega_{FI}(R)$.

Both $\chi(R)$ and $\omega_{FI}(R)$ can be specified rather easily if one looks at the details of the I-D or D-F coupling. I consider the level scheme of Fig. 3a where the I-D coupling enters, but the level scheme of Fig. 3b can be treated in an identical manner if ω_{DI} is replaced by ω_{FD}. Denoting the collisional coupling strength (in frequency units) between composite *atomic* states I and D by $U_c[R(t)]$ and taking the energy of state $|I\rangle$ equal to zero, one finds that the atomic state amplitudes evolve as

$$i\hbar \begin{pmatrix} \dot{a}_I \\ \dot{a}_D \end{pmatrix} = H_c \begin{pmatrix} a_I \\ a_D \end{pmatrix} = \hbar \begin{pmatrix} 0 & U_c \\ U_c & \omega_{DI} \end{pmatrix} \begin{pmatrix} a_I \\ a_D \end{pmatrix} . \tag{5}$$

Diagonalization of the Hamiltonian $H_c[(R(t)]$ at each instant of time t (or internuclear separation R) provides the molecular-state energy levels $[\hbar\omega_I(R)$ and $\hbar\omega_D(R)]$ as well as the molecular-state eigenkets $|I(R)\rangle$ and $|D(R)\rangle$. The diagonalization procedure yields

$$\omega_I(R) = [\omega_{DI} - \omega_{DI}(R)]/2 \tag{6a}$$

$$\omega_D(R) = [\omega_{DI} + \omega_{DI}(R)]/2 \tag{6b}$$

and

$$|I(R)\rangle = \cos[\theta(R)]|I\rangle + \sin[\theta(R)]|D\rangle \tag{7a}$$

$$|D(R)\rangle = -\sin[\theta(R)]|I\rangle + \cos[\theta(R)]|D\rangle , \tag{7b}$$

where

$$\sin[\theta(R)] = \frac{1}{\sqrt{2}}\left[1 - \frac{\omega_{DI}}{\omega_{DI}(R)}\right]^{1/2} \tag{8}$$

and

$$\omega_{DI}(R) = \{\omega_{DI}^{\ 2} + 4[U_c(R)]^2\}^{\frac{1}{2}} \quad . \tag{9}$$

The transition occurs at internuclear separations given by $\Omega = \omega_{FI}(R_o)$ or

$$\Delta = -\omega_I(R_o) \quad . \tag{10}$$

Since R_o is defined by Eq. (10) and since $d\omega_{FI}(R_o)/dR_o = -d\omega_I(R_o)/dR_o$ can be evaluated by using Eqs. (6a), (9), and (10), the only remaining quantity to specify is $\chi(R_o)$.

Explicitly, the coupling parameter is given by

$$\chi(R) = \langle F(R)|\hat{\mu}\mathcal{E}/2\hbar|I(R)\rangle , \tag{11a}$$

where $\mathcal{E}$ is the laser field amplitude (assumed constant during a collision) and $\hat{\mu} = \hat{\mu}_A + \hat{\mu}_{A'}$ is the sum of dipole moment operators for atoms A and A'. Since $\hat{\mu}$ is a sum of operators acting in atoms A and A', respectively, it couples composite atomic states in which only *one* of the atoms changes state. Consequently, $\langle F|\hat{\mu}|I\rangle = 0$, since *both* atoms change state in the $I \rightarrow F$ transition. This feature of the coupling is most easily seen in Figs. 1a and 1b, where the field acts in either atom A or A' to couple intermediate state $|D\rangle$ to final state $|F\rangle$ (for Figs. 1c and 1d, it couples state $|I\rangle$ to $|D\rangle$). Using Eqs. (11a) and (7a), along with the relationships $|F(R)\rangle \simeq |F\rangle$ and $\langle F|\hat{\mu}|I\rangle = 0$, one finds a collisional coupling strength

$$\chi(R) = \chi \sin[\theta(R)] , \tag{11b}$$

where $\chi = \mu\mathcal{E}/2\hbar$ and μ is the F-D dipole moment matrix element. This is the last ingredient needed to calculate σ. Using Eqs. (4)-(11), one can obtain

$$\sigma = \frac{4|\chi|^2\pi^2 R_o^2 \Delta}{|vU_c(R_o)dU_c(R_o)/dR_o|} , \tag{12a}$$

where

$$|U_c(R_o)| = [\Delta(\omega_{DI} + \Delta)]^{\frac{1}{2}} \quad . \tag{12b}$$

For a dipole-dipole collisional interaction,

$$U_c(R) = -C/R^3 , \quad C > 0; \tag{13}$$

Eq. (12a) becomes

$$\sigma = \frac{4}{3} \frac{\pi^2 |\chi|^2 |C|}{v\Delta^{1/2}(\Delta + \omega_{DI})^{3/2}} \quad . \tag{14}$$

If $\omega_{DI} >> \Delta$, as is commonly assumed in conventional "two-state" theories of LICET, then $\sigma \sim \Delta^{-\frac{1}{2}}$. However, if $\Delta \simeq \omega_{DI}$, the quasistatic wing falls off more steeply than $\Delta^{-\frac{1}{2}}$ and the conventional two-state LICET theories are no longer valid. Since $\Delta \simeq \omega_{DI}$ in all high resolution LICET experiments reported to date, it is not surprising that the conventional theory fails to satisfactorily explain the data. On the other hand, Eq. (14), which represents a simple extension of the theory to account for the fact that Δ and ω_{DI} may be of comparable magnitude, provides very good quantitative agreement with experiment.

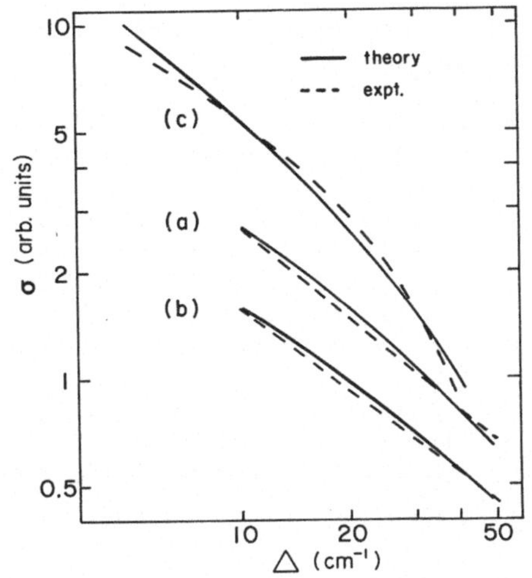

Fig. 4. LICET cross section σ (arbitrary units) as a function of detuning $\Delta(cm^{-1})$ for energy defects (a) $\omega_{DI} = 63cm^{-1}$ (Eu-Sr) (b) $\omega_{DI} = 94cm^{-1}$ (Na-Ca) and (c) $\omega_{DI} = 21cm^{-1}$ (Sr-Li). The point at $\Delta = 10cm^{-1}$ is chosen to normalize the theoretical and experimental curves.

A comparison between theory and experiment is presented in Fig. 4 for Eu-Sr[3], Na-Ca[4], and Sr-Li[5] LICET reactions. The Eu-Sr reaction corresponds to the level scheme of Fig. 1b with the frequency mismatch ω_{DI} (in cm^{-1}) equal to $-63cm^{-1}$; a detuning range $-55cm^{-1} < \Delta < -10cm^{-1}$ was reported.[3] The comparison between theory [Eq. (14)] and experiment [$(-\Delta)^{-0.85}$ dependence] is shown in Fig. 4(a). This experiment was recently repeated by Matera et al.,[8] again showing good agreement with Eq. (14). (b) The Na-Ca reaction corresponds to the level scheme of Fig. 1d with $\omega_{FD} = 94cm^{-1}$; a detuning range $\Delta < 60cm^{-1}$ was reported.[4] As shown in Fig. 4(b), agreement between theory [Eq. (14) with ω_{DI} replaced by ω_{FD}] and experiment [$\Delta^{-0.8}$ dependence] is very good. (c) The Sr-Li reaction corresponds to the level scheme of Fig. 1d with $\omega_{FD} = 21cm^{-1}$; a detuning range $\Delta < 50cm^{-1}$ was reported.[5] A comparison between theory and experi ment is given in Fig. 4(c), where the varying slope of $\log(\sigma)$ vs $\log(\Delta)$ predicted by Eq. (14) is clearly seen in the data. [Equation (14) is not strictly valid for $\Delta < 5cm^{-1}$ since the quasistatic assumption ($\Delta\tau_c >> 1$) used in its derivation fails in this detuning range for the Sr-Li reaction.]

FINAL STATE POLARIZATION

Equation (4) gives the LICET cross section for each of the level schemes of Fig. 1. In other words, a measurement of the *total* LICET cross section does very little to identify which, if any, of the excitation pathways shown in Fig. 1 is providing the dominant contribution to the cross section. The situation changes dramatically if one measures magnetic state coherence in the final state of one of the atoms resulting from the LICET reaction. The final state magnetic sublevel distribution provides valuable additional information that can be used to characterize the LICET process, as well as the interatomic potential.

Previously, polarization effects in LICET were studied for the impact core of the profile ($|\Delta\tau_c| << 1$).[9] A simple physical picture was used to distinguish the different polarizations that arise for the various level schemes shown in Fig. 1. By considering the collisional interaction as an effective unpolarized multipolar field, it is possible to view the over-all excitation as resulting from unpolarized fields (collisional interaction) plus the laser field (which is assumed to be linearly polarized). One can extend some of these arguments to the quasistatic wing of the LICET profile.

Additional insight into polarization effects can be derived from the extensive studies of such effects in the related problem of collisionally-aided radiative excitation (CARE).[7,10] The CARE reaction is, typically,

$$A_i + A'_{i'} + \hbar\Omega \rightarrow A_i + A'_{f'}$$

which, in the composite state picture, is identical to Eq. (1). In CARE, the laser is detuned from a transition in one of the atoms. The collision with the other atom provides translational energy to account for the energy defect associated with the atom-field detuning. The atomic and molecular excitation schemes for CARE are shown in Fig. 5. The nearly-resonant intermediate state $|D>$ present in LICET, is absent in CARE. In the quasistatic line wing, polarization in CARE is produced by excitation at an internuclear radius R_o defined by $\Omega = \omega_{FI}(R_o)$, followed by depolarization in the half-collision following excitation. Depolarization occurs for internuclear separations $R_o < R < R_d$, where R_d is the separation at which the molecular basis is no longer appropriate. Depolarization is neglected in the range $R > R_d$, owing to the weakness of the collisional interaction in this regime.[7]

It would appear that the depolarization rates would be the same for LICET and CARE since the molecular picture of the excitation is nearly identical. This is true to some extent, but there are two important differences between CARE and LICET. First, in contrast to LICET where excitation occurs at separations R_o for which either the initial or final state molecular potentials is flat (see Fig. 3), excitation in CARE occurs at values of R_o for which neither I(R) nor F(R) is flat (see Fig. 5). Second, as has been shown above, the coupling parameter is dependent on both the collisional interaction and the radiation field for LICET; it depends only on the atom-field interaction for CARE (assuming an allowed transition). These differences have important consequences with regard to final state LICET polarizations.

The first feature to note is that there is relatively little depolarization in the half-collision for the level schemes of Figs. 1a and 1b, or, equivalently for the molecular levels of Fig. 3a. Excitation occurs at sufficiently large R_o that the final state molecular level F(R) is flat (negligible collisional interaction) for most of the half-collisions following excitation. Thus one would expect that the final state polarization is determined solely at excitation. When averaged over all collision orientations, the final state relative polarization [i.e., final state magnetic polarization (measured as a difference in magnetic state populations) divided by final state total population], will probably not differ all that much from that obtained in the line core. This is precisely what was seen experimentally[11] for a level scheme corresponding to Fig. 1b - the final state relative polarization was found to be independent of detuning.

In contrast to the level of schemes of Figs. 1a, 1b, and 3a, LICET excitation for the level schemes of Figs. 1c, 1d, and 3b will involve depolarization in the half-collision following excitation. In this regard, the LICET and CARE reactions would be similar. However, the fact that the coupling strength in LICET depends on the collisional interaction can play an important role. Consider, for example, the level scheme of Fig. 1c with $J_i=J_{i'}=J_f=0$; $J_d=J_{f'}=1$, corresponding to a $J_I=0$ to $J_F=1$ transition in

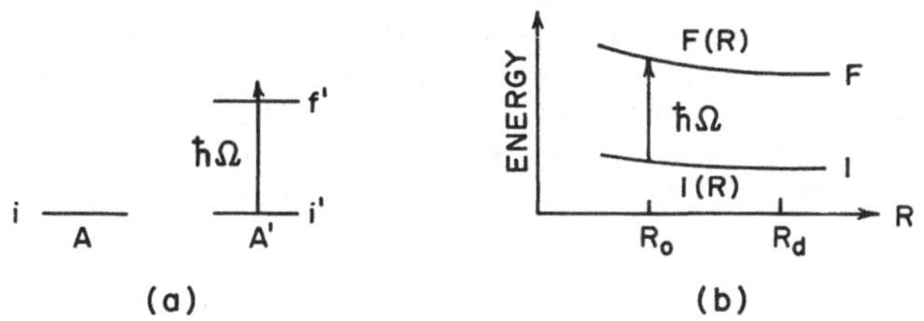

Fig. 5. Excitation scheme in the atomic (a) and molecular (b) pictures for Collisionally-Aided Radiative Excitation (CARE) of the reaction $A_i + A'_{i'} + \hbar\Omega \rightarrow A_i + A'_{f'}$. Since there is no longer a nearly-resonant intermediate state, the I(R) and F(R) molecular energy curves have a similar R dependence. For $R > R_d$, no depolarization occurs.

the composite scheme. In CARE, a laser field polarized in the "z" direction excites only the m = 0 excited state magnetic sublevel and produces a final state relative polarization of unity (which is partially destroyed in the half-collision following excitation). For the actual LICET reaction, however, the final state relative polarization at excitation is much smaller (1/7 with some appropriate definition of relative polarization[9,10]), owing to the dependence of the coupling on the collisional interaction. In other words, the coupling strength is a function of the geometry of the collision and tends to diminish the polarization when an average over collision orientations is taken. Similarly, for the level scheme of Fig. 1d with $J_{i'}=J_{f'}=1$; $J_{d'}=0$, the LICET polarization at excitation is zero,[9] whereas it would be finite for CARE.

Explicit calculations of final state polarizations for LICET are in progress. The ideas expressed above should provide a qualitative guide for understanding the manner in which collisional-radiative reactions lead to final state magnetic polarization in LICET.

CONCLUSIONS

A simple extension of the conventional two-state LICET theory provides a consistent explanation of experimental LICET profiles recorded at non-saturating laser field intensities. The modified theory allows for the fact that the frequency defect of a nearly-resonant intermediate state may be of comparable magnitude with a detuning parameter in the problem. The theory can be extended to saturating field intensities for comparison with recent experimental results.[5] We have also seen how final-state magnetic coherence can be used to obtain additional information regarding the LICET reaction.

With the availability of very high power ($\simeq 10^{14}$W/cm^2) pulsed lasers, it now may be possible to drive LICET reactions *without* a nearly resonant intermediate state [the cross section scales as $(\chi/\omega_{DI})^2$]. There is also a wide range of possibilities for observing LICET-type reactions involving atom-molecule, atom-surface and atom-cluster interactions. Whether or not LICET will live up to its original billing as a means for obtaining laser-controlled atomic and chemical reactions remains to be seen.

ACKNOWLEDGMENTS

I should like to thank Prof. P. Toschek for transmission of his results (Ref. 5) prior to publication. The work on the detuning dependence was carried out in collaboration with Dr. A. Bambini (Quantum Electronics Institute, Florence, Italy) and that for magnetic state polarization with Dr. F. Schuller (LIMHP, Villetaneuse, France). This research is supported by the U.S. Office of Naval Research and, in part, by a National Science Foundation Grant INT-84-13300.

REFERENCES

1. For review of this subject area containing additional references, see see P.R. Berman and E.J. Robinson in "Photon-Assisted Collisions and Related Topics," edited by N.K. Rahman and C. Guidotti (Harwood, Chur, Switzerland,1982), pp. 15-33; S.I. Yakovlenko, Sov. J. Quantum Eletron. 8, 151 (1978); M.G. Payne, V.E. Anderson, and J.E. Turner, Phys. Rev. A20, 1032 (1979).

2. The first experiments in this area were carried out by Harris et al. For a review of this work, see S.E. Harris, J.F. Young, R.W. Falcone, W.R. Green, D.B. Lidow, J. Lukasik, J.C. White, M.D. Wright and G.A. Zdasiuk in "Atomic Physics, Vol. 7," edited by D. Kleppner and F.M. Pipkin (Plenum, New York, 1981) pp. 407-428.
3. C. Bréchignac, Ph. Cahuzac, and P.E. Toschek, Phys, Rev. A21, 1969 (1980); A. Débarre, J. Phys. B15, 1693 (1982).
4. A. Débarre, J. Phys. B16, 431 (1983).
5. F. Dorsch, diploma thesis, University of Hamburg, 1986 (unpublished); F. Dorsch, S. Geltman, and P.E. Toschek, Phys. Rev. A. (submitted, 1987).
6. A. Bambini and P.R. Berman, Phys. Rev. A35, 3753 (1987); A. Agresti, A. Bambini, A. Stefanel and P.R. Berman, Phys. Rev. A (submitted, 1987).
7. For a review of this area, see K. Burnett, Phys. Reps. 118, 339 (1985).
8. M. Matera, M. Mazzoni, R. Buffa, S. Cavalieri and E. Arimondo, in "Photons and Continuum States of Atoms and Molecules," edited by N.K. Rahman, C. Guidotti and M. Allegrini (Springer-Verlag - Heidelberg, 1987) pp. 227-232; Phys. Rev. A36, (1987).
9. P.R. Berman, Phys. Rev. A22, 1848 (1980).
10. See, also, the article by P. Julienne in "Spectral Line Shapes, Vol. 2," edited by K. Burnett, (de Gruyter, Berlin, 1983) pp. 769-785, for an application of this method to LICET.
11. A. Débarre, J. Phys. B15, 1693 (1982).

THEORY OF LASER-PULSE-INDUCED MOLECULAR DYNAMICS: GAS-PHASE MOLECULAR COLLISIONS AND ADBOND DYNAMICS

Hai-Woong Lee
Department of Physics
Oakland University
Rochester, Michigan 48063 USA

Sander van Smaalen
Laboratory of Inorganic Chemisty
Materials Science Center
State University of Groningen
Nijenborgh 16
9747 AG Groningen
The Netherlands

Thomas F. George
Departments of Chemistry and Physics & Astronomy
239 Fronczak Hall
State University of New York at Buffalo
Buffalo, New York 14260 USA

INTRODUCTION

A semiclassical study presented here indicates that a sufficiently short and intense pulse can be much more effective in inducing a collisional radiative transition than cw radiation or a long pulse, although the intensity must not be too high because the Rabi oscillation can bring down the probability. For the situation of a molecule physisorbed on a crystalline surface and irradiated by a laser, a master equation approach, used to describe the time evolution of the population of the vibrational adbond levels, shows that for high intensities laser-induced vibrational excitation is the same for pulsed and cw lasers.

GAS-PHASE MOLECULAR COLLISIONS

Most theoretical and experimental work in the past on the subject of molecular collisions occurring in the presence of a laser field has been concerned with the case in which the field consists of continuous wave (cw) radiation. In this case the colliding molecules are influenced by the laser radiation during both the incoming and outgoing trajectories. If, however, the laser field consists of a pulse whose duration is shorter than the collision time (for a typical collision time of $10^{-11} \sim 10^{-12}$ s, one must have a (sub)picosecond pulse), only a part of the collision process occurs in the presence of the laser field. The outcome of the collision therefore may be significantly different depending whether the laser radiation is cw or pulsed. The difference shows up strongly in particular for collision processes involving an electronic transition that occurs via curve crossing, as we have shown in our earlier work.[1,2] Below we give a brief description of such a transition taking place in the presence of laser radiation, with special attention to the effect of the temporal duration of the laser radiation. An experimental study of this effect has recently been reported.[3]

Let us consider an electric-dipole transition induced by collision that occurs in the presence of laser radiation. Within the semiclassical two-state formalism, the Schrödinger equation yields the following coupled equations for the probability amplitudes $c_1(t)$ and $c_2(t)$ for the two states 1 and 2 being considered:[4]

$$i\hbar \frac{dc_1(t)}{dt} = H_{11}c_1 + H_{12}c_2 \tag{1a}$$

$$i\hbar \frac{dc_2(t)}{dt} = H_{21}c_1 + H_{22}c_2 \quad , \tag{1b}$$

where H_{11} and H_{22} represent adiabatic potential curves in the dressed-molecule representation for the two states 1 and 2 which, we assume, cross at a certain internuclear distance $R = R_c$ (i.e., $H_{22}(R_c)-H_{11}(R_c) = 0$), the interaction term H_{12} is given by $H_{12} = -\vec{\mu}\cdot\vec{E}_0/2$, $\vec{\mu}$ is the transition moment, and $\vec{E}_0$ is the electric field amplitude of the laser radiation. The transition probability $1 \rightarrow 2$ is obtained by solving the coupled equations (1a) and (1b) with the initial conditions $|c_1(0)| = 1$ and $c_2(0) = 0$. These equations, however, cannot be solved analytically in general because the matrix elements H_{ij} vary with time in a complicated manner for a realistic collision system. One therefore must rely on either approximate or numerical solutions. Here we present results of approximate calculations based on the Landau-Zener method along with results of numerical calculations.

Perhaps the simplest approximate solution to Eqs. (1) is provided by the Landau-Zener model. The model assumes that the transition is localized in a narrow neighborhood of the crossing point R_c over which $H_{22} - H_{11}$ varies linearly with R and H_{12} remains constant. The transition probability as the system passes through the crossing point R_c is then given by

$$P = 1 - \exp(-p) \tag{2a}$$

where

$$p = 2\pi[H_{12}(R_c)]^2/\hbar v\gamma[1 - (\frac{b}{R_c})^2]^{\frac{1}{2}} \quad , \tag{2b}$$

b is the impact parameter, γ is the slope of $|H_{22}-H_{11}|$ vs. R at R_c, and v is the relative velocity of the colliding partners at R_c. If the collision occurs in the presence of a short pulse so that the system is illuminated on the way in or out only, the transition probability is given by $P_s = P$. On the other hand, if the system is illuminated both on the way in and out as is the case for cw radiation or a long pulse, the transition probability is given by $P_{cw} = 2P(1-P)$. To exhibit clearly the intensity dependence of the transition probability, we note that p is proportional to I and let $p = \beta I$.

Then the short-pulse probability and cw probability are given respectively by

$$P_s = 1 - \exp(-\beta I) \quad (3a)$$

$$P_{cw} = 2 \exp(-\beta I)[1-\exp(-\beta I)] \quad (3b)$$

where, for the case of a short pulse, a square pulse of intensity I is assumed.

In Table 1 we give the Landau-Zener probabilities P_s and P_{cw} for different values of I and b for the case $R_c = 9.5$, $v = 1.83\times10^{-4}$, $\gamma = 0.002$ and $\mu = 3$ (all quantities are expressed in atomic units), assuming that the laser beam is linearly polarized in the direction of the relative velocity v of the colliding partners and that the transition moment $\vec{\mu}$ is parallel to the internuclear axis as in a $\Sigma - \Sigma$ transition. The probability P_{cw} as a function of the intensity I exhibits the well-known single-peak structure of the Landau-Zener model, whereas P_s monotonically increases with I. As a result, P_{cw} and P_s differ significantly at high I for which $\beta I \gg 1$. Thus, according to the Landau-Zener model, a short pulse can be much more effective than cw radiation or a long pulse in inducing a transition, if the pulse intensity is sufficiently high.

Table 1. The cw probability P_{cw} and short-pulse probability P_s calculated using the Landau-Zener model.

b(a.u.)	$I(W/cm^2)$	10^8	5×10^8	10^9	5×10^9	10^{10}	5×10^{10}	10^{11}	5×10^{11}
1	P_{cw}	0.053	0.23	0.37	0.39	0.13	0.00	0.00	0.00
	P_s	0.027	0.13	0.24	0.74	0.93	1.00	1.00	1.00
5	P_{cw}	0.045	0.20	0.33	0.43	0.18	0.00	0.00	0.00
	P_s	0.023	0.11	0.21	0.69	0.90	1.00	1.00	1.00

In order to check the accuracy of the data presented in Table 1, we also have calculated the transition probability by numerically integrating Eqs. (1). For this calculation, we have assumed that the potential curves satisfy

$$H_{22} - H_{11} = 3 \exp(-0.737R) - 0.0027 \quad , \quad (4)$$

which yields $R_c = 9.5$ and $\gamma = 0.002$. We have also assumed a straight-line constant velocity (= v) trajectory for simplicity of calculation and have chosen $v = 1.83 \times 10^{-4}$ and $\mu = 3$, as before. The result of our calculation is summarized in Table 2. Comparison of Tables 1 and 2 indicates that the Landau-Zener model yields relatively accurate probabilities at low intensities, but the model fails at high intensities ($I > 10^{11}$ W/cm^2). This is mainly due to the inability of the Landau-Zener model to correctly describe the Rabi oscillation of the molecular system at a high laser intensity.[2] Since P_s decreases due to the Rabi oscillation as I is increased beyond ~10^{11} W/cm^2, there is a finite range of the intensity over which a short pulse is much more effective than cw radiation. For the model system being considered, this range is 10^{10} ~ 10^{11} W/cm^2.

Table 2. The cw probability P_{cw} and short-pulse probability P_s calculated by numerical integration of Eqs. (1).

b(a.u.)	I(W/cm^2)	10^8	5×10^8	10^9	5×10^9	10^{10}	5×10^{10}
1	P_{cw}	0.056	0.24	0.39	0.39	0.21	0.15
	P_s	0.029	0.14	0.27	0.73	0.88	0.92
5	P_{cw}	0.034	0.17	0.28	0.41	0.13	0.13
	P_s	0.019	0.091	0.17	0.71	0.93	0.93

b(a.u.)	I(W/cm^2)	10^{11}	5×10^{11}	10^{12}	5×10^{12}	10^{13}	3×10^{13}
1	P_{cw}	0.24	0.39	0.44	0.48	0.48	0.50
	P_s	0.86	0.73	0.68	0.60	0.61	0.54
5	P_{cw}	0.23	0.40	0.46	0.49	0.48	0.47
	P_s	0.87	0.72	0.64	0.58	0.61	0.63

In view of the recent experimental study[3] of the short-pulse effect considered here, we show in Table 3 the total number N_t of transitions per target atom per unit pulse energy, an experimentally observable quantity, for different values of the pulse duration and intensity. Although this quantity was obtained by integrating the Landau-Zener probability over impact parameter, the data shown in the table is expected to be accurate since the Landau-Zener model yields accurate probabilities at the values of I considered here. As can be seen from the table, there is very little difference between long and short pulses at 10^8 ~ 10^9 W/cm^2. The difference, however, shows up at 10^{10} W/cm^2, as a noticable jump in N_t is seen in going from 10 to 1 ps (the collision time of the system being considered is ~ 2.5 ps).

Table 3. Total number of transitions per target atom per unit pulse energy (in arbitrary units).

$I(W/cm^2)$ \ pulse duration (ps)	1000	100	10	1	0.1
10^8	19.8	19.8	19.9	20.2	20.2
10^9	15.1	15.1	15.7	18.5	18.5
10^{10}	1.45	1.57	2.81	8.75	8.84

In conclusion, the results of our calculations summarized in Tables 1, 2 and 3 indicate that a short pulse can be much more effective in inducing a collisional radiative transition than cw radiation or a long pulse. This short-pulse effect becomes significant if the pulse intensity is sufficiently high that the probability of transition at the crossing point is high ($\sim$ 1). On the other hand, the intensity must not be too high because the Rabi oscillation can bring down the probability to a low value at an extremely high intensity.

ADBOND DYNAMICS

Level Populations

We now turn to consider a molecule physisorbed on a crystalline surface and irradiated by a laser. Due to the weak van der Waals bond, the admolecule has a series of vibrational states for the motion perpendicular to the surface. Under certain conditions it is reasonable to assume that a laser, with frequency ω_L, is in reasonance with only one pair of these levels, say $|g\rangle$ and $|e\rangle$. The transition frequency is $\omega_{eg} = \omega_e - \omega_g > 0$; the detuning is defined as $\Delta = \omega_L - \omega_{eg}$. Relaxation of the adbond occurs due to its interaction with the lattice vibrations of the substrate. In the Markov approximation, this is described through the occurrence of rate constants a_{nk} in a set of first-order differential equations for the level populations and coherences of the reduced density operator of the adbond, the so-called master equation.[5-10]

For pulsed lasers with a pulse duration Δt short compared to the relaxation times (= the inverse rate constants) of the adbond, the latter can be neglected during the pulse. Then,[10] the time evolution during the pulse is given by the optical Bloch equations,

$$\begin{aligned}
\frac{dR_1(t)}{dt} &= \Delta R_2(t) \\
\frac{dR_2(t)}{dt} &= -\Delta R_1(t) + \Omega_p(t)R_3(t) \qquad (5) \\
\frac{dR_3(t)}{dt} &= -\Omega_p(t)R_2(t) \quad ,
\end{aligned}$$

where $\Omega_p(t) = \vec{\mu}.\vec{E}_0(t)/\hbar$ is the time-dependent Rabi frequency; $\vec{E}(t) = \vec{E}_0(t)\cos(\omega_L t)$ is the electric field amplitude of the laser, with slowly-varying envelope $\vec{E}_0(t)$; and $\vec{\mu}$ again is the transition dipole moment. The real-valued quantities R_i are defined by the populations of and coherences between the two adbond levels coupled by the laser, in the rotating frame, according to[10]

$$
\begin{aligned}
R_1 &= \tilde{P}_{ge} + \tilde{P}_{eg} \\
R_2 &= -i\,(\tilde{P}_{ge} - \tilde{P}_{eg}) \\
R_3 &= P_e - P_g \ .
\end{aligned}
\tag{6}
$$

The total exciting effect up to time t of a laser is characterized by

$$\theta(t) = \int_{-\infty}^{t} dt'\ \Omega_p(t') \ . \tag{7}$$

Here, we are only interested in the effect of a complete laser pulse, so that the integral can be taken over a single pulse to give a time-independent parameter θ. Let R_1°, R_2° and R_3° be the values just before the pulse.[10] Then, for zero-detuning, the values right after the pulse are given by

$$
\begin{aligned}
R_1(\Delta t) &= R_1^\circ \\
R_2(\Delta t) &= R_2^\circ\cos(\theta) + R_3^\circ\sin(\theta) \\
R_3(\Delta t) &= R_3^\circ\cos(\theta) - R_2^\circ\sin(\theta) \ ,
\end{aligned}
\tag{8}
$$

where the start of the pulse is taken as zero point of time, and Δt is the pulse duration. We shall only consider the situation of maximal excitation, which is obtained for $\theta = \pi$, the so-called π-pulses.

Instead of a single pulse, we shall consider a series of equally spaced π-pulses, with interval time t_p and pulse duration $\Delta t << t_p$. Assume that the system is initially in thermal equilibrium, i.e., $R_1 = 0$, $R_2 = 0$ and $R_3 = R_3(eq)$. Then, after a π-pulse, R_1 and R_2 are zero again, but R_3 is changed into its opposite: $R_3(\Delta t) = - R_3(eq)$. Between two consecutive pulses the adbond evolves in time through its relaxation against the lattice vibrations. This process can be described by a master equation.[5-9] In the present situation only the populations have to be considered. However, unlike the coupling with the laser, transitions to and from all vibrational states are possible. With the result of the laser pulse as initial condition, the formal solution of the master equation is ($\Delta t \leq t \leq t_p$),

$$\underset{\sim}{P}(t) = \underset{\sim}{P}(eq) + e^{-Wt}\{\underset{\sim}{P}(\Delta t) - \underset{\sim}{P}(eq)\} \ , \tag{9}$$

where W is the matrix formed by the transition rate constants a_{nk} and has elements

$$W_{nm} = \sum_k a_{nk}\delta_{nm} - a_{mn} \ . \tag{10}$$

$\underset{\sim}{P}(t)$ is the vector formed by the populations $P_n(t)$, and $\underset{\sim}{P}(eq)$ refers to the thermal equilibrium distribution, given by $W\ \underset{\sim}{P}(eq) = 0$.

After a number of pulses have passed, the adbond will reach a quasi-steady state, wherein the time evolution of the populations is the same in each interval t_p (see Fig. 1). Then, combining Eqs. (8) and (9), two sets

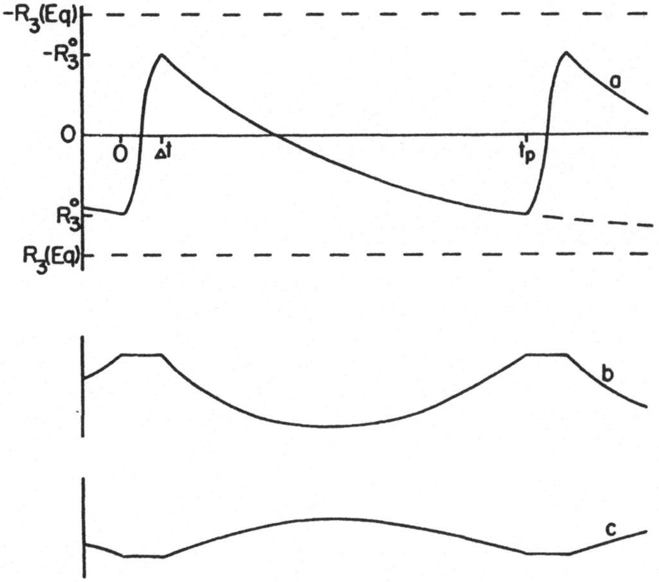

Fig. 1. Schematic drawing of the time evolution of the level occupations for a pulse sequence in the quasi-steady state. Curve a represents R_3; curve b gives $P_g + P_e$; and curve c gives P_n for any other level $n \neq$ e or g.

of equations are obtained, which together determine the constants $P_n(\Delta t)$ in terms of the rate constants a_{nk} and the interval time t_p:

$$\underset{\sim}{P}(t_p) = \underset{\sim}{P}(eq) + e^{-Wt_p} \{\underset{\sim}{P}(\Delta t) - \underset{\sim}{P}(eq)\} \tag{11}$$

$$\begin{aligned} P_n(t_p) &= P_n(\Delta t) = P_n^o \quad (n \neq e,g) \\ P_g(t_p) &= P_e(\Delta t) = P_g^o \\ P_e(t_p) &= P_g(\Delta t) = P_e^o \quad . \end{aligned} \tag{12}$$

For a continuous wave laser, the Rabi-frequency is a constant, Ω_{cw}. Now, the relaxation processes cannot be neglected during the interaction with the laser. The rate constants a_{nk} give rise to additional terms in the optical Bloch equations,[11,12] in a way analogous to their occurence in the master equation. After some time a true steady state will be reached, in which the populations are determined by

$$\sum_k a_{nk} P_n(\infty) = \sum_k a_{kn} P_k(\infty) \quad (n \neq e,g)$$

$$\sum_k a_{gk} P_g(\infty) = \sum_k a_{kg} P_k(\infty) - \frac{1}{2} \frac{\Gamma_{eg}}{\frac{1}{2}\Gamma_{eg}^2 + 2\Delta^2} \{P_g(\infty) - P_e(\infty)\} \tag{13}$$

$$\sum_k a_{ek} P_e(\infty) = \sum_k a_{ke} P_k(\infty) + \frac{1}{2} \frac{\Gamma_{eg}}{\frac{1}{2}\Gamma_{eg}^2 + 2\Delta^2} \{P_g(\infty) - P_e(\infty)\} \quad ,$$

where $\Gamma_{eg} = \sum_k \{a_{ek} + a_{gk}\}$, and the steady-state values are denoted by $P_n(\infty)$.

Pulsed Laser Versus Continuous Wave Laser

To compare the effect of a pulse train with the effect of a continuous wave (cw) laser on the dynamics of some system, a criterion is needed to compare both lasers. Below two possible criteria will be defined. First, it is required that both lasers have equal average power. Because the Rabi-frequency, $\Omega(t)$, is proportional to the electric field of the laser, the intensity is proportional to $\Omega^2(t)$. Performing the time-average for a series of π-pulses then gives:[11]

$$\Omega_{cw}^2 = \pi^2/(\Delta t\ t_p) \quad . \tag{14}$$

A second possible criterion is to require the average dissipation from both lasers to be equal. This leads to[11]

$$(P_g^\circ - P_e^\circ)/t_p = (\Omega_{cw}^2/\Gamma_{eg})(P_g(\infty) - P_e(\infty)) \quad , \tag{15}$$

where P_g°, P_e°, $P_g(\infty)$ and $P_e(\infty)$ follow from Eqs. (11)-(13).

The efficiency of a pulsed laser over a cw laser to induce some process (e.g., desorption or chemical reaction) will be different for each process. Here we restrict ourselves to the effect on the population of the upper one of the two levels coupled by the laser for a two-level system. A more extensive discussion is given elsewhere.[11] Let us define the efficiency ε as the ratio of the average excited level population $P_e(av)$ under pulsed laser irradiation to the steady state population $P_e(\infty)$ under cw laser irradiation. For a two-level system, Eqs. (11)-(13) are easily solved.[11] We are interested in the situation where the laser-induced population is substantially larger than the equilibrium thermal occupation of the levels. Then, neglecting the latter, the efficiency is[11]

$$\varepsilon = \frac{2}{(1 + e^{-\Gamma t_p})} \frac{(1 - e^{-\Gamma t_p})}{\Gamma t_p} \frac{\frac{1}{2}\Gamma^2 + \Omega_{cw}^2}{\Omega_{cw}^2} \quad . \tag{16}$$

For a given integrated intensity of one pulse, the average intensity of the pulse train is proportional to t_p^{-1}. Then a definition for a strong/weak pulsed laser is $\Gamma t_p \lessgtr 1$. Analogously a strong/weak continuous wave laser is defined by (Ω_{cw}^2/Γ^2) being much larger/smaller than one.

Using these definitions, it follows from Eq. (16) that $\varepsilon = 1$ in the strong laser limit. This can easily be understood when it is realized that both for the pulsed laser and the cw laser the system is then continuously driven into saturation ($R_3 = 0$). Using criterion 1 (Eq. (14)) leads to

$$\varepsilon = \frac{1 - e^{-\Gamma t_p}}{1 + e^{-\gamma t_p}} \frac{\Gamma^2 + 2\Omega_{cw}^2}{\Gamma^2} \frac{\Gamma \Delta t}{\pi^2} . \tag{17}$$

In the weak laser limit this gives $\varepsilon = \Gamma\Delta t/\pi^2 \ll 1$, and thus ε can be made arbitrarily small by decreasing the pulse duration Δt. The explanation is that the intensity of the pulsed laser is proportional to $\Omega_p^2\Delta t$, whereas the exciting effect is only proportional to $\Omega_p\Delta t$. Decreasing Δt, while $\Omega_p^2\Delta t$ is held constant, will decrease $\Omega_p\Delta t$ too. For a cw laser both the intensity and the exciting effect are proportional to Ω_{cw}^2. The latter can be regarded as the limiting form of a pulsed laser with $\Delta t \to \infty$, and therefore will give the most efficient form of excitation.

For criterion 2, $\varepsilon = 1$ is obtained, independent of the laser power. This result reflects the fact that the energy flow into the substrate (is absorbed energy) is proportional to the average occupation of the excited level and not on how P_e varies in time.[11,13]

ACKNOWLEDGMENTS

This research was supported in part by the National Science Foundation under Grant Nos. CHE-8512406 and CHE-8620274, the Air Force Office of Scientific Research (AFSC), United States Air Force, under Contract No. F49620-86-C-0009, and the Office of Naval Research. The research of S. van Smaalen has been made possible by a fellowship from the Royal Netherlands Academy of Arts and Sciences.

REFERENCES

1. H. W. Lee and T. F. George, J. Phys. Chem. 83:928 (1979).
2. H. W. Lee and T. F. George, Phys. Rev. A, in press.
3. T. Sizer II and M. G. Raymer, Phys. Rev. Lett. 56:123 (1986).
4. H. W. Lee and T. F. George, Phys. Rev. A 29:2509 (1984).
5. S. Efrima, L. Jedrzejek, K. F. Freed, E. Hood, and H. Metiu, J. Chem. Phys. 79:2436 (1983).
6. Z. W. Gortel, H. J. Kreuzer, P. Piercy, and R. Teshima, Phys. Rev. B 27:5066 (1983).
7. B. Fain and S. H. Lin, Surf. Sci. 147:497 (1984).
8. H. F. Arnoldus, S. van Smaalen, and T. F. George, Phys. Rev. B 34:6902 (1986).
9. H. F. Arnoldus, S. van Smaalen and T. F. George, Adv. Chem. Phys., in press.
10. L. Allen and J. H. Eberly, "Optical Resonance and Two-level Atoms," Wiley, New York (1975).
11. S. van Smaalen and T. F. George, Surf. Sci. 183:263 (1987).
12. W. H. Louisell, Quantum Statistical Properties of Radiation Wiley, New York (1973), Ch. 6.
13. S. van Smaalen, H. F. Arnoldus, and T. F. George, Phys. Rev. B 35:1142 (1987).

RYDBERG STATES IN LASER FIELDS

G. Alber and P. Zoller

Institute for Theoretical Physics
University of Innsbruck, A-6020 Innsbruck, Austria

1. INTRODUCTION

Rydberg states play an important role as intermediate or final states in laser induced processes studied in atoms and molecules. A typical one photon excitation process of Rydberg states $|\epsilon_n\rangle$ from a low-lying atomic state $|g\rangle$ is characterized by three parameters: The spectral width of the laser pulse $\Delta\omega_L \approx \hbar/T_P$ whose pulse duration is T_P, the Rabi frequencies Ω_n associated with the transitions $|g\rangle \longrightarrow |\epsilon_n\rangle$ and the level spacing between adjacent Rydberg states $\Delta\omega_n \approx 2Ry/\hbar(-\epsilon_n/Ry)^{3/2}$. Depending on the relative magnitudes of these parameters we can distinguish between three dynamical cases: (1) The simplest situation arises, if we excite an isolated Rydberg state well below threshold ($\Delta\omega_n > \Omega_n, \Delta\omega_L$). This regime is well understood on the basis of a two-(or few-) level approximation, which treats the coupling between the resonant atomic states nonperturbatively and takes into account the influence of all other nonresonant states perturbatively. On the contrary, cases (2) and (3), which correspond to excitation by a short ($\Delta\omega_L > \Delta\omega_n, \Omega_n$) or intense ($\Omega_n > \Delta\omega_n, \Delta\omega_L$) laser pulse and arise if we are exciting sufficiently close to a Rydberg threshold, are not so well understood. In the extreme situation of excitation directly at a Rydberg threshold both (infinitely) many Rydberg states and part of the adjoining electron continuum are excited and traditional theoretical approaches based on a few-level approximation[1] for the bound-bound transitions $|g\rangle \longrightarrow |\epsilon_n\rangle$ or the pole-approximation[1] as far as the bound-free transitions $|g\rangle \longrightarrow |\epsilon\rangle$ are concerned break down.

Recently we have developed a theory[2], which is capable of describing laser excitation close to a Rydberg threshold. It is based on the observation that the photon absorption process $|g\rangle \longrightarrow |\epsilon_n\rangle$ takes place in a region around the atomic nucleus (≡ reaction zone), which is small in comparison with the extent of highly excited Rydberg states. This *finite range* of the radiative coupling allows us to use concepts of quantum defect theory (QDT)[3] and to derive closed analytical expressions for transition amplitudes, whose time dependence can conveniently be studied with the help of a *multiple scattering expansion*. This expansion expresses these amplitudes as an infinite sum over contributions due to all classical paths of the excited Rydberg electron (Kepler orbits). In particular, for excitation close to a Rydberg threshold, i.e. cases (2) and (3), it leads directly to the physical picture of a (radial) Rydberg wave packet, which is generated inside the reaction zone and moves in the Coulomb potential of the ionic core.

The main purpose of this paper is to discuss laser excitation close to a Rydberg threshold. Therefore in section two we study generation and detection of Rydberg wave packets by short laser pulses and in section three we deal with the excitation of a Rydberg series by an intense laser pulse (generalized Rabi problem).

2. SHORT LASER PULSES

According to our introductory remarks a short laser pulse implies $\Delta\omega_L > \Omega_n$ so that the initial atomic state $|g\rangle$ is undepleted and the excitation of the Rydberg electron can be determined by time dependent perturbation theory in the laser field. For times t long after the interaction with the laser pulse the state of the atom is therefore approximately given by $|\Psi(t)\rangle = e^{-i\epsilon_g t}|g\rangle + |\Psi_g(t)\rangle$ with the state of the excited Rydberg electron described by

$$|\Psi_g(t)\rangle = i\int_{-\infty}^{+\infty} dt'\; e^{-iH_A(t-t')}\, \vec{d}\cdot\vec{\epsilon}\; \mathcal{E}(t')\; e^{-i\bar{\epsilon}t'}\; |g\rangle\,. \tag{2.1}$$

H_A is the atomic Hamiltonian, $\vec{d}$ the dipole operator, $\vec{\epsilon}$ and $\mathcal{E}(t)$ the polarization and slowly varying electric field amplitude of the laser field and $\bar{\epsilon} = \epsilon_g + \omega$ is the mean excited energy (We use Hartree atomic units). From Equ.(2.1) we immediately notice that the laser excitation process affects only a spatial region of the size of $|g\rangle$. This region (≡ reaction zone) is small in comparison with the extent of the highly excited Rydberg states so that the laser excitation process is *localized in space*.

A short laser pulse also implies $\Delta\omega_L > \Delta\omega_n$ so that many Rydberg states are excited. In the time domain this condition states that the interaction time between atom and laser field T_P is small in comparison with the mean classical orbit time $T_{\bar{\epsilon}} = 2\pi(-2\bar{\epsilon})^{-3/2}$ of the excited electron in the Coulomb potential of the ionic core. The laser excitation process is therefore also *localized in time*. Both conditions, namely localization of the photon absorption process in space and time imply that a (radial) Rydberg wave packet is generated[4,5]. Note that due to dipole selection rules only a few angular momentum eigenstates are excited so that only the radial coordinate of the Rydberg electron is localized.

The center of this wave packet performs a periodic motion in the Coulomb potential of the ionic core with period $T_{\bar{\epsilon}}$. If we are exciting a multi-electron atom with one valence electron, for example, the electrostatic potential due to the ionic core deviates from a pure $1/\rho$-Coulomb form inside the core region, which is essentially identical with the small reaction zone where the initial laser excitation process has taken place. Therefore, whenever the excited wave packet returns to the reaction zone it experiences an additional electron-ion scattering due to this modified core potential, which can be characterized by a scattering matrix element $\chi = e^{i2\pi\alpha}$. Due to the fact that inside the reaction zone there is no difference between energy normalized Rydberg- and continuum states close to the Rydberg threshold ($\epsilon=0$) χ is a slowly varying function of energy across threshold. Above threshold ($\epsilon>0$) $\pi\alpha$ is a continuum phase shift and below ($\epsilon< 0$) α is the quantum defect of the excited Rydberg series. This physical picture suggests that $|\Psi_g(t)\rangle$ can be represented by a multiple scattering expansion as a sum over contributions due to all these subsequent (below threshold) electron-ion scattering events, which take place inside the reaction zone and are characterized by a scattering matrix χ (compare with Equ.(3.6)).

The periodic motion of the generated wave packet can be probed by any process which is also localized in space and time. One example is a two-photon Raman process with two time delayed short laser pulses. In such a process a radial electronic wave packet is generated by a first short laser pulse, which excites an atom from an initial state $|g\rangle$. A second short laser pulse then deexcites this wave packet after a time delay $\Delta t \gg T_P$ by inducing a transition to some low-lying bound state $|f\rangle$. As excitation and deexcitation (≡ stimulated recombination) both take place inside a small reaction zone around the atomic nucleus we expect a large Raman transition probability whenever $\Delta t \approx mT_{\bar{\epsilon}}$. This can be seen in Fig.1 where the Raman transition probability is plotted in arbitrary units as a function of $\Delta t/T_{\bar{\epsilon}}$ in the case of excitation of a single Rydberg series. Each peak at $\Delta t \approx mT_{\bar{\epsilon}}$ is associated with a stimulated recombination of the electron-ion complex. The broadening of the recombination peaks reflects the spreading of the generated wave packet. After sufficiently long times (here $\Delta t \gtrsim 5T_{\bar{\epsilon}}$) the wave packet has spread out over the whole extent of its orbit so that contributions due to subsequent returns to the reaction zone overlap in time thus giving rise to a complicated interference pattern.

Additional interesting effects can arise in multi-electron atoms. Figs.2 show the Raman transition probability in the case of excitation of two Rydberg series, which are coupled by configuration interaction (coupling parameter $\tau=0.02$) and converge to different ionization thresholds ($\epsilon_1=4.7\cdot10^{-5}$, $\epsilon_2=0$ a.u.). The first short laser pulse now generates a Rydberg wave packet in each channel with mean classical orbit times T_1 and T_2 (here $2T_2\approx3T_1$). In this case we expect a large Raman transition probability whenever $\Delta t \approx m_1T_1+m_2T_2$. Recombination peaks associated with values $m_1> 0$ and $m_2> 0$ are possible, because with each return to the inner turning point of its orbit the wave packet can be scattered into the other channel by an inelastic (below threshold) electron-ion collision inside the reaction zone, which can be characterized by an off-diagonal matrix element of the approximately energy independent scattering matrix χ. The transition amplitude associated with a particular recombination peak (m_1,m_2) in Figs.2 is proportional to excitation- and deexcitation dipole matrix elements multiplied with scattering matrix elements of χ. So the amplitude of the peak (1,1), for example, is proportional to $(\mathcal{D}^{(-)}_{2f}\chi_{21}\mathcal{D}^{(-)}_{1g} + \mathcal{D}^{(-)}_{1f}\chi_{12}\mathcal{D}^{(-)}_{2g})$ and its value therefore strongly depends on the relative phases of the various photoionization dipole matrix elements $\mathcal{D}^{(-)}_{ig}$, $\mathcal{D}^{(-)}_{if}$ i=1,2. This is apparent from Figs. 2a and 2b. Whereas in Fig.2a $\mathcal{D}^{(-)}_{1g}/\mathcal{D}^{(-)}_{2g} = \mathcal{D}^{(-)}_{1f}/\mathcal{D}^{(-)}_{2f} = 1$ in Fig.2b $\mathcal{D}^{(-)}_{1g}/\mathcal{D}^{(-)}_{2g} = -\mathcal{D}^{(-)}_{1f}/\mathcal{D}^{(-)}_{2f} = 1$ and the peak (1,1) has dissapeared due to destructive interferences between the two contributing quantum paths. Eventually both generated wave packets overlap (here at $\Delta t \approx 2T_2 \approx 3T_1$) and interfere quantum mechanically. Whether this interference is constructive (Fig.2b) or destructive (Fig.2a) also strongly depends on the relative phases of the photoionization dipole matrix elements.

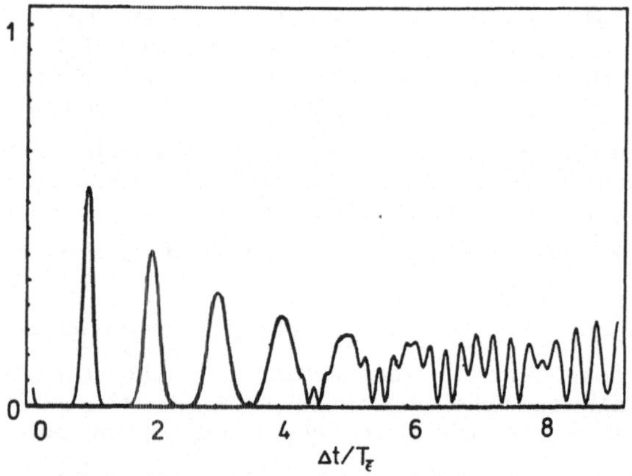

Fig.1: Raman transition probability as a function of time delay $\Delta t/T_{\bar{\epsilon}}$ with T_P=10ps, $T_{\bar{\epsilon}}$=94ps.

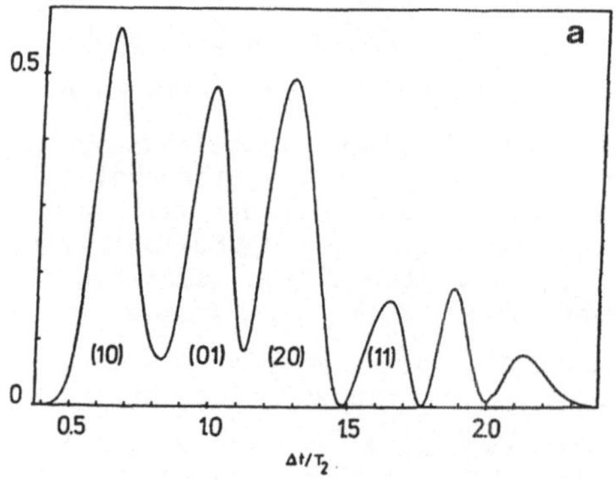

Fig.2a: Raman transition probability as a function of $\Delta t/T_2$ with $T_{P_1}=T_{P_2}$=14ps, $T_2=1.5\cdot T_1$=107ps, $\mathscr{D}^{(-)}_{1g}/\mathscr{D}^{(-)}_{2g} = \mathscr{D}^{(-)}_{1f}/\mathscr{D}^{(-)}_{2f} = 1.$

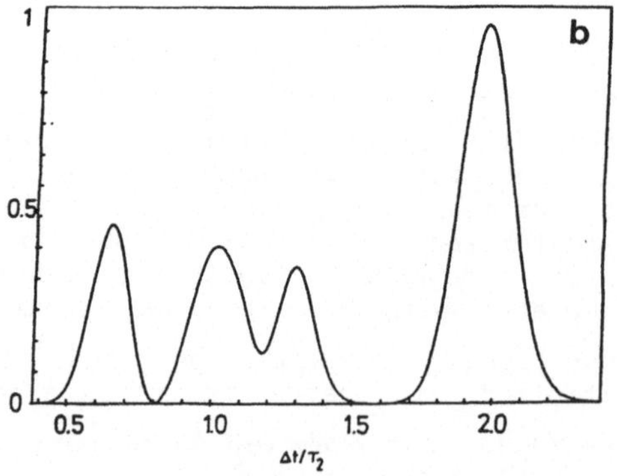

Fig.2b: Same as Fig.2a but with $\mathscr{D}^{(-)}_{1g}/\mathscr{D}^{(-)}_{2g} = -\mathscr{D}^{(-)}_{1f}/\mathscr{D}^{(-)}_{2f} = 1.$

3. INTENSE LASER PULSES

We consider excitation of a Rydberg series with quantum defect α from a low-lying atomic state $|g\rangle$ by an intense laser pulse, which is turned on instantaneously at t=0 and whose electric field envelope $\mathcal{E}(t)$ is constant for t>0. In particular we are interested in situations where the initial state $|g\rangle$ is depleted and many Rydberg states (and possibly part of the adjoining electron continuum) are excited so that the traditional two-(or few-) level approximation is not applicable. The state of the atom at time t is approximately given by $|\Psi(t)\rangle = |g\rangle a_g(t) + |F(t)\rangle$ with $|F(t)\rangle$ characterizing the time evolution of the excited electron. Taking the Laplace transform of $|\Psi(t)\rangle$ and neglecting ionization from the excited Rydberg states to higher electron continua we find in the dipole- and rotating wave approximation the system of equations

$$(z - \epsilon_g)a_g(z) + \langle g|\vec{d}\cdot\vec{\epsilon}^{\,*}\mathcal{E}^*|F(z+\omega)\rangle = i \qquad (3.1)$$

$$(z + \omega - H_A)\ |F(z+\omega)\rangle + \vec{d}\cdot\vec{\epsilon}\mathcal{E}|g\rangle\ a_g(z) = 0$$

with $|\Psi(t)\rangle = 1/2\pi \int_{-\infty+i0}^{+\infty+i0} dz\ e^{-izt}\ |\Psi(z)\rangle$. In the language of QDT these equations describe the coupling between a bound channel $|g\rangle$ and a free channel $|F(z+\omega)\rangle$[3]. According to Equs.(3.1) the Laplace transform $a_g(z)$ is given by

$$a_g(z) = i\ [\ z - \epsilon_g - \Sigma\ (z+\omega)\ ]^{-1} \qquad (3.2a)$$

with the resonant part of the self energy

$$\Sigma\ (\epsilon) = \langle\ g\ |\vec{d}\cdot\vec{\epsilon}^{\,*}\ \mathcal{E}^*\ \frac{1}{\epsilon - H_A + i0}\ \vec{d}\cdot\vec{\epsilon}\ \mathcal{E}|\ g\ \rangle = \qquad (3.2b)$$

$$= \begin{cases} \delta\omega - i\gamma/2 & (\epsilon > 0) \\ \delta\omega + \gamma/2\ \cot\pi(\nu+\alpha) & (\epsilon=-1/2\nu^{-2} < 0) \end{cases} .$$

(The final result for $\Sigma(\epsilon)$ can be obtained with the help of QDT methods or the Poisson summation formula)[2]. $\delta\omega$ is a contribution to the quadratic Stark shift of the initial state $|g\rangle$ and $\gamma = 2\pi\ |\langle\epsilon|\vec{d}\cdot\vec{\epsilon}\ \mathcal{E}|g\rangle|^2$ is the ionization rate according to Fermi's Golden rule and characterizes the radiative coupling between $|g\rangle$ and the excited channel. It is important to note that both parameters are approximately energy independent across the Rydberg threshold (ϵ=0), because inside the reaction zone, where the laser excitation process $|g\rangle \longrightarrow |\epsilon_n\rangle$ takes place and which is of the size of the initial state $|g\rangle$, there is no difference between energy normalized Rydberg- and continuum states close to threshold. This reflects the *localization* of the laser excitation process *in space*.

In order to determine the time evolution of $a_g(t)$ we have to invert the Laplace transform. Traditionally this is done with the help of contour integration in the complex z-plane. This approach leads to a *dressed state representation*, which expresses $a_g(t)$ as a sum over contributions due to all dressed states and a cut contribution from the electron continuum(ϵ>0)[2]. The dressed state energies $\tilde{\epsilon}_n = z_n + \omega = -1/2\tilde{\nu}_n^{-2} < 0$ are associated with the poles of $a_g(z)$ at z_n. From Equs.(3.2) we find in our case the implicit transcendental equation $\tilde{\epsilon}_n = -1/2\,(n - \alpha - \mu(\tilde{\epsilon}_n))^{-2}$ for the dressed states so that the influence of the radiative coupling is characterized by the intensity dependent quantum defect

$$\mu(\epsilon) = -1/\pi \arctan(\gamma/2\,(\epsilon - \bar{\epsilon})^{-1}) \tag{3.3}$$

with $\bar{\epsilon} = \epsilon_g + \omega + \delta\omega = -1/2\bar{\nu}^{-2}$. The appearance of this quantum defect is not surprising, because it is well known that mixing of an isolated bound state into an electron continuum leads to an additional continuum phase shift $\pi\mu(\epsilon)$. In laser excitation such a phase shift gives rise to laser-induced autoionizing-like resonances whereas in the process of laser-assisted electron-ion scattering, which is described by the scattering matrix

$$\tilde{\chi}(\epsilon) = e^{2i\pi(\alpha+\mu(\epsilon))}, \tag{3.4}$$

it manifests itself in a "capture-escape" resonance[6]. Furthermore, continuum phase shifts correspond to quantum defects in the bound state region[3]. So we finally find in the *dressed state representation*[2]

$$a_g(t) = \sum_n e^{-i(\tilde{\epsilon}_n-\omega)t}\, 1/\pi\, \tilde{\nu}_n^{-3} \frac{\gamma/2}{(\tilde{\epsilon}_n-\bar{\epsilon})^2 + (\gamma/2)^2(1+2/\pi\gamma\cdot\tilde{\nu}_n^{-3})} + \tag{3.5}$$

$$i/2\pi\, e^{-i(\bar{\epsilon}-\omega)t}\left\{e^{-\gamma t/2}[E_1(-i\bar{\epsilon}t-\gamma t/2)-2i\pi ST(\bar{\epsilon})] - e^{\gamma t/2}E_1(-i\bar{\epsilon}t+\gamma t/2)\right\}$$

with the exponential integral $E_1(x)$ and the unit step function $ST(x)$. In the limit $\gamma \ll \bar{\nu}^{-3}$ we are exciting an isolated Rydberg state well below threshold and Equ.(3.5) reduces to the two-level result.

In the case of an intense laser pulse, however, $\gamma > \bar{\nu}^{-3}$ and many dressed states contribute to Equ.(3.5). In this limit it is more convenient to expand $a_g(z)$ in terms of the rapidly oscillating function $e^{i2\pi\nu}$ (Note that $2\pi\nu$ is just the classical action along a closed Kepler orbit of energy ϵ). This way we obtain the *multiple scattering expansion*[2]

$$a_g(t) = e^{-i(\bar{\epsilon}-\omega)t}\, e^{-\gamma t/2} + \tag{3.6}$$

$$\sum_{m=1}^{\infty} \int_{-\infty}^{0} d\epsilon\, e^{-i(\epsilon-\omega)t}[i(\epsilon-\bar{\epsilon}+i\gamma/2)^{-1}\mathcal{D}_{\epsilon g}^{(-)}]e^{2i\pi\nu}[\tilde{\chi}(\epsilon)e^{2i\pi\nu}]^{m-1}$$

$$[\mathcal{D}_{\epsilon g}^{(-)} i(\epsilon-\bar{\epsilon}+i\gamma/2)^{-1}]$$

with the complex photoionization dipole matrix elements $\mathcal{D}_{\epsilon g}^{(-)} = -i\, e^{i\pi\alpha} \langle\epsilon|\vec{d}\cdot\vec{e}\mathcal{E}|g\rangle$. If $\gamma \gg \bar{\nu}^{-3}$ we can evaluate the various energy integrals in equ.(3.6) in the stationary phase approximation and the dominant contributions arise from energies $\epsilon_s(m,t)$, which fullfill the stationary phase condition $t = T_{\epsilon_s}(m,t)$.

In the case of laser excitation close to the photoionization threshold by an intense pulse the physical interpretation of Equ.(3.6) is straight forward. For times $t \ll T_{\bar{\epsilon}}$ the stationary phase contributions to Equ.(3.6) are negligible and the initial state probability is exponentially decaying with a rate γ. The laser excitation process is therefore not only *localized* in space but also *in time*, because the characteristic interaction time $1/\gamma$ ($\equiv$ depletion time of $|g\rangle$) is short in comparison with the mean classical orbit time $T_{\bar{\epsilon}}$. This implies that a radial electronic wave packet is generated. The exponential decay reflects the fact that at times $t \ll T_{\bar{\epsilon}}$ this wave packet has not yet "felt" the outer turning point of its Kepler orbit and therefore behaves like in a true ionization process above threshold. For times $t \gtrsim T_{\bar{\epsilon}}$ the stationary phase contributions associated with $m=1,2,\ldots$ become important. They describe the periodic motion of the excited wave packet in the Coulomb potential of the ionic core. In particular, the m-th term is the contribution due to the m-th return of this wave packet to the reaction zone, where it can be deexcited back to the initial state $|g\rangle$. The periodic recombinations of the electron-ion complex inside the reaction zone manifest themselves in pulsations of $|a_g(t)|^2$ with period $T_{\bar{\epsilon}}$. Equ.(3.6) shows that the m-th recombination peak centered at $t \approx mT_{\bar{\epsilon}}$ does not only contain information about the initial excitation- and final deexcitation process but also about the $(m-1)$ intermediate laser-assisted electron-ion scattering events inside the reaction zone, which are characterized by $\tilde{\chi}(\epsilon)$ of Equ.(3.4).

Figs.3 show the initial state probability $|a_g(t)|^2$ as a function of $t/T_{\bar{\epsilon}}$ for $\gamma=10^{-6}$a.u. and different values of $\bar{\epsilon}$. In Fig.3a $1/\gamma > T_{\bar{\epsilon}}$ and we observe the typical Rabi oscillations, which are slightly perturbed due to the presence of the nonresonant Rydberg states. In Figs.3b and 3c the laser excitation process is localized in time, i.e. $1/\gamma < T_{\bar{\epsilon}}$, and we see the characteristic exponential decay for $t \ll T_{\bar{\epsilon}}$ and the pulsations with period $T_{\bar{\epsilon}}$ for $t \gtrsim T_{\bar{\epsilon}}$. As soon as the generated Rydberg wave packet has spread out over the whole extent of its orbit contributions due to successive returns to the reaction zone overlap in time and give rise to a complicated interference pattern. In Fig.3c we are exciting so close to threshold that this already happens at $t \gtrsim T_{\bar{\epsilon}}$.

Until now we have neglected photon absorption from Rydberg states to higher electron continua. Due to the large extent of highly excited Rydberg states it is not obvious that these transitions are also of finite range and it is therefore not clear how to include them into our treatment. However, it has recently been shown by Giusti-Suzor and Zoller[7] that the range of the radiative coupling ρ_R as far as Rydberg-free or free-free transitions are concerned is determined by the frequency ω and the intensity I of the laser pulse and is roughly given by $\rho_R \approx \max\{\omega^{-2/3}, \sqrt{I}\cdot\omega^{-2}\}$ (The a.u. of laser intensity is $I_0=1.4\cdot10^{17}$W/cm^2). This implies that for optical frequencies and not too high laser intensities photon absorption is of finite range even as far as these kind of transitions are concerned. This is an important result, because it allows us to include ionization from the excited Rydberg states in a straight forward manner into our approach and, more generally, to treat *configuration interaction* and *laser coupling* in a *unified way* due to the fact that both take place inside a reaction zone, which is small in comparison with the extent of highly excited Rydberg states. All the physical processes inside this reaction zone can therefore be characterized by a laser-assisted electron-ion scattering matrix $\tilde{\chi}(\epsilon)$, which is a slowly varying function of energy across a Rydberg threshold as inside the reaction zone there is no difference between energy normalized Rydberg- and continuum states close to threshold. With the help of this general scattering matrix it is straight forward to derive multiple scattering expansions for the time evolution of various transition amplitudes and to study the threshold behaviour (compare with Equ.(3.6)).

Fig.3d shows $|a_g(t)|^2$ in the case of an additional decay channel of the Rydberg states (here due to autoionization with channel coupling parameter $\tau=10^{-2}$ and q=20). Whereas in Fig.3c the time averaged "mean" initial state probability (dashed curves in Figs.3c and 3d) reaches a non zero stationary value in the long time limit reflecting population trapping in $|g\rangle$ the additional decay channel in Fig.3d finally leads to complete depletion of the initial state.

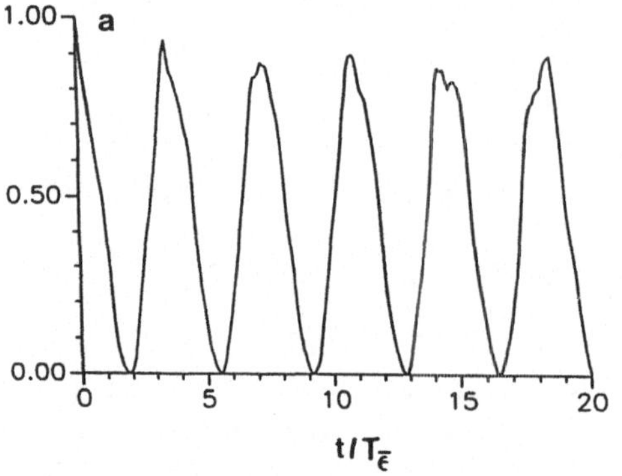

Fig.3a: Initial state probability as a function of $t/T_{\bar{\epsilon}}$ for $\gamma=10^{-6}$a.u. and $\bar{\epsilon}= -2\cdot 10^{-4}$a.u.

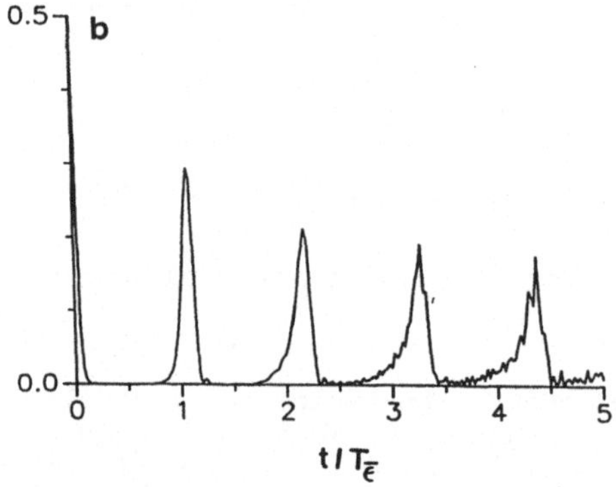

Fig.3b: Same as Fig.3a but with $\bar{\epsilon}= -1.25\cdot 10^{-5}$a.u.

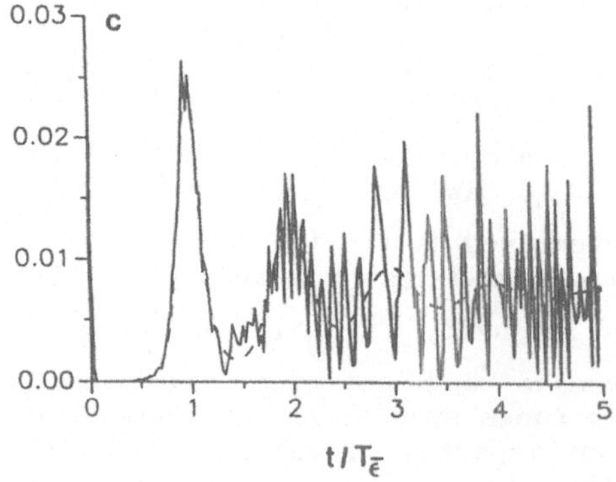

Fig.3c: Same as Fig.3a but with $\bar{\epsilon}= -4\cdot 10^{-6}$a.u.

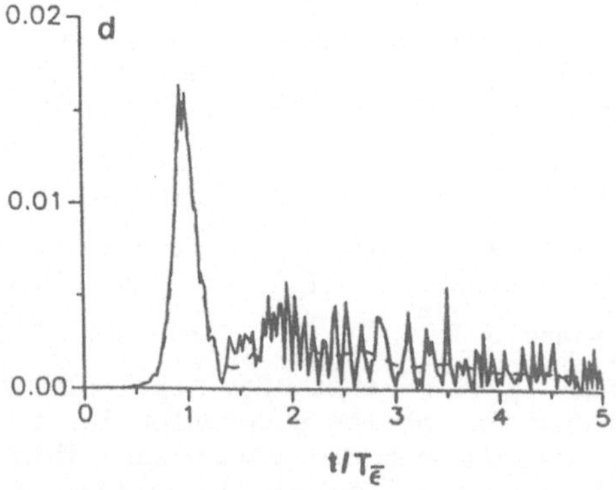

Fig.3d: Same as Fig.3c but with an additional decay channel.

ACKNOWLEDGMENT

This work was supported by the Austrian "Fonds zur Förderung der wissenschaftlichen Forschung" under contract number P6008P.

REFERENCES

1. C.Cohen-Tannoudji, B.Diu and F.Laloe, "Qantum Mechanics" (Wiley, 1977)
2. G.Alber and P.Zoller, Phys.Rev.A (submitted)
3. M.J.Seaton, Rep.Prog.Phys **46**, 167(1983)
4. G.Alber, H.Ritsch and P.Zoller, Phys. Rev. A**34**,1058(1986)
5. W.A.Henle,H.Ritsch and P.Zoller, Phys.Rev.A (in print)
6. L.Dimou and F.H.M.Faisal in "Photons and Continuum States of Atoms and Molecules" edited by N.K.Rahman, C.Guidotti and M.Allegrini (Springer, 1986)
7. A.Giusti-Suzor and P.Zoller, Phys.Rev.A (submitted)

WAVE PACKETS IN VIBRATIONAL MOTION OF SODIUM DIMER

C. Radzewicz* and M.G. Raymer

The Institute of Optics, University of Rochester
Rochester, NY 14627, USA

Wave packets, i.e. quantum states well localized in space, play an important role in our understanding of quantum mechanics. Their time evolution, which is a quantum mechanical equivalent of a classical particle trajectory, provides a useful insight into the motion of microscopic objects. In atomic physics, excitation of wave packets consisting of Rydberg states was proposed as a way to study the motion of an electron in the field of the nucleus in the nearly classical limit.[1-3] The concept of wave packets has also been used for a semi-classical description of molecular spectra[4] and dynamics.[5]

Some processes in laser chemistry are known to depend strongly on internuclear distance r . For example, photodissociation is enhanced in the regions of r where the difference potential shows an extremum. The efficiency of these processes could be controlled if states of well defined r could be prepared. Unfortunately, when a laser pulse longer than the period of molecular vibrations is used for excitation, little information about the initial internuclear distance is preserved. The probability of finding the molecule at a particular r is determined only by the appropriate vibrational wavefunction $\psi_v(r)$ of the excited state. On the other hand, if the exciting pulse is shorter than the vibrational period, a coherent excitation of many vibrational levels leads to formation of a wave packet, which might be concentrated in a relatively small range of r . Since the position of the wave packet changes as it moves in the potential well, a second delayed short pulse can be used for effective excitation of the desired r-dependent transition. Thus this scheme may be useful for controlling the relative efficiencies of competing processes, for example photodissociation versus electronic excitation. To date few experiments have demonstrated control of collision processes by ultrashort laser pulses. One such study showed an enhancement of electronic excitation of sodium atoms during collision with argon atoms (optical collision or collisional redistribution).[6]

In the present study we investigate theoretically the excitation of vibrational wave packets in the sodium dimer molecule, and the subsequent probing of these wave packets by a second ultra-short pulse that dissociates the vibrating molecule (see Fig. 1).

The electric field of the laser pulses is written in the form

$$\vec{\mathcal{E}}_i(t) = \mathcal{E}_i(t)\, \vec{\varepsilon}_i\, e^{-i\omega_i t} + c.c. \quad , \tag{1}$$

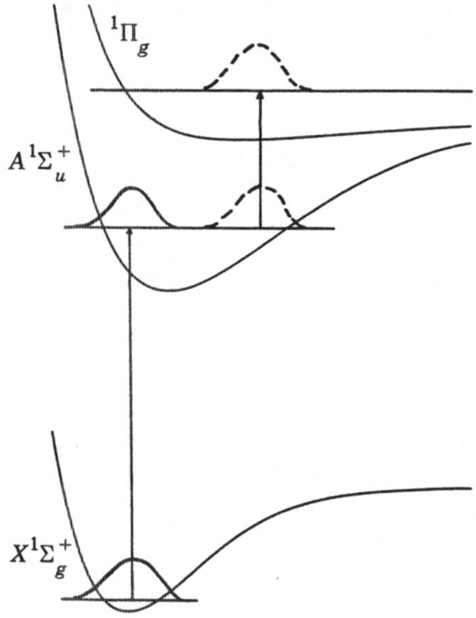

Fig. 1. Schematic diagram of Na_2 potential energy curves for the states involved in a two-step ionization with ultrashort laser pulses. Also shown, schematically, are wave packets in intermediate and final states.

with i=1 denoting the exciting pulse of length τ_1 and i=2 denoting the probe pulse of length τ_2. $E_i(t)$ is the pulse envelope, $\vec{\varepsilon}_i$ is the polarization vector of the electric field, and ω_i is the angular frequency. We assume that $E_1(t)$ is peaked at t=0 and $E_2(t)$ is peaked at $t=t_0$, which is the delay between pulses.

We also assume that initially only the lowest vibrational level of the ground state, with energy E_0, is populated. This can be achieved in an experiment with a supersonic molecular beam. In the following analysis $|\phi_0\rangle$ is the stationary wavefunction of the lowest vibrational level of the ground state, $|\psi_k\rangle$ (k=0,1,...) are stationary wavefunctions of the vibrational levels with energies E_k in the intermediate state and $|\psi_E\rangle$ is the stationary wavefunction of the state with energy E in the dissociation continuum. In the perturbation theory approximation, the time-dependent wavefunction of the intermediate state after the excitation ($t \gg \tau_1$) is given by

$$|\psi_{int}(t)\rangle = \frac{i}{\hbar} \sum_k |\psi_k\rangle\langle\psi_k|\vec{\varepsilon}_1\cdot\vec{\mu}_1|\phi_0\rangle\, \tilde{E}_1(\omega_{k0})\, \exp(-iE_k t/\hbar) \qquad (2)$$

where $\vec{\mu}_1$ is the electronic dipole moment for the transition between ground and intermediate states and $\tilde{E}(\omega_{k0})$ is the Fourier transform of $E_1(t)$ evaluated at $\omega_{k0} = (E_k-E_0)/\hbar-\omega_1$. It can be seen from Eq. 2, that coherent excitation of many vibrational levels in the intermediate state, which is required for the formation of a wave packet, can be achieved only if $\tilde{E}_1(\omega)$ is broader than the energy spacing of $|\psi_k\rangle$ levels. This corresponds to a pulse duration shorter than the vibrational period.

For times after the second pulse ($t > t_0 + \tau_2$) the photodissociation probability $P(t_0)$ is given by

$$P(t_0) = \int dE\, |C_E(t_0)|^2 \quad , \qquad (3)$$

where $C_E(t_0)$ is the amplitude of the state $|\psi_E\rangle$ and has the form

$$C_E(t_0) = \frac{1}{\hbar^2} \sum_k \langle \psi_k|\vec{\varepsilon}_1\cdot\vec{\mu}_1|\phi_0\rangle\langle\psi_E|\vec{\varepsilon}_2\cdot\vec{\mu}_2|\psi_k\rangle \times$$

$$\tilde{E}_1(\omega_{k0})\ \tilde{E}_2\ (\omega_{Ek})\ \exp[-i(E- E_k)t_0/\hbar] \quad (4)$$

where again $\tilde{E}_2(\omega_{Ek})$ is the Fourier transform of $E_2(t)$ and $\vec{\mu}_2$ is the dipole moment for the transition between intermediate and final states.

To calculate $|\psi_{int}(t)\rangle$ and $P(t_0)$ one has to know the stationary vibrational wavefunctions $|\phi_0\rangle, |\psi_k\rangle$ and $|\psi_E\rangle$. We have calculated these functions by numerical integration of the stationary Schrödinger equation for a given internuclear potential using the algorithm of Cooley.[7] Potential energy curves for states $X^1\Sigma_g^+$, $A^1\Sigma_u^+$ and $^1\Pi_g(3s+3d)$ were taken from references 8,9 and 10 respectively. Further, we assume that $|\vec{\mu}_1|$ and $|\vec{\mu}_2|$ do not depend on the internuclear distance r . This is a reasonable approximation for the limited range of r used in the calculations.[11] The laser pulses were assumed to be Gaussian shaped, Fourier limited with the full-width-at-half-maximun duration of 50 fs and frequencies ω_1 and ω_2 corresponding to 16,000 cm^{-1} and 10,000 cm^{-1}, respectively.

The results of the calculations using Eq. (2) are shown in Figs. 2 and 3.

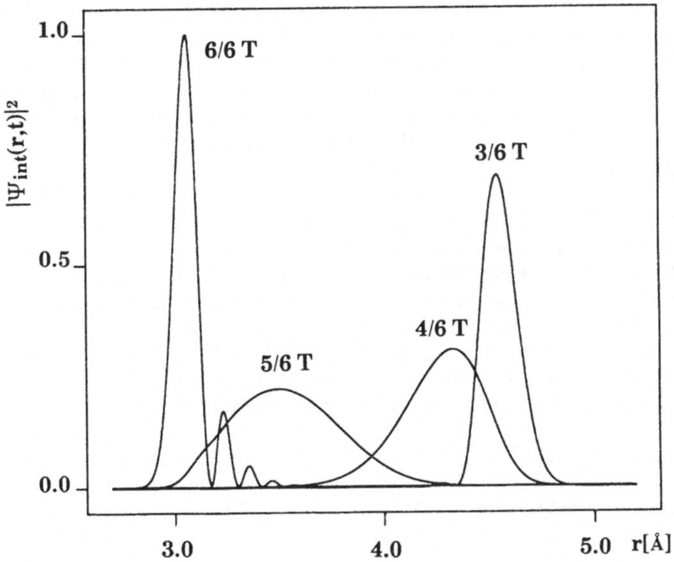

Fig 2. Time evolution of a wave packet in vibrational motion of Na_2 molecule, following excitation by an ultrashort laser pulse. T is the classical vibrational period in $A^1\Sigma_g^+$ state.

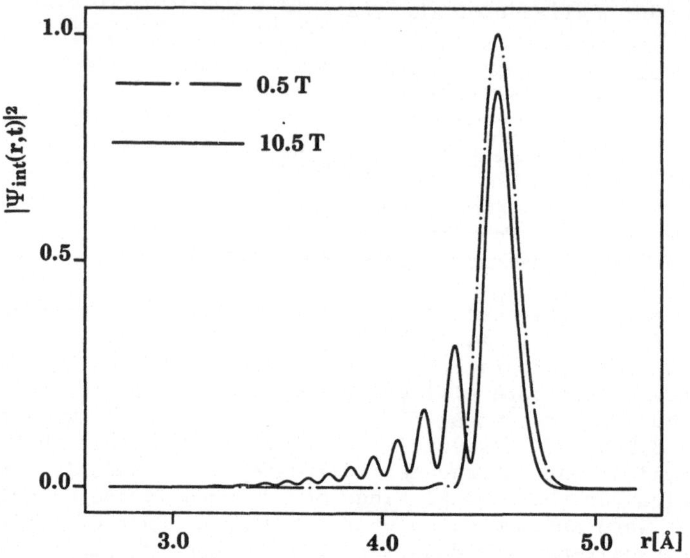

Fig. 3 Wave packet after evolution for times 0.5T and 10.5T.

Fig. 2 shows the time evolution of the vibrational wave packet for short times 0.5T <t<T, where T (~.3 ps) is the vibrational period in the $A^1\Sigma_u^+$ state. The quantity plotted is $|\psi_{int}(r,t)|^2$ v.s. r , i.e. the probability of finding the excited Na_2 molecule at a particular internuclear separation at a given time t . The resemblence of classical motion is remarkable. For example, notice that the velocity of the wave packet is larger on the left side of classical equilibrium point than on the right side, as should be expected in classical motion, due to the fact that the potential curve is steeper near the inner turning point than near the outer one. Fig. 3 shows the wave packet after t=0.5T and t=10.5T (~3 ps). There is a noticable spreading of the wave packet for longer times which can be attributed to the anharmonicity of the potential. In fact, after long enough time the phase factors in Eq. (2) become quasi random, which leads to the loss of coherence and destruction of the wave packet. At a much longer time, however, a near-perfect recurrence of the wave packet will occur, due to the finite number of states excited.

In Fig. 4 the photodissociation probability $P(t_o)$ in the two-step process, versus the delay between the two pulses is plotted. Notice that for the particular potentials used in the calculations the maximum photodissociation probability does not correspond to the classical turning points (it does not reach maximum for t_o = T n/2, n = 1,2,...). There is also a surprising asymmetry in photodissociation probability with respect to whether the wave packet is moving towards smaller or larger r . So far we have no explanation for this result.

The results shown in Fig. 2-4 were calculated with the assumption that only the lowest rotational level with J=0 in the ground state of Na_2 molecule is populated. Additional calculations were performed for higher

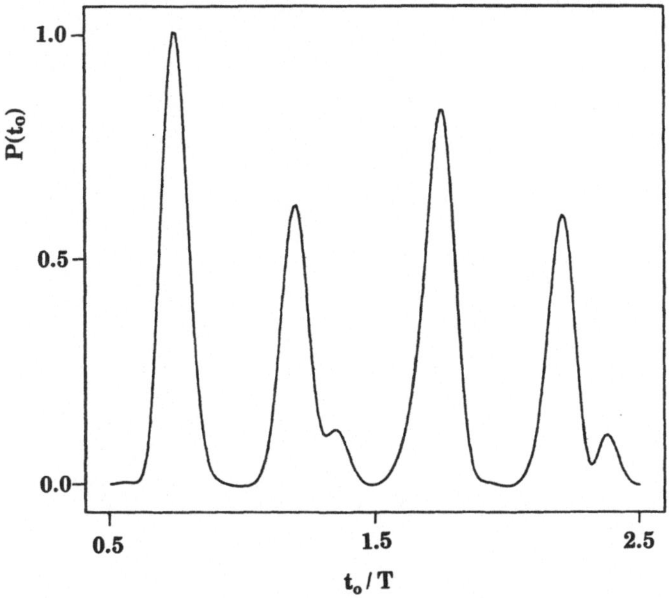

Fig. 4 Photodissociation probability in the two-step process as a function of delay between laser pulses

values of J. We found that including the initial states with rotational numbers, which are characteristic for molecules in a supersonic molecular beam, leads to only slight increase of the width of the wave packet.

In conclusion, it appears that, in a system such as photodissociation of a diatomic molecule, reaction channels can be controlled by using sequences of ultrashort laser pulses. It is an open question whether this idea can be extended to reactions in polyatomic molecules.

References

1. J. Parker and C.R. Stroud, Phys. Rev. Lett. 56, 716 (1986)
2. G. Alber, H. Ritsch, and P. Zoller, Phys. Rev. A 34, 1058 (1986)
3. J.A. Yeazell and C.R. Stroud, Phys. Rev. A 35, 2806 (1987).
4. E.J. Heller, Acc. Chem. Res. 14, 368 (1981).
5. E.J. Heller, J. Chem. Phys. 62, 1544 (1975).
6. T. Sizer II, and M.G. Raymer, Phys. Rev. Lett. 56, 123 (1986).
7. J.H. Cooley, Math. Cumput. 15, 363 (1961).
8. P. Kusch, J. Chem. Phys. 68, 2591 (1978).
9. M.E. Kaminsky, J. Chem. Phys. 66, 4951 (1977).
10. D.D. Konowalow, M.E. Rosenkrantz, and M.L. Olson, J. Chem. Phys. 72, 2613 (1980).
11. M.M. Hessel, E.W. Smith, and R.E. Drullinger, Phys. Rev. Lett. 33, 1251 (1974).

*Permanent address: Warsaw University, Physics Department, Warsaw, Poland

FAST FOURIER TRANSFORM TECHNIQUES IN NUMERICAL SIMULATIONS OF INTENSE PULSE — MOLECULE INTERACTIONS

Paul L. DeVries

Department of Physics
Miami University
Oxford, Ohio 45056 USA

INTRODUCTION

The interaction of an atomic or molecular system with incident light is a principal means by which that system can be probed and physical processes involving the system investigated. With lasers, particularly high intensity tunable ones, the incident light can be used to modify the system as well so as to induce a specific process. Our particular interest is in collision systems in which the incident radiation might be used to control the outcome of a collision event; this is clearly related to the long standing goal of chemists to control chemical reactions by external means.

The present investigation has been motivated by the development of lasers capable of producing pulses of very short duration, as few as three optical cycles. Such pulses are shorter than the collision time so that the motion of the colliding atoms during the laser pulse, when the system can interact with the laser radiation, is minimal. (In contrast, a typical pulse of a nanosecond in duration is long compared to the collision time so that the entire collision takes place in an essentially CW radiation field.) It has been proposed that this additional controllable parameter might be used to further our understanding of the collision process and to produce effects that depend upon the pulse duration. For example, a net excitation of the system can occur by absorption of radiation far from atomic resonance if at that internuclear separation the photon energy coincides with the difference in molecular potential energies. In the case of a long pulse this could happen either as the atoms are approaching one another or as they are receding from one another. Since these possibilities lead to the same physical state, we cannot distinguish between them and our ignorance leads to interference between these pathways. Averaging over thermal distributions, collision vectors, and so on blur the interference patterns, but the interference should result in a smaller net excitation than would have resulted without interference. On the other hand, interference effects should not be present for a "short" pulse since the atoms in a single collision event do not have sufficient time to approach, collide, and recede while the pulse is present. (Of course, an actual observation will involve contributions from many collision events, each of which will be at different stages in the collision — some may be approaching, some receding — but averaging is an incoherent process that does not lead to interference.) It is thus argued that a short pulse should be more efficient in producing excitation that a long pulse — one theoretical study[1] indicates that in the limit of very high intensity the probability for excitation by a short pulse should tend to unity, while for a long pulse it should tend to zero. This motivated some interesting experimental work[2] in which an effect was observed, although it was not as dramatic as had been expected.

Previous theoretical investigations have used straight line classical trajectories to describe the dynamics of the problem. While more sophisticated semiclassical methods could be used, our approach is to complement these studies with a fully quantum mechanical investigation. Since the problem is inherently time dependent, we decided that a time dependent approach would be more intuitive and might lead to a better understanding of the process. This approach provides a substantial computational challenge, however; conventional wisdom would dictate that a time dependent formalism requiring the solution of a set of time dependent close coupled equations would be intractable. Indeed, conventional methods of solution are not suitable. Instead, we have taken the split operator Fourier transform method of Fleck[3] and extended it to the multichannel, short pulse problem. A derivation of the close coupled equations and the Fourier transform technique used in their solution are outlined in the next section, and our application of the method to the sodium-argon collision system is presented in the following section.

THE CLOSE COUPLED EQUATIONS AND THEIR NUMERICAL SOLUTION

The derivation of the time dependent close coupled equations proceeds in much the same way as for the time independent case, except of course that we begin with the time dependent Schrödinger equation and the Hamiltonian contains explicit time dependence through the dipole interaction term

$$H_{interaction} = -e\vec{E}\cdot\vec{r}, \qquad \text{where } \vec{E} = \vec{\epsilon}E_o(t)\,(a + a^\dagger).$$

We use a number state representation for the field so that annihilation and creation operators appear in the interaction term. We should emphasize that the interesting time dependence of this interaction is *not* the $\cos\omega t$ periodic variation of the field, which in our approach is virtually eliminated by the choice of basis set, but rather the temporal evolution of the pulse as described by the envelope function $E_o(t)$. The total wavefunction is expanded as a weighted sum of basis states; elements of the basis contain products of eigenfunctions of the pure radiation field (described in the number state representation), the electronic Hamiltonian, and the angular portion of the kinetic energy operator, and are constructed to be eigenfunctions of $\mathbf{J}^2$ and $\mathbf{J}_z$ as well. The weighting coefficients in this expansion are the (unknown) time dependent nuclear wavefunctions which describe the motion of the atoms. Denoting by α the set of all quantum numbers uniquely describing a basis element, this expansion of the wavefunction leads to the time dependent close coupled equations

$$i\hbar\frac{\partial\psi_\alpha(R,t)}{\partial t} = \frac{\hbar^2}{2m}\frac{\partial^2\psi_\alpha(R,t)}{\partial R^2} + \sum_{\alpha'} V_{\alpha\alpha'}(t)\psi_{\alpha'}(R,t),$$

where the term $V_{\alpha\alpha'}$ contains the electronic potentials, centripetal coupling terms, and all the radiative interactions.

The solution to this set of equations is a formidable task, but can be accomplished using an extension of Fleck's Fourier transform method.[3] In matrix notation, the set of equations take the form

$$i\hbar\frac{\partial\boldsymbol{\psi}(R,t)}{\partial t} = \left[\mathbf{T} + \mathbf{V}(t)\right]\boldsymbol{\psi}(R,t),$$

which has a formal solution given by

$$\boldsymbol{\psi}(R,t+\delta) = e^{-i\int_t^{t+\delta}\left[\mathbf{T}+\mathbf{V}(\tau)\right]d\tau/\hbar}\,\boldsymbol{\psi}(R,t).$$

Recall that the (difficult) time variation of the potential is due to the temporal evolution of the pulse envelope, not the electric field itself. Since we will be using time steps δ on the order of hundredths of a picosecond in order to follow the dynamics of the collision, the envelope (which varies on the tenths of picosecond time scale) can be considered a slowly varying function. The potential in the integral can then be replaced by its average value, yielding the time evolution equation

$$\boldsymbol{\psi}(R,t+\delta) = e^{-i\left[\mathbf{T}+\overline{\mathbf{V}}\right]\delta/\hbar}\,\boldsymbol{\psi}(R,t).$$

We approximate the evolution operator by the symmetric decomposition

$$e^{-i\left[\mathbf{T}+\overline{\mathbf{V}}\right]\delta/\hbar} \approx e^{-i\mathbf{T}\delta/2\hbar}\,e^{-i\overline{\mathbf{V}}\delta/\hbar}\,e^{-i\mathbf{T}\delta/2\hbar}.$$

The unitarity of the operator is preserved by writing it this way, and the error in the approximation is only of order δ^3.

T is diagonal in our representation. Since the exponential of a diagonal matrix is just the matrix of exponentiated diagonal elements, we need to evaluate

$$\psi'_\alpha(R,t) = e^{-i(\hbar^2/2m)(\delta/2\hbar)\partial^2/\partial R^2}\,\psi_\alpha(R,t).$$

This is where the Fourier transform enters: by expressing $\psi_\alpha(R,t)$ in terms of its transform $\phi_\alpha(K,t)$, the exponential of the derivative is easily evaluated as

$$\psi'_\alpha(R,t) = \frac{1}{\sqrt{2\pi}}\int_{-\infty}^{\infty} e^{-iKR}\,e^{-i(\hbar^2/2m)(\delta/2\hbar)(-iK)^2}\,\phi_\alpha(K,t)\,dK,$$

which we recognize as an inverse Fourier transform. This evaluation must be done for each of the N weighting coefficients (e.g., nuclear wavefunctions), and is only practical due to the availability of the Fast Fourier Transform algorithm to actually perform the integrations.

Since $\overline{\mathbf{V}}$ is a hermitian matrix, we have

$$\boldsymbol{\psi}''(R,t) = e^{-i\overline{\mathbf{V}}\delta/\hbar}\boldsymbol{\psi}'(R,t) = \mathbf{S}e^{-i\mathbf{S}^\dagger\overline{\mathbf{V}}\mathbf{S}\delta/\hbar}\mathbf{S}^\dagger\boldsymbol{\psi}'(R,t) = \mathbf{S}e^{-i\mathbf{D}\delta/\hbar}\mathbf{S}^\dagger\boldsymbol{\psi}'(R,t),$$

where $\mathbf{D} = \mathbf{S}^\dagger\,\overline{\mathbf{V}}\,\mathbf{S}$ and $\mathbf{S}$ is a unitary matrix chosen so that $\mathbf{D}$ is diagonal; that is, to diagonalize $\overline{\mathbf{V}}$. The exponential of the diagonal matrix is then easily evaluated, and $\boldsymbol{\psi}''(R,t)$ determined by matrix multiplication. Note that since the potential changes with time, **S** must be reevaluated at each time interval.

The final step in the time propagation is identical to the first, and is likewise performed. While it appears rather complicated, this algorithm is actually quite straightforward to implement and, due to its reliance on the FFT to perform the necessary integrals, it is relatively fast and very stable.

THE SODIUM-ARGON SYSTEM

To investigate the effects of various length gaussian shaped laser pulses on the sodium-argon collision system, we consider the time evolution of wavepackets describing the relative position of the atoms. An initial wavepacket is specified at a particular location and time and allowed to evolve in the presence of the relevant Hamiltonian: suitable electronic potentials and the electric dipole interaction with the incident laser pulse. For this study, a laser of intensity 10^9 W/cm^2 is tuned 45 cm^{-1} to the red side of the atomic D line and is taken to be incoming along the x-axis and linearly polarized in the z-direction. The electronic curves of Saxon, Olson, and Liu[4] are used in the calculations, dressed by the energy of a photon. In the diabatic representation, this gives a curve crossing at 9.4 bohr which represents the resonant interaction between the ground and excited electronic states. That is, at 9.4 bohr the atoms have perturbed one another enough to change the energy difference between the (field-free) ground and excited states by 45 cm^{-1} so that a laser photon can be absorbed. Of course, excitation will only occur if the system is in the vicinity of this curve crossing at the same time that the laser pulse is present. The net excitation is determined by propagating the wavepacket until the pulse has passed and evaluating the integral

$$P_\alpha = \int \psi_\alpha^* \psi_\alpha \, dR.$$

The probabilities we report here are summed over all excited states.

To begin our investigation, a wavepacket was propagated on the ground electronic curve in the absence of radiation. (This and all other calculations presented here were performed with a total angular momentum of 100.) The calculated $<R>$ and $<K>$ thus map a trajectory in phase space; we use this trajectory to choose initial positions for our wavepackets. This is a little complicated, since we must ac-

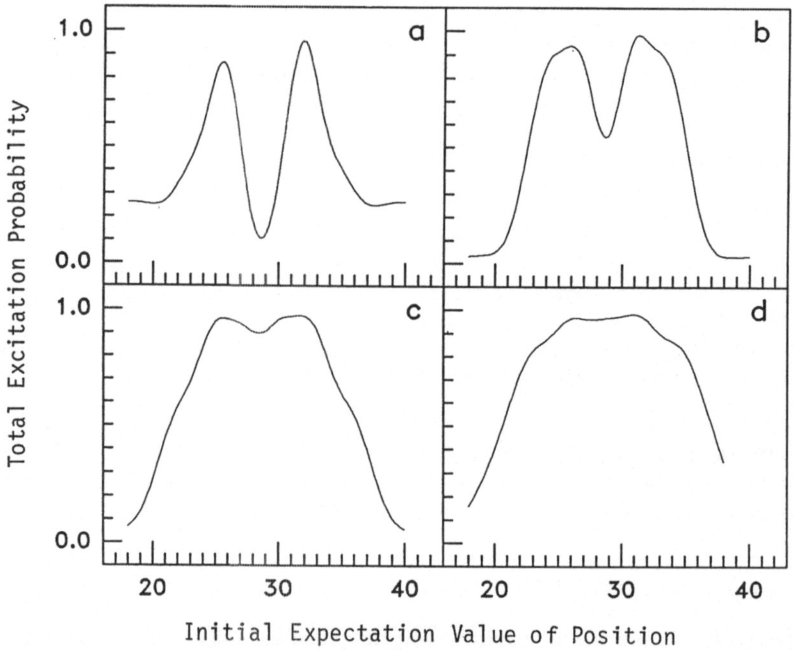

The probability of reaching any excited state is plotted versus the initial position of the wavepacket. The four panels correspond to different pulse lengths: a) τ = 0.4 ps; b) τ = 0.6 ps; c) τ = 0.8 ps; and d) τ = 1.0 ps.

count for the laser pulse arriving at different stages in the collision event. We do this by fixing the arrival time of the pulse at $t = 0$ and starting the wavepackets at different positions. We (somewhat arbitrarily) begin all our wavepacket calculations 2.5 picoseconds before the arrival of the maximum intensity of the laser pulse; from the field-free calculation, we find that atoms closer that about 20 bohr will have sufficient time to collide and rebound into the asymptotic region before the pulse arrives. Such a pair of atoms will thus interact with the laser field as isolated atoms, not as a collision complex — only those atoms at greater initial separation will have the opportunity to interact with the radiation while in the process of colliding. At the other extreme are atoms that are so far removed from one another that the pulse will come and go before the collision actually begins. This of course will depend upon the maximum pulse durations considered; for a 1.0 picosecond pulse, this separation is about 40 bohr. The initial time of 2.5 picoseconds was chosen to insure that in the nearest atom case the separation would still be large enough to be in the asymptotic region, where the atoms can be assumed to be uniformly distributed. A total probability can thus be obtained by integrating over the initial position of wavepackets.

Some of the results of these calculations are presented in the Figure, in which we plot net excitation out of the ground state for various laser pulse durations. In the first panel, the effect of a short pulse duration is clearly present: wavepackets starting near 25 bohr have time to collide and begin to rebound before the pulse arrives, and in fact are near the radiative curve crossing when the pulse is at its maximum and hence have a large probability of making a transition. Meanwhile, those starting near 32 bohr strongly interact with the radiation on the inward leg of the collision. For the intermediate case, such as wavepackets starting near 28 bohr, the excitation is not as large since the wavepackets are near the classical turning point at $t = 0$, an energetically unfavorable position with regards to photon absorption. For this 0.4 picosecond pulse duration, there is a clear distinction between absorption on the inward and outward legs of the collision process. As longer pulse durations are considered, this distinction begins to blur — wavepackets that are near the turning point at $t = 0$ are more readily excited, but is it by interaction on the inward or outward journey? It is, of course, not possible to say and as the distinction blurs the prognosis for observing a pulse duration effect diminishes. Certainly, these results suggest that the temporal region of "short" laser pulses, for the system being considered here, is in the sub-picosecond domain.

Support for this work has been provided by the National Science Foundation under grant PHY-85-06679, and by the Petroleum Research Fund, administered by the American Chemical Society. Computer resources provided by the Pittsburgh Supercomputer Center are gratefully acknowledged.

REFERENCES

1. H. W. Lee and T. F. George, " Molecular Collision Processes in the Presence of Picosecond Laser Pulses," J. Phys. Chem. 83:929 (1979).

2. T. Sizer, II and M. G. Raymer, "Modification of Atomic Collision Dynamics by Intense Ultrashort Laser Pulses," Phys. Rev. Lett. 56:123 (1986).

3. J. A. Fleck, Jr., J. R. Morris, and M. D. Feit, "Time-dependent Propagation of High Energy Laser Beams through the Atmosphere," Appl. Phys. 10:129 (1976).

4. R. P. Saxon, R. E. Olson, and B. Liu, "*Ab initio* calculations for the $X^2\Sigma$, $A^2\Pi$, and $B^2\Sigma$ states of NaAr: Emission spectra and cross sections for fine-structure transitions in Na-Ar collisions," J. Chem. Phys. 67:2692 (1977).

INTERFERENCE IN TWO-PHOTON ABSORPTION AT HIGH FIELD CROSSINGS

George J. Diebold and R.B. Stewart

*Brown University Department of Chemistry
Providence, RI 02912
Applied Biomedical Group
33 Cherry Hill Drive, Danvers, MA 01923

Two-photon absorption at low radiation intensities can be described by second order perturbation theory. The most important features of two photon absorption are a unique set of selection rules, a transition rate proportional to the product of the intensities of the two radiation beams, and an enhancement in the absorption probability when the frequency of one of the beams is nearly coincident with a single photon absorption. Recently it has been shown that when one of the incident light beams is near resonance with an allowed atomic state (the 'intermediate' electronic state), interference effects dependent on the relative polarization of the two incident light beams are produced[1,2]. That is, in a $^1S \leftarrow ^1P \leftarrow ^1S$ energy level configuration, when the two light beams have perpendicular polarizations, the interference is destructive leading to a vanishing two-photon transition probability; when the field polarization vectors of the two beams are parallel, the interference is constructive.

If the sublevels of the intermediate state are perturbed by the application of an external field, the probability amplitudes for population of the upper ^1S state in a destructive interference geometry are altered thereby creating a partially constructive interference. Experiments have been reported[2] where a magnetic field was used to perturb the intermediate ^1P state in atomic Sr. It was shown that measurement of the two-photon absorption rate permits determination of the Landé factor of the intermediate electronic state, in analogy to the Hanle effect.

Here, we calculate the dependence of the two-photon transition rate as a function of the applied magnetic field for the case of a high field crossing in a ^2P intermediate state. A somewhat different formulation of the transition probabililty is given here than was previously presented [1]. Consider an atom whose wavefunction ψ is expressed in terms of energy eigenstates ψ_n as

$$\psi = \sum_n b_n(t)\psi_n e^{-iE_nt/\hbar} \tag{1}$$

where $b_n(t)$ is the time dependent probability amplitude of the state n and E_n is the energy of the state n. The effect of the first light beam with polarization $\hat{\epsilon}_f$ and frequency Ω_1 tuned near resonance with an atomic state whose sublevels are denoted by μ can be found by substitution of matrix elements of the electric dipole operator into the time dependent Schroedinger equation,

$$i\hbar\frac{db_j}{dt} = \sum V_{jk}(t)\, b_k\, e^{i\omega_{jk}t}\ , \tag{2}$$

a solution to which gives the probability amplitude of the intermediate state μ as

$$b_\mu(t) = \frac{H^f_\mu m}{i\hbar}\left[\frac{e^{-i\Omega_1 t+i\omega_{\mu m}t} - e^{-\Gamma\frac{1}{2}t}}{-i(\Omega_1 - \omega_\mu) + \Gamma/2)}\right]\ , \tag{3}$$

where $H^f_{\mu m} = < \mu|\hat{\epsilon}_f \bullet \mathbf{D}|m >, \omega_{\mu m} = (E_\mu - E_m)/\hbar$, Γ is the spontaneous rate of decay of the intermediate state, and m is the atomic ground state. The probability amplitude of the excited state is found by substituting this solution into Eq 2 to give a differential equation for $b_\eta(t)$, the probability amplitude of the excited state. Solution of this differental equation gives

$$b_\eta(t) = \frac{-1}{\hbar^2}\sum_\mu H^g_{\eta\mu} H^f_{\mu m} \frac{\left[e^{i[\omega_{\mu m}-(\Omega_1+\Omega_2)]t} - 1\right]}{\left[-i(\Omega_1 - \omega_\mu) + \Gamma/2\right]\left[-i(\Omega_1 + \Omega_2 - \omega_{\eta m}) + \Gamma'/2\right]} \tag{4}$$

where $H^g_{\eta\mu} = < \eta|\hat{\epsilon}_g \bullet \mathbf{D}\,|\mu >, \hat{g}$ is the polarization of the light beam at frequency Ω_2, $\omega_{\eta m} = (E_\eta - E_m)/\hbar$ and Γ' is the spontaneous decay rate of the upper state.

The factor in brackets in the numerator and the second factor in brackets in the denominator are common to all terms in the summation over the intermediate states μ. Since the rate of population of the upper state R in the low intensity limit is proportional to $|b_\eta(t)|^2/t$ and since

$$\lim_{t\to\infty} \left|\frac{e^{i[\omega_{\eta m}-(\Omega_1+\Omega_2)t]} - 1}{\omega_{\eta m} - (\Omega_1 + \Omega_2)}\right| = \pi t \delta\left[\omega_{\eta m} - (\Omega_1 + \Omega_2)\right] \quad , \tag{5}$$

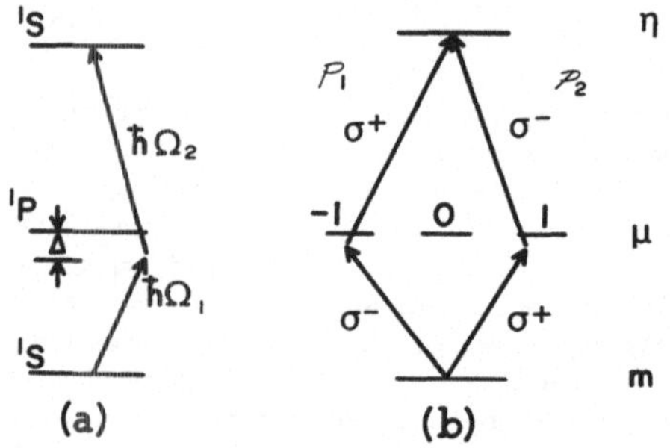

Figure 1: (a) Atomic energy level diagram for the $^1S \leftarrow ^1P \leftarrow ^1S$ two photon absorption treated in the text. The first light beam with frequency Ω_1 is detuned on amount Δ from resonance. (b) Energy level diagram showing the two paths P_1 and P_2 for absorption of radiation.

then the rate of population of the upper state can be written as a sum over the probability amplitudes P_μ as

$$R = |\sum_\mu P_\mu|^2 \tag{6}$$

where

$$P_\mu = \frac{H^g_{\eta\mu}\, H^f_{\mu m}}{(\Omega_1 - \omega_\mu) + i\Gamma/2} \quad .$$

This result is identical to what has been obtained by other authors[3-8] except that the denominator has contributions from the first light beam only since this beam is assumed to be near resonance with the intermediate state with the second beam far from resonance. An additional consequence of the small detuning of the first beam from resonance with the intermediate state is that the sum over all atomic states reduces to a sum over the sublevels of the intermediate state only.

A straightforward application of this result is found in the two-photon excitation of a ^1S state with a ^1P intermediate state as shown in Fig 1*a*. When the light beams have polarizations perpendicular to each other and an external magnetic field H is present, the two-photon absorption rate is given by[1]

$$R = \frac{I_1 I_2 (g\beta H/\hbar)^2}{[(\Delta + g\beta H/\hbar)^2 + \Gamma^2/4][(\Delta - g\beta H/\hbar)^2 + \Gamma^2/4]} \tag{7}$$

where I_1 and I_2 are the intensities of the first and second light beams, g is the Landé factor of the intermediate state, and β is the Bohr magneton. From Eq. 7, it can be seen that the interference is completely destructive when the magnetic field is zero, but becomes partially constructive interference when the degeneracy of the magnetic sublevels of the intermediate state is lifted. That is, the two probability amplitudes P_1 and P_2 in Fig 1*b* have opposite phases and equal amplitudes when $H = 0$ so that absorption rate is zero. When a magnetic field is applied, the relative amplitudes of P_1 and P_2 in Eq 5 are changed as a result of the shifts in ω_1 and ω_2 giving a nonzero value for R.

Consider the case of the $^2S \leftarrow {}^2P \leftarrow {}^2S$ two-photon transition shown in Fig 2 where the population of the $m_j = -\frac{1}{2}$ sublevel of the upper $^2S_{\frac{1}{2}}$ state is monitored. There are three allowed paths for two-photon absorption,

$$P_1 : {}^2S_{\frac{1}{2}}(m_J = -\frac{1}{2}) \leftarrow {}^2P_{\frac{3}{2}}(m_J = -\frac{3}{2}) \leftarrow {}^2S_{\frac{1}{2}}(m_J = -\frac{1}{2})$$

$$P_2 : {}^2S_{\frac{1}{2}}(m_J = -\frac{1}{2}) \leftarrow {}^2P_{\frac{1}{2}}(m_J = \frac{1}{2}) \leftarrow {}^2S_{\frac{1}{2}}(m_J = -\frac{1}{2})$$

$$P_3 : {}^2S_{\frac{1}{2}}(m_J = -\frac{1}{2}) \leftarrow {}^2P_{\frac{3}{2}}(m_J = \frac{1}{2}) \leftarrow {}^2S_{\frac{1}{2}}(m_J = -\frac{1}{2}) \ .$$

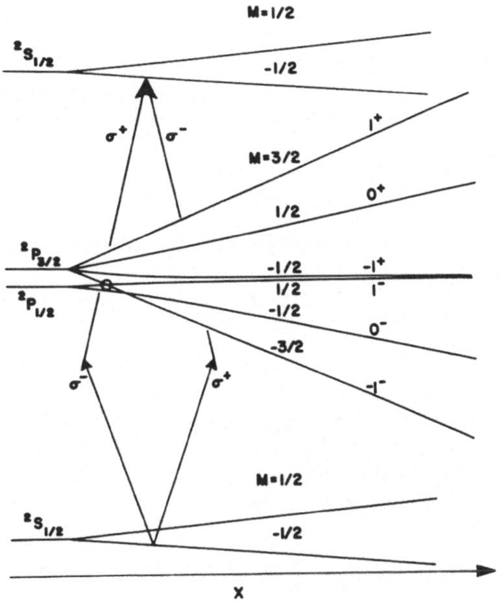

Figure 2: Energy level diagram for a $^2S \leftarrow {}^2P \leftarrow {}^2S$ two-photon absorption.

The absorption rate now becomes

$$R = |P_1 + P_2 + P_3|^2 \tag{8}$$

where

$$P_1 = \frac{<{}^2S_{\frac{1}{2}}\tfrac{1}{2}|D_{+1}|\Psi({}^2P_{\frac{3}{2}} - \tfrac{3}{2})><\Psi({}^2P_{\frac{3}{2}} - \tfrac{3}{2})|D_{-1}|{}^2S_{\frac{1}{2}} - \tfrac{1}{2}> \epsilon^g_{-1}\epsilon^f_{+1}}{\Delta^{\frac{3}{2}}_{-\frac{3}{2}}}$$

$$P_2 = \frac{<{}^2S_{\frac{1}{2}} - \tfrac{1}{2}|D_{-1}|\Psi({}^2P_{\frac{1}{2}}\tfrac{1}{2})><\Psi({}^2P_{\frac{1}{2}}\tfrac{1}{2})|D_{+1}|{}^2S_{\frac{1}{2}} - \tfrac{1}{2}> \epsilon^g_{+1}\epsilon^f_{-1}}{\Delta^{\frac{1}{2}}_{\frac{1}{2}}}$$

$$P_3 = \frac{<{}^2S_{\frac{1}{2}} - \tfrac{1}{2}|D_{-1}|\Psi({}^2P_{\frac{3}{2}}\tfrac{1}{2})><(\Psi({}^2P_{\frac{3}{2}}\tfrac{1}{2})|D_{+1}|{}^2S_{\frac{1}{2}}> \epsilon^g_{+1}\epsilon^f_{-1}}{\Delta^{\frac{3}{2}}_{\frac{1}{2}}}$$

Here, the electric dipole operator and the electric field vector have been written in a spherical basis[4] and the denominator for each term in Eq 5 has been abbreviated as $\Delta^J_{m_J}$ where J and m_J are the total angular momentum and magnetic quantum number of the intermediate state respectively.

As shown by Radford and Evenson[10], the wavefunctions of a given m_J sublevel of a ${}^2P$ state in a magnetic field are described as linear combinations of Russell-Saunders wavefunctions. That is, for a given value of the field parameter,

$$x = \beta H/\delta \ ,$$

where δ is the fine structure splitting, the field dependent wavefunctions are given by

$$\Psi({}^2P_{\frac{3}{2}}\tfrac{1}{2}) = C^U_1|{}^2P_{\frac{1}{2}}\tfrac{1}{2}> + C^U_2|{}^2P_{\frac{3}{2}}\tfrac{1}{2}>$$

$$\Psi({}^2P_{\frac{1}{2}}\tfrac{1}{2}) = C^L_1|{}^2P_{\frac{1}{2}}\tfrac{1}{2}> + C^L_2|{}^2P_{\frac{3}{2}}\tfrac{1}{2}> \ , \tag{9}$$

where

$$C_1 = \{1 + [\alpha + (1-\alpha^2)^{\frac{1}{2}}]x \mp (1 + 2\alpha x + x^2)^{\frac{1}{2}}\}/B$$

$$C_2 = -\{1 + [\alpha - (1-\alpha^2)^{\frac{1}{2}}]x \pm (1 + 2\alpha x + x^2)^{\frac{1}{2}}\}/B \tag{10}$$

$$B = 2\{1 + 2\alpha x + x^2 \mp x[(1-\alpha^2)(1 + 2\alpha x + x^2)]^{\frac{1}{2}}\}$$

$$\alpha = \frac{1}{3} \ ,$$

and when the upper sign is taken for C^U and the lower sign is taken for C^L. Wavefunctions with $m_J = \pm\frac{3}{2}$ are not dependent on the field. Substitution of Eq 9 into Eq 8 thus gives a total of five terms, the matrix elements in which can be determined by standard means[11]. The first term in Eq 8, for example, is given by

$$<{}^2P_{\frac{3}{2}} - \tfrac{3}{2}|D_{-1}|{}^2S_{\frac{1}{2}} - \tfrac{1}{2}> = -(6)^{\frac{1}{2}}\,C(\tfrac{1}{2}\,1\tfrac{3}{2}; -\tfrac{1}{2} - 1 - \tfrac{3}{2})\begin{pmatrix} \frac{3}{2} & \frac{1}{2} & 1 \\ 0 & 1 & \frac{1}{2} \end{pmatrix} <1||D^1||0> \ . \tag{11}$$

For linearly polarized beams with the first beam polarized along the x axis and the second beam inclined at an angle ϕ to the x axis (both propagating in the z direction), the expression for the absorption rate becomes

$$R = \left| \frac{e^{i\phi}}{\Delta^{\frac{3}{2}}_{-\frac{3}{2}}} + \frac{|C_1^L (\frac{2}{3})^{\frac{1}{2}} + C_2^L (\frac{1}{3})^{\frac{1}{2}}|^2 e^{-i\phi}}{\Delta^{\frac{1}{2}}_{\frac{1}{2}}} + \frac{|C_1^U (\frac{2}{3})^{\frac{1}{2}} + C_2^U (\frac{1}{3})^{\frac{1}{2}}|^2 e^{-i\phi}}{\Delta^{\frac{3}{2}}_{\frac{1}{2}}} \right|^2 \tag{12}$$

The energies of each of the relevant states are given by

$$\frac{W^{\frac{3}{2}}_{-\frac{3}{2}}}{\delta} = \frac{1}{3} - 2x$$

$$\frac{W^{\frac{1}{2}}_{\frac{1}{2}}}{\delta} = -\frac{1}{6} + \frac{x}{2} - \frac{1}{2}(1 + \frac{2}{3}x + x^2)^{\frac{1}{2}} \tag{13}$$

$$\frac{W^{\frac{3}{2}}_{\frac{1}{2}}}{\delta} = -\frac{1}{6} + \frac{x}{2} + \frac{1}{2}(1 + \frac{2}{3}x + x^2)^{\frac{1}{2}}$$

where δ is the fine structure splitting. If the first beam is detuned an amount $-p\delta$ from the zero field origin of the ^2P manifold (where $p\delta$ is assumed to be large enough so that the damping terms can be neglected) then the dimensionless detunings are given by

$$\Delta^{\frac{3}{2}}_{-\frac{3}{2}} = -p - \tfrac{1}{3} + 2x$$

$$\Delta^{\frac{1}{2}}_{\frac{1}{2}} = -p + \tfrac{1}{6} - \frac{x}{2} + \tfrac{1}{2}(1 + \tfrac{2}{3}x + x^2)^{\frac{1}{2}} \tag{14}$$

$$\Delta^{\frac{3}{2}}_{\frac{1}{2}} = -p + \tfrac{1}{6} - \frac{x}{2} - \tfrac{1}{2}(1 + \tfrac{2}{3}x + x^2)^{\frac{1}{2}} .$$

It is thus possible to calculate the two-photon absorption rate as a function of field and polarization for a given detuning. The curves in Figs 3– 5 show the absorption rate as a function of the field parameter x for parallel ($\phi = 0$) and perpendicular ($\phi = \pi/2$) electric vectors. The high field crossing between the $^2P_{\frac{3}{2}}$ $m_J = -\frac{3}{2}$ and the $^2P_{\frac{1}{2}}$ $m_J = \frac{1}{2}$ levels is located at $x = \frac{4}{9}$, as can be found by equating the energies of these states in Eq 13. It is clear, however, that the perfectly destructive interference does not occur at this point. This comes as a result of both the unequal transition probabilities between the two states that cross, as well as the effects of the magnetic field mixing of the two levels.

It is thus evident that along with Eq 8, experimental determination of the zero in the two-photon transition rate at several values of the detuning can be used to determine spectroscopic parameters for the intermediate ^2P state. Such measurements of interference in the two-photon transition rate[2] are analogous to Hanle effect[12,13] measurements that have been used in numerous atomic spectroscopic experiments.

The authors are grateful to the U. S. Department of Energy for support of this research.

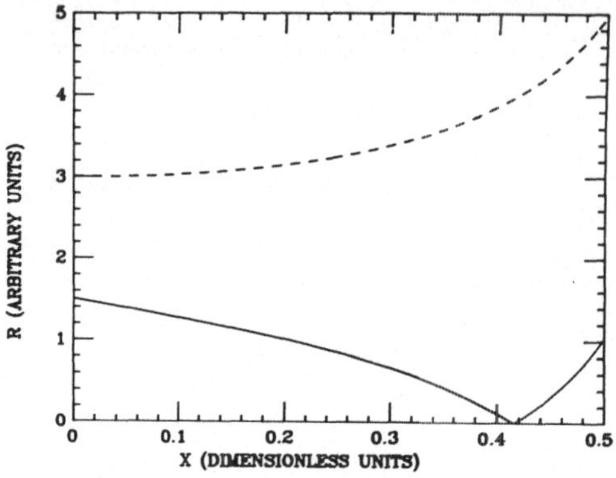

Figure 3: Two-photon absorption rate for parallel (- • -) and perpendicular (—) electric vectors versus the field parameter. The detuning parameter is 1.

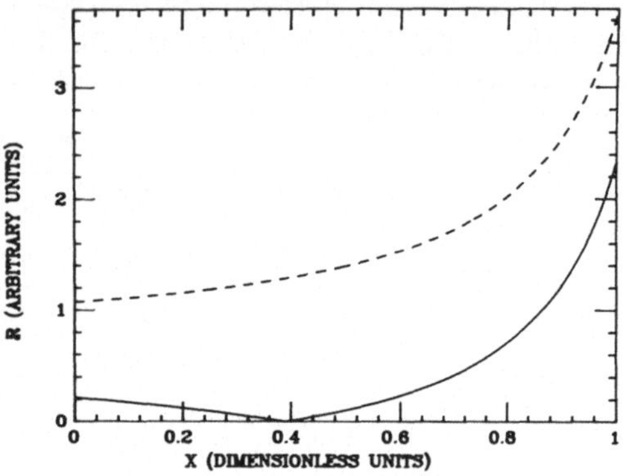

Figure 4: Absorption rate versus the field parameter as in Fig 3 but with $p = 2$.

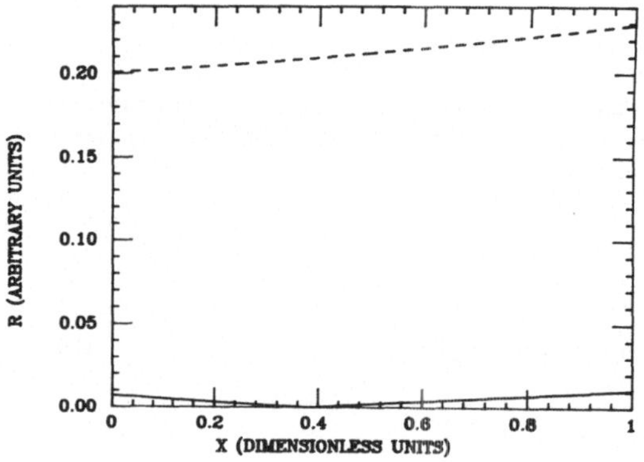

Figure 5: Absorption rate versus the field parameter as in Fig 3 but with $p = 10$.

REFERENCES

1. G. J. Diebold, Phys. Rev. A **32**, 2739 (1985).
2. R. B. Stewart and G. J. Diebold, Phys. Rev. A **34**, 2547 (1986).
3. P. Lambropoulos, in *Advances in Atomic and Molecular Physics*, edited by D. R. Bates and B. Bederson (Academic, New York, 1986), Vol. 12.
4. F. Giacobino and B. Cagnac, in *Progress in Optics*, edited by E. Wolf (North-Holland, New York, 1980).
5. J. S. Bakos, Phys. Reports, **31**, 209 (1977).
6. *Multiphoton Processes*, edited by J. H. Eberly and P. Lambropoulos (Wiley, New York, 1978).
7. A. Schenzle and R. G. Brewer, Phys. Rep. **43**, 455 (1978).
8. L. Allen and C. R. Stroud, Jr., Phys. Rep. **91**, 2 (1982).
9. A. Corney, *Atomic and Laser Spectroscopy* (Oxford University, London, 1979), Chap. 15.
10. H. E. Radford and K. M. Evenson, Phys. Rev. **168**, 70 (1968). We have taken g_s to be 2.0 for simplicity.
11. See, for instance, D. M. Brink and G. R. Satchler, *Angular Momentum*, 2nd ed. (Oxford University, London, 1975).
12. W. Hanle, Z. Phys. **30**, 93 (1924).
13. P. A. Franken, Phys. Rev. **121**, 508 (1961).

TIME-RESOLVED ELECTRONIC RAMAN SCATTERING OF THE NANOSECOND TIME SCALE

J.A. Koningstein

Department of Chemistry
Carleton University
Ottawa, Ontario, Canada K1S 5B6

ABSTRACT

We discuss aspects of time-resolved electronic Raman scattering from ruby. In these experiments information is obtained on the time evolution of population of excited states of Cr^{3+} under conditions of illumination required to observe the excited state electronic Raman process between the E and 2A levels. The population decay of the excited states under these conditions is found to be smaller by orders of magnitude if compared to that of the radiative decay time of the $R_{1,2}$ fluorescence of ruby under weak signal irradiation.

DISCUSSION

In this presentation I intend to give a short review of some recent experiments dealing with the time resolved aspects of electronic Raman scattering. As is well known, the 2A and E excited states from ruby are

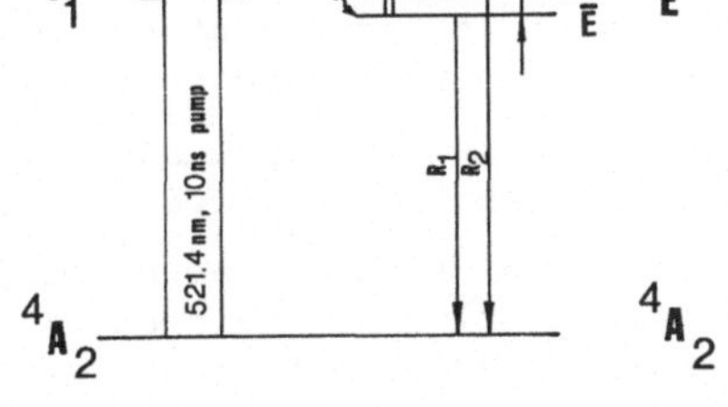

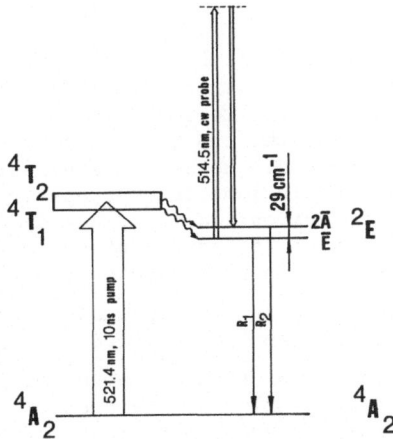

Fig. 1 Part of the energy level diagram of the ruby crystal. Indicated are the position of the 10 ns pulsed optical pump and cw Raman probe.

apart by 29 cm^{-1}. These states can be effectively populated if a single crystal of ruby is exposed to a few hundred mW at 514.5 nm of the cw Argon ion laser. If such an experiment is carried out, not all the radiation of such a laser is used to produce population of the excited states from which the $R_{1,2}$ fluorescence lines originate. Part of the radiation can be used as a Raman probe and the Excited state electronic Raman shift at cm^{-1} is quite easily recorded. Using the radiation from cw lasers information on the time evolution of the population of the excited states cannot be obtained. In order to get some information on that process we decided to carry out a two beam laser experiment. The absorption spectrum of ruby shows a maximum at 530 nm and consequently, population of the lower lying 2A and E levels is most effectively obtained if the wavelength of a tunable laser is in resonance with the above mentioned wavelength. We have taken the 5 ns pulse of a tunable dye laser to produce population of these states while at the same time the Raman spectrum was probed with the radiation of the continuous wave at 514.5 of the argon ion laser. However, the intensity of the Raman probe was reduced to such a degree that the intensity of the electronic Raman shift

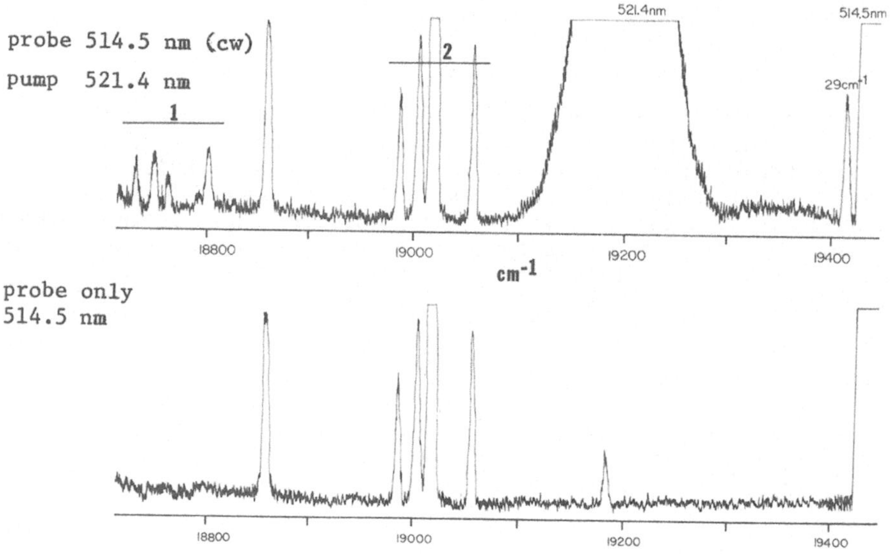

Fig. 2

at 29 cm^{-1} could not be detected with the probe alone. Part of the radiation of the optical pump (repetition rate 25 Hz) was used to trigger a boxcar integrator which is capable to record a pulse with a gate of 3 ns. The radiative lifetime of the red fluorescence of ruby under weak light signals is a few millisecond. Hence, if a nanosecond light pulse at 530 nm is used to induce population, this population should be present after points in time at which the optical pump has died. On the other hand, the Raman probe at 514.5 nm is present at any point in time and this means that by scanning the electronic gate from the point in time t_o at which the pump has reached its maximum intensity, with the monochromator fixed at the wavelength of the Stokes (electronic) Raman shift (514.5 nm - 29 cm^{-1}), the time evolution of the population could in principle be obtained. The population of the excited states occurs in an indirect way via radiationless pathways and it is thought [1] that the leading edge of the population curve is in the picosecond time domain and thus outside the time resolution of our apparatus. On the other hand, the radiative lifetime (for small signal intensity) of the excited states is known to be milliseconds rather than nanoseconds and thus, even with a rate of a few microseconds should the intensity of the signal be observable with the gate tuned away by milliseconds from t_o. The experimental data reveal that this is not the case. The population decay is much faster and we estimate that the decay time is close to 20 μs. Of relevance to this experimental fact is the following. Using only one pass of the cw laser radiation at 514.5 nm through the 5 cm long crystal,

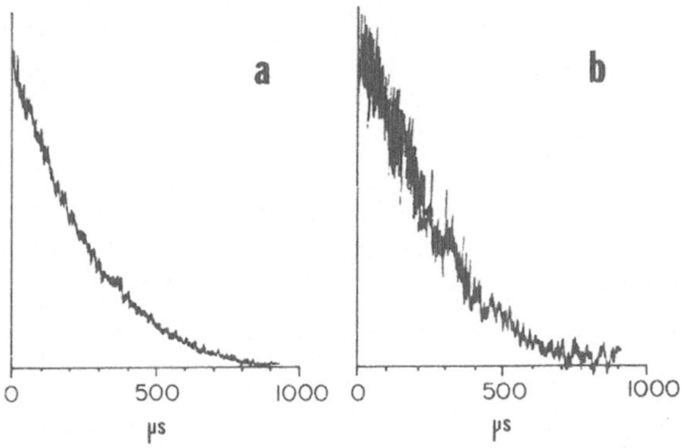

Fig. 3 Time dependence (effective gate width) of the intensity of (a) 10 ns pump laser pulse and (b) excited-state 29 cm^{-1} Raman line (b), under optimized experimental conditions.

laser emission in the $R_{1,2}$ lines of ruby is observed at 77K. Apparently, the quality of our crystal is good enough and the loss in gain for a single pass is small. Such laser action is not observed if the pulsed laser at 530 nm is passing through the same crystal. This is presumably due to the fact that the optical quality of the pulsed laser beam is worse than that of the cw laser. We have previously [2] shown that the time evolution of an excited state E is given by the following expression:

$$\dot{n}_E = -\frac{1}{\phi}\left(\frac{dI(x)}{dx}\right)_A - \left(\frac{dI(x)}{dx}\right)_E - \frac{n_E}{\tau_E} - \frac{1}{\phi} I_F(x)\sigma_E'(n_E - n_A)$$

Here ϕ is the cross section of the light beam and the first and second term on the R.H.S. represent absorptions from the ground state to the excited state E and from this state to a higher lying state of the ion. Both absorptions are taken at the frequency of the pump. If I(x) is the intensity and τ_E the decay of the spontaneous emission and σ_E', is the gain cross secton for stimulated emission. Light amplification in the x-direction of the crystal can occur if $\sigma'_{E,L} > 1/n_E - n_A$. Obviously, the gain cross section for stimulated emission under pumping with cw and pulsed laser radiation is quite different, although population inversion can take place in both cases. For small light signals the equation reduced to the well known expression $n_E = n_o \exp -t/\tau$ however, it is clear that in the present experiment a more intricate relationship exists between the time evolution of the population and the radiative decay time for small signal values.

Our experimental results indicate that the population numbers for the excited states is larger under pumping with the pulsed than with the cw laser source. Presently, this result too is interpreted in terms of the reduction of the stimulated gain in the two cases.

REFERENCES

1. S.K. Gayen, W.B. Wang, V. Petricevic, R. Dorsinville, R.R. Alfano, Appl. Phys. Letters, 47 (1985 455.
2. B. Halperin, J.A. Koningstein, Can. J. Chem., 59 (1981) 2792.

DENSITY MEASUREMENTS OF Hg $^3P_{0,1,2}$ EXCITED STATES IN A DISCHARGE USING SATURATED LASER ABSORPTION

Larry Bigio

General Electric R&D Center
PO Box 8, Bldg. K-1 Rm. 4C34
Schenectady, NY 12301

Much effort continues in understanding the complex physical processes involved in a low-pressure Hg-Ar discharge due to the vast commercial importance of this system to lighting[1]. Computer modeling efforts on such discharges have stimulated the need for experimental determination of parameters such as cross-sections and densities[1-4]. This work describes a laser-based technique to measure the radial density distributions for the $6(^3P_{0,1,2})$ excited states of mercury in the positive column (constant E-field) of a low-pressure Hg-Ar DC discharge in which the polarity is commutated at 60 Hz to avoid axial cataphoresis. The tube diameter is 34 mm, argon partial pressure is 2.8 torr, and a water jacket maintains control of the wall temperature and therefore the Hg vapor pressure (mtorr range). The current is maintained at 400 mA.

The method employed is high-power saturated laser absorption[5,6]. Figure 1 shows the basic experimental arrangement in which a pulsed dye laser beam (20 nsec., .2 cm^{-1} bandwidth), traversing the discharge axially, is scanned through one of the $6(^3P_{0,1,2})$ ---> $7(^3S_1)$ transitions, depending on the lower state of interest. The transmitted signal is detected with a calibrated photodiode (UDT 535 Optometer) and the power-loss on resonance is plotted as a function of the non-resonant input power. Before traversing the tube, the beam is pinhole-filtered and expanded, producing a 1-cm gaussian cross-section beam. A 1-mm aperture is placed before the detector to give the spatial resolution required for radial density measurements. This combination allows for a fairly uniform intensity beam over the roughly 1-mm viewing area of the detector. At sufficiently high powers, the power-loss approaches a constant saturated value which is directly proportional to the initial number density of absorbers in the lower state. This may be seen by writing down the power-loss per unit volume, dI/dz, derived from the density matrix solution for a closed two-level atom excited by a single frequency, ω:[7-9]

$$\frac{dI}{dZ} = \frac{(N_2^o - \frac{g_2}{g_1} N_1^o)\, \hbar\, \omega_o\, T_2\, |\,\Omega\,|^2}{2\,(1 + \Delta\omega^2 T_2^2 + |\,\Omega\,|^2\, T_2\, \tau)} \tag{1}$$

where N_2^o = initial population density in upper level ~ 0

N_1^o = initial population density in lower level

$|\,\Omega\,|^2 = \mu^2 E^2/\hbar^2$ (proportional to laser power)

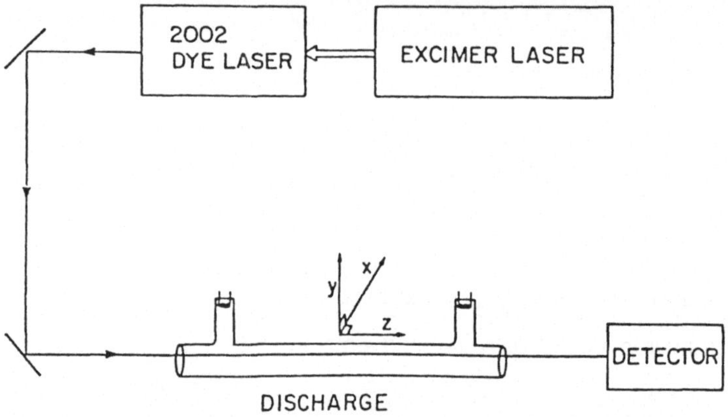

Fig. 1 - Experimental set-up showing excimer pumped dye laser beam passing axially through "V"-shaped discharge tube.

T_2 = dephasing time
τ = lifetime of upper level = 8.3 nsec

For $\Delta\omega$=0 (on resonance), the high power limit ($|\Omega|^2 \gg 1/T_2\tau$) gives:

$$\frac{dI}{dZ} = - \frac{g_2}{g_1} N_1^o \frac{h\omega_o}{2\tau} \tag{2}$$

since N_2^o is less than 1% of N_1^o.

Figure 2 shows three sets of power dependence measurements for three different wall temperatures; 20, 30, and 40°C, corresponding to Hg vapor pressures of 1.3, 3 and 6.4 mtorr, respectively. In order to model this data using Equation (1), it is necessary to choose one of several approximations in order to deal with the fact that the laser is not exactly monochromatic, but rather has an inhomogeneous bandwidth of ~ .2 cm^{-1}.

One method is to take T_2 ~ 25 psec and solve Equation (1) for $\Delta\omega$=0 numerically as[10]

$$IL = \int_o^L \frac{dI(Z)dZ}{dZ} \tag{3}$$

where IL is the total power-loss after traversing the entire length of the plasma. This has the equivalent effect of exciting the transition with a short pulse centered at ω_o whose transform bandwidth is a homogeneous .2 cm^{-1}. The shorter T_2 is, the more power is required to saturate since the power broadened homogeneous bandwidth would have to become broader than the .2 cm^{-1} transform bandwidth.

Another method is to convolute a gaussian (inhomogeneous) laser bandwidth:

$$|\Omega|^2 (Z=0) = |\Omega|_o^2 (Z=0) \frac{1}{\sigma\sqrt{2\pi}} e^{-\frac{1}{2}\left(\frac{\omega - \omega_o}{\sigma}\right)^2} \tag{4}$$

and carry out the following integral numerically:

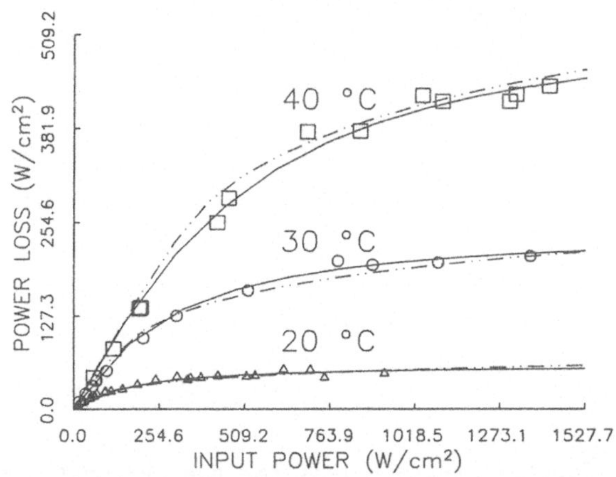

Fig. 2 - Power dependence data sets for three values of the wall temperature: 20, 30, and 40°C corresponding to H_g vapor pressures of 1.3, 3, and 6.4 mtorr, respectively. Solid lines are fits based on Equations (1) and (3) with T_2 + 30 psec. Dash-dot lines are fits based on Equations (1), (4), (5) with r = .04 cm^{-1} and T_2 = .7 nsec.

$$IL = \int_o^L \int_{\omega_-}^{\omega_+} \frac{dI(\omega,Z)}{dZ}\, dZ\, d\omega \qquad (5)$$

where $\omega_{\pm}$ is taken to be $\omega_o \pm 3\sigma$. It can be shown that the gaussian envelope for the laser at Z=0 in Equation (4) is "eaten away" at its center near ω_o as the beam propagates through Z in the medium. A larger laser bandwidth would also tend to push complete saturation to higher powers since, as above, the homogeneous power broadened bandwidth would have to become larger than the inhomogeneous laser bandwidth. In addition to these effects, the approach to saturation depends on the number density of absorbers, N_1^o. The greater this value, the more laser power is required to "burn through" the optically thick discharge, equilibrating upper and lower levels equally throughout its path. This last point is made evident in Figure (2) by comparing the powers at which saturation is attained for the three data sets which differ only in N_1^o. The two lines that fit each data set are derived using the two above approximations, Equations (3) and (5) for comparison.

Though the beam dramatically affects the plasma, it only does so for 20 nsec, during which time the measurement is made and no plasma processes occur. At a rep. rate of 15 Hz, the next pulse interrogates the plasma long after (66 msec) all populations have been re-equilibrated by the plasma.

Thus, this method is useful for making straightforward relative radial density measurements, either by fitting a complete power dependent curve such as in Figure (2), or by merely setting the power in the saturated regime, and measuring the power-loss on resonance as a function of the parameters of interest. Radial distributions were achieved in this latter manner by moving the tube perpendicular to the axial direction and measuring the power-loss in the saturated regime for different values of r, the radial position. Figures (3)-(5) give the results for the three $6(^3P_j)$ states at Hg pressures of 1.3 and 6.4 mtorr. The solid lines shown are 6th order polynomial fits to the data whose + and - r values have been averaged. Figure (4B) also compares results with those of Moskowitz taken for similar conditions[11].

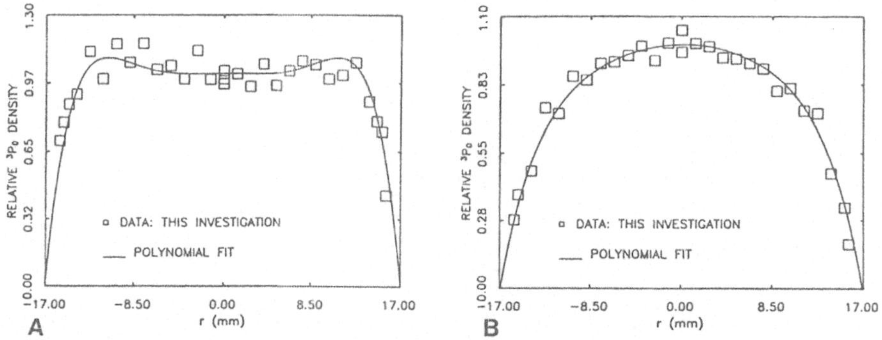

Fig. 3 - Radial density distributions for the $^3P_0$ state at wall temperatures of A) $20^\circ C$ (Hg pp = 1.3 mtorr), and B) $40^\circ C$ (Hg pp = 6.4 mtorr). Absolute densities on axis (r=0) derived using hook method are $1.8 \pm .13 \times 10^{11}$ and $4.8 \pm .25 \times 10^{11}$ cm^{-3}, respectively.

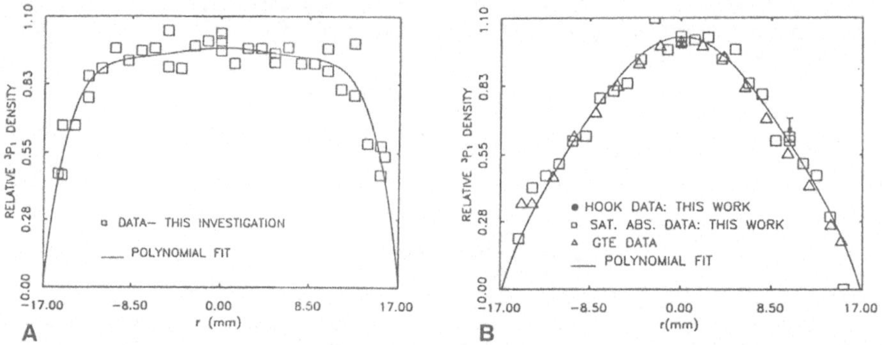

Fig. 4 - Radial density distributions for the $^3P_1$ state at wall temperatures of A) $20^\circ C$ and B) $40^\circ C$. Also shown for $40^\circ C$ is data taken by Moskowitz of GTE[13], and an off-axis hook measurement to compare with the saturated absorption results. Absolute densities on axis (r=0) are $.35 \pm .25 \times 10^{11}$ and $4.4 \pm .1 \times 10^{11}$ cm^{-3} for 20 and $40^\circ C$, respectively.

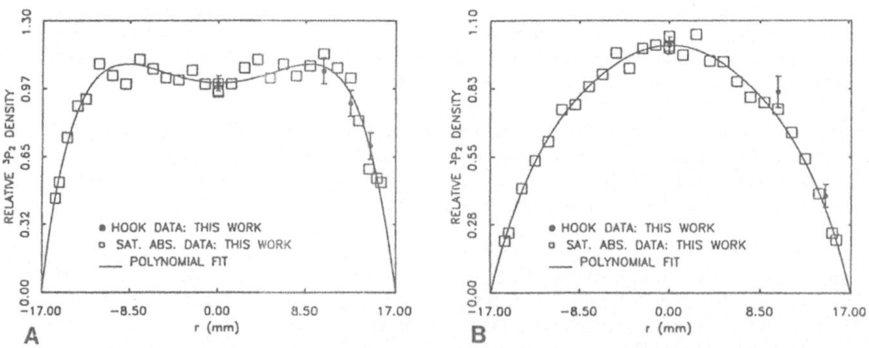

Fig. 5 - Radial density distributions for the $^3P_2$ state at A) $20^\circ C$ and B) $40^\circ C$. Also shown are several off axis hook measurements for comparison. Absolute densities on axis (r=0) are $3.14 \pm .12 \times 10^{11}$ and $8.5 \pm .3 \times 10^{11}$ cm^{-3} for 20 and $40^\circ C$, respectively.

Absolute densities are established in a separate experiment[12] using a well-known interferometric technique known as the hook method[3,13]. Briefly, the interference signal generated as a function of wavelength displays "hooks" at wavelengths where anomalous dispersion undergone by the beam traversing the discharge arm is balanced by the slightly larger optical path length for the beam traversing the reference arm of an interferometer. The dye laser is tuned through the ^3P_j ----> $^3S_1$ transition of interest, and the spectral separation of the hooks is used to determine an absolute density. Several on- and off-axis results are compared with the saturated absorption radial distributions, in Figures 3 and 4.

The saturated laser absorption, or "bleaching" method as some have called it[6], is useful for certain geometries particulary involving optically thick transitions. Furthermore, it doesn't necessarily require an extremely high resolution single mode laser source since, once saturation is complete, the power-broadened bandwidth exceeds all other broadening terms, thus effectively converting the laser into a single mode homogeneously broad source. The higher the resolution, though, the less power it takes to saturate, which may be desirable for optically thinner media. The results presented here are just a sampling of the different radial and absolute density profiles that have been achieved for a variety of discharge conditions, thus enabling a more satisfactory understanding of the complete discharge to emerge.

REFERENCES

1. See, for example; J.T. Dakin, J. Appl. Phys. 60, 563 (1986), and references contained therein.
2. W.J. Van Den Hoek, Philips J. Res. 38, 188, (1983).
3. P. Van de Weijer and R.M.M. Cremers, J. Appl. Phys. 53, 1401 (1982); ibid. 54, 2835 (1983).
4. P.E. Moskowitz, Appl. Phys. Lett. 50, 891 (1987).
5. This is not to be confused with a two-laser sub-Doppler spectroscopic technique exploiting hole-burning which is often referred to as "saturated absorption". For this reason, the word Laser is added as an arbitrary distinction.
6. D.S. Bethune, J.R. Lankard, and P.P. Sorokin, J. Chem. Phys. 69, 2076 (1978).
7. A. Yariv, "Quantum Electronics" John Wiley & Sons, New York, 1975, pp. 149-162.
8. The g_2/g_1 factor implies fast M_J-changing collisions which completely mix the M-levels. Omont and Meunier have measured the alignment-destroying cross-section to be ~ 1.4 X 10^{-3} cm^2 (Ref. 9), which would imply a time between collisions of roughly 40 nsec for 2.8 torr of argon - double the laser pulse duration. Thus it is unclear what factor should really be used, though this doesn't affect relative density measurements.
9. A. Omont and J. Meunier, Phys. Rev. 51, 652 (1968).
10. T_2 has to be $\leq$ 8.3 nsec, the lifetime of the 7($^3S^1$) level. However, it is unlikely that T_2 should be much less than, say, .5 nsec since the argon pressure is fairly low (2.8 torr). Thus, this approximation has the effect of broadening the transition which can be equivalently viewed as broadening the homogeneous width of the exciation source, while leaving the atom at its usual $1/\tau$ width.
11. P.E. Moskowitz, Fourth International Symposium on the Science and Technology of Light Sources, University of Karlsruhe, F.R. Germany, April 7-10, 1986.
12. L. Bigio, to be published.
13. W.C. Marlow, Appl. Opt. 6, 1715 (1967).

ABOVE THRESHOLD IONIZATION IN STRONG FIELDS

AND CLASSICAL BEHAVIOR OF ELECTRONS IN SHORT LIGHT PULSES

Pierre Agostini and Guillaume Petite

Service de Physique des Atomes et des Surfaces
CEN Saclay, 91191, Gif sur Yvette, France

ABSTRACT: Recent experimental results in Above Threshold Ionization in strong fields are reviewed and dicussed. The red shifts observed with picosecond and subpicosecond pulses in the energy spectra are shown to be a direct consequence of the classical behavior of the electrons in space and time dependent electromagnetic fields.

INTRODUCTION

It has been known for some years that an atom can be ionized in an intense laser field by absorption of a number of photons larger than that necessary to simply overcome the ionization potential, a process known as Above Threshold Ionization (ATI)[1] and usually studied[2-5] through the observation of the energy spectrum of the electrons released in the multiphoton ionization process (MPI) (Fig.1).

When the intensity is "low" (a more quantitative criterium is defined hereafter but a typical value to keep in mind is 10^{12}W.cm^{-2} for visible or near IR light), the ATI process is well described and understood within the framework of lowest-order perturbation theory. In brief, the atom ground state is <u>directly</u> coupled to continuum states with energies $-E_g + (N+S)E_p$ where $-E_g$ is the ground state energy, E_p the photon energy, N the minimum number of photons of energy E_p required to ionize the atom and S the number of excess photons absorbed above the threshold. The transition rates to the continuum state labelled by S is given by an (N+S)-order matrix element and is proportional to I^{N+S} (where I is the laser intensity). These transition rates are rapidly decreasing and only a few excess photons are absorbed. Experimental results fully support this picture[6,7,8] and, when the complete calculation is possible, good quantitative agreement is obtained[8].

However, when the intensity is "high" ($I > 10^{12}$W.cm^{-2}), the phenomenology is drastically changed, as observed for the first time in

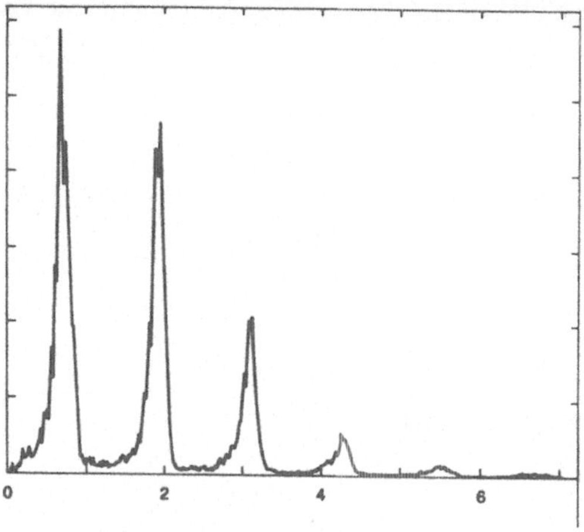

Fig.1- Typical perturbative ATI spectrum.

11-photon ionization of xenon at 10^{13}W.cm^{-2} and 1064 nm : a now famous experiment[2] which has since been repeated many times[3,4,5]. The number of photons aborbed in the continuum can be quite large (several tens), and the corresponding probabilities remain high. Moreover, the low energy peaks are suppressed and the intensity dependence of the rates is at variance with the I^{N+S} law mentioned above [2-5] (Fig.2 and Fig.3). Another key feature of the observations is that the remaining electron peaks were NOT shifted until recent experiments with very short pulses[5,9,10]. It is clear that such spectra cannot be interpreted within the framework of lowest order perturbation theory without pushing it to extreme limits[1].

In this paper we start with a brief discussion of a few simple physical points which are at the center of the understanding of recent experimental results in ATI in strong fields. Two key points to be discussed are final state effects and the energy changes that the electron experiences during its trip to the detector, outside the laser field, especially for short (picoseconde and subpicoseconde) pulses, when, as shown by the new observations, the electron peaks ARE shifted.

FINAL STATE EFFECTS IN ATI IN STRONG FIELDS

The interpretation of the peak suppression[11], at least when space-charge effects are avoided, as given by Muller et al.[11] can be summarized as follows : the atomic energy levels are distorted by the optical Stark-shift, wich affects differently the ground state and the Rydberg states (and continuum states) of the atom[12]. The ground state has a very small shift, while the high-lying states are strongly shifted upwards, as easily found by a second-order perturbation calculation. Neglecting the shift of the ground state, the resulting increase of the ionization potential is given by the so-called ponderomotive potential which is nothing but the classical kinetic energy of an electron oscillating in the electromagnetic field. The electrons which, in the low intensity limit, should be released with an energy smaller than the ponderomotive potential, are not released and the corresponding peaks are suppressed. However, one then expects the other electrons to have kinetic energies red-shifted by the same ponderomotive energy, at variance with the observations with long laser pulses. To explain the observed energies, the electrons were usually said to be accellerated by the "gradient force" whose effect exactly compensated for the energy loss due to the increase of the ionization potential[11], with long pulses. This interpretation raises two main difficulties : (i) the validity of the Stark-shift calculation is questionable[12] since it is based on the high frequency approximation whose validity is not always granted[13] and on an extrapolation of a second-order calculation to 10^{13}W.cm^{-2}. (ii) the work of the total electromagnetic force along the electron trajectory is easily shown[14] to be zero, for long pulses, which makes the standard picture rather inconsistent.

An alternate model has been proposed[14]. It is based on the idea that the final state of electron in the continuum is a Volkov state, perturbed by the Coulomb field rather than a Coulomb state perturbed by the electromagnetic field. The electron ejected in the continuum with a kinetic energy

$$E_N=NE_p-E_i \tag{1}$$

where E_i denotes the ionization potential of the atom, remains coupled both to the Coulomb and laser fields. But, in a very short time, (of the order of the optical period) the electron travels a distance such that the Coulomb energy becomes small compared to the dipole interaction.

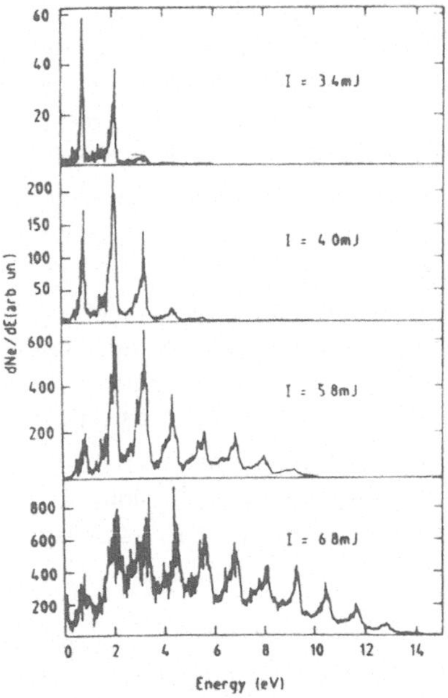

Fig.2- ATI of xenon at 1064 nm for different intensities.

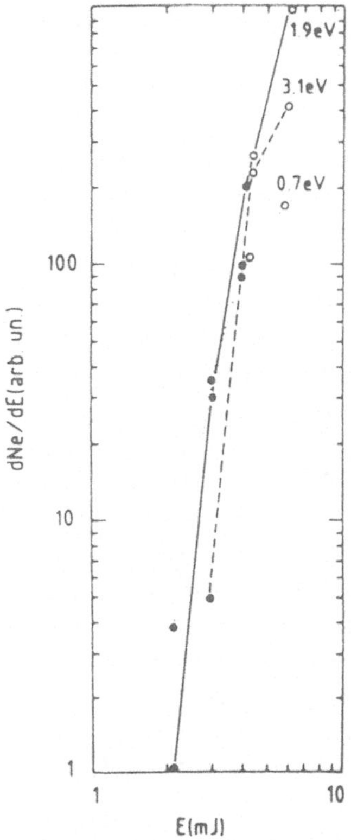

Fig.3- Amplitude of the first three ATI peaks vs pulse energy.

Therefore, the rest of the process is described as transitions between Volkov states induced by the Coulomb interaction, and called Half-Bremsstrahlung[14]. The peak suppression, mentioned above, is the a natural consequence of the energy conservation when one takes into account the coupling of the emerging electron with the external field, that is the oscillatory motion of the electron : the final state, inside the field, is characterized by the energy $m<V_d{}^2+V_{osc}{}^2>/2$ rather than $mV_d{}^2/2$, the drift energy corresponding to E_N and V_{osc} the oscillation velocity of the free electron in the laser field. The <> denote the average over one period of the field. To reach this total energy, the electron must absorb S extra photons in the half-collision process. Hence, the whole ATI process is viewed as, first an N-photon ionization followed by an S-photon inverse Half-Bremmstrahlung process. The idea to describe the final state of the MPI process by Volkov solutions had been used before[15,16], albeit in a different form.

This picture does not suffer from the shortcomings of the AC-Stark shift point of view since the Stark shifts are not a necessary ingredient of the process (although they could be introduced as a corrections if necessary). That is, the peak suppression does not have to be interpreted as due to an increase of the ionization potential, but rather as due to the final state oscillatory motion. In addition, this picture does not need to invoke the gradient force to restaure the electron energy outside the beam, since its total kinetic energy is already incremented by the oscillatory energy. It only requires the oscillatory motion to be converted into drift motion, with no change of the total energy, (for long pulses) as explained in more details in the next section, in full agreement with the classical calculation.

The predictions of the model concerning the intensity dependence of the electron peaks are in excellent agreement with experimental results, especially for the peak labelled S=0, for which an unexpected I^{N-4} dependence is predicted[14] and measured[2,3,5]. For the high order peaks, the model predicts an intensity dependence I^{N+1}, also in good agreement with observations of both the partial[2] and total[17] rates. The model is also consistent with direct measurements of the ionization potential[18] showing an extremely small change of E_i at intensities where the Stark shift should be of the order of 1 eV, when calculated as mentioned above (10^{13}W.cm^{-2} and 1064nm). Finally this point of view allows to set the limit between "low" and "high" intensities as far as ATI is concerned : a weak field is characterized by a ratio of the oscillation average kinetic energy to the drift kinetic energy E_N much smaller than one. As a matter of fact, this condition depends strongly on the light frequency[14] and, if I is expressed in W.cm^{-2} and the wavelength L in micrometers, the strong field condition writes as

$$I.\lambda^3>10^{13}\text{W.cm}^{-2}.\mu\text{m}^3. \qquad (2)$$

that is 10^{13} W.cm^{-2} is a 'high" intensity for a YAG laser while 10^{10}W.cm^{-2} is already a "high" intensity for a CO_2 laser. It should be pointed out that this λ^3 dependence appears also when one evaluates the average number of photons absorbed above the threshold as shown by Shakeshaft[12]. The results of Xiong et al.[24] are consistent with this preiction. Another important feature, seen on Fig.2, is the intensity dependent width of the electron peaks. These widths are much larger than the ionization width of the ground state. The only theoretical treatment accounting for such widths is the "all-order" theory presented in this book[19]. The physical process at the origin of the witdhs is an electromagnetic broadening of the continuum states, due to the continuum-continuum transitions.

As a summary, some of the most prominent features of ATI at high intensity are well understood when the final state of the emerging electron inside the electromagnetic field is described as a Volkov state perturbed by the Coulomb potential. However, the electron is detected and its kinetic energy measured outside the electromagnetic field. The next section is dealing with the classical description of the free electron motion in space and time dependent fields, with particular emphasis on the resulting energy changes when using short light pulses. A more general account of "ponderomotive effects" is given elsewhere[9].

CLASSICAL BEHAVIOR OF FREE ELECTRONS IN SHORT LIGHT PULSES

The energy measured in an ATI experiment is the electron drift energy outside the laser field. This energy is given by the simple equation :

$$E_{out} = E_{in} + \left< \int_{in}^{out} \vec{F}.\vec{dx} \right>_{av} \qquad (3)$$

where E_{in} is the total kinetic energy inside the beam, that is the sum of the drift kinetic energy and the oscillation average energy; $\vec{F}$ is the total electromagnetic force. To solve the non-linear equation of motion, the electron dynamical variables and $\vec{F}$ are separated into a high frequency part corresponding to the oscillation $(x_h, v_h, \vec{F}_h)$ and a low frequency one correponding to the drift $(x_l, v_l, \vec{F}_l)$ following the usual ponderomotive force formalism. If the variations of the low frequency components over one period are neglected, the cross terms in the integral (3) average out so that :

$$E_{out} = E_{in} + \left< \int_{in}^{out} (\vec{F}_h.\vec{v}_h + \vec{F}_l.\vec{v}_l)dt \right> \qquad (4)$$

where

$$\vec{F}_h = -e\vec{E}(x_l,t)\cos\omega t \qquad (5)$$

and

$$\vec{F}_l = -(e^2/4m\omega^2)\vec{\nabla}(E^2(x_l,t) \qquad (6)$$

Eqs. (5) and (6) can be substituted into Eq. (4) to yield :

$$E_{out} = E_{in} - D + H \qquad (7)$$

where D is the classical kinetic oscillation energy and H the work of the gradient force along the average trajectory of the electron :

$$H = \int_{t}^{\infty} \vec{F}_1 \cdot \vec{V}_{drift} \, dt \tag{8}$$

It is easy to see[23] that different situations can arise, depending on the pulse duration. If the pulse is long (compared to the time taken by the electron to escape from the beam waist), then the whole spatial gradient will be experienced. Since most of the electrons are emitted along the laser polarization[3], they will feel mostly the transverse gradient which is parallel to V_{drift}. It turns out that H then exactly compensates for D and the total energy is conserved[14]. The oscillation motion is converted into drift motion when the electron exits the laser focus. The energy peaks are therefore found unshifted : **the quantity measured in this case is the total electron energy after ionization.** On the other hand, if the pulse is short, the electron moving slowly along the polarization direction feels the gradient along the laser propagation direction (since the pulse passes by while the electron is hardly moving), that is F_1 is perpendicular to V_{drift} and, consequently the work of the gradient force is zero. The oscillation energy is lost in the process and the electron peaks appear red-shifted by D : **the quantity measured in this case is just the electron drift energy after ionization.** In intermediate cases, the shift depends on the ratio of the pulse duration to the exit time of the electron[23]. The experimental proofs of this analysis were given recently by using picosecond and subpicosecond pulses[5,23,9,10]. We briefly summarize these results hereafter :

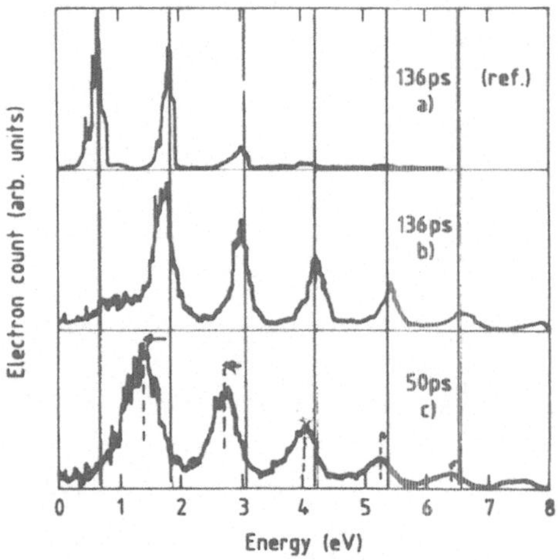

Fig.4- Energy shifts due to the non-adiabatic exit of the electrons.

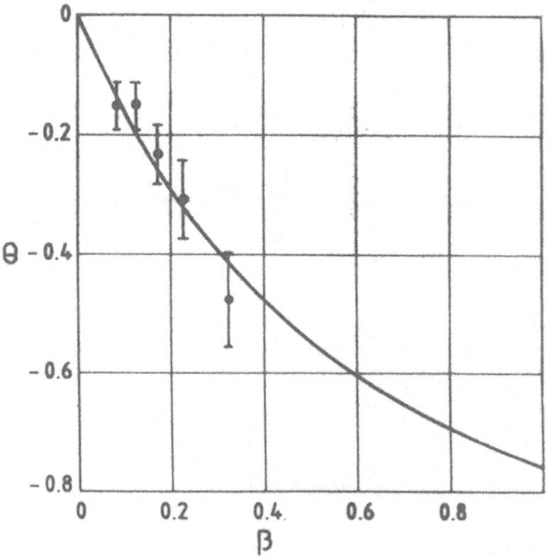

Fig.5- Measured shifts compared to calculation (solid line)[23].

Three electron energy spectra obtained in 11-photon ionization of xenon at 1064 nm are shown on Fig.4. Fig.4a and Fig.4b show spectra taken with a pulse duration of 136 ps and laser intensities of 2×10^{12}W.cm^{-2} and 7×10^{12}W.cm^{-2} respectively. They display the well known phenomenology of ATI in strong fields[2] : peak suppression and no shift. Fig.4c shows a spectrum taken with a 50 ps pulse and the same intensity as on Fig.4b. This time, the peaks appear red-shifted by an amount which rapidly decreases when the order of the peak increases, in agreement with the argument developped above. A more quantitative comparison is represented on Fig.5 where the experimental points correspond to the measurements on Fig.4c and the solid line is the result of an analytical calculation of Eq.(4) for a gaussian beam with an exponential temporal dependence[23]. Added confidence inthe relevance of the model discussed above was obtained by reducing the size of the beam waist and observing the correlated decrease of the red shifts : it is equivalent to increase the pulse length and to reduce the focus size.

The opposite circumstance is met with ultrashort pulses : if the pulse is so brief that even the most energetic electrons travel a very small distance while it lasts, one expects, on the basis of the previous analysis, that all the peaks will be equally red shifted, at variance with the result of Fig.4c. This was done[10] in an experiment on 7-photon ionization of xenon with 0.1 ps pulses at 620 nm. The resulting spectra are shown on Fig.6 for different laser intensities.

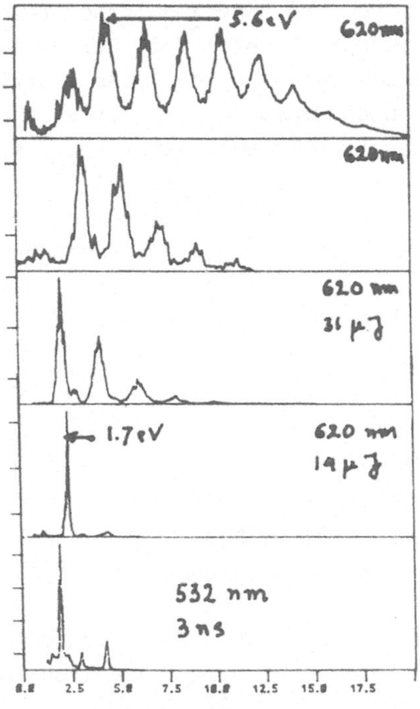

Fig.6- ATI of Xenon at 620 nm, 100 fs pulses for different intensities. The maximum intensity is 10^{14}W.cm^{-2}.

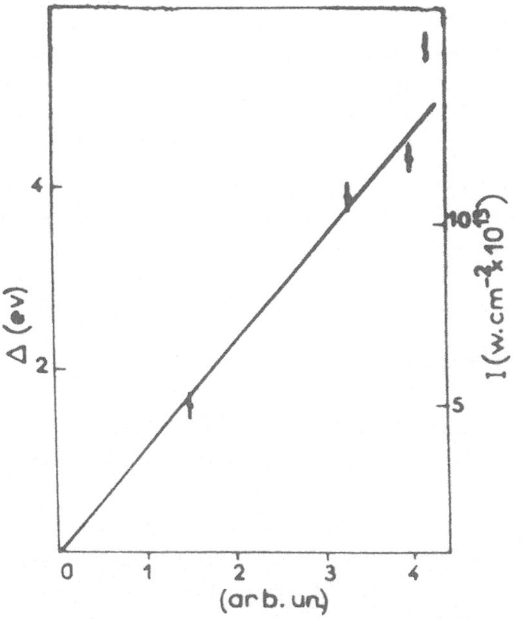

Fig.7- Energy shift as a function of relative intensity. See text for explanation of the right hand scale.

The main difference between the spectra of Fig.6 and the one of Fig.4c is that now the peak separation is constant even when the shifts are quite important (larger than the photon energy, with the consequence that several peaks are suppressed). The red shift as a function of the relative laser intensity is shown on Fig.7. Since the absolute value of the intensity can only be estimated in this experiment, the result shown on Fig.7 is subordinated to the assignement of the peaks of Fig.6. Taking into account the spectrometer calibration and seeking for a linear dependence of the shift on the laser intensity, one is lead to assign the peak recorded at the lowest intensity (Fig.5b) to the 8-photon ionization leaving Xe^+ in the 3/2 state. This assignement yields the linear dependence shown on Fig.7 and absolute values for the shifts and intensity. This question will be discussed in more details elsewhere[10].

CONCLUSION

The phenomenology of ATI under strong fields seems now reasonably established, experimentally. The characteristic features found by Kruit et al.[2] have been confirmed several times. A large fraction of the experimental data can be related to the properties of the final state of the electron inside the electromagnetic field. It is to be noted that it is not necessary to invoke AC-Stark shifts to describe the peak suppression although such shifts could be present and play a role in the transition : recent observations of ATI spectra with subpicosecond pulses have shown [9,10] narrow structures which can be attributed to Stark-shifted resonances. However, the results obtained by Lompre et al.,[24] in Helium and Xiong et al.[24] at 10.6 um are more easily explained by the final state oscillation energy than by AC-Stark shifts which would be much larger than the ionization potential. These results are consistent with the very small shift of the ionization potential when determined directly[17]. On its way to the detector, outside the field, the electron behaves classically and experiences the Lorentz force. Its energy can be changed, depending critically on the ratio of the pulse duration to the exit time of the electron. The resulting red-shift has been demonstrated by using picosecond and subpicosecond pulses.

REFERENCES

1- Y. Gontier and M. Trahin ;J. Phys. B **13**,4381 (1980).
2- P. Kruit ,J. Kimman, M. J. Van der Wiel; J. Phys. B **14**,L597 (1981).
3- H. J. Humpert, R. Hippler, H. Schwier and O. Lutz in Fundamental Processes in Atomic collision Physics, H Kleinpoppen, J. S. Briggs and O. Lutz Ed., (Plenum, New York 1985).
4- P.H. Bucksbaum, M. Bashkansky, R.R. Freeman, T.J. McIlrath and L.F. Dimauro, Phys. Rev. Lett., **56**, 2590 (1986).
5- G. Petite, P. Agostini and F. Yergeau; J. Opt. Soc. Am. B**4**,765,(1987).

6- F. Fabre, G. Petite, P. Agostini and M. Clement; J. Phys. B **15**,1353 (1982).
7- L.A. Lompre, G. Mainfray, C. Manus and J. Kupersztych; J. Phys. B. At. Mol. Phys., **20**, 1009 (1987).
8- G. Petite, F. Fabre, P. Agostini, M. Crance and M. Aymar ; Phys Rev. A **29**,2677 (1984).
9- P. Bucksbaum (this book).
10- P. Agostini, M. Franco, H. Muller, H. B. van Linden van den Heuvell,G. Petite and J. C. Pourny; postdeadline paper, ICOMP IV, Boulder, Colorado (1987) and submitted to Phys. Rev. Lett.
11- H. G. Muller, A. Tip and M. J. Van der Wiel, J. Phys. B **16**, L679 (1983).
12- L. Pan, L. Armstrong, and J. H. Eberly, J. Opt. Soc. Am. B**3**,1319 (1986); see also R. Shakeshaft, J. Opt. Soc. Am. **4**,705(1987)
13- P. Avan, C. Cohen-Tannoudji, J. Dupont-Roc and C.Fabre, J. Phys. (Paris) **37**,993 (1976)
14- J. Kupersztych Europhys. Lett. **4**,23 (1987)
15- L. V. Keldysh, Sov. Phys. JETP **20**,1307 (1965)
16- H. Reiss J. Opt. Soc. Am. B**4**,726 (1987) and ref. therein.
17- L.A. Lompre,G. Mainfray and C. Manus J. Phys. B **13**,85 (1980)
18- L.A. Lompre, G.Mainfray, C. Manus and J. Thebault Phys. Rev A **15**,1604 (1977)
19- Y. Gontier (this book)
20- H. J. Humpert, H. Schwier, R. Hippler and O. Lutz Phys. Rev. A **32**,3787 (1985)
21- D. Feldmann, Invited paper ICOMP IV, Boulder Colorado (1987)
22- Y. Gontier, N. K. Rahman, M. Trahin to be published; K. Rzazewski and R. Grobe, Phys. Rev. Lett. **54**,1729 (1985); A. Maquet and V. Veniard, to be published.
23- P. Agostini,J. Kupersztych, L. A. Lompre, G. Petite and F. Yergeau Phys. Rev. A (in Press).
24- L. A. Lompré, A. L'Huillier, G. Mainfray and C. Manus, J. Opt. Soc. Am. **2**,1902 (1985) ; W. Xiong, F. Yergeau and S.L. Chin (this book)

ABOVE-THRESHOLD IONIZATION AND THE QUANTUM MECHANICS OF WIGGLING ELECTRONS

P. H. Bucksbaum

AT&T Bell Laboratories
Murray Hill, NJ 07974 USA

ABSTRACT

Intense laser fields can dramatically alter the behavior of free and bound electrons. Some effects, such as the scattering of electrons by light, are due to classical wiggling in response to the periodic driving force. Others, such as above-threshold ionization of atoms, depend on the wave nature of the electron, and require quantum mechanics for a complete explanation. This paper reviews some recent experiments that have contributed to our understanding of the physics of light-matter interactions in this new regime.

1. *INTRODUCTION: ATI AND PONDEROMOTIVE FORCES*

Very intense laser light (above 10^{13} W/cm^2), rapidly ionizes most atoms and molecules. These ultra-high intensities can now be produced at many wavelengths between 10μm and 0.2μm. The study of the photoionization fragments, particularly electron energy analysis, has led to a deeper understanding of the physics of high intensity light-matter interactions.[1]

A new feature of photoionization at high intensities is above-threshold ionization (ATI), the absorption of excess photons beyond the minimum necessary to ionize an atom or molecule (see figure 1).[2,3] ATI has now been observed in a wide variety of atoms, ions, and molecules. In some cases, as shown in the photoelectron spectra in figure 1, The threshold channel is greatly suppressed, so that ATI accounts for nearly all of the photoelectrons.

Since ATI occurs only at extremely high laser intensities, there are several practical limitations that profoundly influence all observations. First, ATI experiments invariably take place in tightly focused laser beams. Beam waists of 5 to 50μm are common. This means that a crucial experimental parameter, the intensity where the observations are made, is not well-defined. Depletion of the ground state neutral atoms at the highest intensities (so-called depletion saturation) further complicates matters. In addition, electron dynamics at high intensities are affected by stimulated scattering interactions, called ponderomotive effects.[4] Final state scattering alters the electron energy spectra and angular distributions. Finally, to prevent ionization at low intensity as the laser beam turns on, ATI experiments nearly always involve nonresonant multiphoton ionization, the more photons the better. The most widely studied system is xenon (I.P.=12.127 eV) photoionized by Nd:YAG radiation ($h\nu$=1.165 eV). In principle, CO_2 laser radiation (0.1 eV photons) is even better, but the relatively huge ponderomotive effects associated with intense CO_2 laser beams make electron spectroscopy quite difficult in this case.

2. PONDEROMOTIVE FORCES: THE CLASSICAL PHYSICS OF WIGGLING ELECTRONS

2.1 *Ponderomotive energy and wiggling*

At low intensities, the only observable interaction between free electrons and light is Thomson scattering. In high intensity focused lasers, however, Thomson scattering is utterly negligible. A 1eV electron traversing a tightly focused Nd:YAG laser beam (10^{13}W/cm^2, 10μm beam waist) is likely to Thomson scatter only about one photon out of the beam. The recoil changes the electron's velocity by less than a tenth of a percent. However, although this spontaneous light scattering is a small effect, stimulated scattering is enormous. In fact, the 1 eV electron will not pass unimpeded through the beam, but will actually backscatter from the light!

The classical manifestation of this stimulated scattering is the wiggling motion of the particle under the influence of the electromagnetic wave. "Ponderomotive", refers to this motion. For a periodic electric field with amplitude E_0 and angular frequency ω, the time-averaged energy associated with wiggling is[4]

$$U_p = \frac{e^2E_0^2}{4m_e\omega^2}. \tag{1}$$

U_p is called the ponderomotive energy, and is equal to $0.8\times10^{-13}(I/\hbar\omega^2)$eV, where I is in watts per square centimeter, and $\hbar\omega$ is the photon energy in eV. For 1 μm radiation, the ponderomotive energy is actually equal to the photon energy for $I\approx10^{13}$W/cm^2. This is just the intensity where ATI is observed in xenon, and the equivalence between the quantum energy and the classical wiggle energy has important consequences for ATI.

2.2 *The time-averaged Lorentz force in a laser focus.*

The ponderomotive wiggling motion is rather complicated, particularly when spatial and temporal inhomogeneity of the focused radiation field are taken into account. It is usually possible, however to make two simplifying assumptions: First, laser fields turn on and off slowly compared to the laser period $1/\nu$. This is the adiabatic approximation. Second, the laser intensity varies slowly over the wiggle path of the electron, even in a tight focus.

Under these approximations, it is a simple matter to show that an electron in a spatially inhomogeneous laser focus experiences a net force in the direction of lower intensity. The classical Hamiltonian for a spinless electron in a the radiation gauge is

$$H = \frac{[\mathbf{p}-\frac{e}{c}\mathbf{A}(\mathbf{x},t)]^2}{2m_e}. \tag{2}$$

Here $\mathbf{A}$ is the vector potential for the laser radiation field, with amplitude A_0. The

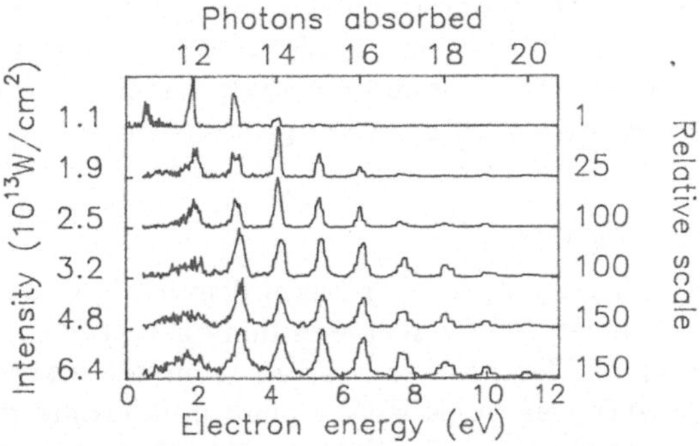

Figure 1: Photoelectron spectra showing above-threshold ionization in xenon by 1.165 eV photons. The laser was linearly polarized and 100 psec in duration, focused to a 12 μm gaussian waist, with peak intensity shown on the left axis. The electrons were detected along the laser polarization direction, and energies were analyzed by time-of-flight. (From reference 3.)

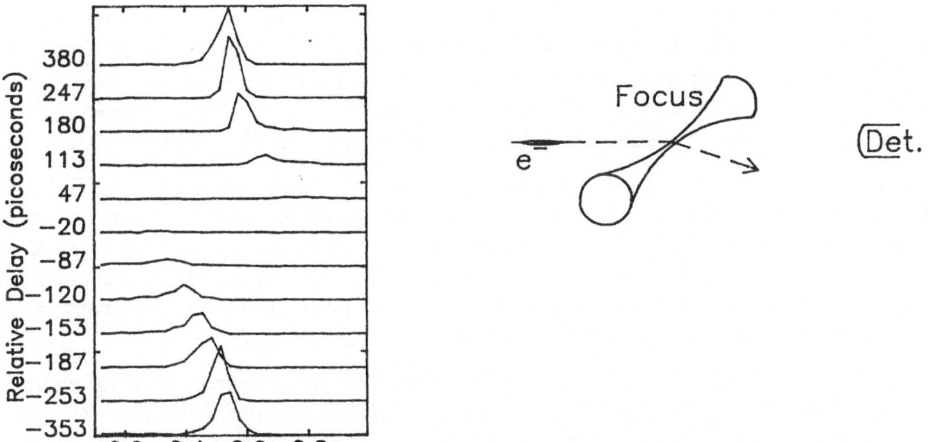

Figure 2: Scattering of electrons by light. 0.54 eV electrons were directed through the focus of a 100 psec laser pulse with peak ponderomotive potential of 8 eV. Those arriving early or late were undeflected. Electrons passing through the leading and trailing edges were accelerated or decelerated, respectively, due to the phenomenon of "surfing", described in the text. Electrons encountering the peak of the pulse were scattered away from the detector. (From reference 5.)

time-averaged hamiltonian is

$$<H> = \frac{p^2}{2m_e} + \frac{e^2 A_0^2(x)}{4m_e c^2}. \tag{3}$$

The second term is equal to the ponderomotive energy U_P in equation 1. Although U_P actually represents wiggle kinetic energy, it acts like, and may be treated as, a conservative potential energy, the "ponderomotive potential." Spatial intensity variations give rise to forces equal to the negative gradient of the ponderomotive potential.

2.3 Electrons scattering from a laser beam

An electron entering a steady state laser beam elastically scatters from the conservative ponderomotive potential. This effect was demonstrated in a recent experiment employing a pulsed beam of electrons scattering from an intense focused Nd:YAG laser.[5] The laser beam, a 100 psec pulse with a peak ponderomotive potential of 8 eV, was between the electron source and the detector, so that the ponderomotive potential could block the electron pulse. The relative timing could be varied to show that the scattering was due to the presence of the light only. (see figure 2)

2.4 Inelastic electron scattering from light

The ponderomotive potential of a pulsed laser is a function of time as well as space. Therefore, the electrons need not conserve total energy. For a time-dependent ponderomotive potential $U_P(t)$, the incremental change in the electron's energy during a time interval δt is $\partial U_P(t)/\partial t \; \delta t$.

The situation is analogous to surfing. A wave can lift a surfer, imparting gravitational potential energy. By taking advantage of the spatial gradients in the wave, the surfer can convert the potential energy into kinetic energy by sliding off the wave. In the same way, an electron overtaken by a light pulse gains ponderomotive potential energy, which can be converted to kinetic energy by accelerating in the ponderomotive potential gradients. The light does work on the electron.

The magnitude of the energy exchanged between the electron and the light pulse is not related to the light quantum $\hbar\omega$, and can be smaller. The quantum mechanical explanation is that inelastic scattering is due to stimulated exchange of photons between different frequency modes of the laser pulse. Likewise, the elastic scattering from a spatial gradient of U_P involves stimulated scattering among different momentum eigenmodes present in a tightly focused laser beam.[4]

In the electron scattering experiment described above, electron surfing was clearly evident as the electrons passed the rising or falling temporal edge of the 100 picosecond laser pulse. Figure 2 shows how the .54 eV electron peak was shifted to higher energy when the electrons gained ponderomotive potential energy by passing through the rising leading edge of the laser pulse. Conversely, when the electrons arrived late in the pulse, they decelerated due to the spatial gradients, and then remained at lower energy as the laser pulse passed by.

2.5 Ponderomotive effects in atoms

Electrons bound to atoms are also affected by ponderomotive forces. A neutral bound state acquires a polarization energy in an oscillating field, which depends on the driving force, the presence of resonances, and the binding strength. The energy acquired by a classical dipole excited well below resonance is inversely proportional to the restoring force constant. By analogy, the ground state of a tightly bound atom such as xenon might be relatively unaffected by ponderomotive forces in a Nd:YAG focus. Loosely bound excited states near the ionization limit, however, certainly shift by the full ponderomotive potential, just like free electrons. Since the ionization limit increases, while the ground state remains nearly unchanged, *more work is required to ionize an atom in an intense field.*

The basic features of the quantum mechanical problem[6] can be seen by considering the Schroedinger hamiltonan for a single electron moving under the combined influence of a laser field and a central potential,

$$H = \frac{[\mathbf{p} - \frac{e}{c}\mathbf{A}(\mathbf{x},t)]^2}{2m_e} + V(r). \tag{4}$$

Two terms depend on the laser: the so-called A^2 term and the $\mathbf{A}\cdot\mathbf{p}$ term. The former is the ponderomotive potential term, and does not depend on the momentum or position of the electron in the potential. It therefore affects all states equally, shifting the ground state by the same amount as a free electron state. The $\mathbf{A}\cdot\mathbf{p}$ term, however, which has no effect on free electrons, tends to cancel the A^2 term for the ground state. To see why this is so, consider the second-order perturbation theory expression for the a.c. Stark shift (due to $\mathbf{A}\cdot\mathbf{p}$ only) of the ground state, for the simple one-electron-central-field problem:

$$U_{\text{Stark}} = \left(\frac{eA_0}{2m_e c}\right)^2 \sum_i \left[|\langle i|\hat{\epsilon}\cdot\mathbf{p}|\phi\rangle|^2 \left(\frac{1}{E_\phi - E_i - \hbar\omega} + \frac{1}{E_\phi - E_i + \hbar\omega}\right)\right] \tag{5}$$

If the photon energy is much less than the separation of any excited state from the ground state, U_{Stark} may be expressed as an expansion in $\hbar\omega/(E_\phi - E_i)$:

$$U_{\text{Stark}} = \left(\frac{eA_0}{2m_e c}\right)^2 \sum_i \left[|\langle i|\hat{\epsilon}\cdot\mathbf{p}|\phi\rangle|^2 \frac{2}{(E_\phi - E_i)}\left[1 + \left(\frac{\hbar\omega}{E_\phi - E_i}\right)^2\right]\right]. \tag{6}$$

The first term in this expansion may be summed using the dipole oscillator strength sum rule.[6] Then we obtain

$$U_{\text{Stark}} \approx -\left(\frac{e^2 A_0^2}{4m_e c^2}\right) \tag{7}$$

thus cancelling the ponderomotive potential. This cancellation may seem fortuitous, but in fact, the separation of the interaction into two terms is somewhat artificial, brought on by our choice of the radiation gauge. The next term in the series is the *static* field Stark effect $U_{\text{D.C.Stark}} = \alpha E_0^2/2$, where α is the polarizability of the ground state. This is much less than the ponderomotive potential in the experiments considered here.

2.6 Energy spectra and angular distributions in intense fields

The increase in the ionization potential should appear in photoelectron spectra, since as the I.P. increases, the residual kinetic energy of the photoelectron must decrease. This effect is masked, however, by the fact that the freed electron possesses ponderomotive potential energy nearly equal to the I.P. shift. This is converted into kinetic energy when the electron leaves the ponderomotive potential, returning the electron's kinetic energy to the value it would have had for ionization in a weak field. So ATI spectra show peaks that are unshifted by the ponderomotive potential.

An interesting situation arises when the ponderomotive potential increases by more than the entire kinetic energy of the threshold electron. In this case, an extra photon is required to ionize the atom, and the lowest spectral peak should disappear

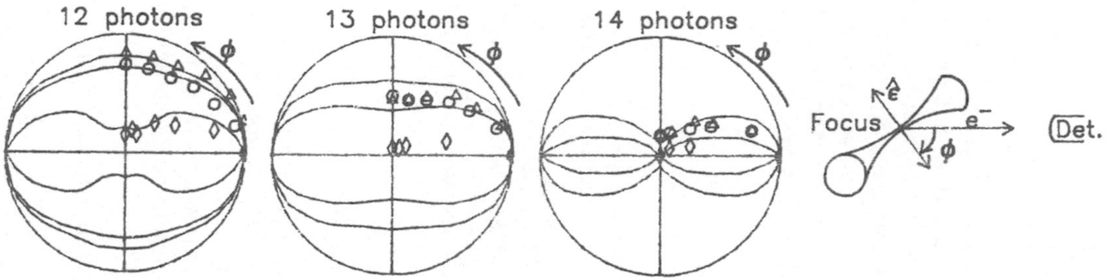

Figure 3: Intensity-dependent ATI angular distributions in xenon, for absorption of 12, 13, and 14, 1.165 eV photons. The peak laser intensity was 1.5, 3.0, and 6.0 $\times 10^{13} W/cm^2$ for diamonds, circles, and triangles, respectively. At low intensities the distributions are sharply peaked along the laser polarization direction. Emission becomes more isotropic at higher intensities due to ponderomotive scattering of the electrons. (From reference 7)

from the electron spectra. In practice, this effect is difficult to observe cleanly, due to the spatial and temporal inhomogeneity of a focused laser pulse; but a strong suppression of the threshold peak is evidence that ionization is occurring at intensities where the threshold channel is cut off (see figure 1).

The ponderomotive shift of the ionization potential may be indirectly observed through measurements of angular distributions of photoelectrons. ATI emission angles with linearly polarized light tend to be sharply peaked in the direction of the laser polarization. However, if the electron velocity is greatly reduced by a ponderomotive increase in the ionization potential, then the electron's direction will be strongly affected by the direction of the ponderomotive force acting on it as it escapes the laser focus. Figure 3 shows several angular distributions from the lowest energy ATI peaks in a xenon photoionization experiment by R. Freeman and coworkers.[7] At low intensities, the electrons are sharply directed along the laser polarization, while at large intensities, the angular distributions become nearly isotropic in the azimuthal plane (normal to **k**). By performing electron trajectory simulations on a model laser focus, they were able to show that the ionization potential shifts for these ATI peaks range from 1 to 3 eV.

3. ELECTRON PHASE-MATCHING: THE QUANTUM PHYSICS OF WIGGLING ELECTRONS

Ponderomotive scattering and surfing are essentially classical effects, that can be understood without resorting to quantum mechanics. However, we know that electrons are actually waves, and ATI cannot be adequately described without a full quantum treatment of the atomic electrons during photoionization.

In this section we consider the quantum-mechanical problem of an electron in a laser field. While we will be forced to skirt most of the difficulties in dealing with the atomic problem, a picture will emerge of quantum wiggling, that emphasizes the importance of phase matching for electrons in a laser field.

3.1 The Volkov state: a quantum-mechanical wiggling electron

We begin by considering Schroedinger's equation for an otherwise free electron in a uniform monochromatic radiation field.

$$\frac{(\mathbf{p}-\frac{e}{c}\mathbf{A})^2}{2m_e}\psi(\mathbf{x},t) = i\hbar\frac{\partial}{\partial t}\psi(\mathbf{x},t) \tag{8}$$

The long wavelength approximation, wherein the vector potential **A** varies only with time, permits direct integration of equation 8. The solutions are known as Volkov states. They are essentially plane waves, whose time dependence has been modified by the electromagnetic vector potential:

$$\psi_{\text{Volkov}}(\mathbf{x},t) = \exp\left(\frac{i}{\hbar}\mathbf{p}\cdot\mathbf{x} - \frac{i}{\hbar}\frac{p^2}{2m_{e_0}}t - \frac{i}{\hbar}\frac{e^2A_0^2}{4\omega m_e c^2}t + \frac{i}{\hbar}\frac{e^2A_0^2}{8\omega m_e c^2}\sin(2\omega t) - \frac{i}{\hbar}\hat{\varepsilon}\cdot\mathbf{p}\frac{eA_0}{2\omega m_e c}\cos(\omega t)\right) \tag{9}$$

Like plane waves, the solutions are eigenstates of the *canonical* momentum with eigenvalue **p**. The extra time-dependent phases are periodic energy fluctuations caused by the oscillating field. The three extra terms are: the average energy change due to the ponderomotive potential; the periodic fluctuations about this value with frequency 2ω; and oscillation at ω (the $\mathbf{A}\cdot\mathbf{p}$ term), due to the component of the driving force in the direction of the drift.

3.2 Photoionization in intense fields

The final state in an ionization process is asymptotically a Volkov state, but it is modified at short distances by the coulomb field of the ion. The initial state has the character of an atomic bound state, but it is also modified, by the radiation field. One would like to find the evolution of an electron wave function under the influence of the full hamiltonian, including both the atomic potential and the radiation field, where the initial wave function is the ground state of an atom. Unfortunately, this problem has yet to be solved analytically, even for the simplest case of a hydrogen atom in a uniform laser beam.

There have been many approaches to this problem,[1] including direct numerical integration, perturbation theory, and analytical approximations of various kinds. One method proposed by Keldysh, and developed by Reiss, Faisal, and others, reproduces several experimental features, providing physical insight into the ionization process in strong fields.[8] The method projects the initial state wave function onto Volkov states. Formally, Reiss has shown that this projection is the lowest order term in an S-matrix expansion using the ion potential as a perturbation.

In this approach, the *canonical* momentum distribution of the final state is the same as the momentum distribution in the initial bound state. In essence, the coulomb potential that formed the ground state is ignored, so that the bound state just evolves outward with time. The *energy* of the final state, however, is quite different from the initial state; after all, the electron has absorbed all those photons in order to ionize! To figure out how many photons were absorbed, one examines the energy structure of the Volkov wave function. It is not an energy eigenstate, since it comes from a time-varying hamiltonian. The energy spectrum is obtained by means of a "Floquet" decomposition, which is simply a projection of the Volkov state on the complete set of plane waves:

$$\psi_{\mathrm{Volkov}}(\mathbf{x},t) = e^{\frac{i}{\hbar}(\mathbf{p}\cdot\mathbf{x} - \frac{p^2 t}{2m_e} - U_p t)} \times \sum_{n=-\infty}^{\infty} \sum_{m=-\infty}^{\infty} J_m\left(\frac{e^2 A_0^2}{8\hbar\omega m_e c^2}\right) J_{n-2m}\left(\hat{\epsilon}\cdot\mathbf{p}\frac{eA_0}{2\omega m_e c}\right) e^{in\omega t} \tag{10}$$

Here J_m's are cylindrical Bessel functions, whose arguments are the coefficients of the oscillating terms in the Volkov phase. The energy spectrum is not continuous, but consists of an infinite series of eigenstates separated by $n\hbar\omega$, where n is an integer. Each state contributes to the total with an amplitude proportional to the Bessel function sum. The interpretation is obvious: the nth component in the Floquet decomposition corresponds to the amplitude for absorption of n photons. Thus ATI is simply the projection of the Volkov final state onto electron states of definite energy, $n\hbar\omega$ above the initial state energy.

Of course, neglecting the coulomb field leads to some serious quantitative disagreements between laser experiments and theoretical predictions. For example, any enhancements in the ionization rate due to intermediate resonances are totally ignored. However, the model naturally accounts for some observations such as channel cutoff, and it has been quite successful in predicting the number and relative magnitude of ATI peaks in Nd:YAG experiments.[9]

The Floquet decomposition shows the absorption of discrete numbers of photons, despite the fact that the electromagnetic field in this problem has not been quantized. The discretization is the result of wave-mechanical interference. Whereas a classical particle exhibits a continuously varying energy as it oscillates up and down in the periodic field, the Volkov wave function is an energy superposition state with an oscillating phase at frequencies ω and 2ω. It's quite easy to show that only energy eigenstates with eigenvalues of $n\hbar\omega$ can be included in this superposition.

Interference phenomena also affect the *directions* of electron propagation. As the electron emission angle changes with respect to the laser polarization, the Volkov phase factor containing $\mathbf{A}\cdot\mathbf{p}$ also changes. This affects the values of the coefficients in the Floquet expansion. The bessel function coefficients may even pass through zero, indicating that ATI electrons with a given energy may not propagate in those directions. Angle-dependent propagation of ATI electrons may be compared to Bragg scattering in crystal. Whereas the Bragg peak is caused by constructive interference of x-ray waves scattering from a spatially periodic potential, the ATI electrons are matter waves scattering from a *temporally* periodic potential. In both cases, wave interference strongly modulates the scattering distributions.

These electron phase-matching effects were recently demonstrated in xenon and krypton ATI experiments utilizing *elliptically* polarized 1.06 μm light.[10] By varying the ellipticity of the light, it's possible to control the variation in $\mathbf{A}\cdot\mathbf{p}$ with angle to optimize the interference effects. In this case, the light was slightly off circularly polarized. Angular distributions in the azimuthal plane are shown in figure 4. The predictions based on the Floquet expansion are also shown. The best match between theory and experiment occurs for a slightly more eccentric retardation in the calculation. The source of this discrepancy is not known, but may be related to coulomb effects neglected in the Keldysh approximation. Nonetheless, the general agreement is quite striking.

4. *ATOMIC STRUCTURE IN INTENSE FIELDS*

The previous analysis ignores atomic structure almost entirely. The atom confines the electron prior to the experiment, and then conveniently disappears the moment the laser turns on. Furthermore, the nonresonant approximation ignores all structure other than the ground state and the ionization potential. This leads to a simple intuitive picture of ionization into wiggling electron states. But in fact, the high-lying bound states may have an important influence on the ionization rate. This is because the loosely bound highly excited states must have a ponderomotive shift comparable to free electrons. For shifts as large as $\hbar\omega$, these states must come into multi-photon resonance with the laser field at some intensity. If parity and angular momentum rules permit, an enhancement in the ionization rate may occur.

Unfortunately, the final state ponderomotive accelerations in the gradients of the laser focus tend to cancel the effect of the level shifts on the energy of the electrons. The resonances therefore do not produce multiple energy peaks, although they may still alter the overall ATI rates and angular distributions. However, under certain circumstances, the final state ponderomotive effects may by effectively turned off. If the laser pulse duration is very short, the ponderomotive force does not act for a long enough time to alter the electron's kinetic energy or momentum. For pulses of about one picosecond or less, the electrons just "kick out" of the ponderomotive potential, giving up their ponderomotive potential energy, but retaining the initial velocity imparted during ionization. Since the ionization potential of the atom increases linearly with the intensity, the electron's energy is a record of the intensity when it was ionized. In this way, intermediate resonances might be seen as peaks in the electron energy distribution.

In a recent experiment, R. Freeman and co-workers have probably observed this phenomenon.[11] A dye laser (wavelength 616 nm) with a variable pulse duration of

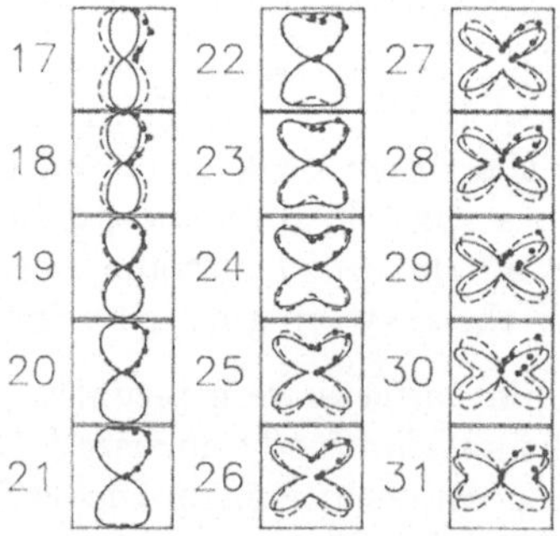

Figure 4: Krypton photoelectron azimuthal angular distributions for ATI peaks corresponding to absorption of 17-31 photons, for nearly circularly polarized light (retardation $\xi = 80°$). Solid lines are the theoretical calculations with the $\xi = 70°$, and $U_P = 2$ eV. (From reference 10.)

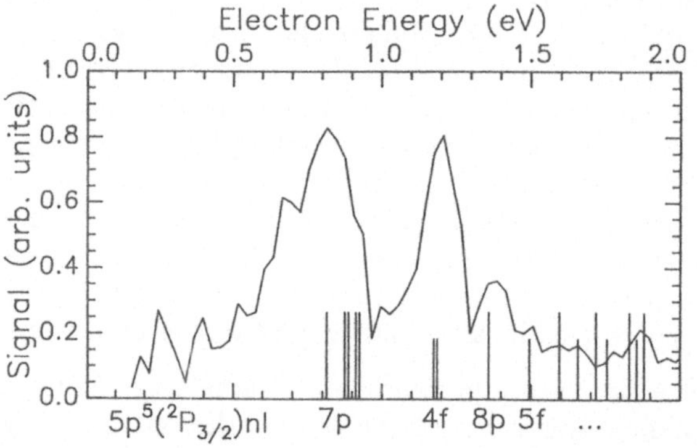

Figure 5: Photoelectron spectrum in xenon for 7 photon ionization by sub-picosecond 616 nm pulses. The peaks line up well with the expected positions of six-photon resonant enhancements, provided that the resonant intermediate states shift by the full ponderomotive potential of a free electron in the laser field. (From reference 11.)

0.5-13 picoseconds was used to ionize xenon. For longer pulsewidths, typical ATI peaks were observed at the appropriate energies. However, when the pulsewidth was reduced below 1 picosecond, the peaks shifted and split into a number of sharp features. Figure 5 shows the short pulse ATI spectrum superimposed on the even parity single-electron excitation spectrum for xenon. Most of the peaks coincide with allowed six-photon resonances, provided the resonant intermediate states have shifted by the same amount as the ionization limit.

5. CONCLUSIONS

ATI experiments have greatly increased our understanding of the interaction of intense fields with electrons. Ponderomotive forces and final state effects are now fairly well understood. Much has also been learned about the quantum nature of the wiggling electron continuum. The newest results suggest that atomic structure may be strongly altered by intense fields. Ultra-short pulsed lasers will be extremely valuable tools in exploring these questions in the future.

It is a pleasure to acknowledge useful discussions with M. Bashkansky and R. R. Freeman.

REFERENCES

1. See, for example, the feature issue on Multielectron Excitations in Atoms, Jour. Opt. Soc. B *4*, ed. by W. E. Cooke and T. J. McIlrath, 701-862 (1987).
2. P. Agostini, F. Fabre, G. Mainfray, G. Petite, and N. Rahman, Phys. Rev. Lett. *42*, 1127 (1979); P. Kruit, J. Kimman, and M. van der Wiel, J. Phys. B *14*, L597 (1981).
3. T. J. McIlrath, P. H. Bucksbaum, R. R. Freeman, and M. Bashkansky, Phys. Rev A *35*, 4611 (1987).
4. T. W. B. Kibble, Phys. Rev. *150*, 1060 (1966); J. H. Eberly, *Progress in Optics VII*, ed. by E. Wolf, Amsterdam: North-Holland, 361 (1969).
5. P. H. Bucksbaum, M. Bashkansky, and T. J. McIlrath, Phys. Rev. Lett. *58*, 2590 (1987).
6. L. Pan, L. Armstrong, Jr., and J. H. Eberly, Jour. Opt. Soc. B *3*, 1319 (1986); M. Mittleman, *Introduction to the Theory of Laser-Atom Interactions*, New York: Plenum Press (1982).
7. R. R. Freeman, T. J. McIlrath, P. H. Bucksbaum, and M. Bashkansky, Phys. Rev. Lett. *57*, 3156 (1986).
8. L. V. Keldysh, Z. Eksp. Teor. Fiz. *47*, 1945 (1964) [Sov. Phys. JETP *20*, 1307 (1964); H. R. Reiss, Phys. Rev A *22*, 1786 (1980); F. H. M. Faisal, J. Phys. B *6*, L89 (1973).
9. H. R. Reiss, J. Phys. B *20*, L79 (1987).
10. M. Bashkansky, P. H. Bucksbaum, and D. W. Schumacher, Phys. Rev. Lett. *59*, 274 (1987).
11. R. R. Freeman, P. H. Bucksbaum, H. Milchberg, S. Darack, D. Schumacher, and M. E. Geusic, Phys. Rev. Lett. *59*, 1092 (1987).

ULTRASHORT PULSE MULTIPHOTON IONIZATION OF XENON AT MODERATE PRESSURES

P.B. Corkum and
C. Rolland

National Research Council
of Canada
Division of Physics
Ottawa, Ontario, Canada
K1A 0R6

ABSTRACT

Multiphoton ionization is investigated in xenon with 620 nm pulses. Near saturated ionization is observed at 2 x 10^{13} W/cm^2 using 0.9 psec pulses and 6 x 10^{13} W/cm^2 using 90 fsec pulses.

INTRODUCTION

Recent experiments have demonstrated that supercontinua are produced by the interaction of high power, ultrashort pulses with gases.[1,2] Similar continua have been observed in condensed media[3] and the origin has been attributed to the third-order nonlinear processes of self-phase modulation and four-wave mixing.[4,5] However, these theories cannot account, even qualitatively, for the experimental observations in gaseous continua.[1,6] Since continua can be produced in such a large variety of materials, understanding their generation is a fundamental problem of nonlinear optics.

In gaseous media, beam self-focusing was shown to be crucial to the onset of continua.[1] Plasma production is an important mechanism that can limit the diameter of the self-focused beam.[7] In addition, plasma production will contribute to the spectral broadening by blue shifting the frequency of the incoming radiation. The intensity required to ionize the gas under our experimental conditions represents a major uncertainty. Indeed, only recently has any work been done on multiphoton ionization (MPI) with femtosecond 0.6 μm pulses.[8,9] In this paper, we will show that the MPI threshold (as defined by nonlinear optics signatures and approximately corresponding to the saturation intensity in conventional MPI experiments) is much higher than that expected by extrapolating MPI results at 0.53 μm and 1.06 μm.[10] Saturation intensities in Xe of 2 x 10^{13} W/cm^2 and 6 x 10^{13} W/cm^2 are measured at 0.62 μm for input pulse durations of 0.9 psec and 90 fsec respectively. These results are the first direct measurements of saturated intensities in Xe using femtosecond pulses and are consistent with recent values deduced from the electron energy spectra obtained in above-threshold ionization experiments.[8,9]

Our diagnostics are based on nonlinear optics signatures of ionization and rely on the change in refractive index ($\delta\eta$) produced by the ionization. At moderate plasma density ($N_e << N_{cr}$), $\delta\eta$ is given by

$$\delta\eta = -\frac{1}{2}\frac{N_e}{N_{cr}} \qquad (1)$$

where N_e is the change in electron density from that present before the pulse and N_{cr} is the critical density

$$N_{cr} = \varepsilon_0 \frac{m\,\omega^2}{e^2} \qquad (2)$$

In eq. (2) ε_0 is the dielectric constant, m the electron mass, e the electronic charge and ω the angular frequency of the input pulse.

In addition to the plasma contribution, the refractive index also varies as a function of the intensity due to the non-resonant electronic contribution $\delta\eta = \eta_2 E^2$ + (high order terms). In xenon, $\eta_2 = (N_0 \times 9.1 \times 10^{-51}) m^2/v^2$ where N_0 is the atomic density.[11] At low enough atomic densities $\eta_2 E^2$ can be made negligible.

In this work, we have used three nonlinear optics signatures of MPI.

(1) Self-defocusing of the beam as a result of plasma production: since the central portion of the input beam sees a lower refractive index than on the edge, the beam will be defocused.
(2) Blue frequency shifting of the transmitted spectrum: the increase in the plasma density due to MPI causes the refractive index to fall (irreversibly on the picosecond/femtosecond time scale) leading to a blue chirp on the transmitted pulse.
(3) Energy extraction from the ionizing beam.

These diagnostics, although unusual in the context of MPI in gases, are used in other areas of physics. Self-defocusing has been used to measure the ionization rate (free carriers production) induced in semiconductors.[12,13] The blue shift caused by plasma production is directly related to the rate of ionization in both condensed and gaseous media.[7,14] Finally, the change in the beam transmission as a function of intensity is an important diagnostic in absorption spectroscopy.

EXPERIMENT

The experiment was performed with two different pulse durations generated from independent dye laser systems. The 90 fsec pulses were generated in a colliding pulse mode-locked dye laser and amplified in an XeCl pumped dye amplifier. The system produces near diffraction- and transform-limited pulses.[15] Figure 1 shows a typical spectrum and autocorrelation trace. The solid line is a fit to the experimental data using a $sech^2$ line shape and a pulse duration of 92 fsec (full width at half-maximum FWHM).

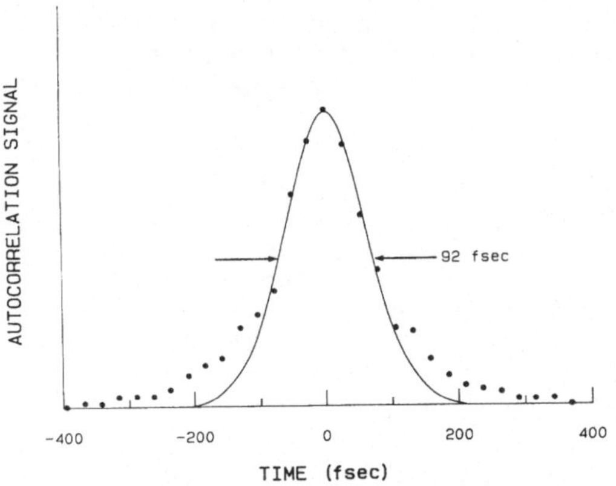

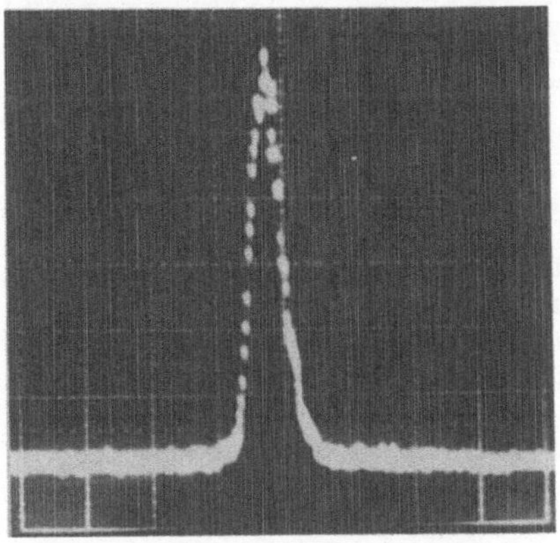

Fig. 1. A typical autocorrelation trace (left) and multishot spectrum (right) of the amplified 90 fsec pulse. The solid line is a fit to the experimental data using a $sech^2$ pulse shape and a FWHM pulse duration of 92 fsec. Note $\Delta t\ \Delta\nu = 0.53$, right trace 154Å/div.

A second dye system uses a synchronously pumped dye laser to produce pulses with an ~1 psec pulse duration. Since these pulses are not transform-limited, they are spectrally filtered using an SF-6 Brewster's angle prism and a diffraction-limited aperture. Single pulses from the train were amplified in a Nd:Yag pumped dye amplifier of similar design to that in Ref. 15. A typical autocorrelation trace and a multishot spectrum of the amplified picosecond pulses are shown

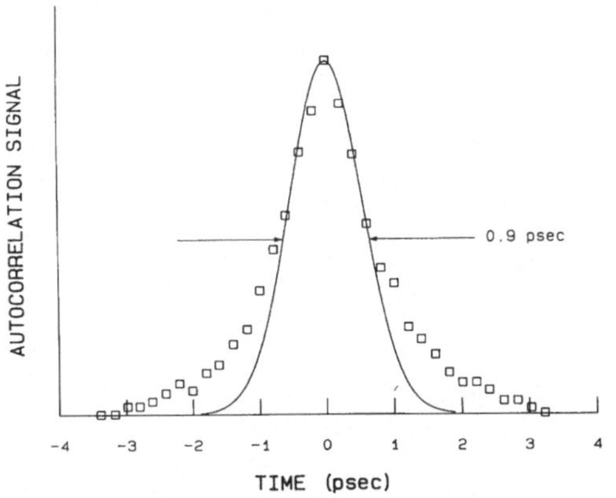

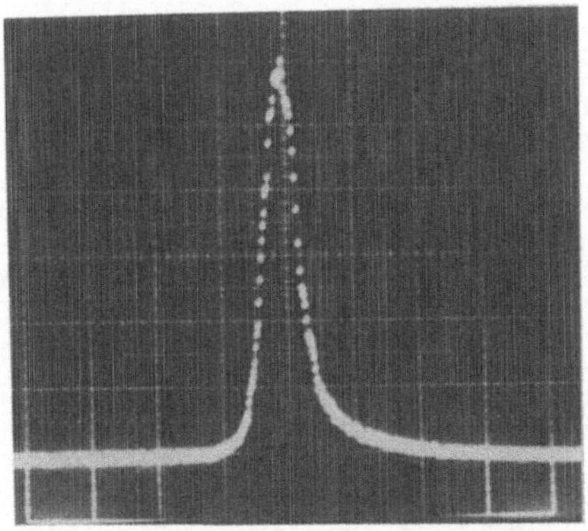

Fig. 2. A typical autocorrelation trace (left) and multishot spectrum (right) of the amplified picosecond pulse. The solid line is a fit to the experimental data assuming a Gaussian pulse shape and a FWHM pulse duration of 0.9 psec. Right trace 12.5Å/div.

in Fig. 2. The amplified pulse duration is 0.9 psec assuming a Gaussian pulse shape. The measured bandwidth of 7.6Å yields a $\Delta t\ \Delta\nu$ product of 0.5 compared with a value of 0.44 expected for a Gaussian pulse.

Both beams were spatially filtered with a vacuum spatial filter using an undersized pinhole. The undersized pinhole ensured that the beam divergence after the spatial filter is dominated by the aperture. Thus the beam divergence of both the picosecond and femtosecond pulses are identical. The spatially filtered radiation was focused to a diffraction-limited spot of $\omega_o = 21\ \mu m$ (radius at e^{-2} intensity level) as measured by scanning a 10 μm diameter aperture in both the vertical and horizontal directions.

The gas cell containing xenon is 38 cm long and has 5 mm thick quartz windows. Quartz has a relatively low η_2 (1.1×10^{-13} esu) and for an intensity of $I < 10^{10}\ W/cm^2$, the spectrum and the beam divergence are unaffected by their presence. The absence of nonlinear phase distortion was verified by monitoring the spectrum of the pulse transmitted through the evacuated cell as a function of the input power. The experiment was performed with a gas pressure of ~50 Torr, well below any significant η_2 contribution to the refractive index at intensities that we could achieve experimentally.

To observe ionization we gradually increased the incident intensity while monitoring changes to the spatial distribution of the beam as measured at the output of the cell, the frequency spectrum of the pulse and the energy transmitted through the cell.

Results and Discussion

For the purpose of this paper we will define the ionization threshold as the intensity at which we start to observe beam defocusing. For plasma to modify the beam by one diffraction limited beam divergence, a $\sim\pi/2$ phase change must be impressed on the wave front.

The phase advance $\delta\phi$ due to free carriers is given by

$$\delta\phi = \frac{2\pi \int \delta\eta\ dz}{\lambda} = \frac{\pi \int N_e\ dz}{\lambda\ N_{cr}} \qquad (3)$$

where λ is the wavelength of the radiation propagating along the z axis. For $\lambda = 0.6\ \mu m$ and $\delta\phi = \pi/2$, $\int N_e\ dz = 10^{17}\ cm^{-2}$.

Given the above experimental parameters, we can estimate the degree of ionization necessary for observable beam defocusing. Near the beam waist, the intensity of a Gaussian beam remains constant within a factor of 2 over a distance of $2\ z_o = 2\pi\omega_o{}^2/\lambda = 5$ mm. Assuming that the ionization is restricted to this interaction length, a plasma density of $N_e \sim 2 \times 10^{17}\ cm^{-3}$ is required for a $\pi/2$ phase change. Our detection threshold due to beam deflection corresponds to ~10% ionization at a gas pressure of 50 Torr (the atomic density is $\sim 2 \times 10^{18}\ cm^{-3}$).

Beam deflection due to plasma production is illustrated in a series of optical multichannel analyzer (OMA) traces taken near the output of the cell. Similar traces were obtained for both the picosecond and femtosecond pulses. Only the picosecond results are shown in Fig. 3. An ~0.5 mm thick slit was used to select a slice of the beam passing through the beam axis. Figure 3 (a) was obtained with an evacuated cell and shows the unperturbed beam. A similar beam distribution is observed for low intensity pulses when the cell is filled with 50 Torr of Xe. Figure 3 (b) shows a typical beam profile obtained with the cell filled with 50 Torr of Xe and with the pulse energy approximately twice the ionization intensity. Finally, Fig. 3 (c) shows the differential beam profile (Fig. 3 (b) - Fig. 3 (a)) which demonstrates that radiation is moved from the center of the beam to the edges.

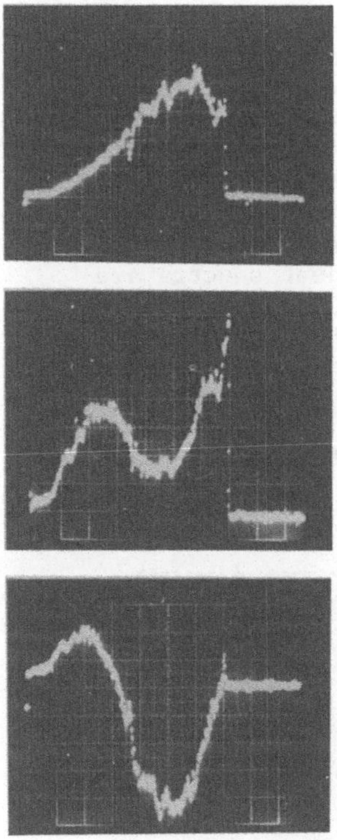

Fig. 3. Oscilloscope traces illustrating plasma beam deflection: (top), the undeflected beam profile; (middle), the beam profile deflected by plasma produced at a pulse intensity in excess of the saturation intensity; (bottom), the differential beam profile (b minus a). Note, radiation has been arbitrarily truncated on the right side of the traces for base line calibration. All distributions are, in reality, symmetric.

We have defined the threshold intensity as the intensity at which beam defocusing just becomes observable. For the 90 fsec pulse, the ionization threshold is 6 x 10^{13} W/cm^2. For the 0.9 psec pulse, a threshold intensity of 2 x 10^{13} W/cm^2 is obtained. In the case of the picosecond pulse, the center frequency was changed from 610 nm, 616 nm and 618 nm with no affect on the threshold intensity.

At such high intensities, the energy required to produce sufficient plasma to deflect the beam accounts for a small fraction of the beam energy, even for the 90 fsec pulse. Given a plasma density of $\int N_e dz = 10^{17} cm^{-2}$ and an absorbed energy of 14 eV per ionized atom, the absorbed fluence is ~0.2 J/cm^2 compared with a threshold fluence of 5.4 J/cm^2 with 90 fsec pulses. Experimentally we could not observe changes in the energy transmission through the cell with or without xeñon. If the ionization threshold were 10^{13} W/cm^2, substantial beam absorption would have occurred. This simple observation confirms the high ionization threshold for femtosecond pulses.

The frequency spectrum of the light transmitted through the plasma generation region is a clear signature of ionization. The frequency shift is given by the differential of Eq. (3):

$$\delta\omega = \frac{\pi \int (dN_e/dt) dz}{\lambda N_{cr}} \tag{4}$$

which can be simplified for a seven photon process (the minimum number of photons required to ionize Xenon) to

$$\delta\omega = \frac{N_o \pi \sigma \int I^7 dz}{\lambda N_{cr}} \tag{5}$$

where σ is the generalized cross-section, I is the intensity and $dN_e/dt = N_o\sigma I^7$. Since dN_e/dt is a monotonically increasing function of time (recombination is insignificant in our time scale), the spectrum must be blue shifted. Figure 4 shows differential spectra (that is, the spectrum with Xenon minus the spectrum obtained with an evacuated cell) obtained on an OMA for a laser intensity in excess of the saturation intensity. Figure 4(a) was obtained with the 90 fsec pulse and Fig. 4(b) with the 0.9 psec pulse. For comparison the incident spectra are shown in Figs. 1 and 2. As expected, the radiation is shifted from the red to the blue side of the spectrum. It should be noted that this technique for measuring ionization thresholds is not as sensitive as beam defocusing, since only that fraction of the radiation present during the rise in the electron density gets shifted. The proportion of the radiation that is shifted is too small to be detected at intensities below the saturation intensity.

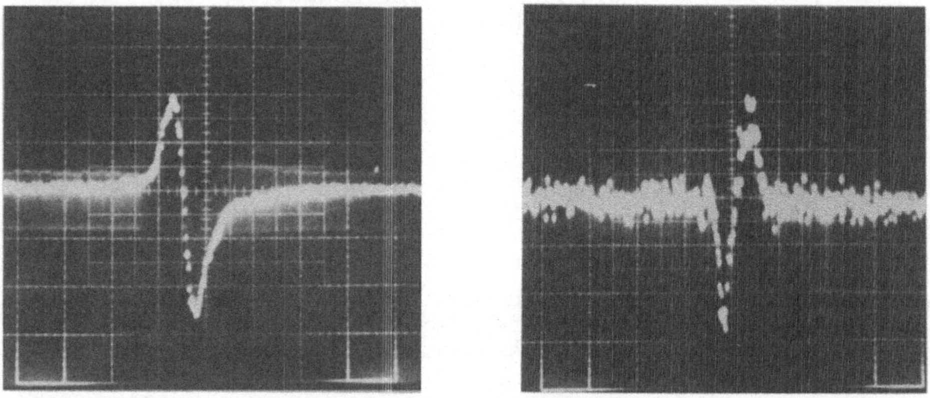

Fig. 4. OMA traces obtained by subtracting the spectrum of the radiation transmitted through the evacuated cell from the spectrum when the cell is filled with 50 Torr of Xe: left trace 0.9 psec pulse, 12.5Å/div. (increasing frequency to left) right trace - 90 fsec pulse, 154Å/div. (increasing frequency to right).

The high ionization thresholds of Xe are surprising in the context of previous 1.06 μm and 0.53 μm MPI experiments[10] but are consistent with the conclusions in Refs. 1 and 16. Inhibited ionization has been observed in resonant media[17] where it results from interference effects between different optical harmonics.[18] We consider this process unlikely in our case since there are no near resonant states in xenon for 1-5 photons and we have observed a similarly enhanced ionization threshold in krypton (the only other gas investigated).

Our motivation for performing this experiment was to find the intensity above which plasma contributions to the refractive index exceed those due to electronic nonlinearities. Assuming $dN_e/dt = N_o\sigma I^7$ and using the 10% ionization threshold measured in this experiment, plasma production becomes significant ($\pi/2$ phase change) at intensities of only 3×10^{13} W/cm^2 in 10 atm Xenon (10 atmospheres is a typical pressure at which continua are observed). Experimentally the difference in ionization threshold for picosecond and femtosecond pulses exceeds that expected for $dN_e/dt = N_o\sigma I^7$. A factor of only 1.4 is due to the difference in pulse durations while we measure a factor of 3. We believe that a nonperturbative dependence of the ionization rate is indicated for Xe and thus even intensities lower than 3×10^{13} W/cm^2 may produce substantial phase change in high pressure gases.

In conclusion, we report the first direct measurement of the ionization threshold in Xe with picosecond and femtosecond visible pulses. The high ionization thresholds and the pulse duration scaling of MPI imply that, the lowest order term in the perturbation expansion is insufficient for describing the interaction of ultrashort pulses with xenon.

References

1. P.B. Corkum, C. Rolland and T. Srinivasan-Rao, Phys. Rev. Lett. 57, 2268, (1986).
2. J.H. Glownia, J. Misewich and P.P. Sorokin, J. Opt. Soc. B 3, 1573, (1986).
3. R.R. Alfano and S.L. Shapiro, Phys. Rev. Lett. 24, 592 (1970).
4. G. Yang and Y.R. Shen, Opt. Lett. 9, 510 (1984).
5. J.T. Manassah, M.A. Mustafa, R.R. Alfano and P.P. Ho, IEEE J. Quantum Electron. QE-22, 197 (1986).
6. P.B. Corkum, C. Rolland and T. Srinivasan-Rao, in Ultrafast Phenomena V, G.R. Fleming and A.E. Siegman editors, Springer Series in Chemical Physics 46, Springer-Verlag, New York, p. 149, 1986.
7. W.Lee Smith, P. Liu and N. Bloembergen, Phys. Rev. A15, 2396 (1977).
8. P. Agostini et al. this issue.
9. R.R. Freeman, P.H. Bucksbaum, H. Milchberg, S. Darack, D. Schumacher, and M.E. Geusic, Phys. Rev. Lett. 59, 1092 (1987).
10. A. l'Huillier, L.A. Lompre, G. Mainfray and C. Manus, Phys. Rev. A27, 2503 (1983).
11. Note: in Ref. 1 there is an error in the atmospheric value of η_2 for Xenon. It should be $\eta_2 = 2.2 \times 10^{-16}$ esu.
12. S. Guha, E.W. Van Stryland and M.J. Soileau, Opt. Lett. 10, 285 (1985).
13. T.F. Boggess Jr., A.L. Smirl, S.C. Moss, I.W. Boyd, and E.W. Van Stryland, IEEE J. Quantum Electron. QE-21, 488 (1985).
14. P.B. Corkum, IEEE J. Quantum Electron. QE-21, 216 (1985).
15. C. Rolland and P.B. Corkum, Optics Comm. 59, 64 (1986).
16. P.B. Corkum and C. Rolland, XV International Quantum Electronics Conference, Postdeadline Paper PD21 (1987).
17. J.C. Miller, R.N. Compton, M.G. Payne and W.W. Garrett, Phys. Rev. Lett. 45, 114 (1980).
18. D. Jackson and J.J. Wynne, Phys. Rev. Lett. 49, 543 (1982).

DISCUSSION OF LIMITING DEPOSITION POWER DENSITIES

IN LASER TARGET MATERIALS

A. McPherson, K. Boyer, H. Jara, T. S. Luk
I. A. McIntyre, R. Rosman, J. C. Solem*, and C. K. Rhodes

Laboratory for Atomic, Molecular, and Radiation Physics
Department of Physics, University of Illinois at Chicago
P. O. Box 4348, Chicago, Illinois 60680

ABSTRACT

The availability of terawatt KrF^* laser systems is expanding capabilities to explore new atomic processes at high intensity in the ultraviolet. By utilizing the nonlinear absorption of laser energy into a target material and the relativistic motion of electrons in the high charge state plasma produced by intensities exceeding 10^{19} W/cm^2, it should be possible to exceed the deposition power density necessary to achieve amplification in the x-ray region.

INTRODUCTION

The availability of small ultraviolet laser systems capable of producing peak powers in excess of a terawatt is expanding the horizon for studying atomic processes at high intensity. With the advent of KrF^* (248 nm) excimer systems producing joule level subpicosecond pulses of low divergence, it will be possible to subject atomic systems to intensities in excess of 10^{19} W/cm^2.[1-3] Since the principal difficulty in the production of coherent short wavelength radiation is the controlled deposition of energy into a target material at high specific power,[4] these lasers are natural candidates for pumping x-ray laser systems.[5]

The atomic processes under investigation lead to the production of ions, electrons, and photons by the general multiphoton reaction

$$N\hbar\omega + A \rightarrow A^{q+} + qe^- + \gamma. \qquad (1)$$

For the present discussion the detailed path leading to the production of the photons (γ), whether it be by harmonic production or fluorescence, will not be considered. It is sufficient to state that at least four processes in the target material are possible: (a) harmonic generation in the neutral gas or plasma; (b) direct atomic and ionic excitation; (c) electron collisions leading to excited high charge states; and (d) recombination in the plasma. That high charge states, which can yield short wavelength radiation, are produced is illustrated[6] in Fig. (1). The Xe target was produced by a pulsed valve delivering ≃ 10 – 100 Torr pressure over a region characterized by a few hundred micrometers. In this spectrum of Xe,

---

* on leave from the University of California, Los Alamos National Laboratory

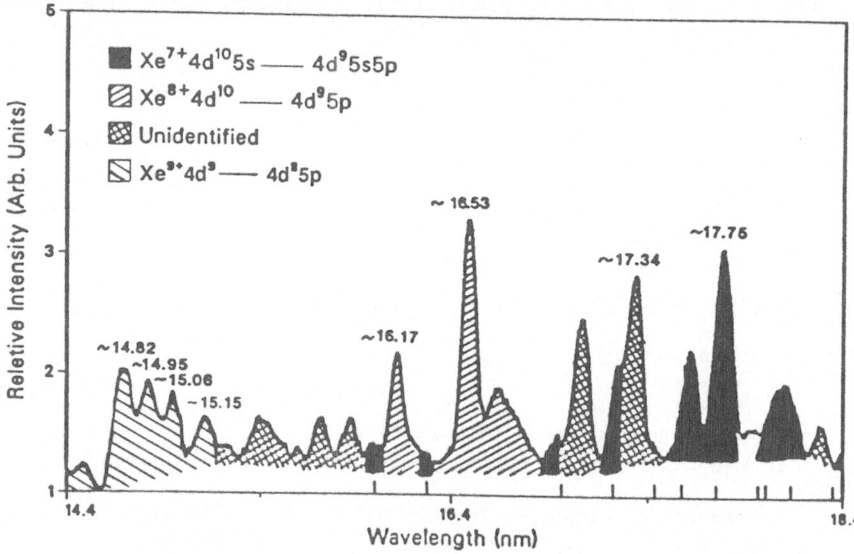

Fig. (1). Xe spectrum obtained with 450 fs pulses of 248-nm radiation from a KrF* laser system. The focused intensity into the target was approximately 3 x 10^{15} W/cm². Three charge states are identified, each of which yielded significant amounts of short wavelength radiation.

produced by focusing 450 fs pulses of 248-nm radiation into the Xe target, three charge states are identified, each of which yielded significant amounts of radiation. However, the specific mechanism leading to the fluorescence is not yet firmly established.[6]

Each of the four processes listed above leads to the same general question, regardless of details, namely, how much energy can be deposited into an atomic system to ensure the generation of radiation? This is not an easy question to answer because, even at laser intensities of ~ 10^{15} W/cm², the details of multiphoton absorption are not well understood. However, it is possible to determine an achievable upper bound to the energy deposition rate by making reasonable approximations of the physical quantities based on existing experimental evidence.

In the absence of collisionally induced effects (collisional processes can only increase energy deposition), the deposition power density Φ (W/cm³) can be written in the form

$$\Phi = I \sigma \rho, \tag{2}$$

where I is the pump laser intensity, σ is the energy transfer cross section and ρ is the target density. The maximization of each of these quantities will in turn be considered in the following discussion. It has been estimated[4] that the production of significant x-ray amplification at a quantum energy of ~ 1 keV requires a state selective deposition power density of at least 10^{14} W/cm³.

ENERGY TRANSFER CROSS SECTION

For laser intensities exceeding 10^{16} W/cm^2, atomic systems are expected to respond in a manner different from that corresponding to exposure at much lower field intensities. For 248-nm irradiation at intensities above a few times 10^{19} W/cm^2 the electronic motions not only will become relativistic, but ordered many-electron motions of the outer-shell electrons may become increasingly significant. It has been suggested that such an ordered motion could lead to enhanced rates of energy deposition.[7-10]

An estimate of the maximum attainable energy transfer cross section can be made by considering: (a) threshold measurements for low stages of ionization in the inert gases for low laser intensity ($\simeq 10^{14}$ W/cm^2); (b) the experimentally determined average cross section for total energy transfer at an intermediate intensity ($\simeq 10^{16}$ W/cm^2); and (c) an extrapolation into the high intensity regime. Figure (2) presents the information stemming from these three considerations.[11]

From collision-free studies of threshold ion production,[12] low laser intensity measurements yield an energy transfer cross section of $\simeq 3 \times 10^{-22}$ cm^2 for krypton and xenon at a laser intensity of $\simeq 3 \times 10^{13}$ W/cm^2. At the higher intensity of $\simeq 7 \times 10^{15}$ W/cm^2 an average cross section for energy transfer can be defined as

$$\langle \sigma_{N\gamma} \rangle \simeq \frac{\langle \varepsilon \rangle}{I\tau} \quad , \tag{3}$$

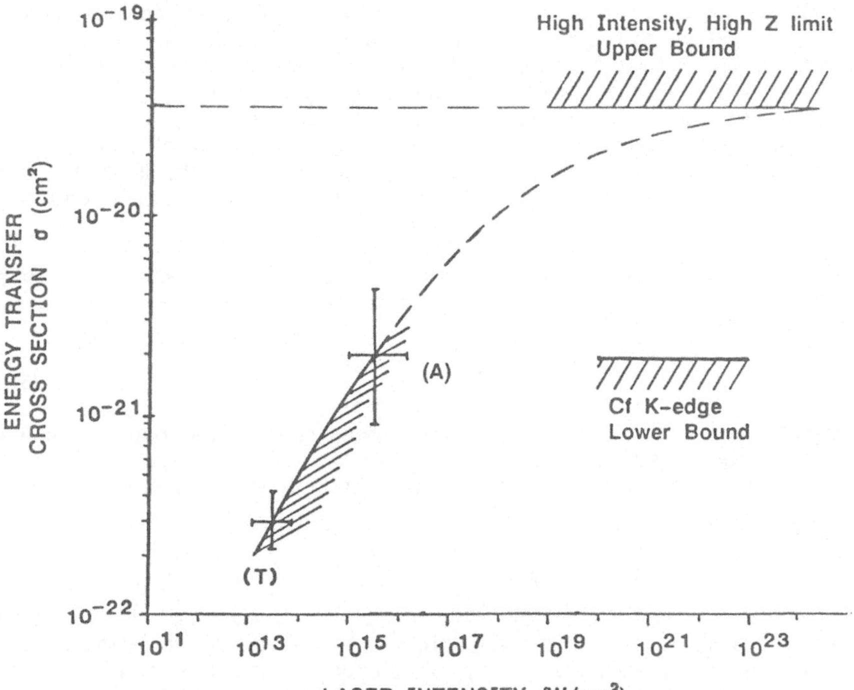

Fig. (2). Energy deposition cross section versus subpicosecond 248-nm laser intensity for collision-free conditions. Typical threshold (T) values for Kr and Xe are shown as well as the average (A) over all observed charge states for Xe. Extrapolation into the high field region is shown by the dashed line.

where $<\varepsilon>$ is the average energy determined from known atomic ionization potentials and the measured abundances of the ionic species formed.[13] I and τ denote the laser intensity and pulse width, respectively. At this higher intensity, which allowed the production of up to Xe^{8+}, the average energy transfer cross section for Xe was found to be $\simeq 2 \times 10^{-21}$ cm^2. Extrapolation of earlier estimates[9] on the high field behavior leads to two results.[11] The first is that an approximate lower bound for the energy transfer cross section can be obtained by using the known atomic total photoionization cross sections[14] for the K edge of Cf which is approximately 2×10^{-21} cm^2. The second is the establishment of a corresponding upper bound. The upper bound can be established by assuming that all of the atomic electrons move collectively, the matrix elements for the cross section tend to a_0/Z, and the quiver velocity of the electron approaches the speed of light.[11] The maximum ability of the atomic system to absorb energy is then given by

$$\sigma_m \simeq 8\pi\alpha^2 a_0^2 = 8\pi \lambda\!\!\!^{-}{}_c^2 \simeq 3.6 \times 10^{-20} \text{ cm}^2 \tag{4}$$

where $\lambda\!\!\!^{-}{}_c$ is the Compton wavelength of the electron, α is the fine structure constant, and a_0 is the Bohr radius. An interesting result of these estimates is the observation, based on Fig. (2), that the variation of the cross section with intensity is quite weak.

LASER INTENSITY

For laser systems delivering target intensities significantly greater than 10^{14} W/cm^2, nonlinear dispersive properties of the medium will begin to dominate the laser beam propagation through the target material at sufficiently high density. The polarizability and refractive index of the medium is changing as the material is progressively ionized in the laser focus and these dynamic changes will influence the achievable deposition power density. Of particular interest is the dynamic self-focusing of the propagating laser beam since: (a) this is a physical response which automatically leads to the concentration of energy; and (b) it is a process which may be unavoidable at our considered domain of power density and target density.

Self-focusing leading to filament formation in a plasma produces a filament with a diameter of about a wavelength.[15] Therefore, the maximum laser intensity in such a filament is given by

$$I \simeq P/\lambda^2 \tag{5}$$

where P is the pump laser power. If such a focusing condition can be obtained, the increased intensity due to self-focusing can be expected to enhance the tendency to produce high charge states and, hence, the production of short wavelength radiation.

TARGET DENSITY

For a sufficiently high target density, the plasma frequency will exceed the pump laser frequency and propagation through the medium will cease. It is desirable to have the plasma frequency, which determines the maximum target density, as close as possible to the laser frequency. At sufficiently high laser power the laser field will drive the electrons in the plasma at relativistic velocities. Thus the maximum target density considered, which will allow beam propagation, is given by[16]

$$\rho = \frac{m\omega_p^2}{4\pi\zeta e^2}\left(1 + \frac{2Pe^2}{\pi m^2 c^5}\right)^{1/2} \tag{6}$$

where e and m are the electron's charge and mass, c is the speed of light, ω_p is the plasma frequency, and ζ is the number of free electrons per atom.

DEPOSITION POWER DENSITY

A combination of Eqs. (2), (5), and (6) enables formulation of the deposition power density. For these conditions, the quantity Φ is given by

$$\Phi = \frac{\pi P \sigma m c^2}{\lambda^4 \zeta e^2} \left(1 + \frac{2Pe^2}{\pi m^2 c^5} \right)^{1/2} . \tag{7}$$

From the wavelength dependence it is evident that a KrF^* laser system could provide more than 300 times the deposition power density of an equivalently powerful neodymium-glass system.[17]

Figure (3) illustrates deposition power density curves, determined from Eq. (2) and using $\sigma_m = 10^{-20}$ cm^2 as an estimate of the high field limit cross section, as a function of both peak laser power and target density. As an example, the region of no beam propagation was determined for Xe taking into account the relativistic motion of the electrons at high laser power and the incomplete ionization at low

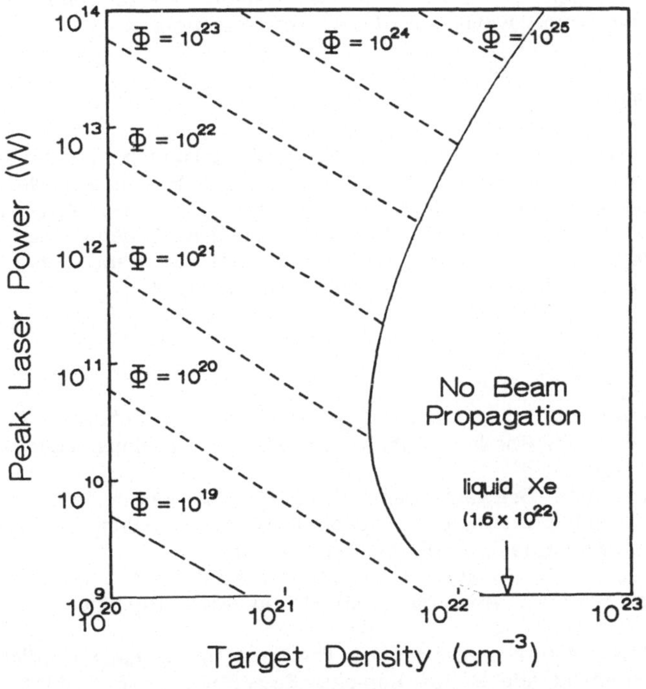

Fig. (3): Deposition power density Φ (W/cm^3) curves as a function of peak laser power and target density. The region of no beam propagation was determined for the plasma frequency cutoff in Xe.

laser power. According to Fig. (3), a deposition power density exceeding 10^{22} W/cm^3 should be possible in less than liquid density Xe. Not only would this correspond to an energy absorption rate in excess of 1 Watt per atom, but would greatly exceed the deposition rate necessary for the generation of significant amplification in the keV region.

GAIN MEDIUM

In order to produce an x-ray laser, it is not sufficient to achieve either $\Phi \geqslant 10^{14}$ W/cm^3 or high charge states. Rather the target material needs to have a linear dimension considerably in excess of a few hundred micrometers. This can be achieved by using three target designs. Solid metallic targets have been the most studied to date. Such targets have shown modest gains with the physical pumping process being recombination in the dense plasma. A variation on this target design is the use of solid inert gases in which a laser is used to evaporate the solid from a cryostat surface. The pump laser can then be focused into the resulting high density gas. Such a target design appears to be easy to produce and yields a reproducible surface for firing at a rate of 1 Hz. The most promising target design we have investigated is the slotted nozzle pulsed valve. A high Mach number design should be able to deliver target densities approaching 10^{20} cm^{-3} with linear dimensions up to 1 cm in length.

CONCLUSION

A multiterawatt 248-nm laser system which drives a dynamically self-focused filament through a target material approaching 1 cm in length should produce a deposition power density exceeding that necessary to produce significant short wavelength gain. Both nonlinear absorption of laser energy into the target material and the relativistic motion of electrons in the ensuing plasma combine to produce conditions which favor short wavelength lasers for excitation.

ACKNOWLEDGEMENTS

The authors wish to acknowledge the technical assistance of Rick Bernico, Rick Slagle, and Jim Wright. This work is supported by the U. S. Office of Naval Research, the U. S. Air Force Office of Scientific Research, the Innovative Science and Technology Office of the Strategic Defense Initiative Organization, the U. S. Department of Energy by the Lawrence Livermore National Laboratory, the National Science Foundation, and the Los Alamos National Laboratory.

REFERENCES

1. A. P. Schwarzenbach, T. S. Luk, I. A. McIntyre, U. Johann, A. McPherson, K. Boyer, and C. K. Rhodes, Subpicosecond KrF* excimer-laser source, Opt. Lett. 11:499 (1986).
2. I. A. McIntyre, A. P. Schwarzenbach, T. S. Luk, A. McPherson, K. Boyer, and C. K. Rhodes, High power subpicosecond KrF* laser system, in "SPIE 664, High Intensity Laser Processes", SPIE, Bellingham (1986).
3. H. Jara, K. Boyer, U. Johann, T. S. Luk, I. A. McIntyre, A. McPherson, and C. K. Rhodes, Dynamic Absorption Effects in KrF* Amplifiers, Appl. Phys. B 42:11 (1987).
4. A. V. Vinogradov and I. I. Sobel'man, The Problem of Laser Radiation Sources in the Far Ultraviolet and X-Ray Regions, Sov. Phys.-JETP 36:1115 (1973).
5. C. K. Rhodes, Multiphoton Ionization of Atoms, Science 229:1345 (1985).
6. A. McPherson, G. Gibson, H. Jara, U. Johann, T. S. Luk, I. A. McIntyre, K. Boyer, and C. K. Rhodes, Studies of Multiphoton Production of Vacuum Ultraviolet Ratiation in the Rare Gases, J. Opt. Soc. Am. B4:595 (1987).

7. A. Szöke, Interpretation of Electron-Spectra Obtained from Multiphoton Ionization in Strong Fields, J. Phys. B18:L427 (1985).
8. C. K. Rhodes, Studies of Collision-Free Nonlinear Processes in the Ultraviolet Range, in "Multiphoton Processes", P. Lambropoulos and S. J. Smith, editors Springer-Verlag, Berlin, (1984) p. 31.
9. K. Boyer and C. K. Rhodes, Atomic Inner-Shell Excitation Induced by Coherent Motion of Outer-Shell Electrons, Phys. Rev. Lett. 54:1490 (1985).
10. A. Szöke and C. K Rhodes, Theoretical Model of Inner Shell Excitation by Outer-Shell Electrons, Phys. Rev. Lett. 56:720 (1986).
11. K. Boyer, H. Jara, T. S. Luk, I. A. McIntyre, A. McPherson, R. Rosman, and C. K. Rhodes, Limiting Cross Sections for Multiphoton Coupling, submitted to Revue de Physique Appliquée.
12. T. S. Luk, U. Johann, H. Egger, H. Pummer, and C. K. Rhodes, Collision-Free Multiple Photon Ionization of Atoms and Molecules at 193 nm, Phys. Rev. A32:214 (1985).
13. U. Johann, T. S. Luk, I. A. McIntyre, A. McPherson, A. P. Schwarzenbach, K. Boyer, and C. K. Rhodes, Multiphoton Ionization in Intense Ultraviolet Laser Fields, in "AIP Conference Proceedings N0. 147, Optical Science and Engineering Series 7, Short Wavelength Coherent Radiation: Generation and Applications", D. Attwood and J. Boker eds., AIP, New York (1986) p. 202.
14. E. F. Plechaty, D. E. Cullen, and R. J. Howerton, Tables and Graphs of Photon Interaction Cross Sections from 0.1 keV to 100 MeV Derived from the LLL Evaluated-Nuclear-Data Library, UCRL-50400, V6, Rev. 3 Nov 1981.
15. Heinrich Hora, "Physics of Laser Driven Plasma", Wiley-Interscience, New York (1981) p.220.
16. Prediman Kaw and John Dawson, Relativistic Nonlinear Propagation of Beams in Cold Overdense Plasmas, Physics of Fluids 12:472 (1970).
17. K. Boyer, H. Jara, T. S. Luk, I. A. McIntyre, A. McPherson, R. Rosman, C. K. Rhodes, and J. C. Solem, X-Ray Amplifier Excitation in a Dynamically Self-Trapped Ultraviolet Beam, submitted to Optics Letters.

ABOVE THRESHOLD IONIZATION IN Xe WITH A CO_2 LASER

F. Yergeau+, W. Xiong+, S.L. Chin+ and P. Lavigne*

+Dép. de Physique, Université Laval, Québec, Canada G1K 7P4

*INRS-Energie, Varennes, Canada J0L 2P0

INTRODUCTION

Above Threshold Ionization has been the subject of many investigations since its discovery nearly ten years ago[1]. Soon after initial observation, an experiment showed that the first peak in the electron spectrum could disappear at high enough intensity[2]. A wealth of explanations appeared in the litterature concerning both the presence of multiple peaks and their suppression. Some of these involved the so-called ponderomotive force, that is, the force on the photoelectron due to the gradient of intensity of the laser field[3]. Many investigations of the coupling between ATI and the ponderomotive force have been undertaken up to now[4,5,6], but all experiments involve wavelengths in the visible of near-infrared regions of the spectrum. This poses a limitation on the range of ponderomotive potentials attainable, because atoms irradiated at such wavelengths are not ionized at intensities higher than about 10^{14} W/cm^2 (that is, the ionization process is saturated before the intensity reaches higher values). The ponderomotive potential however, depends inversely on the square of the laser frequency, so that using a wavelength ten times larger than that of the Nd:YAG laser, for instance, should yield a hundred-fold increase in the potential if similar intensities are used. This, in turn, sets the lower limit for the energies of the photoelectrons higher and leads to the production of high energy electrons in this ionization process. In this paper, we report observation of electrons with kinetic energies up to 350 eV in the ionization of Xe by a CO_2 laser at λ = 10.6 μm.

EXPERIMENTAL

The laser is an TEA-CO_2 oscillator-amplifier chain producing 1.1 ns pulses with tens of joules of energy at 10.6 μm. A detailed description can be found in Yergeau et al.[7]. The Xe sample is leaked into a vacuum chamber at a pressure of 10^{-6} torr, the background being 10^{-8} torr. A 250 mm focal length lens focuses the beam between the two first plates of a retarding potential electron energy analyzer (fig. 1). An accelerating field of about 35 V/cm is applied between these two plates, and a retarding potential is set on plates 3 and 4. Those electrons that can pass through this barrier are then detected by an electron multiplier tube (EMT).

The number of electrons collected is read as the amplitude of the output of the EMT, and plotted against V_r, the retarding potential corrected for the acceleration incurred between the first two plates. V_r is thus the initial energy of an electron that can barely pass through the barrier. The measurements were corrected for shot-to-shot laser energy variations by taking data for a few shots at each value of V_r and, using a fitting procedure, estimating the "effective" number of electrons at a target laser energy E_0 straddled by the actual values obtained from the laser. Our target energy was 5.2 J, which corresponds to a peak intensity of about $5x10^{13}$ W/cm^2 (see reference 7 for the intensity calibration). At this intensity, only single ionization occurs, as was verified by operating the electron analyzer in reverse, as a time-of-flight mass spectrometer. Figure 2 shows two spectra obtained in this manner, at two different times. The difference between the two is due to sample purity, as will be explained later.

DISCUSSION

The shape of the spectrum can be explained as follows. We know from earlier work[7] that the xenon atoms are all ionized at (almost) the same intensity I_s, which corresponds to the saturation intensity of the process. Thus at a given instant, ionization is occuring only in a thin shell surrounding the focal point where $I=I_s$. As it emerges from the atom, the electron feels the ponderomotive force and is rapidly ejected from the focal volume. In our conditions of wavelength, intensity, pulse duration and focal size, it is a simple matter to show that the transit time of the electron through the focal volume is much shorter than the pulse duration, so that the electron effectively "feels" a frozen field distribution[8]. Then the force is conservative and the kinetic energy acquired is equal to the value of the associated potential at the point of ionization, which depends only on the local intensity through the well known formula:

$$U = e^2E^2/4m\omega^2 \qquad (1)$$

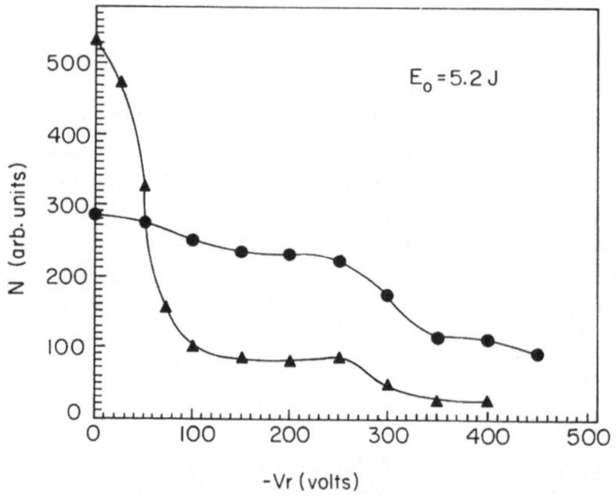

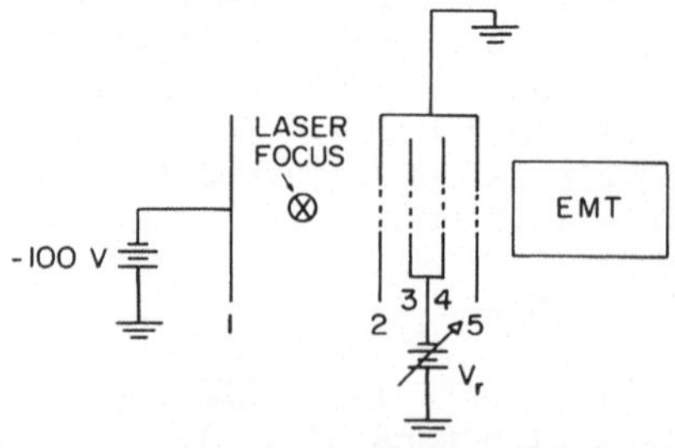

Figure 1. Schematic of the retarding potential electron spectrometer.

Figure 2. Two spectra of xenon at about $5x10^{13}$ W/cm^2. (▲): contaminated sample. (●): purer sample.

where E is the electric field amplitude associated with intensity I and the other symbols assume their usual meanings. But we know that the intensity at the point of ionization is just I_s, the saturation intensity, by virtue of the high nonlinearity of the process. Two estimates of this value, one experimental[7], and another one theoretical[9], yield the value $I_s=1.88\times10^{13}$ W/cm^2, which, introduced in eq. 1, gives 190 eV as the minimum kinetic energy of the photoelectrons.

This energy is a minimum since the electrons can have some initial energy before being accelerated by the ponderomotive force. Indeed our spectra show a broad peak starting at about 250 eV and centered around 300 eV, which implies that more than the ponderomotive potential is involved. The latter can only give a well-defined energy of about 190 eV to the electrons, since ionization occurs at the well-defined intensity I_s. All the energy over and above this threshold must be provided by some other mechanism, and is thus in the realm of ATI. To avoid empty discussions, we must make clear that in this context, we understand ATI as any mechanism occuring during ionization that provides the photoelectron with more energy than it needs to overcome the ionization threshold.

We can think of many reasons to explain the difference between the foot of our peak at 250 eV and the estimated ponderomotive potential of 190 eV. First of all, the calibration of the energy scale of the spectra is good to only about 10 eV, because of an uncertainty as to the value of the correction for acceleration that must be applied, due to inexact knowledge of the position of the focal point between the first two plates of the analyzer. Second, the intensity calibration is not perfect, and the estimate of the ponderomotive potential depends directly on it. Third, it is quite possible that peaks are suppressed in the ATI process to an energy higher than the minimum imposed by the ponderomotive potential.

The two spectra of fig. 2 differ in the purity of the gas sample used. The spectrum with a large peak at low energy was obtained when the gas was introduced through a regulator, which could not be baked out and cleaned properly. Ion spectra taken with this sample show many impurity peaks apart from the xenon ions. When the regulator was removed and the gas line baked out, much cleaner ion spectra resulted and the second electron spectrum of fig. 2 was obtained. This shows that the large, low-energy peak of the bad spectrum comes from low-intensity ionization of impurities introduced with the sample.

CONCLUSION

We succeeded in obtaining the photoelectron energy spectrum of ionization of Xe by an intense CO_2 laser. The impressively high energy of these electrons can be well explained by acceleration due to the ponderomotive force together with some ATI-like mechanism.

This work was supported in part by the Natural Science and Engineering Research Council of Canada and by Le Fonds FCAR de la Province de Québec. One of us (W.X.) acknowledges scholarships from Université Laval and LROL.

REFERENCES

1. P. Agostini, F. Fabre, G. Mainfray, G. Petite and N.K. Rahman, Phys. Rev. Lett. **42**, 1127 (1979).

2. P. Kruit, J. Kimman and M.J. van der Wiel, J. Phys. B **14**, L597 (1983).

3. T.W.B. Kibble, Phys. Rev. **150**, 1060 (1966).

4. M.J. Hollis, Opt. Commun. **25**, 395 (1978).

5. P. Kruit, J. Kimman, H.G. Muller and M.J. van der Wiel, Phys. Rev. A **28**, 248 (1983).

6. P.H. Bucksbaum, R.R. Freeman M. Bashkanski and T.J. McIlrath, J. Opt. Soc. Am. B **4**, 760 (1987).

7. F. Yergeau, S.L. Chin, and P. Lavigne, J. Phys. B **20**, 723 (1987).

8. This is the "high-intensity" case of reference 4.

9. M. Crance, J. Phys. B **17**, L355 (1984); J. Phys. B **17**, 3503 (1984); J. Phys. B **17**, 4333 (1984); J. Phys. B **17**, 4823 (1984).

MULTIPHOTON MULTIELECTRON EXCITATION AND IONIZATION OF H_2

Munir H. Nayfeh, J. Mazumder, David Humm,
Thomas Sherlock and Kwan Ng

Department of Physics
Department of Mechanical Engineering
University of Illinois at Urbana-Champaign
1110 W. Green Street
Urbana, IL 61801

Introduction

Considerable discussion and controversy has concerned the interaction of short pulse radiation with atoms.[1-5] Are they best explained as processes occurring in steps, or as collective phenomena? These arguments were triggered by recent measurements in which a 0.5 ps pulse of 192 nm laser radiation was focused on Xe producing similar numbers of both singly and multiply stripped ions.[2] Collective emission occurring on a time scale of 1 fs was originally postulated to explain this abundance of multiply ionized Xe, but later theoretical study concluded that successive stripping of single electrons was the dominant mechanism.[5]

More recently, these processes were investigated in the diatomic systems N_2 and O_2 exposed to a 0.6 ps, 600 nm, $3x10^{15}$ W/cm^2 laser field[4]. Fast (38 eV) N^{+3} ions were observed by a time-of-flight mass spectrometer. These were interpreted as resulting from the collective four-electron ionization of N_2^{+2} (formed from N_2 by steps) to form N_2^{+6}, which dissociates into two N^{+3} ions sharing equally the Coulomb repulsion energy of 76 eV. However, it was deemed that the process is molecule specific at the intensities used, since the excitation was not observed in oxygen.

We now have preliminary observations of the multiphoton ionization of molecular hydrogen in a 28 ns, 308 nm, $\sim 10^{11}$ W/cm^2 laser beam.[6] The radiation is in near three photon resonance with the B (v = 6) state and near four photon resonance with the H_2^+ (v = 2) state, as shown in Figure 1, which results in the production of these species at the interaction point. We analyzed the energy of the ions produced in the process with a stopping voltage. Our measurements show that ions of kinetic energies as high as 12.5 eV are produced in addition to other lower energies. Direct multiphoton dissociative ionization[3] of the B state and multiphoton dissociation of the H_2^+ ground state can explain the low energy ions. However, we found it necessary to postulate direct multiphoton excitation of two electrons from the ground state to the dissociative ionization limit of the H_2^+ ground state or to the two electron continuum in order to explain the high energy ions. Molecular hydrogen has only two electrons and two possible ions, H^+ and H_2^+, so theoretical treatment is much easier than in other systems, and the dynamics of possible ionization mechanisms are better understood. Study of this system will hopefully provide the answer to the question of whether direct multiphoton multielectron ionization can compete with stepwise ionization processes.

Experiment

Hydrogen molecules are observed in a molecular beam formed when they diffuse through a multicollimator assembly from a region of H_2 gas at a pressure of several mtorr. This loosely collimated beam has a density of about $10^{12}/cm^3$. It passes between the lower two of three parallel electrodes in a diffusion pumped cell. The lower two electrodes are horizontal, flat metal plates separated by 8 mm, with a 3x5 mm slot covered by fine mesh (the analyzing grid) in the middle electrode to allow the passage of ions to the detector. They are biased at values between 50 volts retarding and 100 volts accelerating in order to analyze the energy of the produced ions. The upper electrode is a wire mesh with an extraction voltage that is kept at -5 kV. It is connected to a metal tube leading directly to an eighteen-stage Venetian-blind electron multiplier capable of single ion detection.

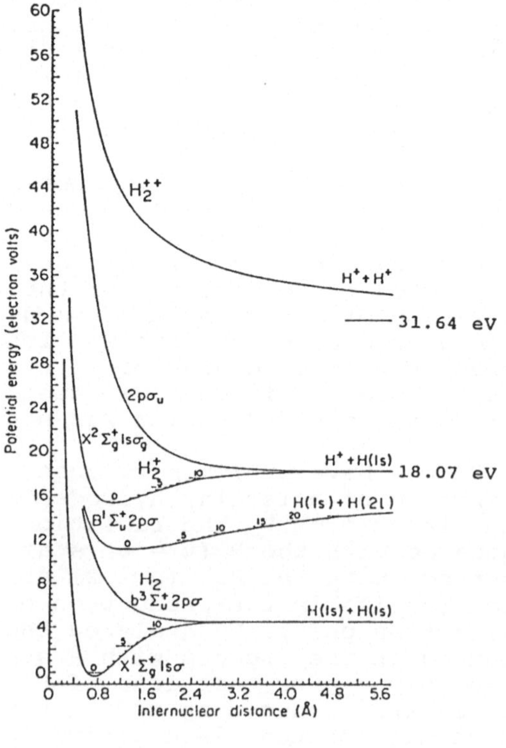

Figure 1

This gives the energy levels of the H_2 molecule, including the ground state, the B state, the one-electron continuum threshold (the H_2^+ ground state), the $2p\sigma_u$ repulsive ground state of H_2^+, and the two-electron continuum threshold (H_2^{++}). Our radiation is in near three photon resonance with the B state and near four photon resonance with the H_2^+ ground state.

The laser radiation used in the experiments is the 308 nm unpolarized fundamental of an XeCl excimer laser which gives 6.5 J per 28 ns pulse. Most of the highly divergent fraction of the beam departs over the 4 meter path it travels before it is focused into the cell, halfway between the analyzing electrodes and directly beneath the analyzing grid. The pulse energy entering the cell is 1.5 J and focuses to a fine line, giving power densities estimated to be on the order of 10^{11} W/cm^2. The energy of each pulse is monitored by a photodiode, which measures the light reflected by the window of the cell.

The energy of the ions is determined by measuring the voltage necessary to stop them. Unfortunately, this procedure provides analysis only of the velocity component parallel to the stopping field. Ions are produced with velocities in other directions as well. As a result, for an ion with energy E_0, a distribution of energies from 0 to E_0 will be observed, because a lesser stopping potential will suffice for ions with only a component of their velocity in the direction of the stopping field.

When there are several species of ions, each with a characteristic energy, the lower energy ions cannot be distinguished from higher energy ions with large transverse velocities. Since this effect depends on how many ions with significant transverse velocities are accepted by the detector, it can be alleviated by reducing the acceptance solid angle of the detector (by changing the analyzing grid area or the distance of the focal point from it). In spite of the large solid angle we were able to acheive some resolution of the lower energy ions by utilizing the power dependence of the processes involved. We quenched the higher energy ions by lowering the intensity of the laser radiation, but we realize they will be quenched only if the intensity dependence of the ion yield varies monotonically with the ion energy. We may be missing low-energy ions produced by high order transitions, and such will be studied in further detail in the future. Finally, we do not believe that space-charge effects can change the energy distribution, because our pulse time is long, and the ions are fast, so they should not accumulate.

Figures 2-4 give the number of ions reaching the detector as a function of stopping potential for three selected laser intensities of increasing magnitude. In Figure 5 we show an enlargement of the high energy portion of Figure 2. At highest power we measure stopping potentials of up to 12.5 volts, corresponding to ions with energies as high as 12.5 eV. At the next lower power, the fast ions are gone, but we see ions with energies of 5 eV. At the lowest power, the 5 eV ions disappear, and ions of 3 eV are observed.

Analysis

We now discuss the various mechanisms responsible for the energy distribution of the ions we observed (see figs. 2-5). First of all, it is clear that the excitation scheme we used populates the v = 6 vibrational level of the B state through a near three photon resonance, as well as the H_2^+ (v = 2) bound ground state by near four photon resonance. We believe

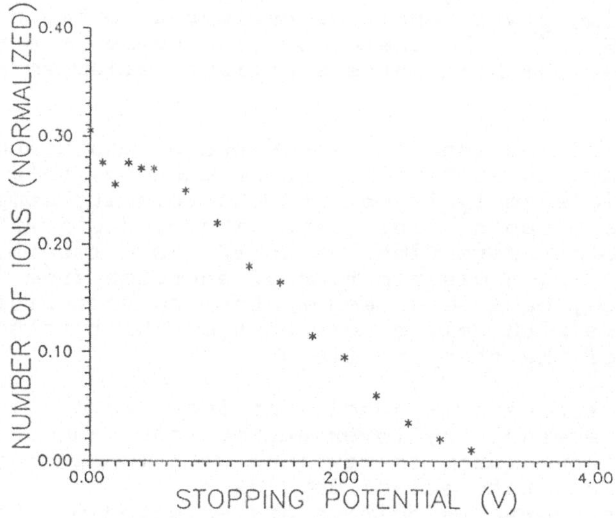

Figure 2. Laser intensity of I_0.

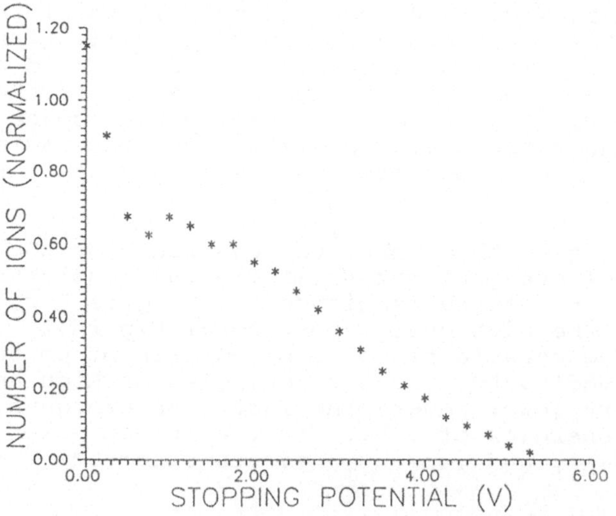

Figure 3. Laser intensity of $1.7I_0$.

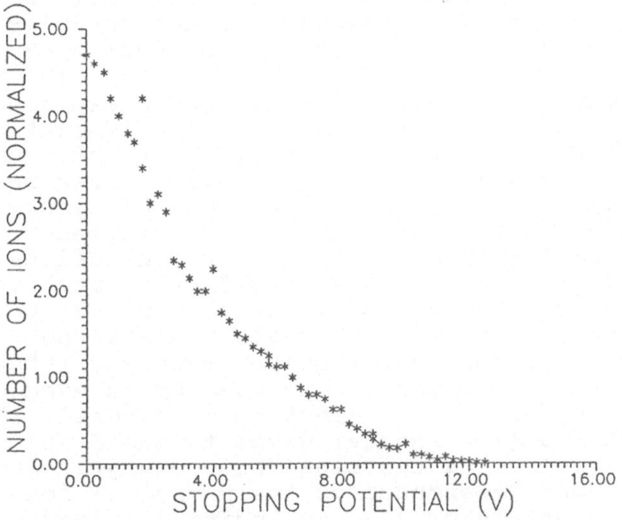

Figure 4. Laser intensity of $2.9I_0$.

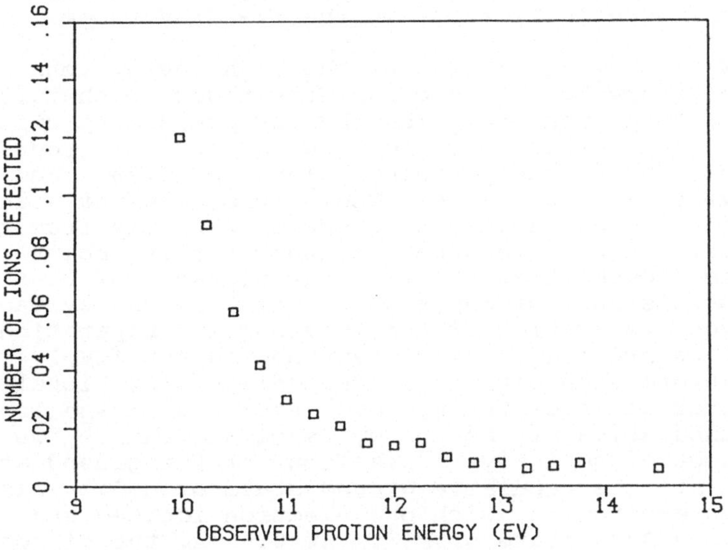

Figure 5. Laser intensity of $2.9I_0$.

that the 5 eV ions arise through a three photon resonant excitation from the H_2^+ (v = 2) state (at internuclear separation 1.00 Å) to the repulsive $2p\sigma_u$ ground state of the ion. At this internuclear separation, the energy of the $2p\sigma_u$ state is 28 eV: as a result the dissociation of the H_2^+ into H(1s) and H^+ fragments liberates 10 eV of kinetic energy shared by the products (5 eV each). Selection rules inhibit a similar process using two photons to produce 3 eV ions.

The three (and also five) eV ions can be explained by three (and four) photon direct dissociative ionization (via the $2p\sigma_u$ of the H_2^+) of the B state of H_2 at internuclear separations of 1.28 Å (and 1.00 Å). This process results in the simultaneous ejection of one electron and dissociation of the H_2^+ repulsive state into H(1s) and H^+ fragments. The energy of the $2p\sigma_u$ at these internuclear separations is 24 eV and 28 eV, producing ions of 3 eV and 5 eV respectively, consistent with our observations. We should note that the dissociative ionization of the B state, unlike the dissociation of H_2^+, does not require resonance interaction with the laser radiation and can proceed by an even or odd number of photons, because an electron is ejected. This will tend to produce red (low energy) wings to the 5 and 3 eV energies without altering the blue (high energy) wings. We inspected the power dependence of the number of ions detected for stopping potentials of 2 V, excluding all ions with energies less than 2 eV, and 9 V, excluding all ions of less than 9 eV. We found the exponent to be approximately 3 for the former and 8 for the latter, but these measurements are difficult and our results preliminary, so the only conclusion we draw is that the high-energy ions are produced by a much higher-order process than the low-energy ions. We will attempt to accurately measure the power dependence for all ion energies, especially those in the 12-13 eV range.

We now discuss the origin of the high energy ions observed in Figure 5. It is clear from Figure 1 that it is not possible to produce such ions by any process starting from the B (v = 6) state nor from the H_2^+ (v = 2) ground state because their inner turning points, at 0.78 Å and 0.77 Å respectively, are too large. These ions, however, can be accounted for if the excitation proceeds directly from the ground state. This is because the inner turning point of the ground state occurs at an internuclear distance of 0.65 Å, the smallest distance of any turning point in the system. In order to avoid relaxation of the internuclear separation as the excitation proceeds, the process should not involve resonance interaction with any intermediate vibrational highly excited state of the system. From Fig. 1 and from numerical tabulation of the energy level diagram[6,8], we can examine direct dissociative ionization of the ground state (at 0.65 Å) via the repulsive ground state of H_2^+. This is a two-electron excitation, with one electron ionized and one excited to the $2p\sigma_u$ state of H_2^+. At 0.65 Å, the dissociate ionization threshold occurs at 37.7 eV which is 19.6 eV above the dissociation limit, thus giving ions of 9.8 eV each. It is a 10 photon excitation with the ejected electron acquiring 2.5 eV kinetic energy. The ion energy of 9.8 eV, however, is less than the maximum of the energies in the experimental results of Fig. 5.

Another possible mechanism which can give large ion energies is the direct two electron excitation to the two electron continuum. In this process, two electrons are stripped simultaneously, followed by Coulomb explosion of the two protons. As an example, with excitation from near the smallest turning point (at 0.65 Å), the two electron ionization threshold occurs at 53.8 eV. Using 31.64 eV for the H^+-H^+ system at infinite distance gives 11.1 eV for the kinetic energy of each of the resulting protons. At this turning point the process proceeds via the absorption of 14 photons, and so the ejected electrons share a kinetic energy of ~2 eV. Similar calculations at other R values also show that the two electron ionization processes yield ions of energies higher than the dissociative ionization case. In fact, this is due to the electronic energy of the p orbital of the H_2^+ repulsive state. This energy is shown in Fig. 6 along with the energy of other orbitals.

Since Fig. 5 shows that ions of energies larger than 11.1 eV are observed, namely ions in the range 11 - 12.5 eV, then excitation from the classically forbidden region of the ground state (R < 0.65 Å) must have taken place. Although the vibrational wavefunction in this region is exponentially decaying, it is not vanishingly small. We calculated the internuclear separations at which the two electron ionization and the dissociative ionization processes produce ions of a given energy; these are illustrated in Fig. 7. The lower horizontal axis gives the energy and we use the upper horizontal axis to illustrate the corresponding R values for the two processes. As expected, the R values approach each other as they decrease (as the ion energy increases). However, for a given R, the two electron ionization process still gives ions of a higher energy than those produced by the dissociative ionization process. Note that the time scale of the process is short. It takes about 1 fs for the protons to move from an initial internuclear separation of 0.6 Å to one of 1.0 Å, so we are not only exciting from the forbidden region, but also examining very short time scales.

Fig. 7 also shows that with excitation from the forbidden region, marked as F_1 and F_2 on the energy axis, which correspond to an R value of 0.65 Å in both processes, the fastest ions observed in our experiment can be explained by either one of the processes discussed above. However, to get the same energy the dissocative ionization process requires the excitation to occur from shorter R values. Thus, it may be less probable than the two electron ionization process, due to the exponential drop of the proton wavefunction in this region. To see the importance of the decay, we calculated the wave function and its square, the probability density, of the ground state in the well and in the inner forbidden region. We converted this to a probability of observing ions of a certain energy, assuming that the excitation takes place at constant internuclear separation. This probability is of course different for the two different processes, which yield ions of different energy from the same internuclear separation, and it is plotted as solid lines for both in Fig. 7. The curves clearly show that in the region of interest, around 12.5 eV, the dissociative ionization

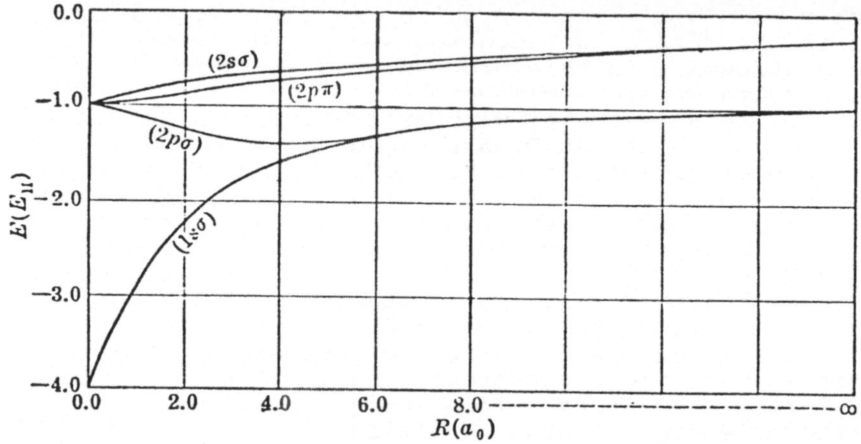

Figure 6. Electronic energy of the $2p\sigma_u$ state and others.

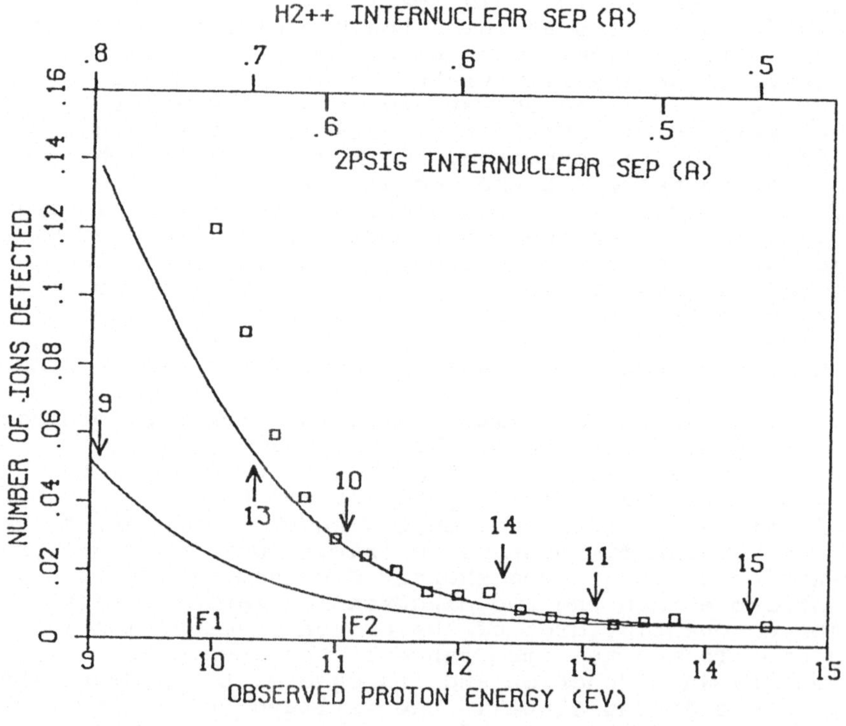

Figure 7. The two upper scales give the internuclear separations that would lead to the observed energies given on the bottom scale, for the two different processes. The arrows give the observed proton energy for a given number of photons absorbed if the electron(s) do not carry off any energy (threshold of ionization). There are 9, 10, or 11 photons for the dissociative ionization process and 13, 14, or 15 photons for the two-electron ionization process. F1 and F2 give the observed proton energy for which the protons are initially at the edge of the forbidden region of the ground state vibrational wavefunction (0.65 Å) in each process.

process is less probable than the two electron ionization by a factor of about 3.5. In the same figure we present the experimental data, normalized to the probability density at a two electron ionization energy of 11 eV. The agreement is quite excellent at the high energy side, but deviations are seen at energies less than 9 eV. Of course, we got similar conclusions when we normalized the results to the probability density evaluated at dissociative ionization energies. This is expected since the two probability density curves have nearly the same energy dependence except for a shift in the curve along the energy axis.

The detection efficiency in our experiment is less for dissociative ionization because events in which the neutral atoms are ejected towards the detector (relevant to the case of dissociative ionization) will be missed unless the laser manages to ionize the atoms with 100 percent efficiency before they leave the interaction region. Under the conditions of our experiment we expect to ionize only a few percent of the hydrogen atoms. This conclusion is based on some measurements we made where the H_2 beam was converted to an H beam by a DC discharge. Our measurements indicated that 10 percent or less of the H atoms are ionized by the laser beam. This percentage drops further in the actual experiment because the H atoms produced in the dissociative ionization process are not thermal; they are produced with a kinetic energy of 12 eV. At this energy they have velocities of 4.8×10^6 cm/s, and cross the beam in ~2 ns. Since the laser pulse width is 28 ns, we expect the rate of ionization to drop by a factor of 7 from that of the thermal atom.

Thus, as a result of consideration of vibrational wavefunctions and detection efficiencies we project that the two-electron ionization process is favored by a factor of 7 over the dissociative ionization process. Moreover, there are other factors which may influence the branching ratios of the two processes. These include: the order of the multiphoton process, the photoionization cross section, and the polarization of the laser radiation.

The two processes each involve a "large" number of photons (12-14) and no resonant intermediate states. In this case, the parameter $4\pi r_0 F/\Delta^2$ is useful, where r_0 is the classical radius of the electron, and Δ is an effective average detuning of all the intermediate states involved.[9] Using simple arguments based on an earlier analysis of atoms by Bebb and Gold,[9] one finds that the ionization cross section is proportional to

$$(4\pi r_0 F/\Delta^2)^N \int \psi_f{}^* e^{iN\underline{k}\cdot\underline{r}} \psi_i d^3\underline{r}$$

where N is the number of photons, and ψ_i, ψ_f are the initial and final state wavefunctions respectively. If $\Delta^2 = 4\pi r_0 F$, the first factor, coming from the number of photons, will be the same for both processes. At the powers we are using, we find that the average detuning required for the condition to be achieved is 1/2 eV which is not totally unrealistic. The second factor favors the two-electron ionization process. Since $N\underline{k}\cdot\underline{r} \approx .03$, the dipole approximation is valid, and the integral reduces to an overlap integral between the initial and final two-electron wavefunctions, which can be broken down into a sum of products of one-electron overlap integrals. In the case of dissociative ionization, every term in the sum contains a factor of the one-electron overlap integral between the $1s\sigma_g$ ground state and the $2p\sigma_u$ excited state, which is zero from parity considerations. This does not happen in the two-electron ionization, where all the overlap integrals are between ground state electrons and free electrons. Thus, we expect the dissociative ionization to be suppressed. The suppression, of course, may not be as strong if the number of photons is not high enough.

Also because of the large number of photons involved, we expect no polarization effects when the laser radiation is unpolarized as in our experiment. However, we would expect some effects for a completely polarized beam.

We display the location of various multiphoton processes on the energy axis of Figure 7. It is interesting to see that there is evidence in the data of the 14 photon resonance in the two electron ionization process. This resonance occurs in excitation from the forbidden region of R. Also one can see evidence for the 13 photon resonance of the same process. Both of these resonances cause deviation of the data from the predictions based on the R dependence of the wavefunction of the ground and final state. On the other hand, it is not clear if we have any evidence for the 11 photon resonance of the dissociative ionization process, which would account for ion energies of 13 eV. There might be some evidence for the 10th order resonance of this process occuring at 11.1 eV; however, it is not as strong as the rise in the yield at the 13 photon resonance of the two electron ionization process.

In conclusion, we have seen multiphoton two-electron processes, and we believe that we have evidence for the simultaneous ionization of two electrons. It is remarkable that these processes can be investigated, and such short time scales examined, with a laser of such low power and long pulse duration. Future work will involve detailed studies of the branching ratios, power dependencies and polarization effects, as well as attempts to achieve better energy resolution.

References

1. A. L. Robinson, Science 232:1193 (1986).
2. U. Johann, T. S. Luk, H. Egger, and C. K. Rhodes, Phys. Rev. A 34:1084 (1986).
3. K. Codling, L. J. Frasinski, P. Hatherly, and J. Barr, to be published.
4. L. J. Fransinski, K. Codling, P. Hatherly, J. Barr, I. N. Ross, and W. T. Toner, Phys. Rev. Lett. 58:2424 (1987).
5. P. Lambropoulos, Phys. Today, 40, No. 1:S25 (1987).
6. T. E. Sharp, J. Atomic Data 2:119 (1973).
7. D. Humm, T. Sherlock, K. Ng, J. Mazumder, and M. H. Nayfeh, to be published.
8. D. R. Bates and R. H. G. Reid, Adv. Atomic and Mol. Phys. 4:13 (1968).
9. A. Gold and H.B. Bebb, Phys. Rev. Lett. 14:60 (1965).

RESONANCE IONIZATION PROCESSES IN SMALL MOLECULES

Philip M. Johnson

Department of Chemistry
State University of New York
Stony Brook, N.Y. 11794

INTRODUCTION

Resonance ionization spectroscopy has been proven to be a valuable tool in the study of the structure and dynamics of molecules[1]. In addition, the resonance ionization process provides insight into the interaction between intense coherent light fields and either isolated molecules or molecules in the process of undergoing a collision. Here we would like to explore what can be learned about molecules with this technique and how nonlinear effects in the intense laser fields affect the molecules and their spectra.

If one examines what is known about the excited states of small molecules, it is found that high resolution information has been gained primarily from two sources--absorption spectra from the ground state and emission spectra between the various states excited either optically or electrically. Other techniques such as electron scattering and photoelectron spectroscopy also can be very valuable but do not usually have the rotational resolution necessary for the most detailed analysis of excited state structure. This means that states are not very well known in detail if: 1) they are not optically connected upward from the ground state or downward from an emitting state and 2) they do not emit themselves. Techniques are therefore valuable which either change the optical selection rules to connect a new set of states or enable absorption spectroscopy to be done on an excited metastable state. Because of multiphoton selection rules on the one hand and great sensitivity on the other, resonance ionization spectroscopy has historically been heavily involved in the detection and characterization of new excited states. In some cases the mass selectivity inherent in the technique is vital in seeing the transitions of a species of interest in the presence of another less interesting molecule. This situation often occurs in radical spectroscopy where the concentration of some parent molecule can overwhelm a small amount of radical created from it. In fact interfering absorptions are the primary reason that higher nonemitting states (such as Rydbergs) are known for very few radicals.

While many of the advantages of resonance ionization spectroscopy can be invoked with single photon transitions, in its most general form the ionization process involves some multiphoton transitions. These multiphoton transitions require high light intensities because of the well-known nonlinear relationships involving the laser power and the number of photons involved in the transition. The large fields produced at the focus of the laser have the capability of producing other nonlinear effects besides absorption. These other processes such as third harmonic generation and the ac-Stark effect can greatly perturb the appearance of the spectrum, making the extraction of molecular data much more difficult.

The most apparent perturbations in spectra taken at pressures above one Torr are produced by third harmonic generation. This collective effect scales as the square of the pressure, modulated by phase matching conditions, and therefore is not apparent in spectra involving low pressures such as in molecular beams. However in higher pressure spectra the high energy photons created by third harmonic generation can probe one photon transitions with their higher transition probabilities. These high energy transitions can become more probable than multiphoton transitions at certain pressures, thus dominating the spectra. Once they are understood, the changes in spectra created by high intensity light interacting with the medium have the potential of providing additional information about the molecules and their interaction with light.

NONLINEAR EFFECTS IN MOLECULAR CHLORINE

The particular case of a three-photon study of molecular chlorine serves to illustrate some of the spectral changes produced by nonlinear effects. In this study light around 400nm was directed into a proportional counting ionization cell containing chlorine at pressures ranging from 0.2 to 1.0 Torr[2]. This optical region encompasses the wavelength at one-third the photon energy of the 134.33 and 133.22 nm $2^1\Pi_u \leftarrow X^1\Sigma_g^+$ (1,0-0) rovibronic excitations, and ionization occurs with a fourth photon. Also in this region are several three photon atomic chlorine transitions and one of four photons.

Some spectra of chlorine at various pressures and orientation of light vectors are shown in figure 1. It is immediately seen that the spectrum changes dramatically as conditions are changed. In order to extract molecular information from these spectra it is necessary to examine the source of these changes. However first we must explore a strange phenomenon which occurs when a transition is both one-photon and three-photon allowed. Then, at higher pressures, supposedly strong transitions are missing from the spectrum.

It was first noticed in xenon gas that the 6s←X transition which is seen so strongly in one photon does not appear in the three-photon resonant ionization spectrum at pressures higher than 10^{-2} Torr[3,4]. Later it was discovered that the transition is recovered when the laser light is redirected back through the sample cell so there are counterpropagating beams[5]. From the work of Wynne and Jackson[6], and of Payne and Garrett[7], this phenomenon arises from the fact that the polarization of the medium set up by the third harmonic wave is exactly out of phase with the polarization created by the fundamental light, producing a cancellation of the polarization of the medium and thus no net transition strength if the sample is optically thick at the third harmonic wavelength. At lower pressures the third harmonic wave does not appear and the transition can take place. With counterpropagating beams, two photons can be absorbed from one beam and one from the other, creating an excitation pathway which cannot be canceled by the third harmonic wave. It should also be noted that third harmonic generation cannot take place with circularly polarized light because of symmetry restrictions, but neither are many three photon transitions allowed for the same reason. The ability to use counterpropagating beams to bring back three photon transitions enables us to distinguish whether a line is missing because of polarization cancellation or from some more fundamental reason (like dissociation competing with ionization).

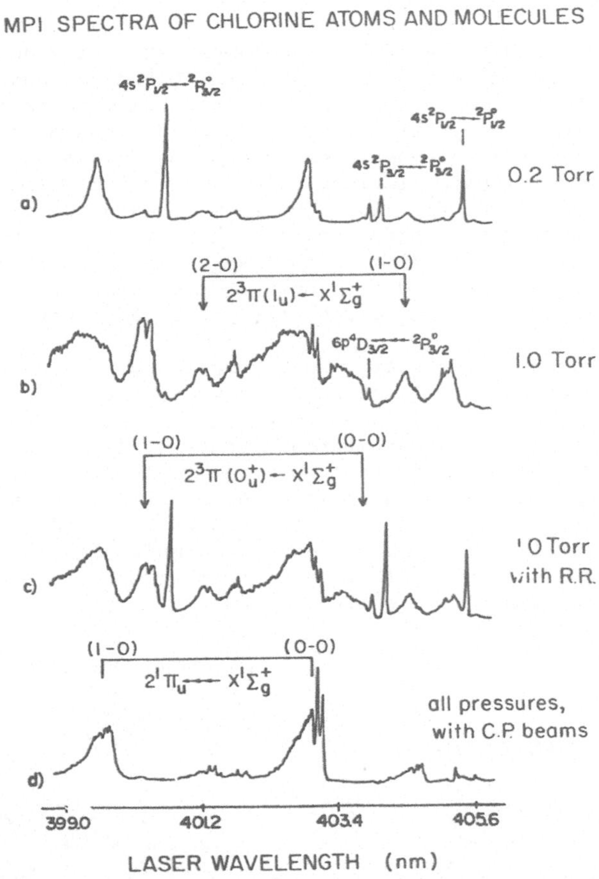

Figure 1. MPI spectra of chlorine: (a) Linearly polarized light and traveling beam excitation at 0.2 Torr of chlorine gas; (b) same as above but at 1.0 Torr; (c) same pressure and polarization as b but with counterpropagating beam excitation; (d) circularly polarized light and traveling beam excitations at all pressures ($P \leq 1$ Torr) (Ref. 2).

With that background the behavior of the chlorine spectrum begins to become clear. The complexity is due to the interaction between the transitions of chlorine atoms and chlorine molecules. The sharp peaks in Fig. 1a are due to three photon transitions of chlorine atoms. The broader peaks are molecular transitions. These remain when the pressure is raised, Fig. 1b, but the atomic transitions go away because of polarization cancellation. In their place, slightly to the blue, some new structure appears. These new transitions turn out to be singlet-triplet transitions of the molecule. They are enhanced relative to their natural three-photon intensity because they are in fact one photon transitions. The vacuum ultraviolet photons used to produce these ionizations arise from third harmonic generation by the chlorine atoms. It is characteristic that third harmonic generation occurs strongly to the blue of atomic resonances because that is where phase matching can occur in a normally dispersive medium.

In order to show that the atomic resonances are being quenched by polarization cancellation, a spectrum is shown in Fig. 1c where the light is retroreflected back through the sample. The atomic peaks are again prominent and the third harmonic induced molecular resonances remain. To demonstrate that the triplet resonances are indeed due to third harmonic light one can use circularly polarized light with which third harmonic generation is not possible, Fig. 1d. The atomic resonance also do not appear because of symmetry restrictions and the spectrum becomes relatively independent of pressure. There are some changes in the molecular rotational contours because of differences in the rotational branch intensities between parallel and circularly polarized light.

This whole sequence of spectra can be explained in terms of the energy diagrams of molecular and atomic chlorine, Fig. 2. At the one photon level in the molecule is a dissociation continuum which leads to the production of ground state atoms. In the excitation process this dissociation competes with three photon excitation so some molecular excited states are produced as well as substantial quantities of atoms. These atoms can then undergo a three photon resonant ionization process or produce third harmonic light which is rapidly absorbed by the molecules to produce new molecular ionization channels. Thus the interplay between atoms and molecules neatly explains the spectra.

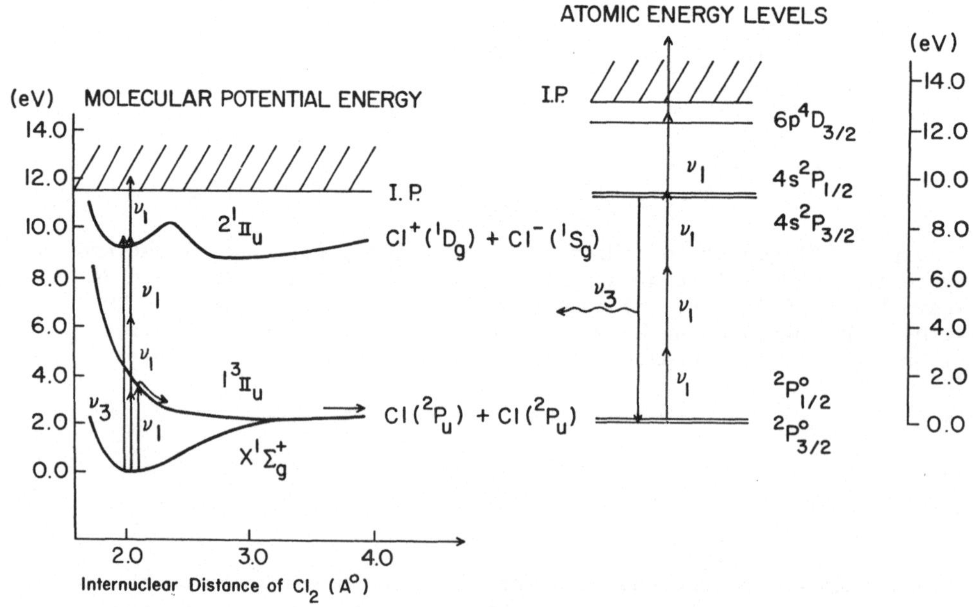

Figure 2. Selected atomic and molecular energy levels of chlorine and photoprocesses involved in creating the spectra of Fig. 1 (Ref. 2).

Although the polarization cancellation effects dramatically alter the intensities of atomic transitions, as indicated in Fig. 1, it is of interest to examine what effect, if any, this is going to have on molecular spectra. Certainly there are no dramatic effects on the molecular transitions in going from Fig. 1a to 1b. There are some changes however.

It turns out that one can categorize molecular transitions into three classes, which in diatomic molecules can be correlated with angular momentum changes in the molecule. Class I includes the transitions which are allowed in one photon--the P, Q, and R branches of Σ_u^+ and Π_u transitions from a Σ_g^+ ground state. Class II includes those transition which have angular momentum changes requiring three photon--e.g. N, O, S and T branches and Δ_u and Φ_u electronic states. Class III does not occur for small molecules. Since polarization cancellation requires a transition to be both one photon and three photon allowed, it is immediately clear that only Class I transitions will be affected. The transition moment in this class has two tensor components called weight-1 and weight-3, of which only the weight-1 component i sensitive to polarization cancellation. The ratio of weight-1 strength to weight-3 strength in a transition depends upon the intimate details of the wavefunction, but in general even a Class I transition cannot go to zero because of polarization cancellation. In certain cases such as that recently recognized in acetylene, the weight-1 component dominates the oscillator strength and three photon interference is a large effect. As seen here in chlorine, the more general case is that the intensity ratios between the branches are changed because the P, Q, and R branches are Class I and the N, O, S and T branches are Class II. This perturbation caused by harmonic cancellation must be taken into account when analyzing linear/circular polarization ratios in three photon transitions, as is commonly done in order to assign symmetries to the states involved in transitions.

SPECTRA OF COLD RADICALS--CARBON CHLORIDE

When one goes to the rarefied density of a supersonic beam, the interesting pressure effects described above are lost. This may be viewed by some as a blessing, but there are other distinct advantages in doing MPI in supersonic beams, foremost among them the possibilities of mass selection, cold rotational temperatures and a collisionless environment enabling the production of unstable or reactive species. A good example of these advantages coming into play is in a study of the CCl radical, which is of current interest in atmospheric chemistry because of the possible effect of halocarbons on the ozone layer.

The CCl radical is relatively easily made in a supersonic beam[8]. A pulsed jet of argon, seeded with CCl_4, is subjected to a pulsed high-voltage electric discharge during adiabatic cooling. This electrical pulse of about 40 kV is imposed between a wire in the gas flow and the pulsed valve body for about one microsecond and produces a glow discharge in the entire expansion cone. Most of the radical formation appears to take place in the early parts of the expansion, as evidenced by the low rotational temperature of the dissociation products (about 15 K).

Previous to our work[9] only two excited state bands were known for this molecule, assigned as the 0-0 and 1-0 transitions to the $A^2\Delta$ state from the $X^2\Pi_{1/2}$ ground state[10]. The spectrum of the 1-0 transition is seen in Fig. 3, along with a computer simulation of the spectrum using a Boltzmann distribution of 15 K. Measurement of the 2-0 band allowed the determination of the vibrational constants of the A state as ω_e = 876.2 and $\omega_e x_e$ = 12.0 cm^{-1}.

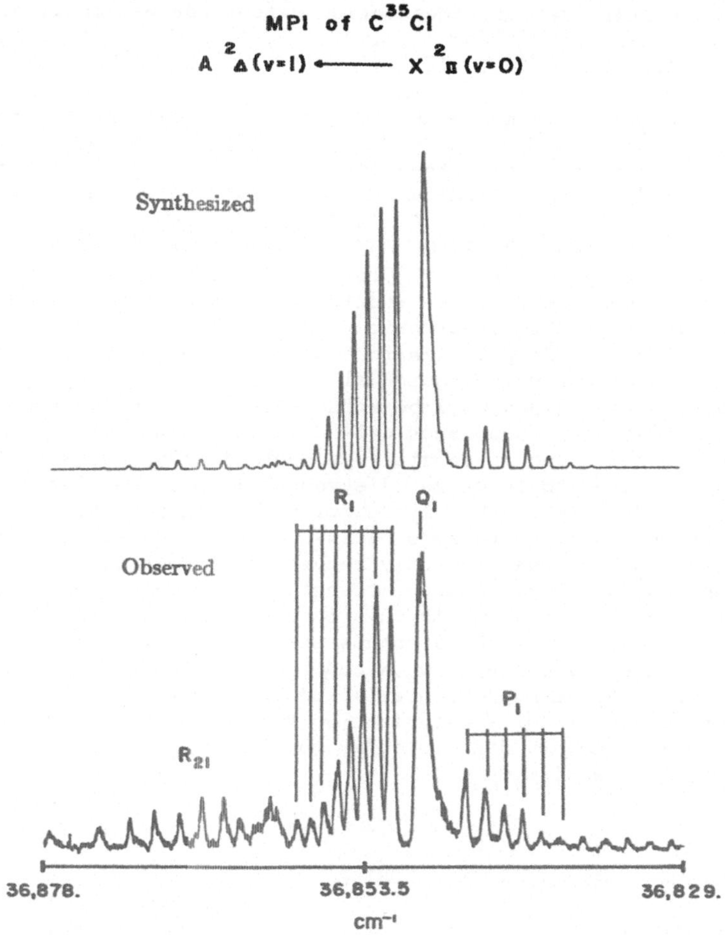

Figure 3. Experimental and computer-synthesized spectra of the (1-0) band of $C^{35}Cl$ with a temperature of 15 K (Ref. 9).

The higher excited states of most radicals are not very well known because the absorption bands are obscured by some precursor. Mass selectivity gets around this problem in an obvious way if the photodissociation of the precursor at the wavelength being used does not produce large quantities of the ions at the mass of the radical. Usually mass interference is not a large problem or only creates a continuous background which is easily subtracted. In fact quite often dissociation of the precursor competes with ionization and very few ions are seen from the parent molecule as compared to those from radicals produced in a discharge or by other means. That means that three photon spectroscopy is an excellent tool for looking at the Rydberg structure of radicals with visible light.

Fig. 4 shows the three photon Rydberg spectrum of CCl. Several series are evident, giving the opportunity to determine the ionization potential of the radical to fairly high accuracy. The most prominent series has a quantum defect of 0.6, indicating that it is a p-series. It converges at 10.6 eV. This is in surprising disagreement with the value 8.9±0.2 eV determined from multiphoton photoelectron spectroscopy[12]. There, however, the radical was formed from the photodissociation of CCl_2F_2 and CCl_3F and it was necessary to assume that the dissociation occurred in the neutral molecule and that the radical was formed in its ground state. Apparently one of these assumptions is not correct. On the other hand the Rydberg value for the ionization potential agrees exactly with one derived from the reaction threshold technique[13]. Other series have quantum defects of 0.79 and 0.11.

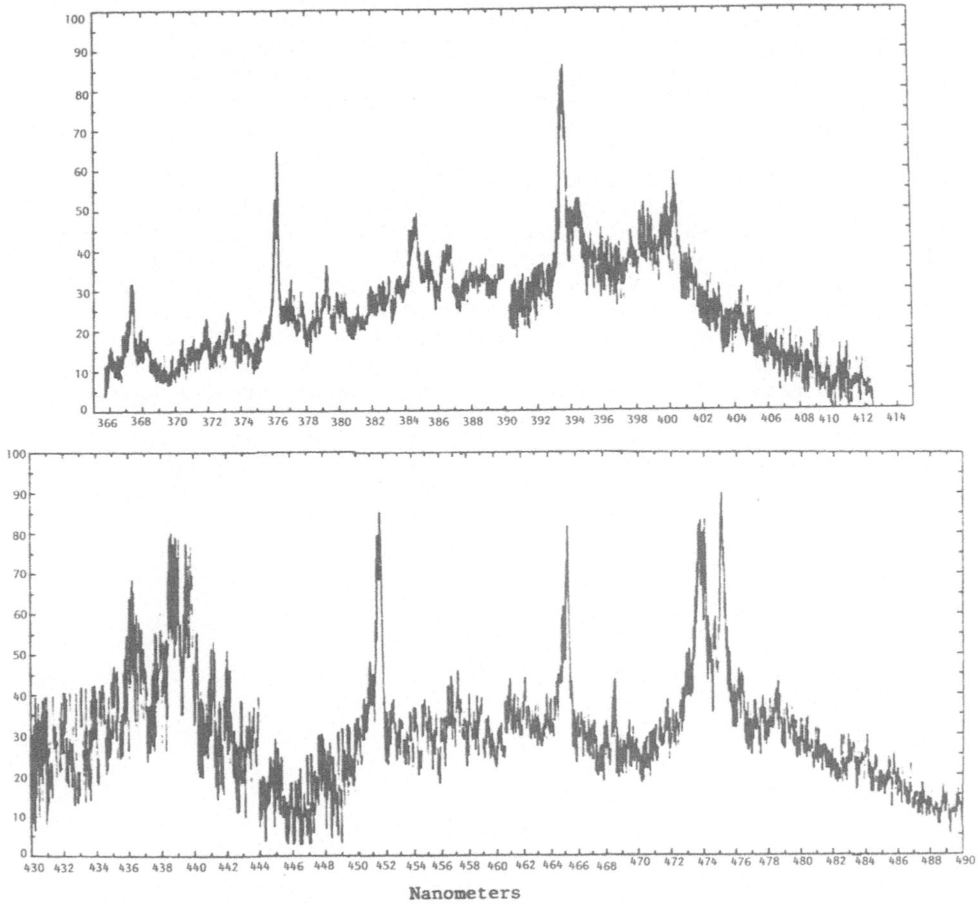

Figure 4. The three photon Rydberg spectrum of CCl radical.

These spectra of CCl illustrate some of the possibilities in getting information about radicals from resonance ionization. The status of this area has been recently reviewed by Hudgens[14]. Our recent work on benzene[15] and on nitrogen molecule[16] also has shown there is a future in resonance ionization of metastable triplet states. We have prepared these by the electric discharge in the supersonic beam in the same way as describe above for the CCl formation.

Even though resonance ionization spectroscopy is now becoming a fairly mature technique, it is hoped that the above illustrates that there are still interesting new phenomena to explore and new molecules to put to the test.

ACKNOWLEDGMENTS

I would like to give credit to my coworkers Steven Sharpe and Leping Li, who did much of the work reported here. This work has been supported by the National Science Foundation and the Department of Energy.

REFERENCES

1. P. M. Johnson, Acct. Chem. Res., **13**, 20 (1980); P. M. Johnson and C. Otis, Ann. Rev. Phys. Chem., **32**, 139 (1981).
2. L. Li, M. Wu, and P. M. Johnson, J. Chem Phys., **86**, 1131 (1987).
3. K. Aron and P. M. Johnson, J. Chem. Phys., **67**, 5099 (1977).
4. R. N. Compton, J. C. Miller, A. E. Carter and P. Kruit, Chem. Phys. Lett., **45**, 114 (1980).
5. J. H. Glownia and R. K. Sander, Phys. Rev. Lett., **49**, 21 (1985).
6. D. J. Jackson, J. J. Wynne, and P. H. Kes, Phys. Rev. A, **28**, 781 (1983); J. J. Wynne, Phys. Rev. Lett., **52**, 751 (1984).
7. M. G. Payne, W. R. Garrett, and H. C. Baker, Chem. Phys. Lett., **75**, 468 (1980); M. G. Payne and W. R. Garrett, Phys. Rev. A, **26**, 356 (1982).
8. S. Sharpe and P. Johnson, Chem. Phys. Lett.,**107**, 35 (1984).
9. S. Sharpe and P. Johnson, J. Mol. Spectrosc. **116**, 247 (1986).
10. R. D. Verma and R. S. Mulliken, J. Mol. Spectrosc. **6**, 419 (1961).
11. R. D. Gordon and F. W. King, Canad. J. Phys., **39**, 252 (1961).
12. J. W. Hepburn, D. J. Trevor, J. E. Pollard, D. A. Shirley, and Y. T. Lee, J. Chem. Phys., **76**, 4287 (1982).
13. L. C. Frees, P. L. Pearl, and W. S. Koski, Chem. Phys. Lett.,**63**, 108 (1979).
14. J. Hudgens, *Advances in Multiphoton Processes and Spectroscopy*, S. H. Lin, ed. (World Scientific, Singapore, 1987).
15. S. Sharpe and P. Johnson, J. Chem. Phys., **81**, 4176 (1984).
16. S. Sharpe and P. Johnson, J. Chem. Phys., **85**, 4943 (1986).

THIRD HARMONIC GENERATION AND MULTIPHOTON IONIZATION SPECTROSCOPY

C. Fotakis*, J.A.D. Stockdale** and M.J. Proctor

Research Center of Crete
Institute of Electronic Structure and Laser
P.O. Box 1527, Iraklion, Crete, Greece

INTRODUCTION

Nonlinear phenomena, such as multiphoton excitation (MPE) processes leading to ionization and third-harmonic generation (THG) in gases have become the issue of extensive studies over recent years. Studies in rare gases in particular have resulted in several interesting observations in this respect [1]. Thus, it has been established that three-photon resonantly enhanced multiphoton ionization (MPI) in the vicinity of states which are single photon optically coupled to the ground state may occur in efficient competition with THG. A characteristic example is the competition between the MPI and THG in xenon when tuning through the 6s 3/2 J=1 three-photon allowed intermediate state [1]. Several novel aspects related to these effects have been treated theoretically by Garrett et al. [2] and by Jackson and Wynne [3] in a series of papers. In these papers, the THG in the negatively dispersive side of three-photon resonances is treated quantitatively. Thus, the dependence of the wavelength range of the third-harmonic radiation profile on gas pressure, gas composition, and exciting beam intensity may be predicted with a high degree of accuracy.

Finally, in a recent study of four-photon non resonant excitation in krypton in the autoionizing region above the first ionization threshold, it has been shown that the presence of third-harmonic radiation may result in a strong enhancement of the observed signal [4]. Recent calculations by Lambropoulos and his co-workers are consistent with these observations [5].

The present work demonstrates several new effects which may appear when third-harmonic radiation is present during MPI.

EXPERIMENTAL SETUP

For the purposes of this work a six-way cross ionization cell, capable of background pressures smaller than 10^{-6} Torr, was used. The output of a

---

*Also Department of Physics, University of Crete, Greece.

**Permanent address: Health and Safety Research Division, Oak Ridge National Laboratory, Oak Ridge, Tennessee 37831-6125, U.S.A. (Oak Ridge National Laboratory is operated by Martin Marietta Energy Systems, Inc., under contract DE-AC05-84OR21400 with the U.S. Department of Energy).

KrF pumped dye laser was focused in the ionizing region by means of a 3.8 cm focal length lens placed in adjustable positions inside the cell. The dye laser beam could be focused to a spot of ~10 μm giving a power density of ~100 GW cm^{-2}. Along the laser beam axis, the cell was coupled via a LiF window either to a 0.2 m vacuum ultraviolet (VUV) monochromator equipped with a solar blind photomultiplier or to a second ionization cell for single photon excitation studies using the third-harmonic radiation generated in the first cell.

LINEWIDTH AND LINESHAPE EFFECTS

The MPI spectra of high lying even parity Rydberg states in xenon and krypton have been reported by Blazewicz et al. [6] . The p and f series and the p' and f' series leading to the $P_{3/2}$ and $P_{1/2}$ ionization limits have been identified and were attributed to a direct four-photon excitation. More recently, Proctor et al. by recording simultaneously MPI spectra and third-harmonic radiation have shown that in the case of high pressures of krypton, THG arising from the negatively dispersive region of the 5S state, i.e., at pump laser wavelengths shorter than 349.4 nm, may contribute significantly to the observed ionization signal [4]. In Fig. 1 the 9f' line of krypton is shown in greater detail. This figure clearly demonstrates that:

a. The linewidth decreases with increasing pressure, and

b. The asymmetric line profile is reversed by increasing the pressure.

These observations may be interpreted on the basis of the participation of third-harmonic radiation in the excitation process. At low krypton pressures, the phasematching conditions do not favor THG and the observed MPI spectra are attributed to a direct four-photon excitation process which occurs with the highest probability in the region of strongest focusing. For the laser intensities used in our experiments field strengths of ~10 V cm^{-1} are expected at the focal point. These high field strengths should in turn give rise to intense Stark broadening of the observed lines. As the pressure increases phasematching favors THG. In this work, THG was initially observed by means of the VUV monochromator and the solar blind photomultiplier. In later experiments THG was observed by using the second ionization cell (cf experimental setup) which could be filled with NO, CO or another rare gas. In Fig. 2 are shown the single-photon ionization spectra of nitric oxide due to the THG produced in the krypton gas in the first cell. These spectra clearly demonstrate the increase of intensity of THG together with a shift to shorter wavelengths of the THG profile with increasing pressure of krypton. These observations are consistent with the theoretical predictions of Garrett et al. [2]. It should be noted that the observed structure can be attributed to the existence of autoionizing states of nitric oxide, which have not been reported before in this wavelength region. Similar experiments with CO or Ne instead of NO resulted in characteristic structureless THG profiles.

The increasing intensity of THG with pressure leads to an increasing contribution to the observed ionization signal of a two-photon excitation process, involving one third-harmonic photon and one dye laser photon. The efficiency of this two-photon process is expected to be high even in regions outside the strong focusing region, where the Stark broadening effect is weaker. This is compatible with the observed line narrowing with increasing pressure shown in Fig. 1. Additional support for this qualitative interpretation is provided by recent theoretical work, where it is estimated that the cross sections for the two-photon process involving one THG photon and one dye laser photon under our experimental conditions is four orders of magnitude higher than the direct four-photon excitation cross section [5].

The change in asymmetry of the lineshape could also be examined in

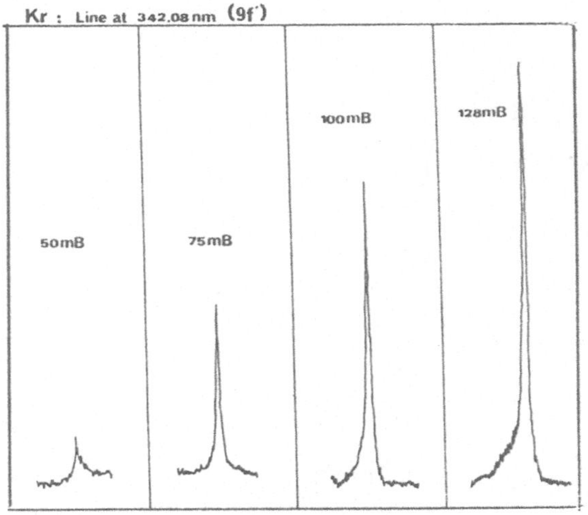

Fig. 1. Line asymmetry reversed

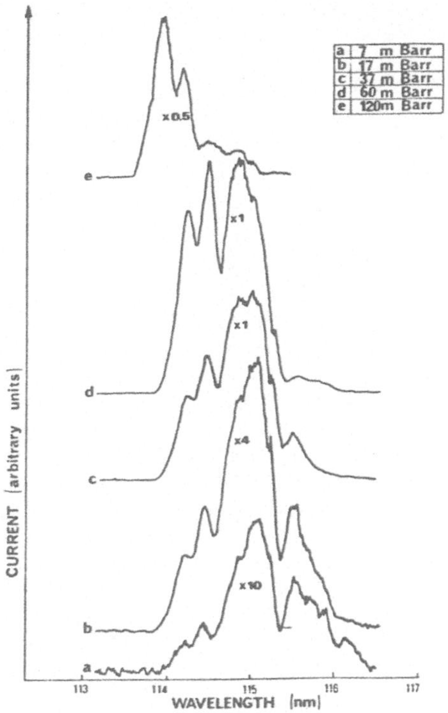

Fig. 2. Single-photon ionization spectrum of nitric oxide reflecting the shift in THG with Kr pressure.

terms of THG involvement in the ionization process, in the sense that the Fano "q" factors, which determine the asymmetric lineshapes of autoionizing states [7], may in principle involve different matrix elements for four- and two-photon ionization resulting in different "q" values. Recent preliminary calculations by Lambropoulos and his co-workers, however, do not reproduce the lineshape asymmetric reversal effect [8].

OBSERVATION OF RARE-GAS DIMERS AND THG PROFILES

In regions where the third-harmonic radiation produced in a rare gas is strongly absorbed by rare gas dimers, intense molecular MPI may occur. This has been demonstrated by Proctor et al. for the III* band of Kr_2, which has been observed by one VUV photon resonant, two-photon ionization [4]. By altering the pressure or using another buffer gas, shift in the THG profile may be induced allowing the probing of different parts of the upper potential surface. A typical two-photon excitation spectrum of the Kr_2 III* band obtained in this way is shown in Fig. 3.

In other cases, however, the THG profiles do not overlap with absorption features of the dimer. Structureless broad features may then be observed in the MPI spectra, due to two-photon nonresonant excitation in the medium, reflecting the THG profiles, as shown in Fig. 4 for xenon. The broad features at 342 nm and 333 nm correspond to THG in the vicinity of the 7s$[1\frac{1}{2}]$ J=1 and 6d$[1\frac{1}{2}]$ J=1 states of xenon correspondingly.

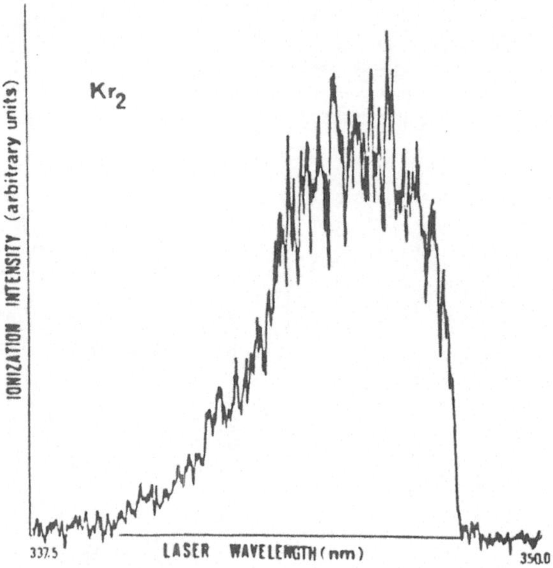

Fig. 3. Kr_2 two-color two-photon ionization spectrum

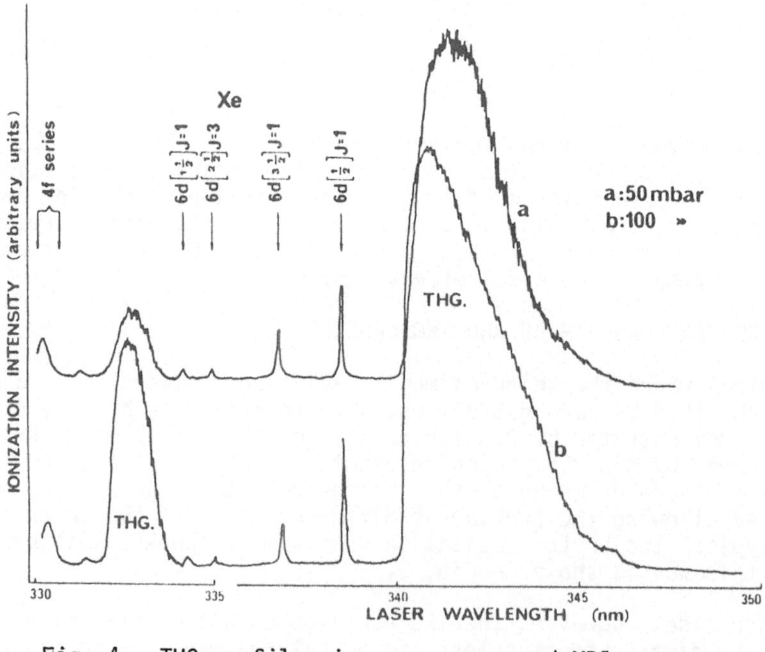

Fig. 4. THG profiles in pure xenon and MPI spectra

CONCLUDING REMARKS

It has been demonstrated that THG in rare gases may influence drastically the expected intensity, linewidth, lineshape, and the observed spectral features in the MPI spectra. Such effects are expected to be important in a wider range of atomic and molecular systems and their potential contribution should be considered when deducing spectroscopic or lifetime information from MPI spectra.

ACKNOWLEDGEMENTS

The authors are indebted to Professor P. Lambropoulos for many stimulating discussions. Support under NATO Collaborative Research Grant No. 86/423 is acknowledged.

REFERENCES

1. J. C. Miller, R . N. Compton, M. G. Payne, and W. R. Garrett, Resonantly enhanced multiphoton ionization and third harmonic generation in Xe gas, Phys. Rev. Lett. 45:114 (1980).
2. M. G. Payne, W. R. Garrett, and W. R. Ferrell, Influence of third harmonic fields on MPI of noble gases in unfocused laser beams, Phys. Rev. A 34: 1143; and references therein (1986).
3. D. L. Jackson, J. J. Wynne, and P. H. Kes, Resonance-enhanced multiphoton ionization interference effects due to harmonic generation, Phys. Rev. A 28:781 (1983).
4. M. J. Proctor, J. A. D. Stockdale, T. Efthimiopoulos, and C. Fotakis, Third harmonic generation and ionization processes in Kr, Chem. Phys. Lett. 137:223 (1987).
5. X. Tang and P. Lambropoulos (to be published).
6. P. R. Blazewicz, J. A. D. Stockdale, J. C. Miller, T. Efthimiopoulos, and C. Fotakis, Four-photon excitation of even parity Rydberg series in Kr and Xe, Phys. Rev. A 35:1093 (1987).
7. U. Fano, Effects of configuration interaction on intensity and phase shifts, Phys. Rev. 124:1866 (1961).
8. P. Lambropoulos (private communication).

EXPERIMENTAL INVESTIGATION OF

MULTIPHOTON FREE-FREE TRANSITIONS

Antonio Weingartshofer and Barry Wallbank

St. Francis Xavier University
Antigonish, Nova Scotia
Canada B2G 1C0

Electron-atom collisions in the presence of a strong laser field have attracted considerable theoretical attention in recent years not only because of the importance of these processses in applied areas (plasma heating by lasers) but also in view of their interest in fundamental collision theory. Cross-sections can change significantly when a large field is present and laser-photons are exchanged in the event. Moreover, laser-photons can absorb energy and momentum and since the coupling and dynamics of photons are particularly simple the laser is considered an attractive "third body" in a collision. The presence of laser parameters in the collision problem justify the expectation that new collision physics may evolve from this kind of "three body" interaction. There is now a wealth of theoretical calculations (Ferrante 1985, Faisal 1987) but new experiments are very much needed. In this presentation the emphasis will be on experiment.

Experimental efforts so far have been limited to the investigation of the most elementary collision process, i.e., "elastically" scattered electrons from an atomic target in an intense laser field. Since continuum states of the electron-atom system are involved, such radiative transitions are called "free-free" (FF). They have been directly observed in carefully designed three-beam experiments that were initiated ten years ago by two groups: one conducted experiments in the low-power regime (10^4 W cm^{-2}) examining the one-photon FF process with a c.w. CO_2 laser while the second looked for multiphoton FF processes at higher laser powers (10^8 W cm^{-2}) with a pulsed CO_2 TEA laser. These early experiments have been described in recent review papers (Andrick, 1980, Weingartshofer and Jung, 1984).

The low-power regime was demonstrated to be in good agreement with theory, i.e., first order perturbation expansion with respect to the laser field in the soft-photon limit, $\hbar\omega \rightarrow 0$. The expression simply relates the one-photon FF cross-section as a fraction of the corresponding scattering cross-section without the laser field, i.e. the elastic scattering cross-section $d\sigma_{e\ell}/d\Omega$ (Kruger and Schulz 1976).

$$\frac{d\sigma}{d\Omega}_{FF}(1) = \frac{p_f}{p_i}\left(\frac{\lambda}{2}\right)^2 \quad \frac{d\sigma_{e\ell}}{d\Omega} \qquad (1)$$

where $\lambda^2 = 1.944 * 10^{-12} \lambda_o^4 F_o E_i \left[\frac{\vec{\varepsilon} \cdot (\vec{p}_i - \vec{p}_f)}{2p_i} \right]^2$

where λ_o is the laser wavelength in microns, F_o is the laser intensity in W cm^{-2}, E_i is the incident electron energy, $\vec{\varepsilon}$ is the laser polarisation normalised according to $\vec{\varepsilon}.\vec{\varepsilon}=1$ and $(\vec{p}_i-\vec{p}_f)$ is the change in the electron momenta. If the elastic scattering cross-section is known we have here an important practical tool to calculate FF cross-sections. It is therefore, very useful to test the range of validity of the soft-photon approximation.

A scattering resonance might be thought to have a strong influence on FF transitions since the lifetime of the negative ion is up to 10^4 times longer than a normal collision. If the soft-photon approximation, which assumes the radiative transitions to occur outside the region of the electron-atom interaction, is valid then this is not the case. The experiments of Andrick and Langhans (1977) confirmed the validity of the soft-photon approximation.

For flux densities of the order of 10^8 W cm^{-2} a perturbation expansion with respect to the laser field no longer applies. In the case of a CO_2 laser, however, a semiclassical soft-photon approach (Kruger and Jung 1978) can be applied, which yields the following cross-section formula for multi-photon FF transitions with a net absorption/emission of n laser-photons

$$\frac{d\sigma_{FF}(n)}{d\Omega} = \frac{p_f}{p_i} J_n^2(\lambda) \frac{d\sigma_{e\ell}}{d\Omega} \quad \text{and} \quad \sum_{n=-\infty}^{+\infty} \frac{d\sigma_{FF}(n)}{d\Omega} = \frac{d\sigma_{e\ell}}{d\Omega} \tag{2}$$

Here $J_n(\lambda)$ is the Bessel function of the first kind and order n and λ is defined as before. The equation to the right represents the Sum Rule that can be derived from the equation to the left and the well-known properties of Bessel functions. We note that $n < 0$ corresponds to emissions and $n > 0$ corresponds to absorption of a net number of photons ($n\hbar\omega$).

The Sum Rule, Eq. (2, right) is one of the most interesting results that was carefully measured in the early experiments (Weingartshofer and Jung 1984) and which provided theorists (Jung 1980) with a solid basis to carry out realistic speculations that have stimulated significant progress in this field.

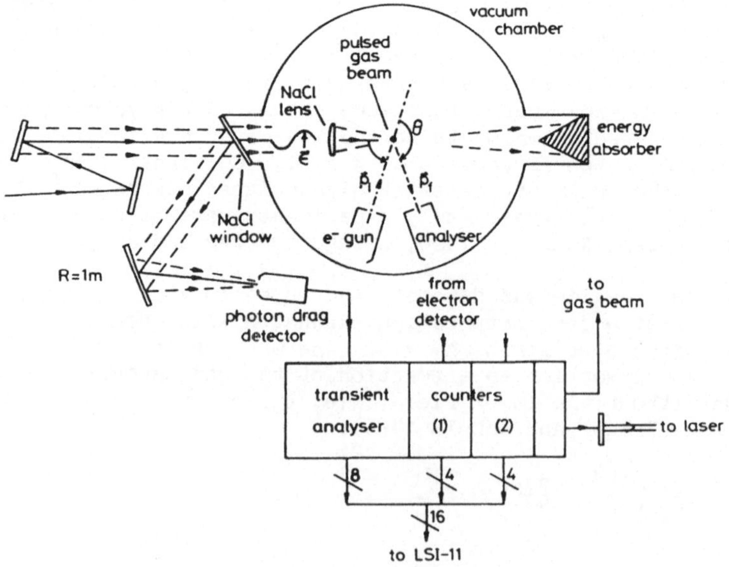

Fig. 1. Schematic diagram of the new three-beam apparatus

The early experiments were carried out with a pulsed CO_2 TEA laser (peak power ca. 10^8 W cm^{-2}) operated in the multimode optical configuration. Considerable effort was devoted to the quantitative confirmation of the Sum Rule, Eq. (2, right). Although this important result is not in disagreement with Eq. (2, left) it does not provide conclusive confirmation. The question of the range of validity of the soft-photon approximation has not been resolved.

The presence of the Bessel function J_n in Eq. (2) is due to the assumption of an ideal model of the radiation field: pure single-mode, spatially homogeneous and linearly polarised. It has been shown that the modifications to the cross-section due to the presence of the pulsed laser are strongly dependent on the model adopted to represent the field of the focused laser pulses in the target region (Daniele et al 1983). Considerable experimental and theoretical work has been done on laser statistics and bandwidth effects on collisionless multiphoton processes (interaction of strong radiation fields with isolated atoms), but in collision theory this kind of investigation is at its very beginning. This study is very essential to understand effects of field correlation on the parameters of atomic collisions in a strong radiation field to compare with experiments.

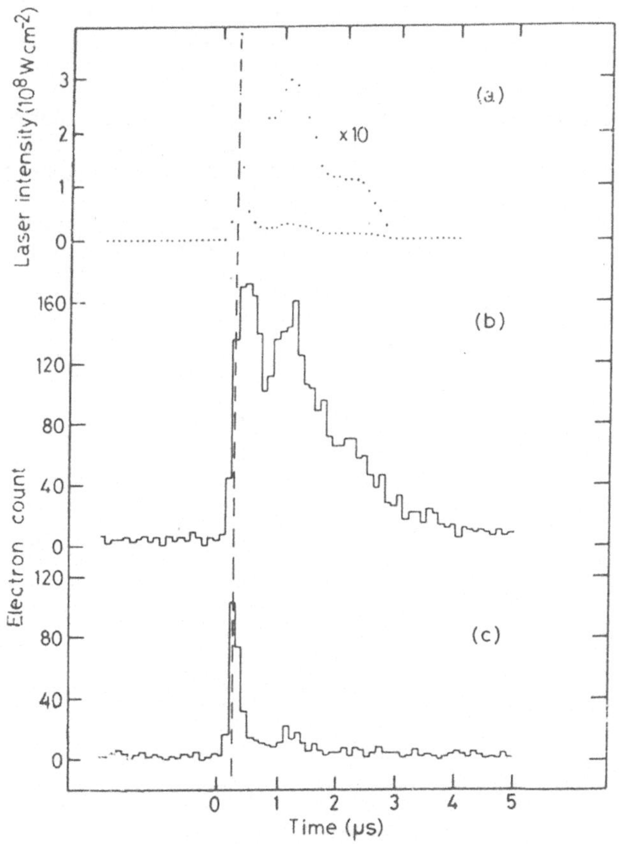

Fig. 2. Experimental electron distributions for electrons that have emitted 1 photon (b) and 2 photons (c) obtained with multimode laser pulses, whose average pulse profile is shown in (a)

Jung (1980) carried out the first practical analysis of experimentally obtained multiphoton FF transition cross-sections and provided a useful tool to compare theory with experiment and this was used to evaluate the data measured up to 1983 (Weingartshofer and Jung 1984). It is indeed very encouraging to find well-established links between theory and experiment but Jung's method of evaluation and also the Sum Rule do not provide a sensitive test for a detailed examination of basic theory, unless the statistics of the experiments can be improved considerably.

The original apparatus has been gradually updated in recent years and important changes have been made. A novel technique has been developed to measure, for the first time, experimental differential FF cross-sections as a function of laser intensity. The new system and details of the technique have been published (Wallbank et al 1987) and only the important aspects of this work shall be reported here. The present arrangement is shown in Fig. 1. The effusive gas beam has been replaced with a pulsed supersonic beam producing very high atom densities in the scattering region. The previous CO_2 laser operating in the multimode optical configuration has now been replaced by a system consisting of a hybrid oscillator, to produce the single mode pulse, followed by two single pass amplifiers to increase the energy in the shaped pulse to ca. 10 J. The scattered electrons are detected and recorded with a home-built counter which also simultaneously digitises and stores (transient analyser) a small fraction of the laser pulse. The data obtained for each laser pulse are aligned with respect to the peak of the laser pulse before being added to the accumulated data stored in a LSI-11 computer system to eliminate variations in timing.

The first experiments were conducted with multimode laser pulses (by not triggering the low-pressure section of the hybrid oscillator) in order to demonstrate the validity of the new experimental arrangement. Data from such a pulse, presented in Fig. 2, are consistent with the experiments of 1983 described above. To be noted in Fig. 2 (a) is the evolution in time of the laser pulse as measured with the transient analyser and in Figs. 2 (b)

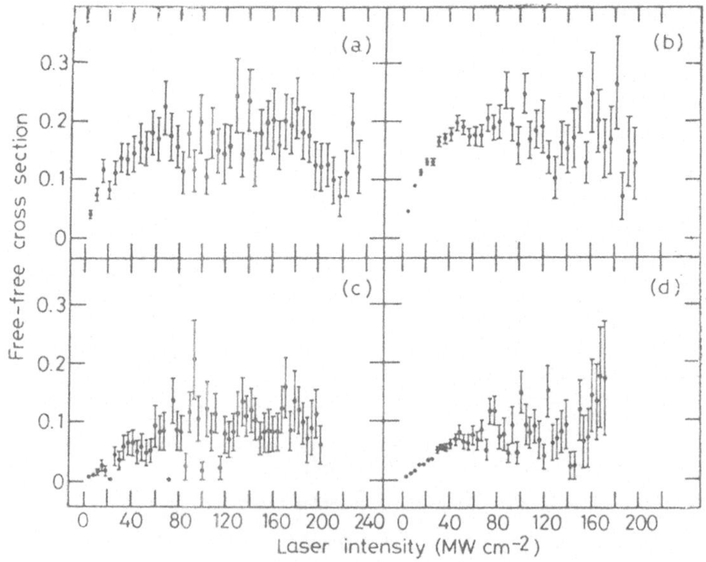

Fig. 3. Experimental free-free cross-section as a function of laser intensity for electrons that have emitted 1 photon ((a),(b)) and emitted 2 photons ((c),(d)) obtained with the "flux sorting" technique.

and (c) the correlated electron distributions for the one- and two- photon emission respectively. The dip in the tail of the laser pulse is clearly reproduced in the one- and two-photon data and at powers a little higher the one-photon cross-section appears to saturate.

A typical example of the results obtained with single-longitudinal-mode pulses is shown in Fig. 3. For these measurements we have to take into account any pulse-to-pulse variation in the laser pulse shape. The technique consists in analyzing the electron counts as a function of intensity for each laser pulse before adding to the accumulated data. It is these "flux-sorted" data which allow us to measure directly the differential cross sections for FF transitions as a function of laser intensity. Fig. 3 shows the experimental FF cross-sections for the one- and two-photon processes. No oscillations of the Bessel function type can be detected in the data. This is the effect of the laser inhomogeneities and the overlap between the electron beam and the laser (the electron beam is at least twice as wide as the laser beam). A complete analysis (Wallbank et al 1987) shows that the general form of the observed cross-sections can be described by using typical theoretical laser models as a guide but they predict smooth functions and offer no explanation for the distinct structures.

A difficult variable to determine is the actual laser intensity experienced by the collision partners. But here again a link can be found between theory and experiment. This value can be extracted by analysing the one-photon cross-section at low flux density values and comparing this with the calculated value from Eq. (1). Very reasonable and consistent values can be determined with this procedure.

References

Andrick, D., 1980, Free-free processes in electron-atom scattering: experiment, in: "Electronic and Atomic Collisions," N. Oda and K. Takayanagi, ed., North-Holland, Amsterdam.

Andrick, D. and Langhans, L., 1977, Measurement of free-free processes in e^--Ar scattering, ICPEAC X Abstracts, 1:350.

Daniele, R., Faisal, F. H. M., and Ferrante, G., 1983, Photon correlation effects in charged-particle scattering in a radiation field, J. Phys. B: At. Mol. Phys., 16:3831.

Faisal, F. H. M., 1987, "Theory of Multiphoton Processes," Plenum Press, New York.

Ferrante, G., 1985, Particle-atom collisions in strong laser field, in: "Fundamental Processes in Atomic Collision Physics," H. Kleinpoppen, J. S. Briggs and H. O. Lutz, ed., Plenum Press, New York.

Jung, C., 1980, Laser pulse-shape independence of the mean energy absorption per electron in multiphoton free-free transitions, Phys. Rev. A, 21:408.

Kruger, H., and Jung, C., 1978, Low frequency approach to multiphoton free-free transitions induced by realistic laser pulses, Phys. Rev. A, 17:1706.

Kruger, H., and Schulz, M. J., 1976, Laser-induced free-free electronic transitions near a resonance, J. Phys. B: At. Mol. Phys., 9:1899.

Wallbank, B., Holmes, J. K., and Weingartshofer, A., 1987, Experimental differential cross-sections for multiphoton free-free transitions, J. Phys. B: At. Mol. Phys., to be published.

Weingartshofer, A., and Jung, C., 1984, Multiphoton free-free transitions, in: "Multiphoton Ionization of Atoms," S. L. Chin and P. Lambropoulos, ed., Academic Press, New York.

THEORY OF MULTIPHOTON IONIZATION

Abraham Szöke

Lawrence Livermore National Laboratory
Physics Department
Livermore, CA 94550

A consistent, quantum-mechanical, many-body theory of multiphoton excitation and ionization is reviewed. The laser pulse is considered to be a classical field consisting of many optical cycles of a single, slowly varying frequency. The results are useful both for numerical calculations and for qualitative interpretation of experiments.

1. INTRODUCTION

The earliest calculation of multiphoton ionization probabilities is that of Bebb and Gold.[1] The most prominent feature of their calculations is the presence of non linear, or multiphoton resonances, which occur when the energy of N photons becomes equal to an atomic energy difference. At intensities of 10^8 - 10^{10} W/cm^2 (considered high at that time but modest today), it is much more probable to excite or ionize an atom close to these resonances than at other wavelengths. Over the following years, experimentalists have been observing these multiphoton resonances and theoreticians have been calculating them, using increasingly sophisticated techniques of diagrammatic perturbation theory. Recent publications by Lambropoulos and associates,[2] Delone and Krainov,[3] and l'Huillier and Wendin[4] exemplify the achievements of these efforts. Setting aside fundamental questions of convergence, perturbation theory does not appear to be the most direct or illuminating way to understand some recent experiments. I refer in particular to the observation of an abundant number of photoelectrons whose energies correspond to the absorption of many more photons than the minimum number needed to ionize a particular atom or ion. Also, photoionization between the resonances is sizeable at intensities of the order of 10^{14} - 10^{16} W/cm^2, so much so that some of the resonances seem to disappear.

A parallel development, started by Keldysh,[5] emphasized the importance of the oscillatory motion of the outgoing electron in the radiation field.[6-9] The wave function of the continuum electron is no longer a simple plane wave, but a frequency modulated wave; in the language of quantum electronics, it is dressed by a large number of photons. This effect is pronounced at long wavelengths, but it can be dominant even in the visible part of the spectrum. In consequence, we expect to see that a large fraction of the photoelectrons absorbs more than the minimum number of photons required to liberate them, and that the ionization rate depends on the wavelength and ionization energy rather than on the detailed atomic structure.

A third theoretical line, usually called Floquet theory, was developed recently by Chu and Reinhardt.[10] It calculates the atomic wave function non-perturbatively by finding stationary states of the atom in a periodic, arbitrarily strong laser field. These states have complex eigenvalues, whose imaginary parts are ionization rates. Recently Chu[11] has incorporated some of the dynamics associated with the pulsed nature of the laser in his calculations.

The rest of this paper will review a more comprehensive theory[12] which reduces to all the previous ones in the proper limits. The theory is described in Section 2. Section 3 contains some useful formulas, and Section 4 presents model calculations that are compared with experiments. Section 5 outlines some work yet to be done, and a general outlook.

2. THEORY

Physical Ideas

The central goal of the present theory is to include as much of the motion of bound and free electrons as convenient into non-perturbative formulas and to do perturbation theory on the rest. The first mathematical tool used is the two-time expansion technique. Its applicability follows from the observation that all lasers used for multiphoton ionization are pulsed, and that the relative bandwidth of the laser is a small parameter; thus the laser amplitude and frequency are slowly varying quantities. The theory identifies a natural basis set: it is the set of Floquet states of the atom in a monochromatic laser field, a similar set of Floquet states describing an electron scattered by a singly ionized atom in the laser field, etc. These basis sets are defined in the framework of the theory of rearrangement collisions. The dynamics of multiphoton ionization is described as follows. The atom, which is in one of its states (usually its ground state) at the start of the laser pulse, stays initially in the Floquet state adiabatically connected to it. As the laser gets stronger, the quasi energies of all these Floquet states change and avoided level crossings can occur, corresponding to multiphoton resonances. During all this time, the atom gets ionized through transitions to the continuum, in a picture that closely resembles the usual picture of single photon ionization; in fact in most cases it can be adequately described by very similar, first order (Fermi golden rule) formulas. The multiphoton character of the transition is incorporated into the bound- and free-electron wave functions, which become dressed by many photons.

Formal Theory

We are interested in solving the Schrodinger equation,

$$i\hbar \frac{\partial\Psi}{\partial\tau} = H\Psi \qquad (2.1)$$

subject to the initial condition, $\Psi \to \Psi_0$, as $t \to -\infty$. Usually Ψ_0 is the atomic ground state. We will use a semiclassical Hamiltonian,

$$H = \frac{1}{2m}\sum_{i=1}^{K} [-i\hbar\bar{\nabla}_i - \frac{e}{c}\bar{A}(\bar{r}_i,t)]^2 + \sum_{i=1}^{K} V(\bar{r}_i) + \frac{1}{2}\sum_{i\neq j}^{K} \frac{e^2}{|\bar{r}_i - \bar{r}_j|} + H', \qquad (2.2)$$

where $A(r_i,t)$ is the vector potential of the laser pulse, $V(\bar{r}_i)$ is a central potential, $\bar{r}_i$ are the electron coordinates, and H' may include spin-orbit interaction or relativistic corrections. It is convenient to split up the Hamiltonian into two parts,

$$H = H_0 + H_1 \tag{2.3}$$

where H_0 is e.g. a Hartree - Fock Hamiltonian in the external field; the interaction with the external field is always incorporated into H_0. In the dipole approximation, for a (single color) narrow band laser, of band width $\Delta\omega$ the vector potential can be written as

$$\bar{A}(t) = \bar{A}_0(t) \cos[\int^t \omega(t')\,dt'] \tag{2.4}$$

where the amplitude, $\bar{A}_0(t)$ and the instantaneous frequency $\omega(t)$ are both slowly varying functions. The two - time formalism is introduced by defining two dimensionless time-like variables: the fast time,

$$\tau_o = \int^t \omega(t')\,dt' \tag{2.5}$$

which is really the phase of the optical field, and the slow time,

$$\tau_1 = \varepsilon\tau_0 \quad ; \quad \varepsilon = \overline{\Delta\omega}/\bar{\omega} = \text{const.} << 1 \tag{2.6}$$

The vector potential, Eq. 2.4, has the form

$$\bar{A}(\tau_0, \tau_1) = \bar{A}_0(\tau_1) \cos \tau_0 \tag{2.7}$$

and using the chain rule,

$$\frac{\partial}{\partial t} = \omega(\tau_1) \left[\frac{\partial}{\partial \tau_0} + \varepsilon \frac{\partial}{\partial \tau_1}\right] \tag{2.8}$$

we get from Eq. 2.1 the "two-time" Schrodinger equation,

$$i\hbar\omega(\tau_1) \left[\frac{\partial}{\partial \tau_0} + \varepsilon \frac{\partial}{\partial \tau_1}\right] \Psi(\bar{r}_i, \tau_0, \tau_1) = H(\bar{r}_i, \tau_0, \tau_1)\, \Psi(\bar{r}_i, \tau_0, \tau_1). \tag{2.9}$$

So far this looks like empty formalism. The essence of the method is to solve this equation for arbitrary values of τ_0, τ_1 not necessarily constrained by their connection to the physical time (but usually close to it). When finally these times are made to satisfy the constraints (Eqs. 2.5, 2.6), solutions of Eq. 2.9 reduce to those of the real Schrodinger's equation, 2.1. The advantage of the formalism is that for a constant value of τ_1, the Hamiltonian is strictly 2π periodic in τ_0.

If we stop at this point, we can rederive Floquet theory.[12] In its customary form[10,11] a complex coordinate transformation is introduced in order to deal with convergence difficulties of the wave function. These difficulties stem from the fact that in the presence of the radiation field the atom is unstable: its ionization rate is the very effect we try to calculate. An alternative and more general way is to use the projection operator method of Feshbach, as described in Newton.[13]

We define a complete, orthogonal set of projection operators, P_i, $i = 0, \ldots K$, which project onto the unionized, singly ionized, etc. arrangement channels. These channels are defined in terms of the boundary conditions on the wave function: the unionized channel is L^2 normalizable; the first ionized channel has a single electron going to infinity, which in our context means ingoing, (-) type, boundary conditions; etc. We decompose H_0 as

$$H_o = \left(\sum_i P_i\right) H_o \left(\sum_i P_i\right) = \sum_i P_i\ H_o\ P_i + \sum_{i \neq j} P_i\ H_o\ P_j \quad . \tag{2.10}$$

We now consider the set of equations

$$\left[\sum_i P_i\ H_o\ P_i - i\hbar\omega \frac{\partial}{\partial \tau_o}\right] \Psi = 0 \quad . \quad i = 0, \ldots, K \tag{2.11}$$

These are the Floquet equations in the individual arrangement channels. These equations contain the slow time, τ_1 as a parameter, but they have the correct mathematical properties: they are strictly 2π periodic, their solutions are normalizable, (L^2-, or delta-function- normalizable, for $i = 0$, and $i > 0$ respectively.) Their solutions are of the Floquet form,[14]

$$\Psi_\lambda^{(i)}(\bar{r}_i, \tau_o; \tau_1) = \exp[- i\ e_\lambda^{(i)}(\tau_1)\ \tau_o]\ \phi_\lambda^{(i)}(\bar{r}_i, \tau_o; \tau_1)$$

$$\phi_\lambda^{(i)}(\bar{r}_i, \tau_o + 2\pi; \tau_1) = \phi_\lambda^{(i)}(\bar{r}_i, \tau_o; \tau_1), \tag{2.12}$$

where $e_\lambda^i = E_\lambda^i/\hbar\omega$ is the dimensionless Floquet exponent, or quasi - energy. The physical interpretation of these functions gives insight into the multiphoton ionization process: the individual eigenstates (steady states) which are the states of the atom in the strong field, account for the largest part of the effect of the field on the atom.

Now we consider the dynamics. We will do this in an elementary way;[15] a more advanced, formal solution can be found in Newton.[13] The wave function is written in a general form, as a superposition of solutions of Eq. 2.11, which are of the form, Eq. 2.12

$$\Psi(\bar{r}_i, \tau_o, \tau_1) = \sum_{i=o}^{K} \sum_\lambda b_\lambda^{(i)}(\tau_o, \tau_1)\ \phi_\lambda^{(i)}(\bar{r}_i, \tau_o, \tau_1) \tag{2.13}$$

Here, we expanded in terms of the periodic parts of Eq. 2.12 and we wrote out the time- and space-dependence of the quantities explicitly. After defining the adiabatic Floquet exponent,

$$b_\lambda^{(i)}(t) = a_\lambda^{(i)}(t)\ \exp\left[- i \int^t \omega_\lambda^{(i)}(t')\ dt'\right] \tag{2.14}$$

where the quasi energy is

$$\omega_\lambda^{(i)}(t) \equiv \frac{E_\lambda^{(i)}(t)}{\hbar} = \omega(t) \left[e_\lambda^{(i)}(t) - i\ \varepsilon \left\langle \phi_\lambda^{(i)*} \frac{\partial \phi_\lambda^{(i)}}{\partial \tau_1} \right\rangle - \frac{i}{\hbar} \left\langle \phi_\lambda^{(i)*}\ H_1\ \phi_\lambda^{(i)} \right\rangle \right] \tag{2.15}$$

we get the equations,

$$\frac{da_\lambda^{(i)}}{dt} = - \sum_j \sum_\mu (1 - \delta_{ij}\ \delta_{\lambda\mu})\ a_\mu^{(j)}\ \exp\left[- i \int^t (\omega_\lambda^{(i)} - \omega_\mu^{(j)})\ dt'\right]$$
$$\left\langle e^{-i(n_\mu - n_\lambda)\ \tau_o} \left[\varepsilon\ \omega\ \phi_\lambda^{(i)*} \frac{\partial \phi_\mu^{(j)}}{\partial \tau_1} + \frac{i}{\hbar} (1 - \delta_{ij})\ \phi_\lambda^{(i)*} P_i H_o P_j \phi_\mu^{(j)} + \frac{i}{\hbar}\ \phi_\lambda^{(i)*} P_i H_1 P_j \phi_\mu^{(j)} \right] \right\rangle . \tag{2.16}$$

The pointed brackets, < >, are the usual scalar product of Floquet theory,[14] defined as

$$\langle \phi_\mu^* \phi_\lambda \rangle \equiv \frac{1}{2\pi} \int_0^{2\pi} d\tau_o \int d\bar{r}_i \, \phi_\mu^* (\bar{r}_i, \tau_o; \tau_1) \, \phi_\lambda (\bar{r}_i, \tau_o; \tau_1) = \delta_{\lambda\mu} \qquad (2.17)$$

The adiabatic quasi energy, Eq. 2.15, contains the shift due to H_1 and the Berry phase.[16]

The grand conclusion of this development is that these equations are identical to the formulas of standard time dependent perturbation theory, where the unperturbed wave functions are solutions of the Floquet Hamiltonian,

$$H_F = \sum P_i H_0 P_i \qquad (2.18)$$

giving the natural basis set

$$\Psi_\lambda^{(i)}(\bar{r}_i, t) = \exp\left[- i \int^t \omega_\lambda^{(i)}(t') \, dt'\right] \phi_\lambda^{(i)}(\bar{r}_i, t) \qquad (2.19)$$

with the perturbation Hamiltonian represented symbolically as

$$H' \equiv (1 - \delta_{ij} \, \delta_{\lambda\mu}) \left[- i \hbar \varepsilon \omega \frac{\partial}{\partial \tau_1} + (1 - \delta_{ij}) H_o + H_1\right] \qquad (2.20)$$

This is a very satisfactory result; it shows that the usual quantum mechanical formalism applies to these strong field problems almost unchanged if the correct basis set is used. Recent work by Baranger[17] has shown how to build a diagrammatic perturbation theory on a time dependent basis.

A more formal and general solution starts with H_F of Eq. 2.18 as the unperturbed Hamiltonian and H' of Eq. 2.20 as the perturbation. The formal development of Newton[13] can be taken over in its entirety; e.g. a Lippman-Schwinger equation can be written in terms of Floquet states.

Discussion of the Theory

This section presents some of the qualitative predictions of the theory which can be discussed without detailed calculations.

i.) The dynamics within the unionized arrangement channel can be described as follows. When the laser slowly turns on, the atom stays in the Floquet state that is adiabatically connected to its initial state. In particular, if the laser is weak, the atom returns to its ground state when the laser turns off. As the laser gets stronger, the quasi energies of all states change; it is called the AC Stark effect. At some intensity, some of these quasi energies may become equal. This corresponds to a multiphoton resonance in ordinary quantum mechanics. In the presence of strong fields, there is always some interaction between levels (unless it is forbidden by symmetry) resulting in an avoided crossing. As the laser's intensity or frequency changes, these resonances are traversed at varying rates. The simplest way to account for the dynamic effect of isolated avoided crossings is to use the Landau - Zener formula,[18] which assumes a completed passage during the laser pulse:

$$P = 1 - e^{-w} \quad ; \quad w = \frac{2\pi}{\hbar} \frac{\Delta E_{min}^2}{|d(\Delta E_d)/dt|} \qquad (2.21)$$

In the case of weak coupling, i.e. an unsaturated resonance, the formula reduces to the perturbation theory result: the total transition to the excited state is proportional to the square of the matrix element and to the time of the passage. When the coupling is strong and the time of passage is long, an adiabatic passage results to the excited state. The behavior of the ion in the first ionized channel (etc.) is entirely similar.

ii.) The coupling to the next higher channel can be solved to lowest order using the same arguments that lead to Fermi's golden rule.[13] The underlying assumptions are, in addition to the narrow bandwidth of the laser, that the ionization rate is low compared to the optical frequency and that the frequency dependence of the transition rate is weak, i.e. there are no overlapping resonances, or "Fano profiles." The ionization rate is then,

$$W_N = \frac{2\pi}{\hbar} \left| < \phi_\mu^{(i+1)} \left| P_{i+1} H' P_i \right| \phi_\lambda^{(i)} > \right|^2 \rho(E_\mu)\, \delta[\hbar(\omega_\mu^{(i+1)} - \omega_\lambda^{(i)}) - N\hbar\omega] \tag{2.22}$$

Note that the multiphoton character of the ionization is incorporated into the dressed channel wave functions. The density of states is modified from the field free case.[12] Also, the delta function expresses the conservation of the quasi energy, thus it correctly includes the outgoing electron's quiver energy. It is very important that the outgoing electron wave function is a solution of the P_1HP_1 Hamiltonian, therefore there is no interference among the outgoing electron waves. This is not true about any other outgoing waves, especially not plane waves in the absence of the electromagnetic field.

iii.) The low intensity limit is obviously correct, as all the quantities that appear in Eq. 2.22 reduce to the perturbation theory result. In particular, the formula for single photon ionization and for the Auger effect is correct. Eq. 2.22 does not include Fano profiles of overlapping resonances; the treatment can be modified to do that. It is relatively easy to see that N photon ionization, at sufficiently low intensities, is <u>always</u> proportional to I^N, where I is the laser intensity. It is more difficult to see that in the off-resonant case, Eq. 2.22 incorporates the sum of all second order diagrams that are needed to obtain the correct frequency dependence of the atomic polarizability.

The central assumption of most perturbation treatments of multiphoton ionization has been that only the atomic electrons are perturbed by the laser field. This is equivalent to the assumption that the wave function in the ionized channel is unperturbed, and that the perturbed atomic wave function is calculated by diagrammatic perturbation theory. It is clear that perturbation theory is a special case of Eq. 2.22. That equation also shows that perturbation theory should work well close to resonances and at high laser frequencies, where the polarizability of the bound electrons is higher than that of the free electrons.

The diametrically opposite assumption is that the bound electrons are not perturbed by the field, and that the laser photons dress only the free electrons. This, the Keldysh approximation, is clearly another special case of Eq. 2.22. (Recently that approximation has been modified to include the atomic potential.[19,20]) Above- threshold ionization, i.e. the absorption of a larger than the minimum number of photons by an electron, can occur in both the resonant (perturbation) and the non resonant (Keldysh) limit. The essential difference is that in the former limit both the ionization rate and the electron energy spectrum are expected to be sensitive functions of both the laser wavelength and the atomic specie; while in the latter limit the ionization rate depends smoothly on the atomic ionization energy and the laser wavelength, and the electron energy spectrum depends on the laser wavelength and intensity alone.

Floquet theory, especially in its new form by Chu,[11] comes closest to the present treatment. If only ionization rates are needed and there are no overlapping resonances, the calculation of the imaginary part of the complex pole is a more accurate and simpler calculation than our prescription. If electron energy distributions, or angular distributions are calculated, this simplicity disappears, and in the presence of overlapping resonances Floquet theory becomes problematic. This is not different from the situation in other areas of collision physics.

iv.) The above theory can be put into an even more general, formal and intimidating form. The amplitude and frequency of the laser field constitute a two-dimensional space that will be taken as the continuous part of the base space. The Hilbert space of the quantum mechanical wave functions at each point in the base space provide a manifold. The projection operators cut this manifold into K ordinary Hilbert spaces, where K is the number of electrons in the atom. Thus each of these Hilbert spaces is a fiber on this two dimensional continuous space, plus the K discrete points, and the whole multiphoton ionization problem is defined on a fiber bundle. The time evolution of the wave function is equivalent to finding a connection on this space: the wave function is represented by a point in this space and, as the laser amplitude and frequency change, it is transported in this fiber bundle. The advantage of looking at it this way (in addition to its snob appeal) is that some powerful theorems may be evoked; e.g. the Berry phase appears quite naturally.

3. SOME USEFUL FORMULAS

The multiple scale method can be used to obtain some known and useful results.[21] It can be shown that to first order a short laser pulse focuses similarly to a long laser pulse: the optics provide a "path", and the laser pulse propagates on it. Also the classical motion of an electron in the electromagnetic field can be integrated in a series of successfive approximations. The zeroth approximation reproduces the electron's "quiver" motion, its "figure eight" motion, and proves the presence of the ponder-motive potential.[22] Once the canonical equations are solved, the action function (Hamilton's first principal function) can be obtained by quadrature[23] and the WKB wave function can be written down. The (unnormalized) wave function is

$$\psi = \frac{1}{\sqrt{p}} \exp[\frac{i}{\hbar}(\bar{p}\cdot\bar{x} - E_F t - \hbar n_f \sin\omega t + \frac{1}{2}\hbar\, n_{osc} \sin 2\omega t)] \tag{3.1}$$

where the Floquet energy is

$$E_F = \hbar\omega(n_k + n_{osc}) = \frac{p^2}{2m} + \frac{e^2 A^2}{4mc^2} \tag{3.2}$$

and

$$n_f = \frac{e\bar{p}\cdot\bar{A}}{mc\hbar\omega} = (8 n_k n_{osc})^{1/2} \cos\theta \tag{3.3}$$

These formulas define the important parameters determining the motion of an unbound electron. In particular, $n_f + n_{osc}$ is the approximate number of photons dressing the free electron. In a single-electron ionization this is one ingredient in the wave function $\phi_\mu^{(i+1)}$ in Eq. (2.22).

Another simple, exactly soluble problem is a one-dimensional harmonic oscillator in the strong electromagnetic field.[25] In a steady field, the solutions are again of the Floquet form, they allow one to define a bound-electron parameter,

$$n_b^2 = n_{osc} \frac{4\omega_o \omega^3}{(\omega_o^2 - \omega^2)^2} \tag{3.4}$$

where n_{osc} is the same as in Eq. 3.2, and ω_0 is the harmonic oscillator resonance frequency. This equation shows clearly that, as expected, the dressing of the bound electron dominates multiphoton ionization near resonance. If an electronic transition is modeled as a harmonic oscillator, with oscillator strength, f, the condition $n_f < n_b$ implies

$$\frac{(\omega_0^2 - \omega^2)^2}{\omega_0 \omega^3} < \frac{f}{2n_k} \qquad (3.5)$$

i.e., the laser has to be close to resonance.

4. COMPUTER MODELS OF MULTIPLE IONIZATION

Two simple computer models were designed to explore some aspects of multiphoton ionization. The first one is a one dimensional, cutoff harmonic oscillator in a strong, steady electromagnetic field. An exact, analytic solution can be found both in the inner- and the outer regions; the former in terms of parabolic cylinder functions, the latter in terms of Volkov states. A superposition of many such solutions is needed to ensure continuity of the wave function and of its derivative at the boundary of the inner and outer regions. When it is demanded that in the outer region the wave function have outgoing waves only, an eigenvalue equation results with complex eigenvalues; the resulting states are decaying states. The computer program follows the state adiabatically connected to the ground state, as the electromagnetic field intensity is increased in steps. It was shown explicitly that the spectrum has many ATI peaks, that the decay rates can be calculated both by the outgoing flux (i.e. explicitly) and by the imaginary part of the appropriate eigenvalue. The most interesting finding is that level crossings (multiphoton resonances) occur commonly under realistic values of excitation energy, of ionization energy and of laser frequency, and intensity. Unfortunately, the way the computer program is written, it is very difficult to follow the states beyond the level crossing. Some numerical results are presented in Fig. 1.

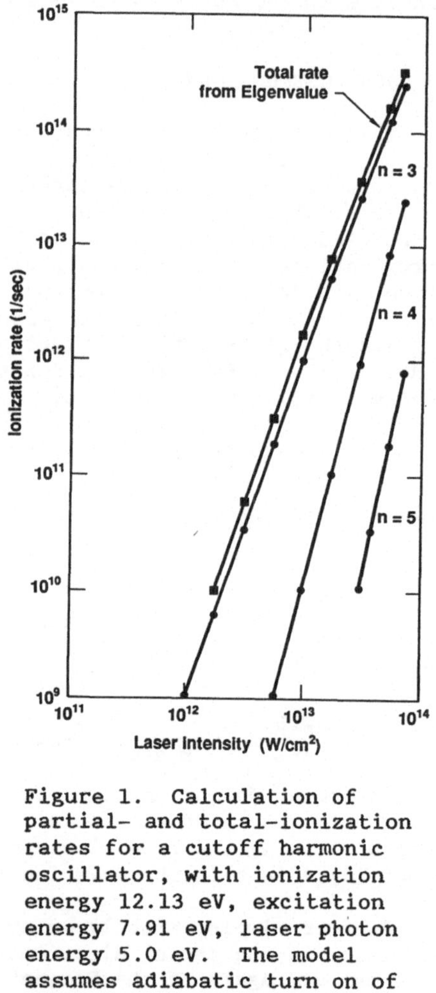

Figure 1. Calculation of partial- and total-ionization rates for a cutoff harmonic oscillator, with ionization energy 12.13 eV, excitation energy 7.91 eV, laser photon energy 5.0 eV. The model assumes adiabatic turn on of the laser; the partial rates are labeled by the number of photons absorbed. The highest intensity corresponds to a multiphoton resonance.

The second computer program evaluates a modified Keldysh approximation. The starting point is the exact Fermi golden rule expression, Eq. 2.22, for multiphoton ionization. The model is three dimensional, but it treats only a single electron. The bound electron wave function is hydrogenic, with the appropriate ionization energy, but unmodified by the laser field. The free electron wave function is modeled as a plane wave in the laser field (i.e. a Volkov state) in a constant potential whose depth is equal to the ionization energy. The model results in the following formula for the ionization rate,

$$w = \sum_{n=n_o}^{\infty} \frac{32\omega(n-n_b-n_{osc})\, n_b^{5/2} n^2}{(n-n_{osc})^{1/2}(n+n_b-n_{osc})^4} \int_0^1 J_n^2\left(n_f\mu, -\frac{n_{osc}}{2}\right) d\mu \qquad (4.1)$$

$$n_b = \frac{E_b}{\hbar\omega}, \quad n_{osc} = \frac{e^2A^2}{4mc^2\hbar\omega} \qquad n_f = (8(n-n_{osc})n_{osc})^{1/2}$$

where E_b is the atom's (or ion's) binding energy and $J_n(a,b)$ is a modified Bessel function, as defined by Reiss.[9] The individual terms are the correct expression for the differential cross section of absorption of N photons and emitting an electron in a given direction. The program calculates photoelectron angular distributions for each electron peak, corresponding to the absorption of N photons, beginning with the minimum number required for ionization. In a linearly polarized laser field a narrow peaked distribution is obtained, occupying about 10^{-2} of the sphere. The angular distributions are integrated to get the individual ATI peaks, and they are in turn added up to get the ionization rate of each ionic species at a given intensity. The total intensity of the individual ATI electron peaks is in good agreement with measurements. The program then integrates the rate equations for a given laser pulse shape, and tabulates the number of ions of each species at the end of the laser pulse. Finally, the total number of ions is calculated for a Gaussian laser beam. It should be stressed that the model has no adjustable parameters, once the measured ionization energies are introduced, and the laser wavelength and pulse length are fixed. Unexpectedly good agreement is obtained with our own measurements of ion yields in Ar, Kr, Xe at 586 nm (visible) wavelength, with 2 psec laser pulses. Figure 2 shows the Ar data and Fig. 3 is an overall plot of the threshold intensity of all these gases plotted against their ionization energy.

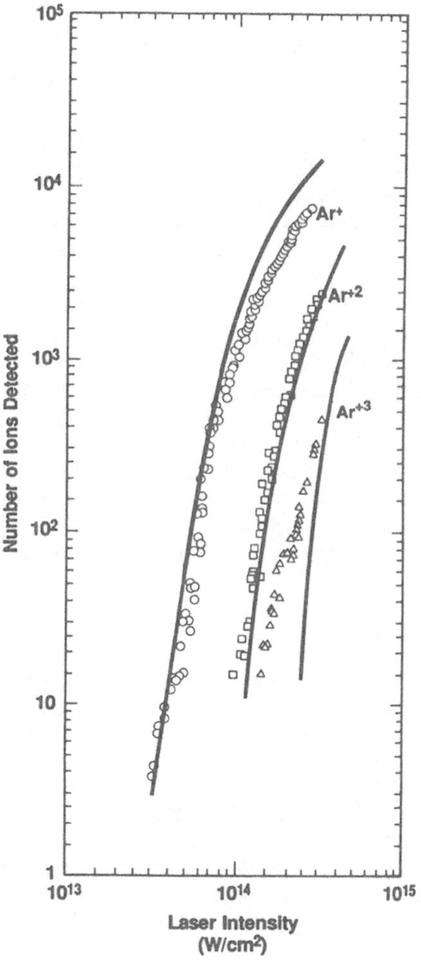

Figure 2: Comparison of argon ion yields, from Perry et al. (Ref. 19), and the modified Keldysh model (Eq. 4.1). Laser wavelength is 586 nm, pulse length is 2 psec.

Also, good agreement with experiment is obtained for 248 nm measurements in He, Ne; but not in Ar Kr Xe - apparently resonance effects are

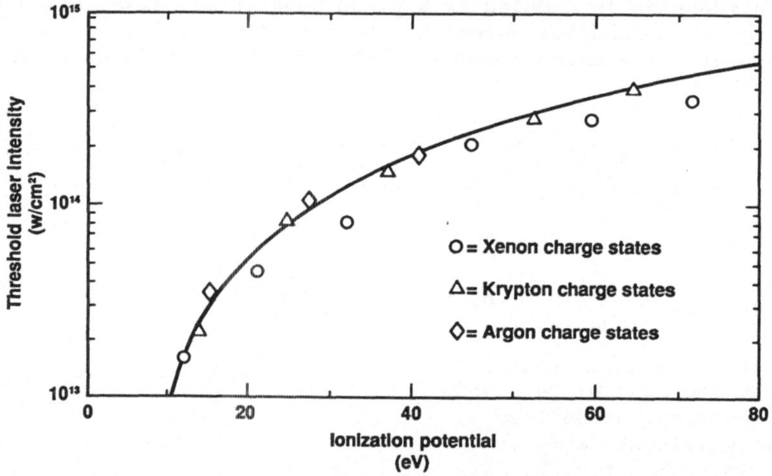

Figure 3: Threshold intensities for production of multiply ionized Ar, Kr, Xe plotted against the ionization potentials of the parent species. The smooth dependence suggests a Keldysh-type mechanism. (From Perry et al., Ref. 19.

important in the latter. The angular- and ATI peak-distributions of the photoelectrons are well modeled for 1.06 micron wavelength in Xe, but the absolute rates are high by about 300. This is probably a reflection that the classical quiver motion of the electron is getting of the order of the atomic radius at these long wavelength, therefore it is incorrect to model the Coulomb potential as a constant pseudopotential. These calculations delineate both the importance of the non-resonant contribution to multiphoton ionization and its limitation in the presence of the resonant part, which depends in an essential way on the structure of the atom.

5. OUTLOOK

The theory presented here explains qualitatively what is known today about multiphoton ionization of atoms. It can be easily extended to incorporate the time-dependent Hartree-Fock method; the fundamental observation being that if each single-electron wave function is of the Floquet form, so is the Slater determinant and the Hartree-Fock (averaged) potential. This gives the hope of doing computer calculations on the slow time scale, rather than on the fast one. A second interesting possibility is to use time-propagation methods to determine the stationary (Floquet) states[26] and trying to follow as few of them as needed to describe the time evolution of the wave function.

Work performed under the auspices of the U.S. Department of Energy by the Lawrence Livermore National Laboratory under contract number W-7405-ENG-48.

REFERENCES

1. H. B. Bebb and A. Gold, Phys. Rev. 143, 1 (1966).
2. X. Tang and P. Lambropoulos, Phys. Rev. Letters 58, 108 (1987).
3. N. B. Delone and V. P. Krainov, Atoms in Strong Light Fields Springer, New York, 1985).
4. A. L'Huillier and G. Wendin, J. Phys. B 20, L37 (1987).
5. L. V. Keldysh, Sov. Phys. JETP 20, 1307 (1965).
6. N. M. Kroll and K. M. Watson, Phys. Rev. A 8, 804 (1973).
7. F. H. M. Faisal, J. Phys. B6, L89 (1973).
8. M. H. Mittleman, Phys. Lett 47A, 55 (1974); G. Pert, J. Phys. B 8, L173 (1975).
9. H. R. Reiss, Phys. Rev. A 22, 1786 (1980), J. Phys. B 20, L79 (1987).
10. S-I. Chu, Adv. Mol. Phys. 21, 197 (1985).
11. S-I. Chu, Adv. Chem. Phys. (1987).
12. A. Szöke, Lawrence Livermore National Laboratory, Livermore, CA, UCRL 95294 (1986).
13. R. A. Newton, Scattering Theory of Waves and Particles, 2nd Edition (Springer, New York, 1982).
14. J. H. Shirley, Phys. Rev. 138, B979 (1965); Ya. B. Zeldovich, UFN 110 (1973) [Sov. Phys. Usp. 16, 427 (1973)]; H. Sambe, Phys. Rev. A7, 2203 (1973).
15. L. I. Schiff, Quantum Mechanics, 3rd Edition (McGraw-Hill, New York, 1968).
16. M. V. Berry, Proc. Roy. Soc. (London) A392, 45 (1984); J. C. Garrison, Lawrence Livermore National Laboratory, Livermore, CA, UCRL-94267 (1986).
17. M. Baranger, I. Zahed, Phys. Rev. C29, 1005, 1010 (1984).
18. R. D. Levine, Quantum Mechanics of Molecular Rate Processes, (Oxford, 1969).
19. M. D. Perry, O. L. Landen, A. Szöke, and M. Campbell, Phys. Rev. A., Submitted 1987; Phys. Rev. Letters, submitted 1987.
20. A. Szöke, J. Phys. B, submitted 1987.
21. A. Szöke, J. Garrison UCID-21103.
22. T. W. B. Kibble, Phys. Rev. 150, 1060 (1966); P. Mulser, J. Opt. Soc. Am. B2, 1814 (1985); W. Becker, R. R. Schlicher, M. D. Scully, J. Phys. B19, L785 (1986); P. H. Bucksbaum, M. Bashkansky and T. J. Mclrath, Phys. Rev. Letters 58, 349 (1987).
23. V. I. Arnold, Mathematical Methods of Classical Mechanics, (Springer, 1980).
24. L. Rosenberg, Adv. At. Mol. Phys. 18, 1 (1982); M. H. Mittleman, Introduction to the Theory of Laser-Atom Interactions, (Plenum, 1982).
25. E. H. Kerner, Can. J. Phys 36, 371 (1958).
26. J. N. Bardsley, B. Sundanam, L. A. Pinnaduwage and J. E. Bayfield, Phys. Rev. Letters 56, 1007 (1986).

ELECTRON ENERGY SPECTRUM

AFTER MULTIPHOTON IONISATION

Michèle Crance

Laboratoire Aimé Cotton
C.N.R.S.II Bat 505
91405 Orsay Cedex France

I INTRODUCTION

The energy spectrum of electrons emitted after Multiphoton Ionisation has been observed for several years in alkali atoms and noble gases (Fabre et al, 1982; Kruit et al, 1983; Petite et al, 1984; Compton et al, 1984; Humpert et al, 1985; Rhodes, 1985; Lompré et al, 1985; Yergeau et al, 1986; Bucksbaum et al, 1986). When an atom is irradiated by a strong electromagnetic field, an electron may be emitted with energy

$$E_k = (n+k)\hbar w - E_I \qquad (1)$$

for $E_k > 0$. w is the frequency of the field, E_I is the ionisation potential of the atom. According to perturbative interpretation, one expects that the number of electrons N_k with energy E_k results from absorption of n+k photons and thus varies as the (n+k)-th power of light intensity I. Actually, this has been observed for Caesium and noble gases irradiated by Neodymium YAG laser light at moderate intensity (Fabre et al, 1982; Petite et al, 1984). For large intensities, the behaviour of the N_k's departs from power laws. The envelope of the peaks in electron energy spectrum is bell-shaped, the maximum of the curve shifts to high energy as the intensity is increased. The most striking feature in this evolution is the vanishing of first peaks. However, the magnitude of high energy peaks is still governed by power laws. In the framework of perturbation theory, applied at minimum non vanishing order, it has been shown that, for a given light intensity, the ratio between two successive peaks is roughly constant: $N_{k+1}/N_k = N_k/N_{k-1}$. When the envelope of the peaks is bell shaped, this feature is still observed for high energy peaks that is after the maximum. Such an observation suggests a statistical distribution of energy which will be the basis of the treatment presented in section II.

There has been numerous attempts to interpret the vanishing of first peaks in electron energy spectrum. Some of them are based on model atom (Bialinicka Birula, 1984; Deng et al, 1985; Mittleman, 1984; Muller et al, 1983; Edwards et al, 1985; Keldysh, 1965; Reiss, 1980; Kupersztych, 1987). Some invoke a ponderomotive potential which would shift the ionisation threshold (Mittleman 1984). Another type of approach consists of a non perturbative treatment of the atom plus light interaction. Chu and Reinhardt (1977), Crance and Sinzelle (1985), Chu and Cooper (1985) and more recently Giusti-Suzor and Zoller (1987) have solved the stationary problem of finding perturbed energy for a dressed atom or equivalently for a Floquet atom-field Hamiltonian. None of them obtains a reversal of peaks height (that is first peak lower than second peak possibly lower than third peak...) except with an ad hoc introduction of ponderomotive potential (Chu and Cooper 1985). The latter calculations all treat a one electron atom: Hydrogen in (Chu and Reinhardt, 1877; Chu and Cooper, 1985; Giusti-Suzor and Zoller, 1987), Lithium in (Crance and Sinzelle, 1985). Another group of works follows from the treatment initially proposed by Keldysh (1965) (Reiss, 1980; Kupersztych, 1987). For continuum states, the atomic potential is implicitly assumed to be short range, so that the electron is no longer subjected to Coulomb interaction once it has been ejected. The latter treatments reproduce the bell shape of the electron energy spectrum. The comparison between results obtained with and without Coulomb field is not conclusive. Computations in the former case are much more difficult and cannot be carried out beyond an intensity of about 10^{-2} a.u. It is thus difficult to know whether the difference in results is due to the restriction of intensity range or to the nature of the potentials. We shall discuss this problem is section IV.

In a previous work (Crance, 1984a), we have presented a statistical interpretation for the energy spectrum of electrons ejected after multiphoton ionisation which reproduces the power laws obtained for high energy peaks but not the vanishing of first peaks. In the latter work we did not investigate how an electron ejected from an atom leaves the light beam and reaches the detector or, in other words, how the energy spectrum of ejected electrons transforms into the energy spectrum of detected electrons. In section II, we recall the basic principles of statistical interpretation of multiphoton ionisation. Then we use classical arguments to show how an ejected electron leaves (or not) the interaction volume. This leads us to describe the spectrum of detected electrons.

II STATISTICAL DESCRIPTION OF MULTIPHOTON IONISATION

When studying the energy spectrum of electrons ejected after multiphoton ionisation from a theoretical point of view, one has to find a way to determine transition probabilities when an atom is irradiated by an intensity I during an interaction time t. One has to define the ionisation probability $P(I,t)$ and the probability $P_k(I,t)$ for an electron to reach continuum number k. We focus on the case of noble gases irradiated by a laser beam. The outershell consists of m electrons which have comparable orbital size and extraction potential. We consider that all m electrons have an equal probability to absorb a photon. An electron will be ejected if it absorbs a sufficiently large number of photons, say n. Suppose that p photons are available for absorption, we calculate the possible distributions of p photons among m electrons and deduce the ionisation probability $Q(p)$ as the probability that an electron has absorbed n photons or more. In such a procedure, we treat on a same footing the absorption of one, two, three, ...n, ...p photons. However, absorption of less than n photons is a virtual absorption while absorption of n photons or more may be real. It is thus justified to compare the probability of virtual and real absorption only if we consider an interaction time short enough for Heisenberg uncertainty principle to be satisfied. The energy defect is at most E_I so the interaction time to be defined is $T = h/E_I$. The size of the atom can be characterised by the geometrical area $S = \pi\langle r\rangle^2$ ($\langle r\rangle$ is the mean electron nucleus distance in ground state). For a light beam of intensity I the number of photons available for absorption during a time T is thus p = IST. Then, the ionisation probability Q(p) can be reinterpreted as the ionisation probability P(p/ST, T) for an atom irradiated by an intensity I during a time T.

A detailed examination of the distribution of p photons among m electrons allows us to define the probability $Q_k(p)$ that an electron has absorbed n+k photons when p photons are available for absorption (Crance, 1984a). As Q(p) can be reinterpreted in terms of total ionisation probability, the Q_k's can be reinterpreted as the probability $P_k(I,T)$ for an electron to be ejected with energy E_k and thus contribute to the peak number k when an atom is irradiated by an intensity I during a time T. For large value of p and moderate values of m and n, $Q_k(p)$ is approximately equal to $(1+(m-2)/p)^{-(n+k)}$ (Crance, 1984b). So we find the power laws obtained experimentally for high energy peaks. For low intensity, it is clear that n is defined as the smallest integer larger than E_I/Hw. For high intensity such a definition is questionable. In relation with ponderomotive forces, it has been argued that an electron with too weak a kinetic energy could not leave the atom. However, the concept of ponderomotive potential holds only when the light field envelope is time independent and does not apply to picosecond pulses used in recent experiments. In the following section, we examine with classical arguments how an electron just escaping from an atom leaves the interaction volume, in order to deduce an estimate of n.

III LEAVING THE INTERACTION VOLUME

An electron which has just escaped from the Coulomb potential of an ion is still in the light field. It is subjected to the Lorentz force

$$\vec{F} = e\,(\vec{E} + \vec{V} \times \vec{B}) \tag{2}$$

which determines its trajectory. e is the electron charge, V is its velocity. Note that the second term is quite small for non relativistic electrons and can be neglected in calculation of trajectories (Gontier 1987). For the sake of simplicity, we consider a linearly polarised light field and we assume that the electron is emitted with a velocity parallel to the polarisation. The trajectory of the electron remains in a plan perpendicular to the light beam axis. We consider a monochromatic field with slowly varying time and space envelope. The electric field can be written $E(z,t)\cos(wt+\varphi)$ with $E > 0$. At time t = 0, an electron is ejected at a point of coordinates x_0, y_0, z_0 with a velocity V_0. The electron is then subjected to a force e $E(z,t)\cos(wt+\varphi)$ and it moves according to the equation

$$m\ddot{\gamma} = eE(r,t)\cos(wt+\varphi) \tag{3}$$

Successive integrations by part lead to

$$mV = mV_1 + \frac{eE\sin(wt+\varphi)}{w} + \frac{dE}{dt}\,\frac{e\cos(wt+\varphi)}{w^2} + \ldots . \tag{4}$$

The choice of V_1 will be discussed later. d/dt means the total time derivative along the electron trajectory. For time independent amplitude, it reduces to the product of electron velocity and field gradient. This is the starting point for the concept of ponderomotive potential. From the structure of Eq.3, it is clear that the electron motion can be analysed as the superposition of a low-frequency and a high-frequency one. One calculates the low-frequency component for the rate of change of kinetic energy by averaging $mV\gamma$ (Eq.2 and 3) over a few periods. The resulting quantity is then interpreted as resulting from a low frequency force deriving from a potential. Such an interpretation remains, in fact, ambiguous. Starting only from energetic considerations, how can one distinguish between a slow variation of the high frequency component and a noticeable variation of the low-frequency component ?

The slowly varying envelope assumption means that

$$dE/dt \ll wE \ , \quad d^2E/dt^2 \ll w^2E \ ... \tag{5}$$

We thus have written an expansion in power of $1/w\tau$ where τ is the time which characterises the time variation of E. It is clear that in situations currently encountered in experiments of multiphoton ionisation (picosecond pulses and electrons with non relativistic velocity) $w\tau$ is much larger than one. It is then justified to write the above expansion for the electron velocity. For each type of term (constant, in phase and in quadrature with the electric field) we have kept the leading term. An important point is to choose V_1, that is how to match the velocity at $t=0_-$ with the velocity at $t=0_+$. The velocity consists of two oscillating terms and a constant one. The electron has entered the light field adiabatically, it is thus consistent to ensure the continuity of the slowly varying component of the velocity and write $V_0 = V_1$. It is clear from the variation of V(t) that if the electron ever leaves the light beam, its final velocity will be V_0. This indicates that the peaks in detected electron energy spectrum have the positions defined by E_k. The trajectory can be obtained from the velocity by integration. If $mV_0 < eE/w$, in the first cycle of electric field, the electron is thrown back to the ion where it has the possibility to absorb additional photons. Consequently, only the electrons with an energy larger than $e^2E^2/2mw^2$ leave the interaction volume freely. Electrons with weaker energy will absorb additional photons. The slower their initial velocity is, the larger the probability to be trapped in the ion potential. This simple argument explains why slow energy peaks are redistributed towards higher order peaks. The derivation given above does not allow one to define a value of n, "the minimum number of photon absorptions required for ionisation to occur". In fact, n depends on both the initial electron velocity and the position of the parent atom in the light beam. This conclusion is closely related to the ones deduced from considerations of ponderomotive potentiel. However, it explains, at least quantitatively, why first peaks are not switched off suddenly but decrease smoothly when the light intensity is increased.

We stress that the result obtained here does not depend on the pulse length provided that the slowly varying envelope assumption remains valid. Note that we do not need to invoke ponderomotive potential. It has been observed quite recently (Agostini et al, 1987; Freeman et al, 1987) that the position of the peaks in electron energy spectrum is no longer described by the E_k's when the pulse length becomes shorter than a picosecond. When we consider the order of magnitude of dE/dt in such experiments, it appears that the validity of the slowly varying envelope assumption might be questionable. The latter assumption is in fact the basis of any calculation of ionisation probability carried out up to now with the remarkable exception of explicit time integration performed recently on the simple case of Helium (Kulander 1987).

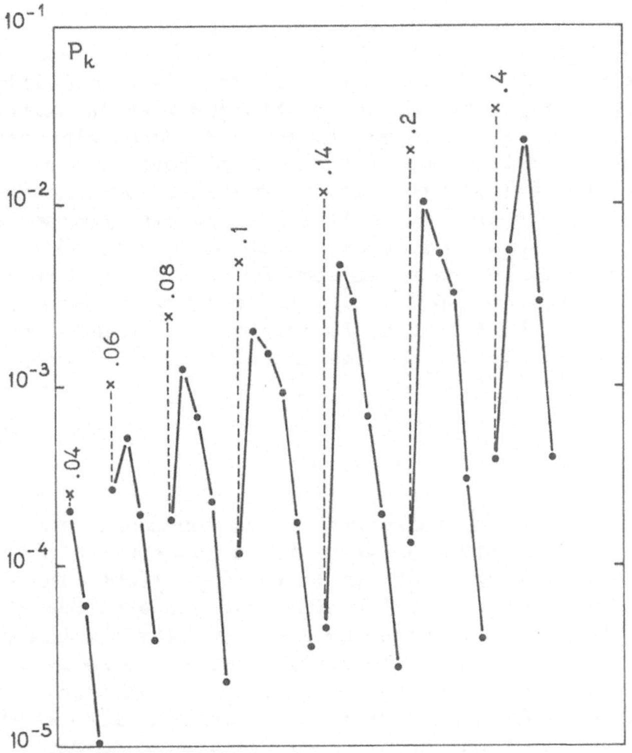

Fig. 1

For various field amplitudes (in a.u. the number next to the star), in logarithmic scale, a star indicates the ionisation probability per unit time, successive points indicate the probability per unit time P_k $(k=0,1,2,3\ldots)$ to obtain an electron after absorption of k+2 photons.

IV FLOQUET TREATMENT OF ABOVE THRESHOLD IONISATION

The complex dilatation method (Junker, 1982) allows one to study ionisation as a resonance in the dressed atom. Starting from an atomic state $|0\rangle$ dressed by M photons, one builds a non hermitian Hamiltonian on a finite basis of complex functions and calculates the energy E of the perturbed state which follows from $|0\rangle$ by continuity when the field intensity is increased. The energy E is complex, its imaginary part is one half of the ionisation probability of state $|0\rangle$. One can also deduce the expansion of the perturbed state on the finite basis as

$$|\varphi\rangle = \sum_N |f_N\rangle\ |N\rangle \tag{6}$$

f_N is an atomic wavefunction, $|N\rangle$ is a field state. The probability for an electron to escape in multiplicity N can be obtained from f_N since it is proportional to the weight of $|\varphi\rangle$ on multiplicity N. Such calculations give accurate values of probabilities when the number of basis states is large enough. As a consequence, calculations are tractable only for moderate intensities. Using this method, we have calculated the electron energy spectrum for Hydrogen at frequency .3 a.u. with intensity up to 10^{-1} a.u. For weak intensity, the heights of peaks decrease monotonically with their number. When the intensity reaches .002 a.u., the first peak becomes smaller than the second one. When the intensity is increased, the envelope of the peaks becomes bell shaped. Typical results are plotted in Figure.

V CONCLUSION

We have presented two different points of view on energy spectrum of electrons ejected in multiphoton ionisation which reproduce the essential features observed experimentally. The statistical description gives an insight of the mechanism. It shows that, in fact, the light increases the temperature of electrons (power laws for high energy peaks) and at the same time stabilises the Coulomb well formed by each ion. The quantum mechanical description gives the projection of the system on a subspace of the Hilbert space. It is a non perturbative treatment based on a stationary picture of atom field interaction. The interest of the calculation is to show that the bell shaped envelope of electron energy spectrum widely demonstrated with short range potential still occurs for Coulomb potential provided that the light intensity is strong enough. This calculation shows also that it is possible to reproduce experimental features without invoking semiclassical concepts such as the ponderomotive potential.

References

Agostini, P., Kupersztych, J., Lompré, L.A., Petite, G., Yergeau, F., 1987, Phys.Rev.A, to be published

Bialinicka Birula, Z., 1984, J.Phys.B, 17, 3091

Bucksbaum, P.H., Bashkansky, M., Freeman, R.R., McIlrath, T.J., DiMauro, L.F., 1986, Phys.Rev.Lett., 56, 2590

Chu, S.I., Cooper, J., 1985, Phys.Rev.A, 32, 2769

Chu, S.I., Reinhardt, W.P., 1977, Phys.Rev.Lett., 39, 1195

Compton, R.N., Stockdale, J.A.D., Cooper, C.D., Tang, X., Lambropoulos, P., 1984, Phys. Rev.A, 30, 1766

Crance, M., Sinzelle, J., 1985, Lecture Notes in Physics, 229, 290

Crance, M., 1984, J.Phys.B17 4333

Crance, M., 1984b, J.Phys.B, 17, L355

Deng, Z., Eberly, J.H., 1985, J.Opt.Soc.Am.B, 2, 486
Edwards, M., Pan, L., Armstrong Jr, L., 1985, J.Phys.B, 18, 1927
Fabre, F., Petite, G., Agostini, P., Clement, M., 1982, J.Phys.B, 15, 1353
Freeman, R.R., Bucksbaum, P.H., Milchberg, H., Darack, S., Schumacher, D., Gensic, M.E., 1987, Phys.Rev.A to be published
Gontier, Y., 1987 private communication
Giusti-Suzor, A., Zoller, P., 1987, to be published
Humpert, H.J., Schwier, H., Hippler, R., Lutz, H.O., 1985, Phys.Rev.A, 32, 3787
Junker, B.R., 1982, Adv. At.Mol.Phys., 18, 207
Keldysh, L.V., 1965, J.E.T.P., 20, 1307
Kruit, P., Kimman, J., Muller, H.G., Van der Wiel, M.J., 1983, Phys.Rev.A, 28, 248
Kulander, K.C., 1987, Proceedings of International Conference on Multiphoton Processes IV, Boulder Colorado
Kupersztych, J., 1987, Europhysics Lett. to be published
Lompré, L.A., L'Huillier, A., Mainfray, G., Manus, C., 1985, J.Opt.Soc.Am., B2, 1906
Mittleman, M.H., 1984, Prog. Quantum Electron., 8, 165
Muller, H.G., Tip, A., Van der Wiel, M.J., 1983, J.Phys.B, 16, L679
Petite, G., Fabre, F., Agostini, P., Crance, M., Aymar, M., 1984, Phys.Rev.A29, 2677
Reiss, H., 1980, Phys.Rev.A22, 1786
Rhodes, C.K., 1985, Fundamentals of laser interaction Lecture Notes in Physics, 229, 111, Springer Verlag
Yergeau, F., Petite, G., Agostini, P., 1986, J.Phys.B, 19, L663

ALL-ORDER ABOVE THRESHOLD IONIZATION AND RELATED PROBLEMS

Y. Gontier and M. Trahin

Service de Physique des Atomes et des Surfaces
Centre d'Etudes Nucléaires de Saclay
F-91191 Gif sur Yvette Cedex, France

INTRODUCTION

It is now well-known that in Above-Threshold ionization processes, any bound electron of an atom can be carried into a continuum state of high energy. This gain in energy comes from the absorption by the electron of more photons than the minimum number required to ionize the atom[1]. From the pressure low observed in experiments it appears that A.T.I. is essentially a one-step process where the electron absorbs several additional photons in the potential of its parent ion. Under these conditions both energy and momentum are conserved during the interaction of the electron with the photon field. The resulting spectrum of the electrons consists in peaks spaced from each other by multiples of the photon energy. The behaviour of this spectrum with respect to the intensity and the polarization of the radiation field and to the interaction time has been studied in many experiments[2].

One observes that, at low intensity, the amplitudes of the peaks decrease when their order increases i.e. the amplitude of the S-th peak is greater than that of the (S+1)-th, etc... At higher intensity a lowering of the first few peaks occurs which can lead to the suppression of one, two, n peaks.

On the other hand the effect of pulse shape is to modify the position and the width of peaks. The peaks are broadened and shifted towards lower energies when the pulse width becomes smaller and smaller. The very early theories which predicted some of the features of A.T.I. involved the lowest-order perturbation theory (LOPT)[3]. Thus, at intensity values just below those for which the perturbation series diverges, one observed a significant lowering of the first two A.T.I. peaks in H. Nevertheless LOPT is not satisfactory in several respects if one wishes realistic comparisons with actual experiments. Firstly, in this theory the peaks are infinitely sharp and are note shifted. Secondly, at intensities one must consider, higher-order terms comming from absorption-emission of photons, are note negligible and must be taken into account. We note in passing that these terms give rise to a shift and a broadening of the peaks. The preceding arguments prove the necessity of working within an "all-order model" involving the search of a non-perturbative solution of the Schröedinger equation. To this end one can resort to several techniques. In one of them the Schröedinger equation is solved by using a finite base of "essential" states. After little algebra the A.T.I. transition amplitudes are obtained by solving a set of integro-differential equations[4,6]. The only differences between the different authors lie in the approximation introduced to reduce the integrodifferential equations to algebraïc ones.

ALL-ORDER THEORY

An alternate method consists in using the exact operator expressions of the transition amplitudes within a more engranded basis of states. It is this last view point we adopt here.

The formalism of the resolvent together with the technique of projectors is used. By this way one generalizes to the continuum states the framework used for discrete states. In this model the transition amplitudes are obtained from the resolvent operator G(z) through the inversion integral

$$U(t) = \frac{1}{2\pi i} \int e^{-izt} \, G(z) \, dz \tag{1}$$

As a result of the re-summation of the perturbation series one obtains the expression of G(z) corresponding to the absorption or the emission of N-photons

$$G^{(N)}_{abs.}(z) = (G^{-}(z) \, V^{-})^{N} \, G^{0}(z) \tag{2a}$$

$$G^{(N)}_{emis.}(z) = (G^{+}(z) \, V^{+})^{N} \, G^{0}(z) \quad . \tag{2b}$$

The G-operators are expressed in terms of $V^{\pm}$, the atom-field operators describing the emission and the absorption of a photon as

$$G^{-}(z) = \frac{1}{z - H_o - V^{-} \, G^{-}(z) \, V^{+}} \tag{3a}$$

$$G^{+}(z) = \frac{1}{z - H_o - V^{+} \, G^{+}(z) \, V^{-}} \tag{3b}$$

$$G^{0}(z) = \frac{1}{z - H_o - V^{+} \, G^{+}(z) \, V^{-} - V^{-} \, G^{-}(z) \, V^{+}} \quad . \tag{3c}$$

It is more convenient, for practical uses to put forward explicitly the states which are expected to be appreciably populated. To this end one resorts to projectors. In this case the resolvent is expressed in terms of an operators R(z). Let P and Q be the projectors onto and outside the subspace spanned by these particular states. The expression of G(z) to be used for a transition taking place between states belonging to this subspace is

$$\tilde{G}(z) = PG(z)\,P = \frac{1}{Z - H_o - P\,R(z)\,P} \quad , \tag{4}$$

where the diagonal and the non-diagonal expressions of R(z) are

$$R(z) = V^{+}\, \tilde{G}^{0}(z)\, V^{-}\, \tilde{G}^{-}(z) + V^{-}\, \tilde{G}^{0}(z)\, V^{+}\, \tilde{G}^{+}(z) \tag{5a}$$

and

$$R_{\substack{abs.\\ emis.}}(z) = V^{\pm}\, \tilde{G}^{0}(z)[V^{\mp}\, \tilde{G}^{\mp}(z)]^{N+1} + V^{\mp}\, \tilde{G}^{0}(z)\, [V^{\mp}\, \tilde{G}^{\mp}(z)]^{N-1} \tag{5b}$$

respectively. The tilded operators are obtained from Eqs. (3) by making the substitution $V^{\pm} \to Q\, V^{\pm}$ everywhere.

As an example we consider the case where one particularizes the ground state |0> and the continuum states |1E>, |2E> and |3E>,

$$P = |0\rangle \langle 0| + \sum_{i=1}^{3} dE\ |iE\rangle \langle iE| \ . \tag{6}$$

One finds that for the transition $|0\rangle \to |1E\rangle$

$$G_{1E,0}(z) = \frac{R^{(3)}_{1E,0}}{(z-H^{0}_{1E,1E}-R^{(3)}_{1E,1E})\ (z-H^{0}_{0,0}-R^{(3)}_{0,0}) - R^{(3)}_{1E,0}\ R^{(3)}_{0,1E}} \tag{7a}$$

with

$$R^{(3)}(z) = R^{(2)}(z)+R^{(2)}(z) \int \frac{|2E'\rangle \langle 2E'|}{z-H^{0}_{2E',2E'}-R^{(2)}(z)}\, dE'\ R^{(2)}(z) \tag{7b}$$

$$R^{(2)}(z) = R^{(1)}(z)+R^{(1)}(z) \int \frac{|1E'\rangle \langle 1E'|}{z - H^{0}_{1E',1E'} - R^{(1)}(z)}\, dE'\ R^{(1)}(z) \tag{7c}$$

$$R^{(1)}(z) = R(z) \quad . \tag{7d}$$

By substitutions one obtains from Eqs. (7) the six irreducible terms corresponding, up to arbitrary order, to the channels which the electron can take to reach the continuum state |E>.

As a further simplification we only specify the ground state and the continuum $|\alpha E\rangle$. To calculate the probability one must know the pole of $G_{\alpha E,0}$ as given by Eqs. (7). By labelling the continua by the number of photons involved in the transition one has to calculate the poles of

$$G_{\alpha E,0}(z) = \frac{R_{\alpha E,0}(z)}{(z-E+\alpha E_{p}+R_{\alpha E,\alpha E}(z))(z-E_{0}-R_{0,0}(z)) - R_{\alpha E,0}(z)\ R_{0,\alpha E}(z)} \tag{8}$$

which are given by the solution of the equation

$$(z-E+\alpha E_{p}+R_{\alpha E,\alpha E}(z))\ (z-E_{0}-R_{00}(z)) - R_{\alpha E,0}(z)\ R_{\alpha E,0}(z) = 0 \tag{9}$$

One finds that around $z \simeq E_{0}$ this equation provides two solutions:

$$z^{\pm}(E) = E^{\pm}_{\alpha}(E) - i\ \Gamma^{\pm}_{\alpha}(E)$$

which enable to calculate the transition amplitude according to

$$U_{\alpha E,0}(t) = \frac{1}{2\pi i} \int \frac{R_{\alpha E,0}(z)}{(z - z_\alpha^+(E))\,(z - z_\alpha^-(E))}\, e^{-izt}\, dz$$

$$= \frac{1}{z_\alpha^+(E) - z_\alpha^-(E)} \left(R_{\alpha E,0}(z^+)\, e^{-iz_\alpha^+ t} - R_{\alpha E,0}(z^-)\, e^{-iz_\alpha^- t}\right) \quad (10)$$

With the notations $z_\alpha^\pm(E) = E_\alpha^\pm(E) - i\,\Gamma_\alpha^\pm(E)$ the probability density reads

$$W_\alpha(E,t) = \frac{1}{(E_\alpha^- - E_\alpha^+)^2 + (\Gamma_\alpha^- - \Gamma_\alpha^+)^2} \{(\Delta_{\alpha,0}^{+2} + \Gamma_{\alpha,0}^{+2}) e^{-2\Gamma_\alpha^+ t} + (\Delta_{\alpha,0}^{-2} + \Gamma_{\alpha,0}^{-2})\, e^{-2\Gamma_\alpha^- t}$$

$$- 2\,[(\Delta_{\alpha,0}^+\,\Delta_{\alpha,0}^- + \Gamma_{\alpha,0}^+\,\Gamma_{\alpha,0}^-) \cos\,(E_\alpha^+ - E_\alpha^-)\, t$$

$$+ (\Delta_{\alpha,0}^+\,\Gamma_{\alpha,0}^- - \Gamma_{\alpha,0}^+\,\Delta_{\alpha,0}^-) \sin\,(E_\alpha^+ - E_\alpha^-)\, t]\, e^{-(\Gamma_\alpha^+ + \Gamma_\alpha^-)\, t} \}\,, \quad (11)$$

where

$$\Delta_{\alpha,0}^\pm = \mathrm{Re}\,\{R_{\alpha E,0}\,(E_\alpha^\pm, \Gamma_\alpha^\pm)\} \quad (12a)$$

$$\Gamma_{\alpha,0}^\pm = \mathrm{Im}\,\{R_{\alpha E,0}\,(E_\alpha^\pm, \Gamma_\alpha^\pm)\}\;. \quad (12b)$$

Formally this expression is identical to that obtained in the theory of resonant transitions between discrete states. The difference lies in that all the quantities are now functions of the energy E.

Some features of the spectrum are predicted by this model. For example if we neglect the last term in Eq. (9) and if we make the single pole approximation, one has

$$z_\alpha^+ \to E_\alpha^+ - i\Gamma_\alpha^+ = E - \alpha\, E_p + \Delta_\alpha(I) - i\;\Gamma_\alpha(I) \quad (13a)$$

$$z_\alpha^- \to E_\alpha^- - i\Gamma_\alpha^- = E_0 + \Delta_0(I) - i\;\Gamma_0(I) \quad . \quad (13b)$$

The real contributions to the poles z^+ and z^- are

$$E_\alpha^+ = E - \alpha\, E_p + \Delta_\alpha(I) \quad (14a)$$

$$E_\alpha^- = E_0 + \Delta_0(I) \quad (14b)$$

respectively.

From Eq. (11) the maximum probability will be obtained when $E_\alpha^+ = E_\alpha^-$. Thus in each pile of half-planes representing the infinitely degenerated continua it exists a particular curve corresponding to an energy

$$E = E_0 + (N + S - 1) E_p + \Delta_0(I) - \Delta_{N+S+1}(I) \tag{15}$$

which crosses the curve $\varepsilon_0 = E_0 + \Delta_0(I)$.

Aside this particular curve there are other possible crossings with curves of energies around E. They will provide non negligible enhancement of the probability contributing to a broadening of the resonance peaks. This outlines the importance of calculating accurately the perturbations undergone by the atomic energy levels. In doing so for continuum states one encounters a well-known difficulty, which needs the use of a special technique recently proposed[7].

SECOND ORDER FREE-FREE TRANSITIONS

The quantity one has to calculate is

$$R_{\alpha E,\alpha E} = \frac{I}{I_0} \sum_j \left(\frac{D^*_{\alpha E,j} D_{j,\alpha E}}{E - E_j + \omega} + \frac{D_{\alpha E,j} D^*_{j,\alpha E}}{E - E_j - \omega} \right) \tag{16a}$$

or

$$R_{\alpha E,\alpha E} = \frac{I}{I_0 \omega^2} \left\{ \varepsilon^* . \varepsilon + \sum_j \left(\frac{Q^*_{\alpha E,j} Q_{j,\alpha E}}{E - E_j + \omega} + \frac{Q_{\alpha E,j} Q^*_{j,\alpha E}}{E - E_j - \omega} \right) \right\} \tag{16b}$$

according to whether the gauge "$\vec{r}.\vec{E}$" or "$\vec{A}.\vec{P}$" is used respectively. In Eqs. (16), the matrix elements $D_{\alpha E,j} = \langle \alpha E | \vec{\varepsilon}.\vec{r} | j \rangle$ and $Q_{\alpha E,j} = \langle \alpha E | \vec{\varepsilon}.\vec{P} | j \rangle$ diverge when $|j\rangle \equiv |\alpha E\rangle$. This can be shown from the well-known relations between momentum and position operators[8]

$$[P,H]_{-\alpha' E',\alpha E} = -(E'-E) \vec{P}_{\alpha' E',\alpha E} = -i(E'-E)^2 \vec{r}_{\alpha' E',\alpha E} \tag{17a}$$

and

$$[P,H]_{-\alpha' E',\alpha E} = -i \langle \alpha' E' | \vec{\nabla} \frac{Z}{r} | \alpha E \rangle = -i Z \langle \alpha' E' | \frac{\vec{r}}{r^3} | \alpha E \rangle \tag{17b}$$

where use has been made of

$$i(E'-E) \vec{r}_{\alpha' E',\alpha E} = \vec{P}_{\alpha' E',\alpha E} \tag{18a}$$

and

$$[\vec{P},H]_- = [\vec{P}, \frac{Z}{r}]_- \tag{18b}$$

By identifying Eqs. (17a) and (17b) one finds that

$$(\vec{\varepsilon}.\vec{r})_{\alpha' E',\alpha E} = \frac{Z}{(E'-E)^2} \langle \alpha' E' | \frac{\vec{\varepsilon}.\vec{r}}{r^3} | \alpha E \rangle = \frac{Z}{(E'-E)^2} \rho_{\alpha' E',\alpha E} \tag{19}$$

The matrix element of the operator ρ thus defined is finite. Eq. (19) shows, the reason why the matrix element of $\vec{\varepsilon}.\vec{r}$ diverges when E'=E: there exists a second-order pole which does not appear explicitly.

To surmount this difficulty we calculate the matrix elements $R_{\alpha E,\alpha E}$ of Eqs. (16) in the following way. Let us consider the "$\vec{A}.\vec{P}$" gauge, and the following relations obtained from Eqs. (17a) and (19)

$$(E - E_j)\, Q_{\alpha E,j} = - i\, \rho_{\alpha E,j} \qquad (20a)$$

$$(E_j - E)\, Q_{j,\alpha E} = i\, \rho_{j,\alpha E}$$

One obtains from Eq. (16b)

$$R_{\alpha E,\alpha E} = \frac{I}{I_0 \omega^2} \{\vec{\varepsilon}^*.\vec{\varepsilon} - \sum_j \left(\frac{\rho^*_{\alpha E,j}\, \rho_{j,\alpha E}}{(E-E_j+\omega)(E-E_j)^2} + \frac{\rho_{\alpha E,j}\, \rho^*_{j,\alpha E}}{(E-E_j-\omega)(E-E_j)^2}\right) \} \qquad (21)$$

Since the components of the momentum operator commute, $[Q^*,Q]_- = 0$ and we add to the right hand side of Eq. (21) the vanishing quantity

$$\frac{1}{\omega} \langle\alpha E|\ [Q^*,Q]_-\ |\alpha E\rangle = \frac{1}{\omega} \sum_j (Q^*_{\alpha E,j}\ Q_{j,\alpha A} - Q_{\alpha E,j} Q^*_{j,\alpha E}) \qquad (22)$$

After replacing the matrix elements of Q by those of ρ according to Eqs. (20) together with their complex conjugate, one obtains

$$R_{\alpha E,\alpha E} = \frac{I}{I_0 \omega^2} \{\vec{\varepsilon}^*.\vec{\varepsilon} + \frac{1}{\omega} \sum_j \left(\frac{\rho^*_{\alpha E,j}\, \rho_{j,\alpha E}}{(E-E_j+\omega)(E-E_j)} - \frac{\rho_{\alpha E,j}\, \rho^*_{j,\alpha E}}{(E-E_j-\omega)(E-E_j)}\right)\} \qquad (23)$$

The real and the imaginary parts of $R_{\alpha E,\alpha E}$ are given, to second order, by calculating the integrals of Eq (23) in the complex plane containing only first order poles. This method which can be generalized to higher order couplings enables one to put into a fully calculable form the matrix elements of the so-called shift operator R(z) up to arbitrary order.

CONCLUSION

Numerical analysis shows that the quantities Γ_α which, according to Eq. (11), play an important role in the behaviour of the probability, have a very typical energy dependence. From the continuum limit (E=0), they increase with energy, they reach a maximum and then decrease monotonically to zero. The height of the maxima as well as the width of the curves depend on the orbital quantum number and thus on the order of the peaks. The consequence is that the amplitude and the width of the A.T.I. peaks are deeply affected by the behaviour of Γ_α. This strongly reinforces the intensity effects which favor the growth of high order peaks and is one of possible mechanisms responsible of the "disappearance" of low order peaks.

REFERENCES

1. Y. GONTIER and M. TRAHIN, in "Photons and continuum states of atoms and molecules", ed. by N.K. RAHMAN, C. GUIDOTTI and M. ALLEGRINI, Springer-Verlag, 91 (1987).
2. F. YERGEAU, G. PETITE and P. AGOSTINI, in: "Photons and continuum states of atoms and molecules", ed. by N.K. RAHMAN, C. GUIDOTTI and M. ALLEGRINI, Springer-Verlag, 78 (1987).
3. Y. GONTIER, M. POIRIER and M. TRAHIN, J. Phys. B, 13:1381 (1980).
4. Z. DENG and J.H. EBERLY, J. Opt. Soc. Am. B, 2:486 (1985).
5. M. EDWARDS, L. PAN and L. ARMSTRONG Jr., J. Phys. B 18:1927 (1985).
6. Z. BIALYNIKA-BIRULA, J. Phys. B 16:4351 (1983).
7. Y. GONTIER and M. TRAHIN, to be published in Phys. Rev. A.
8. H. A. BETHE and E.E. SALPETER, "Quantum mechanics of one and two electron atoms", Academic Press, New York (1957).

NUMERICAL SIMULATIONS OF MULTIPHOTON PROCESSES IN MANY ELECTRON ATOMS

Kenneth C. Kulander and Charles Cerjan

Theoretical Atomic and Molecular Physics Group
Lawrence Livermore National Laboratory
Livermore, California 94550

INTRODUCTION

The effects of high intensity, short pulsed lasers on many electron atoms are actively being studied in a number of laboratories. Measurements of ion yields and electron energy and angular distributions as functions of wavelength, intensity and pulse length have provided detailed information about the process of multiphoton ionization. The pulse lengths and intensities currently used are in the regime where standard perturbative methods are invalid. However, direct solution of the time dependent Schrödinger equation for single electron atoms has been shown to be possible.[1] In this way, arbitrary pulse shapes can be treated exactly as long as the intensity is high enough that ionization occurs rapidly so that long integration times are not required. In this regime, the highly non-linear process encountered can be accurately described. Extensions of these method are now under development, with applications to the simplest multi-electron atom, helium, having been accomplished.[2] In this report, we will present some details of two numerical methods which we have been pursuing for treating the multi-electron case. Both involve a numerical representation of the time dependent Hartree Fock electronic orbitals and a direct time integration of the coupled equations. Using these methods we will be able to determine the dynamics of the energy absorption process (preionization dynamics and the mechanism of ionization, i.e. whether it is direct multi-electron emission or sequential).

In the following section, a brief description of the two computational methods will be presented. The final section contains some recent results obtained for the multiphoton ionization of helium along with a discussion of future directions for this research.

NUMERICAL METHODS

We represent our wave function (or orbital) on a spatial grid and solve a set of coupled equations which gives the values of the wave function at these points at each integration time step. The first method used to accomplish the time propagation will be presented for the single particle, one-dimensional case; the generalization to three dimensions and many electron atoms is discussed below.

We wish to solve the semi-classical time dependent Schrödinger equation for an electron in an electromagnetic field,

$$i\hbar \frac{d}{dt} \psi(x,t) = \left[-\frac{\hbar^2}{2m} \left(\frac{d}{dx} - \frac{e}{c} A(x,t) \right)^2 + V(x) \right] \psi(x,t) \qquad (1)$$

where V(x) is the static, field-free potential and A(x,t) is the vector potential. In the dipole approximation, which is valid for all the wavelengths and intensities we have considered, $A(x,t) = A_0 f(t)$. Since we wish to treat arbitrary pulse shapes, we make no assumptions about the periodicity of the vector potential (as required by Floquet analysis) and solve for the actual time dependent field. We assume the laser field is turned on over several oscillations so that we have $f(t)=f_0(t)\cos wt$, where f_0 is a slowly varying function of time, generally chosen to be constant after some turn-on period so that an ionization rate for a particular intensity can be determined.

The method for solving Eq. (1) uses a pseudo-spectral technique. The application of pseudo-spectral methods to the solution of partial differential equations is well-known and extensively developed.[3] The use of a special case of these methods leads to an efficient and accurate technique for the solution of the explicitly time dependent Schrödinger equation.[4,5,6] The essence of this technique is to obtain a global representation of the Laplacian differential operator on a grid while remaining in configuration space so that the potential function is just a multiplicative operator. It should be emphasized that the spectral analysis applies only to the spatial derivative operators. The time dependence of the partial differential equation is handled separately--usually by a differencing method. This approach offers some advantages over standard finite element or finite difference schemes since its extension to many dimensions is automatic and can require fewer grid points than other approaches. The pseudo-spectral method chosen to solve this partial differential equation is the so-called Fast Fourier Transform (FFT) method. The transform is applied to the spatial derivative using the identity:

$$\frac{d^n}{dx^n} \psi (x,t) = \frac{i^n}{2\pi} \int_{-\infty}^{+\infty} e^{ikx} \; k^n \psi (k,t)dk \qquad (2)$$

where the circumflex denotes Fourier transformation for suitably bounded functions $\psi(x)$. Algorithmically, the technique is to transform the wave function on a grid at a given time, multiply by a suitable power of k determined by the order of the spatial derivative and then synthesizing to obtain the spatial derivative on the original grid. Obviously, this method requires two Fourier transforms for every time step. As mentioned above, the time evolution is handled by forward differencing,[4] or by exponentiation[5] (split-operator technique). The solution of Eq. (1) in the dipole approximation involves only a first and second order spatial derivative of the wave function.

This method can be generalized easily to higher dimensions and to some other co-ordinate systems. For example, cylindrical or spherical symmetry produces Hankel transforms with the obvious reduction in dimensionality of the problem. The difficulties associated with the method derive from the small time steps needed for numerical stability and similarly for the number of grid points required. The generality of the technique, though, makes this approach a viable and attractive alternative to standard differencing schemes.

The second approach uses a finite difference representation of the kinetic energy operator. In describing this method, we develop the equations for the time dependent Hartree Fock (TDHF) model of helium in a classical electromagnetic field. In this case we use the E•r rather than the p•A interaction shown in Eq. (1). We assume the laser is linearly polarized in the z-direction so that the Hamiltonian has cylindrical symmetry. The two electron, Hartree Fock wave function is given by:

$$\Psi(\underline{r},t) = 2^{-1/2} \varphi(\underline{r}_1,t) \varphi(\underline{r}_2,t) \{\alpha_1\beta_2 - \beta_1\alpha_2\} \tag{3}$$

in the usual notation. Note that in the TDHF model, both electrons occupy the same spatial orbital at all times. This is the major approximation in this treatment, the consequences of which have been discussed elsewhere.[1] Putting Eq. (3) into the time dependent Schrödinger equation, we obtain

$$i\hbar \frac{\partial}{\partial t} \varphi(\underline{r},t) = \{ (-\hbar^2/2m)\nabla^2 - Ze^2/r + e^2\int d\underline{r}' \rho(\underline{r}',t)/|\underline{r}-\underline{r}'| + eE(t)z\sin wt \} \varphi(r,t) \tag{4}$$

where

$$\rho(r,t) = | \varphi(\underline{r},t)|^2.$$

Taking advantage of the cylindrical symmetry and that the initial state, the static Hartree Fock ground state is $1s^2$, we can write

$$\varphi(\underline{r},t) = (2\pi\rho)^{-1/2} g(\rho,z,t). \tag{5}$$

We define a two dimensional grid in the ρz-space and evaluate the second derivatives in the Hamiltonian using a three point formula to obtain the following coupled, time-dependent equations for the value of the orbital on the grid points:

$$i\hbar \frac{\partial}{\partial t} g_{jk} = (Hg)_{jk} + (Vg)_{jk} \tag{6}$$

with

$$(Hg)_{jk} = - (1/2\ \Delta^2) (g_{jk+1} + g_{jk-1} - 2g_{jk}) + \frac{1}{2} W_{jk}g_{jk} \tag{7}$$

$$(Vg)_{jk} = - (1/2\ \Delta^2) (c_j\ g_{j+1k} + c_{j-1}g_{j-1k} - 2g_{jk}) + \frac{1}{2} W_{jk}g_{jk} \tag{8}$$

Here Δ is the grid spacing assumed to be the same in both dimensions,

$$c_j = j/(j^2 - 1/4)^{1/2} \tag{9}$$

and W is the total, laser plus atomic, potential. The time propagation is accomplished using the Peaceman-Rachford (PR) alternating directions implicit method given by:

$$g^{(n+1)} = [I+i\tau V]^{-1} [I-i\tau H] [I+i\tau H]^{-1} [I-i\tau V] g^{(n)} \tag{10}$$

where I is the unit matrix, $g^{(n)}$ is the orbital at $t=n\Delta t$, and $\tau=\Delta\tau/2\hbar$. Since the Fock Hamiltonian depends on the time through the Coulomb term

$$V_c(\underline{r},t) = \int d\underline{r}' \ \rho(\underline{r}',t)/|\underline{r}-\underline{r}'| \tag{11}$$

and this depends in turn on the orbital, we use a two step, predictor-corrector type integration. First we propagate to t_{n+1} to obtain $\bar{W}^{(n+1)}$ which is the interaction evaluated using the extrapolated $g^{(n+1)}$, then use

$$W^{(n+1/2)} = \frac{1}{2} \left[W^{(n)} + \bar{W}^{(n+1)} \right] \tag{12}$$

to propagate the next time step. We have found this to give accurate wave function evolution with reasonable time steps. The propagator defined in Eq.(10) is unitary, second order in time and involves either the multiplication by or the inversion of tri-diagonal matrices. These are accomplished by a few vector multiplications so that the calculations are very efficient.

The evaluation of the Coulomb integral is a major part of the calculation. From Eq. (11) it can be seen that it must be determined for each grid point for each time integration step. We do this by first performing a moment expansion of the density using Legendre polynomials which is then used to explicitly evaluate V_c for the points on the grid boundary. Values for the balance of the grid points are determined by solving the differential equation

$$\nabla^2 V_C(\underline{r},t) = -4\pi\rho(\underline{r},t) \tag{14}$$

using an iterative procedure similar to the PR algorithm.

Thus, the calculation consists of choosing a laser frequency, a pulse shape and a maximum intensity; then solving for the time propagation using Eq. (10) subject to the initial condition of the atom being in its ground state. The wave function evolution is followed until an ionization rate can be determined in cases where our exponential decay of the initial state occurs. Since the final state generally corresponds to free electrons, some accommodation for the finite size of the numerical grid must be made. We do this by including an imaginary part to the potential defined only on a band around the edge of the grid which absorbs all flux which reaches the boundary. Therefore all probability which is excited enough to reach the boundary is assumed to be ionized. Then the ionization rate is computed by determining the change in the norm of the total wave function due to the absorption of this flux. We check the sensitivity of the results to the size of the grid, i.e. the distance from the nucleus to the boundary, to assure the box is large enough.

We have performed a number of calculations on the helium system for various wavelengths and intensities. A sample of the results are presented in the next section.

RESULTS AND CONCLUSIONS

We have determined the multiphoton ionization rates for helium at wavelengths commonly employed in these experiments--1064, 532 and 248 nm. Intensities in the range from 10^{14} – 2×10^{15} W/cm^2 were used. We chose a pulse shape [see Eq. (4)] defined by $E(t) = E_{max}\, f(t)$, where

$$f(t) = \begin{cases} 1/t_{max}, & t<t_{max} \\ 1, & t>t_{max} \end{cases} \quad (15)$$

t_{max} is chosen to be a small number of periods of the laser, but large enough that the turn-on for these wavelengths is adiabatic. In these cases we found an exponential decay of the wave function during the constant intensity interval.

The calculated rates are shown in the accompanying table. We note two interesting facts about these results. First, the rates are comparable in magnitude although the order of the ionization process (i.e. the minumum number of photons required to create an ion) change significantly from 22 to 5. As the intensity increases, we find the rates become independent of wavelength. Second, the rates do not increase as I^n where n is the order of the process as would be predicted by perturbation theory. Clearly, these results are in the regime where perturbation theory is invalid. This should be expected since the electric field due to the laser is becoming comparable to the Coulomb interactions in the atom so that the electromagnetic field cannot be considered a weak perturbation.

These results illustrate the computational techniques described above. More extensive results have been presented and discussed elsewhere.[2,4] The methods are being generalized to treat the multiphoton ionization of Xe. In this case, we explicitly solve for the outer valence shell ($5s^2 5p^6$) electrons with the core represented by a local pseudopotential. Exchange terms are present in this case and are included also by a local density approximation. In this way we can examine the multi-electron character of the pre-ionization dynamics.

A second direction of development is the inclusion of a second configuration in the helium calculation so that the electrons can interact with the field independently. The restriction of a single determinantal wave function means we cannot investigate the role of direct double ionization relative to sequential ionization for producing completely stripped helium.

We conclude by emphasizing that the methods described here show great promise for detailed investigation of the dynamics of many-electron atoms in intense laser fields.

Table 1. Multiphoton ionization rates (s^{-1}) for helium as functions of wavelength (nm) and intensity (W/cm^2)

I\λ	1064	532	248
$5x10^{14}$	$4.7x10^{11}$	$9.0x10^{11}$	$3.9x10^{12}$
$1x10^{15}$	$1.5x10^{13}$	$1.9x10^{13}$	$2.5x10^{13}$
$2x10^{15}$	$1.2x10^{14}$	$1.4x10^{14}$	$1.8x10^{14}$

ACKNOWLEDGEMENT

Work performed under the auspices of the U. S. Department of Energy by the Lawrence Livermore National Laboratory under contract number W-7405-ENG-48.

REFERENCES

1. K. C. Kulander, "Multiphoton Ionization of Hydrogen: A Time Dependent Theory," Phys. Rev. A 35 445 (1987).
2. K. C. Kulander, "Time Dependent Hartree Fock Theory of Multiphoton Ionization: Helium," Phys. Rev. A 36 xxxx (1987).
3. D. Gottlieb, M. Y. Hussaini and S. A. Orszag, "Theory and Applications of Spectral Methods" in Spectral Methods for Partial D. Differential Equations, ed. by R. G. Voigt, D. Gottlieb and M. Y. Hussaini, SIAM publications (Philadelphia, 1984).
4. C. Cerjan and R. Kosloff, "Non-Perturbative Treatment of Particle Dynamics in a Semi-Classical Photon Field," J. Phys. B 20 xxxx (1987).
5. M. D. Feit, J. A. Fleck and A. Steiger, "Solution of the Schrödinger Equation by a Spectral Method," J. Comp. Phys. 412, 47 (1982).
6. H. Tal-Ezer and R. Kosloff, "An Accurate and Efficient Scheme for Propagating the Time Dependent Schrodinger Equation," J. Chem. Phys. 3967, 81 (1984).

NUMERICAL ATI WAVE FUNCTIONS AND THEIR IMPLICATIONS

J. Javanainen, Q. C. Su and J. H. Eberly

Department of Physics and Astronomy
University of Rochester
Rochester, N.Y. 14627

INTRODUCTION

In high-intensity photoionization of atoms several peaks separated by photon energy are observed in the energy spectrum of the liberated electrons, as if an electron continued to absorb photons even after it has been released from the atom. This phenomenon, now known as Above-Threshold Ionization(ATI), has spurred a large body of experimental[1-3] and theoretical[4-6] work. Nonetheless, the understanding of ATI is at present far from complete, mostly because it is extraordinarily difficult to work out theories to the point where a comparison with an experiment can really distinguish between them. For instance, the existing theories typically ignore either complicated features of real experiments such as temporal and spatial shape of the laser pulse, or most of the atomic structure.

The idea of our work is to make ATI experiments stripped of complications by integrating numerically the time-dependent Schrödinger equation in space and time, to thereby obtain clean material for comparison with theoretical concepts. We emphasize that we do not aim at reproducing real experiments and their perplexities but rather at making unique experiments of our own, the analogs of which at the moment mostly cannot be carried out in real laboratories.

We have earlier[7] used this approach (i) to point out that the number of ATI peaks in our simulations is a function of the ratio of the ponderomotive potential to the photon energy, rather than a function of the intensity or the field frequency alone; (ii) to demonstrate explicitly threshold shift and channel closing[6] as a function of the field intensity, basically as it has also been observed in recent experiments with short pulses[2,3]; and (iii) to compare two Keldysh type models[4] with our simulations, finding bad quantitative agreement but some degree of qualitative similarity.

In this paper we discuss two new developments. First, we present a comparison of the results obtained with square pulses and smooth pulses of the laser field. The agreement is usually good, provided the square-pulse calculations have been done judiciously. In particular, the d·E rather than the p·A form of the dipole interaction should be used. Nonetheless, we have met with a case where the square pulse produces unexpected "energy-nonconserving" peaks in the electron spectrum. These peaks are our second main topic.

CALCULATION METHOD

Our numerical experiments are based on the Hamiltonian of a one-dimensional atom in an external field,

$$H(t) = -\frac{1}{2}\frac{\partial^2}{\partial x^2} - \frac{1}{\sqrt{1+x^2}} - xE(t)\sin\omega t. \tag{1}$$

The atomic potential has a Coulomb tail, hence a Rydberg series structure of high-lying states. However, the potential has no singularities. We can treat the atom on the whole real axis, parity is a good quantum number, and bound states do not have permanent dipole moments. We shall either employ a square pulse with $E(t) = E_0$, or the particular form of a smooth pulse $E(t) = E_0 \sin^2(\pi t/T)$.

We integrate the time-dependent Schrödinger equation with the Hamiltonian (1) numerically in space and time. To interpret the results we resort to the eigenvalues and eigenfunctions of the bare atomic Hamiltonian (without dipole coupling to the field). We always use as the initial wave function the ground state of the atom with energy $W_0 = -0.6698$, and most often our objective is the photoelectron energy spectrum at time T, $P(W) = |\langle W|\psi(T)\rangle|^2$.

In practice we replace the x coordinate with discrete points $x_n = n\delta x$, $n = -N/2, \ldots, N/2$, the wave functions with N+1 dimensional vectors, and the second-derivative operator with the finite difference,

$$\frac{\partial^2}{\partial x^2}\psi(x) \rightarrow \frac{1}{(\delta x)^2}(\psi_{n+1} - 2\psi_n + \psi_{n-1}). \tag{2}$$

Finding the eigenstates and eigenvectors of the atom reduces to a tridiagonal matrix problem, which we solve with the QL and inverse iteration algorithms.[8] The wave function is propagated in time according to[9]

$$\psi(t+\delta t) = [1-\frac{i\delta t}{2}H(t+\frac{\delta t}{2})]\,[1+\frac{i\delta t}{2}H(t+\frac{\delta t}{2})]^{-1}\psi(t), \tag{3}$$

which is nothing but the time-honored Crank-Nicholson algorithm for partial differential equations. Every time step requires solving a linear set of equations, but the matrix is tridiagonal and the CPU time only grows linearly with the matrix size. Typical computation parameters for the results given below are $N = 32K$, $\delta x = 0.07$ and $\delta t = 0.08$.

Throughout the computations we use reflecting boundary conditions for the wave functions, i.e., $\psi_{-N/2-1} = \psi_{N/2+1} = 0$. Such an atom only has discrete states, so the counterpart of the continuous electron spectrum has to be defined. We do it by first arranging the energy eigenvalues W_i in ascending order, and then assigning to the photoelectron spectrum the value

$$P[\frac{1}{4}(W_{i-1}+W_i+W_{i+1}+W_{i+2})] = \frac{|\langle W_i|\psi\rangle|^2}{W_{i+1}-W_{i-1}} + \frac{|\langle W_{i+1}|\psi\rangle|^2}{W_{i+2}-W_i} . \quad (4)$$

at the energy appearing on the left-hand side. This convention averages over the alternating even- and odd-parity states, which may have vastly different populations. Eq. (4) also provides an extension of the electron spectrum to negative-energy states, which is particularly natural if the interaction time T does not yet permit resolving the Rydberg levels.

For square pulses (and for smooth pulses at times less than the pulse duration T) the electron whose energy spectrum is to be calculated still dwells in the driving field. On top of its drift motion it executes a forced quiver motion: in the field $\sin\omega t$ the velocity oscillates as $\cos\omega t$. We have found that electron spectra with ATI peak spacing equal to the photon energy are obtained only at times corresponding to $n \pm 1/4$ cycles of the field, i.e. when $\cos\omega t = 0$. Obviously only the drift motion remains at such times. Away from these times, for high multiphoton orders (five-photon ionization or more) the spectrum oscillates violently already in modest fields ($E_0 \sim 0.02$), and rapidly becomes seemingly meaningless. To take the spectrum at $n \pm 1/4$ cycles of the field not only is well motivated in that the quiver motion should not contribute, but also is a practical necessity.

PULSE RESPONSE

To take the electron spectra at times corresponding to $n \pm 1/4$ cycles of the external field may nonetheless sound somewhat artificial. To dispel the associated doubts we have carried out comparisons between spectra obtained with various length square and smooth pulses. An example with $\omega = 0.148$ (five-photon ionization) and $E = 0.1$ (4×10^{14} W/cm^2) is presented in parts a, b and c of Fig. 1 which show the ATI spectrum for 4 1/4 and 32 1/4 cycle square pulses, as well as after a 32 cycle smooth pulse.

With the square pulses the peak positions remain unchanged to within 0.005; and, as far as the peak positions for the smooth pulse can be defined, they too are qualitatively the same. Interestingly enough, the population that for the short square pulse of Fig. 1a resided in the Rydberg levels seems to have preferred accumulating in the lowest above-threshold peak in Fig. 1b. Except for this, the number and relative heights of the above-threshold peaks are very much the same in Figs 1a, b and c.

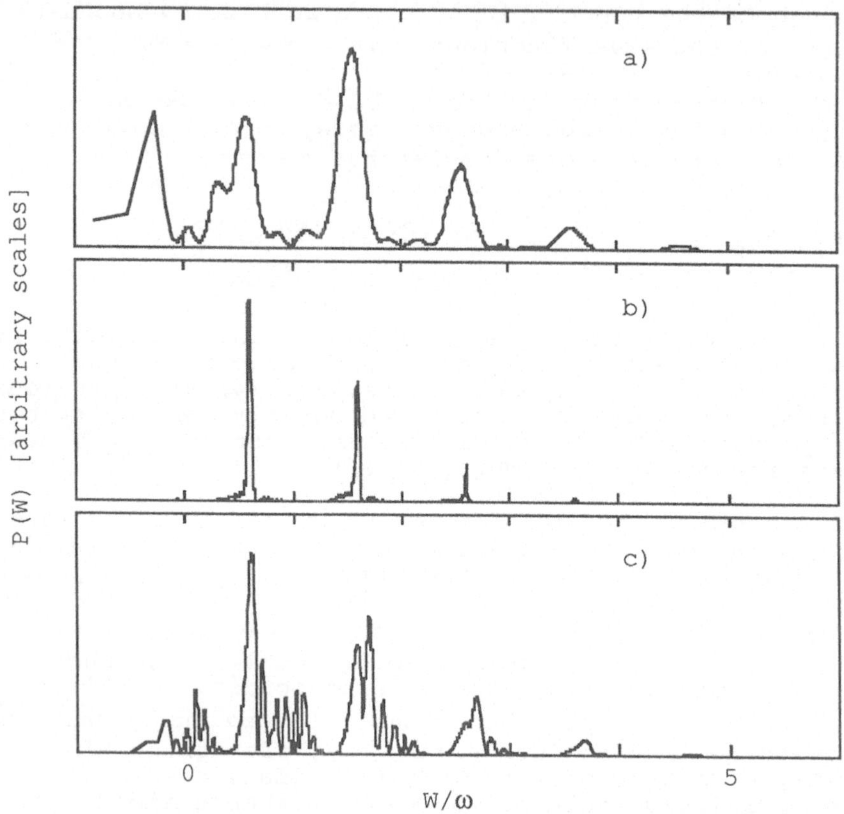

Fig. 1. Photoelectron spectra for (a) 4 1/4 cycle and (b) 32 1/4 cycle square pulses, as well as (c) for a 32 cycle smooth pulse. The fixed parameters are the field frequency $\omega = 0.148$ and the maximum amplitude $E_0 = 0.1$.

Figures 1a and 1b happen to be our only runs so far where the long-term change of the relative heights of the ATI peaks with varying length of the square pulse (after the spectrum is initially formed during typically 2-3 field cycles) is apparent even to the eye. Results on the positions and number of the peaks would usually be practically the same for short square pulses and long smooth pulses. However, for the smooth pulses (i) the number of the problem parameters has increased, because the pulse shape has to be described; (ii) the computer runs cost a lot more than for short square pulses, e.g., 4000 seconds on a Cyber 205 with two vector pipes for Fig. 1c; and (iii) the results may be complicated, exhibiting physics that depends on the time-varying field strength. A smooth pulse is not necessarily the preferred choice for ionization studies.

One physical complication of smooth pulses is immediately

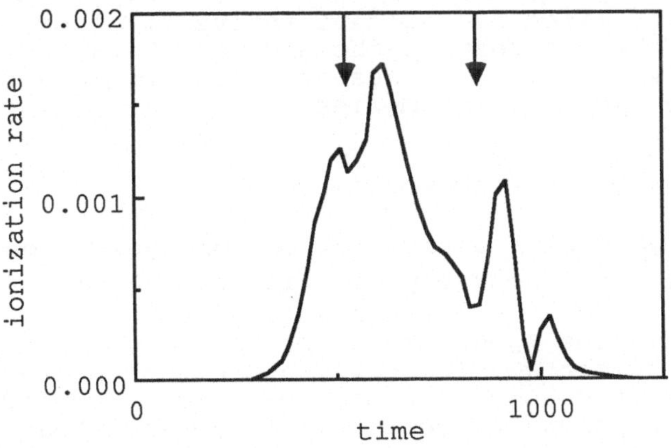

Fig. 2. Ionization rate of the atom as a function of time for a 32 cycle smooth pulse with $\omega = 0.148$ and $E_0 = 0.1$. The arrows mark times when the transition between the ground state and the first excited state is Stark-shifted to three-photon resonance, as indicated by maxima in the excited-state population.

obvious in Fig. 1c: because the peaks shift as a function of intensity, a smooth pulse gives a mixture of peaks at different positions. That is, the peaks broaden.

Another possible problem can be detected by plotting the populations of the ground state and of the first excited state as a function of time. It turns out that at times $t = 520$ and 840, with the field strength $E(t) = 0.087$, the dynamic Stark shifts have made a three-photon resonance with frequency 0.4440 out of the zero-field frequency 0.3949 between these states. The transition rate, i.e., the time derivative of the total population in the continuum, is plotted in Fig. 2. It shows the resonance too, although not as pronounced as one might expect. Tempting as it is to speculate that the bound-level resonance is responsible for some of the peak splittings in Fig. 1c, our attempts to substantiate any such claim have failed.

Given that the $n \pm 1/4$ cycle times and the rapid variation of the peak positions on the time scale of the field cycle reflect the physical quiver motion of the electron, we make one last remark concerning the gauge of the electromagnetic fields. Our Hamiltonian (1) is written in the $d \cdot E$ gauge. We have repeated some of the square-pulse computations for the $p \cdot A$ form of the dipole interaction, using the same energy eigenstates. Contrary to the common belief, the photoelectron spectrum is in general quite different for the $d \cdot E$ and the $p \cdot A$ interactions. In fact, the positions of the peaks in the $p \cdot A$ gauge *do not* oscillate as a function of time. Worse still, $A(t)$ and $A(t)+c$ give the same electric field so the physical results should be the same, but the frozen peak positions depend on the constant c. Obviously the "unperturbed" eigenfunctions of the time-independent Hamiltonian represent physically observable energies only in the $d \cdot E$ gauge.[10] Unless the field is turned on and off slowly (and the turn-on/off is included in the calculations!), or the gauge transformation to the $d \cdot E$ form of the interaction is explicitly utilized in the data analysis, electron spectra obtained from the p·A form of the dipole interaction are as good as random. This is simply because an inappropriate form of the kinetic energy operator is implied.

TURN-ON TRANSIENTS IN ELECTRON SPECTRA

In Fig.3 we plot a portion of the photoelectron spectrum for the 62 1/4 cycle square pulse with $E_0 = 0.05$ and $\omega = 0.52$. As the entire energy range shown is less than one photon energy, the structure in the spectrum cannot be ascribed to ATI. To find a clue to its origin we also give part of the energy level diagram of the atom and a few possible light-induced resonance transitions, all drawn to scale. It is obvious that the middle peak in the spectrum corresponds to two-photon ionization of the ground state, while the rest of the peaks (including the largest one) come from one-photon ionization of the excited bound states. Here the laser nearly matches the frequency from |0> to |2>, but this transition is dipole-forbidden and no resonance ensues.

We discuss these results in a simple semi-quantitative model. We start by assuming that the light first couples the ground state |0> to the other bound states |n> (n = 1, 2, ...), and in the second step the excited bound states |n> to the continuum {|W>}. The respective dipole moment matrix elements are denoted by d_{0n} and d_{nw}, and the coupling energies by Ω_{0n} ($= d_{0n}E_0/2$) and Ω_{nw}. The rotating-wave approximation (RWA) is the second ingredient of our model. We firstly replace the level energies with their detuning from resonance; re-defining the ground-state energy as zero, the excited state |n> will have the energy $\Delta_n = W_n - W_0 - \omega$, and the continuum state |W> the energy

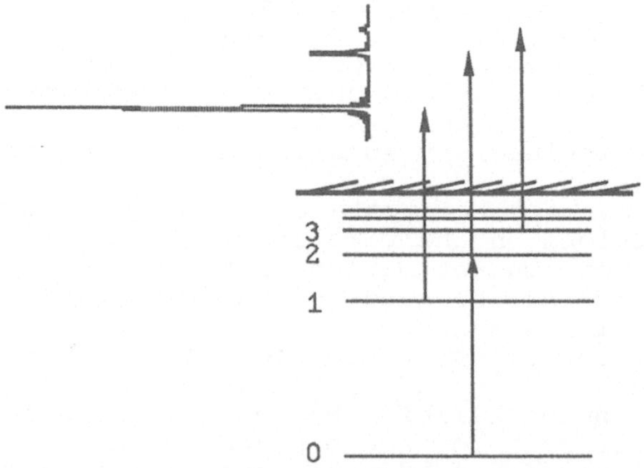

Fig. 3. Part of the energy level diagram of the atom, a few possible transitions for the photon energy ω = 0.52, and the photoelectron spectrum for a square pulse with $\omega = 0.52$, $E_0 = 0.05$ and $T = 752.17$ (upper left corner). All energies are drawn to scale.

$\Delta_W = W - W_0 - 2\omega$. Secondly, the couplings between the levels become time-independent.

Let us treat the light-induced coupling between the bound states as a small perturbation. To first order the ground state acquires a small admixture of excited-state character and the excited states a small ground-state component. The standard time-independent perturbation theory gives as the "dressed" energy eigenstates

$$|0'\rangle = |0\rangle + i\sum_n \frac{\Omega_{0n}}{\Delta_n} |n\rangle \ ; \qquad |n'\rangle = |n\rangle - i\,\frac{\Omega_{0n}}{\Delta_n} |0\rangle. \qquad (5)$$

In second order the levels also shift, but this is of no concern here.

The coupling to the continuum causes an exponential decay of the dressed states and accumulation of electron energy peaks in the continuum. The peaks corresponding to ionization of $|0'\rangle$ and $|n'\rangle$ will have the bare energies $2\omega + W_0$ and $\omega + W_n$, respectively, and Fermi's Golden rule gives the transition rates

$$\Gamma_0 = 2\pi\left[\sum_n \frac{\Omega_{0n}\,\Omega_{n,\,2\omega+W_0}}{\Delta_n}\right]^2 \ ; \qquad \Gamma_n = 2\pi\,\Omega^2_{n,\,\omega+W_n}. \qquad (6)$$

Γ_0 is easily recognized as the usual two-photon ionization rate from time-dependent perturbation theory (within the RWA).

In the numerical experiment of Fig. 3 the field is turned on suddenly at the initial time $t = 0$. We therefore decompose the ground state $|0\rangle$ in terms of the dressed states, whose populations to the lowest order in the perturbation strength are

$$|\langle 0'|0\rangle|^2 = 1 \ ; \qquad |\langle n'|0\rangle|^2 = \frac{\Omega^2_{0n}}{\Delta^2_n}\ . \qquad (7)$$

Subsequently the dressed states decay exponentially. The total populations in the electron spectrum as a function of time are

$$P_0 = 1 - \exp(-\Gamma_0 t)\ ; \qquad P_n = \frac{\Omega^2_{0n}}{\Delta^2_{0n}}\,[1 - \exp(-\Gamma_n t)]\,. \qquad (8)$$

After a long enough time when all exponentials have decayed out, the ratio of the populations in the "energy-nonconserving" (P_n) and "energy-conserving" (P_0) peaks is proportional to the small parameter of the perturbation theory.

A more interesting situation occurs when the relevant dressed levels have not yet decayed significantly,

$$\Gamma_0 t,\ \Gamma_n t \ll 1. \qquad (9)$$

Then (6), (8) give

$$P_0 = 2\pi \left[\sum_n \frac{\Omega_{0n}\,\Omega_{n,2\omega+W_0}}{\Delta_n} \right]^2 t \; ; P_n = 2\pi \left[\frac{\Omega_{0n}\,\Omega_{n,\,\omega+W_n}}{\Delta_n} \right]^2 t. \tag{10}$$

As long as the inequality (9) is not violated, the ratio of the peak heights remains unchanged both as a function of time and when the intensity of the field is varied. If the coupling matrix elements were slowly varying functions of continuum energy so that the difference between $2\omega + W_0$ and $\omega + W_n$ would not be crucial, the energy-nonconserving peaks would give the incoherent decomposition of the two-photon transition rate. Also, in the true perturbation theory limit $E_0 \to 0$ the extra peaks become infinitely long-lived. This is the reason we have dubbed them "energy-nonconserving".

In our numerical experiments we find that up to T = 752.17 (62 1/4 cycles of the field, ~ 20 fs) and E_0 = 0.05 (~ 10^{14} W/cm^2) the relative peak heights remain reasonably well unchanged with changing interaction time and field strength. This is as expected in view of the fact that $\Gamma_1 T$= 1.8 and $\Gamma_3 T$ = 0.3 hold true for the upper limits of T and E_0.

The structure of the spectrum in Fig. 3 is thus a turn-on

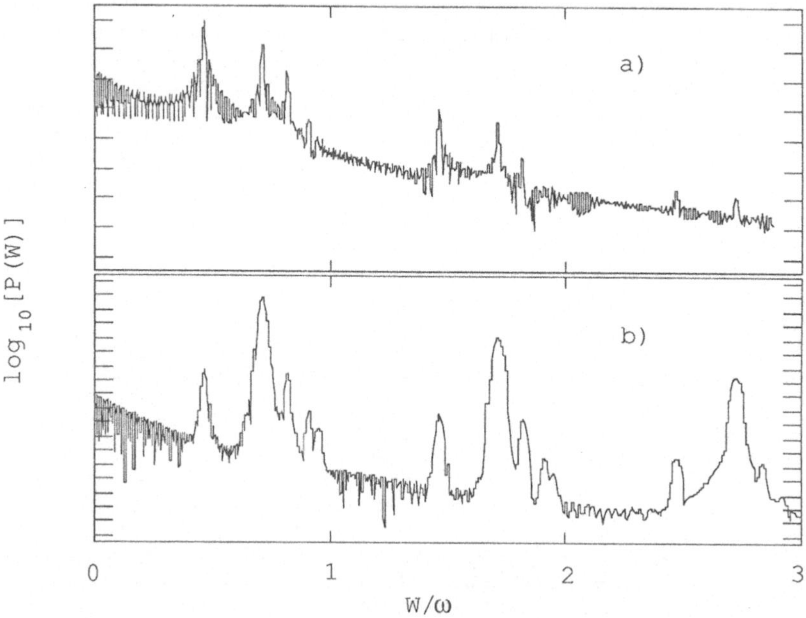

Fig. 4. Photoelectron energy spectra for the 62 1/4 cycle square pulse (a) and 96 cycle smooth pulse (b), with ω = 0.52 and E_0 = 0.05. Notice the different logarithmic scales for traces a and b. ATI images of the peak multiplets can also be seen.

transient. If the field is turned on slowly, the state |0> transforms into |0'> adiabatically and no other dressed states are populated. The energy-conserving peaks are the only ones to remain. This is clearly seen in Fig. 4, where the electron spectrum is plotted for the 62 1/4 cycle square pulse and 96 cycle smooth pulse.

The extra peaks in the spectrum are an example of a rather esoteric way how time-dependent perturbation theory may fail, and in principle at least they also provide a method to obtain information about the bound states of the atom. Nonetheless, in real experiments the pulses turn on smoothly, so it may be doubted whether such effects can be observed. To this end we note that we have studied two-photon ionization basically because then our programs allow us to run a large number of field cycles. However, in real experiments two-photon ionization may be the worst case for turn-on transients. For large multiphoton orders the time scales are automatically longer because the field period is longer. Even more importantly, the (virtual or real) intermediate levels are higher up in the Rydberg ladder and may have neighbors close by that can be excited non-resonantly. With the quest toward shorter pulses the possibility of transients in the electron spectra should be borne in mind.

DISCUSSION

Our discovery of energy-nonconserving peaks suggests that, even after a fair amount of work, our ATI simulations may still have surprises forthcoming. Nonetheless, for all we know now, after the characteristic ATI spectrum with peak spacing equal to photon energy has been established, the positions and the number of the peaks do not change qualitatively when either the square pulse is made longer or it is replaced by a smooth pulse.

Our earlier computations of 5- and 10 photon ionization did not show energy-nonconserving peaks.[7] One plausible reason is that the level spacing around the last intermediate state is generally too small to be resolved during the interaction times we can simulate in the computations.

In real experiments the situation may be different. Apparent bound-state structure attributed to intermediate-state resonances was recently reported with sub-picosecond pulses.[3] However, in Fig. 2 we see that the occurrence of an intermediate resonance does not necesarily lead to an apparent enhancement of ionization. Therefore we would like to propose the energy-nonconserving peak multiplets as a possible alternative mechanism leading to the experimental observations. It should be noticed, though, that the fine structure of the peaks in Fig. 1c cannot be of this origin. In fact, regarding to Fig. 1c or the experiments of Ref. 3 as well, at the moment even more intriguing hypotheses such as coherent interference of ATI electrons produced at different field strengths, might deserve serious consideration.

ACKNOWLEDGMENT

We are grateful for grants for computer assistance from the John von Neumann Center and the Allied Corporation.

REFERENCES

1. P. Agostini, F. Fabre, G. Mainfray, G. Petite, and N. K. Rahman, Phys. Rev. Lett. 42, 1127 (1979); P. Kruit, J. Kimman, H. G. Muller, and M. J. van der Wiel, Phys. Rev. A 28, 248 (1983); L. A. Lompré, A. L'Huillier, G. Mainfray, and C. Manus, J. Opt. Soc. Am. B 2, 1906 (1985); H. J. Humpert, H. Schwier, R. Hippler, and H. O. Lutz, Phys. Rev. A 32, 3787 (1985); U. Johann, T. S. Luk, H. Egger, and C. K. Rhodes, Phys. Rev. A 34, 1084 (1986); M. Bashkansky, P. H. Bucksbaum, and D. W. Schumacher, Phys. Rev. Lett. 59, 274 (1987).
2. G. Petite, invited talk in the 4th International Conference on Multiphoton Processes (ICOMP IV); P. Agostini, J. Kupersztych, L. A. Lompré, G. Petite, and F. Yergeau (unpublished).
3. R. R. Freeman, invited talk in ICOMP IV; R. R. Freeman, M. Geusic, S. Darack, and H. Milchberg (unpublished).
4. L. V. Keldysh, Sov. Phys. JETP 20, 1307 (1965); H. R. Reiss, Phys. Rev. A 22, 1786 (1980); M. Lewenstein, J. Mostowski, and M. Trippenbach, J. Phys. 18, L461 (1985); H. R. Reiss, J. Phys. B 20, L79 (1987); A.Dulcic , Phys. Rev. A 35, 1673 (1987); W. Becker, R. R. Schlicher, and M. O. Scully, J. Phys. B 19, L785 (1986).
5. Z. Bialynicka-Birula, J. Phys. B 17, 3091 (1984); M. Edwards, L. Pan, and L. Armstrong, Jr., J. Phys. B 18, 1927 (1985); Z. Deng and J. H. Eberly, J. Opt. Soc. Am. B 2, 486 (1985).
6. H. G. Muller, A. Tip, and M. J. van der Wiel, J. Phys. B 16, L679 (1983); M. Mittleman, J. Phys. B 17, L351 (1984); S.-I Chu and J. Cooper, Phys. Rev. A 32, 2769 (1985); A Szöke, J. Phys. B. 18, L427 (1985); L. Pan, L. Armstrong, Jr., and J. H. Eberly, J. Opt. Soc. Am. B 3, 1319 (1986).
7. J. Javanainen and J. H. Eberly, to appear in the proceedings of ICOMP IV, ed. S. J. Smith (Cambridge University Press, Cambridge, 1987); and J. Phys. B (submitted).
8. W. H. Press, B. P. Flannery, S. A. Teukolsky, and W. T. Vetterling, "Numerical recipes: the art of scientific computing", Cambridge University Press, Cambridge (1986).
9. A. Goldberg, H. M. Schey, and J. L. Schwartz, Am. J. Phys. 35, 177 (1967).
10. K. H. Yang, Ann. Phys. (N.Y.) 101, 62 (1976); C. Cohen-Tannoudji, B. Diu and F. Laloe, "Quantum Mechanics", Hermann/Wiley, Paris (1977); R. R. Schlicher, W. Becker, J. Bergou, and M. O. Scully, in "Quantum Electrodynamics and Optics", A. O. Barut, ed., Plenum, New York (1984).

IONIZATION OF ONE-DIMENSIONAL HYDROGEN ATOMS WITH INTENSE MICROWAVE AND LASER PULSES

R.V. Jensen and S.M. Susskind

Mason Laboratory, Yale University
New Haven, CT 06520, USA

INTRODUCTION

In the last five years the development of intense, short-pulse-length lasers has stimulated a tremendous amount of research on the behavior of atoms and molecules in oscillating electric fields which are comparable to the Coulomb binding fields. A wide variety of interesting and often puzzling physical phenomena have been observed experimentally and predicted theoretically in this novel, intense field regime of atomic and molecular physics where the traditional theoretical methods based on perturbation theory begin to fail. The present workshop on Atomic and Molecular Processes with Short Intense Laser Pulses provides a forum for the presentation of these exciting new results and for the extensive discussion of a number of controversial issues regarding the theoretical understanding of these intense field phenomena.

The purpose of this paper is, first, to briefly describe recent experimental and theoretical progress on the closely related problem of the behavior of highly excited hydrogen atoms in intense *microwave* fields.[1–2] In particular, we will show that numerical studies of a simple, one-dimensional model of a hydrogen atom in a strong, oscillating electric field can successfully account for many of the features of the experiments with real, three-dimensional hydrogen atoms.[2–4] Second, since researchers studying the interaction of intense lasers with atoms and molecules may be unfamiliar with this work, we will show that many of the physics issues are very similar (although there are some important differences) and that our detailed understanding of the microwave excitation and ionization of hydrogen atoms may have important applications in resolving some of the controversy surrounding the ionization of ground-state atoms with intense laser pulses. As a specific illustration, we will use the one-dimensional numerical model (which has been so successful in the microwave problem) to study the nonperturbative behavior of the two-photon ionization of atoms in an intense laser field.

Since the computer can be used to generate highly accurate solutions for the corresponding Schrödinger equation, it provides a very convenient "laboratory" for examining the role of intermediate states, the effects of the temporal behavior of the laser pulse (i.e. sudden versus adiabatic switching of the field), the appearance of multiple, Above-Threshold Ionization (ATI) peaks in the energy spectrum of the ionized electrons, and the effects of the ponderomotive potential on the location of these peaks.[5–7] In addition, these numerical "experiments" can be used to assess a number of assumptions and approximations which have been introduced in an effort to develop a theory for the multiphoton ionization of atoms in intense fields, such as the importance of continuum-continuum interactions[8] (since the corresponding matrix elements can be turned on or off in the numerical calculation) and the differences between the solutions of the Schrödinger equation using different gauges.[9]

Before describing the fascinating experiments on the microwave ionization of highly excited hydrogen atoms, which were pioneered by Bayfield and Koch[1] in 1974, and the very successful theoretical analysis of these experiments, which has been jointly developed by a number of research groups around the world,[3-4,10-12] we will briefly review the relevant physical parameters of these experiments and compare them with those for the intense laser ionization of ground state atoms.

The most recent experiments of Koch et al.[2] have examined the ionization of highly excited hydrogen atoms which were carefully prepared in states with principal quantum numbers ranging from n=32 to 90 and exposed to $\sim$ 300 oscillations of a 9.92 GHz field. In general it is convenient to characterize the oscillating perturbation in terms of the scaled field strength, n^4F, which is proportional to the ratio of the peak microwave field, F, to the Coulomb binding field, $\sim 1/n^4$ measured in atomic units, and the scaled frequency, $n^3\Omega$, which is the ratio of the angular frequency of the microwave perturbation and the Kepler frequency of the atom, $1/n^3$. For the microwave ionization experiments the scaled frequencies ranged from $n^3\Omega \sim .05$ to 1. for $n = 32$ to 90 and the scaled threshold fields for the onset of ionization ranged from n^4F =.025 to .125.

By comparison, the typical experiments on the ionization of ground state xenon have an effective principal quantum number of $n \sim 1$ and the laser frequencies for the 1064, 532, and 193 nm light used in the experiments correspond to scaled frequencies of $n^3\Omega = .043$, .086 and 0.24, respectively and the laser intensities ranging from 10^{13} to 10^{15} W/cm^2 correspond to scaled fields between $n^4F = .017$ and 0.17.[5-7] Consequently, both the microwave and laser ionization experiments involve the same range of the perturbation parameters. In addition, the recent interest in sub-picosecond laser pulses[13] means that the interaction times in the laser experiments can also be of the order of several hundred field oscillations or less.

There are, however, two significant differences between the two types of experiments. The first is that between 45 and 320 microwave photons must be absorbed to ionize the highly excited hydrogen atoms as compared to 3-12 for the ionization of ground state xenon atoms with 193 to 1064 nm light. Second, the dynamics of the microwave ionization involves the interaction of many bound states (in numerical calculations 100 or more principal quantum numbers must be represented) while the laser ionization involves the coupling of relatively few states (at most 10 or 20 principal quantum numbers). As a consequence the microwave ionization problem is much harder to analyze since conventional theoretical methods based on perturbation theory are likely to be hopeless. Nevertheless, despite these differences, the behavior of both systems is governed by the same Hamiltonian with comparable perturbation parameters and one might hope that new methods of analysis and the associated improvements in our physical understanding of one problem may be applied to the other.

MICROWAVE IONIZATION OF HIGHLY EXCITED HYDROGEN ATOMS

The most remarkable feature of the original experiments of Bayfield and Koch[1]was the observation that the microwave ionization of highly excited hydrogen atoms exhibits a very strong dependence on the intensity of the radiation and a relatively weak dependence on the frequency, just the opposite of the usual photo-electric effect. The experiments are performed with a beam of highly excited hydrogen atoms which pass through a microwave cavity where they are exposed to a microwave field for $\sim$ 300 microwave periods. (Since the atoms see a rising field as they enter the cavity and a falling field as they exit, the perturbing field is conveniently modeled as a pulse with a $\sim$ 50 cycle rise and fall.) For weak fields no ionization is observed during the transit time through the cavity. However, when a threshold field strength is exceeded the entire beam of atoms is rapidly ionized.

The open boxes in Fig. 1 show the experimental values for the scaled threshold field at which 10% of the beam is ionized as a function of $n^3\Omega$ for the experimental studies by Koch et al. with n ranging from 32 to 90. (Since the ionization probability generally increases sharply with increasing field, these 10% thresholds provide an experimental upper bound on the threshold fields for the onset of ionization.) One of the most striking features of this experimental data is the peaks in the threshold field at scaled frequencies with values close to low order rationals, $n^3\Omega$ =1, 2/3, 1/2, 2/5, 1/3, 1/4, 1/5. Since the Kepler frequency, $1/n^3$ is approximately equal to the transition frequency between neighboring levels for large n, a rational frequency, $n^3\Omega = p/q$, corresponds very closely to the resonant frequency for a q photon transition between two levels with $\Delta n = p$. Perturbative arguments would suggest that the atom would be less stable and more likely to ionize at these resonant frequencies; however, the experimental results shown in Fig. 1 provide yet another surprise – the prominent peaks in the ionization threshold near these resonant frequencies indicate that the atom is more stable and a stronger perturbation is required for the onset of rapid ionization. In fact the rising curve near $n^3\Omega = 1.0$ shows that $n = 82$ with an $n^3\Omega = 0.83$ requires a much larger microwave field to ionize than the more weakly bound $n = 90$ state.

These remarkable experimental results have provided a challenging puzzle for theorists. Both the sharp onset of ionization and the curious dependence of the threshold fields on the scaled frequency defy conventional theoretical analysis. In fact, the first successful account of the onset of ionization in this quantum system was based on a purely classical theory of the nonlinear electron dynamics in which the onset of ionization is associated with the onset of chaos in the classical phase space.[2-3,14-16] The results of classical calculations for the critical fields for the onset of chaotic ionization in a simple *one-dimensional* classical model of the experiment are shown in Fig. 1 by the solid line.[2,16] (More extensive three-dimensional classical simulations show even better agreement.[17]) The excellent agreement in this comparison (with no adjustable parameters) indicates very clearly that this quantum system can exhibit the effects of classical chaos and that the classical theory provides a remarkably good explanation of both the sharp onset of ionization as well as the peaks at rational frequency ratios. In particular a detailed examination of the classical theory reveals that these peaks are associated with classical nonlinear resonances, "islets of stability", in the classical phase space, which inhibit the ionization. Moreover, the stabilizing effects of these islands is significantly enhanced by the slow turn-on and turn-off (~ 50 microwave periods) of the microwave "pulse".[18] (The interested reader is refered to Refs.2-3,16-18 for a detailed account of the comparison of the classical theory with the microwave ionization experiments.)

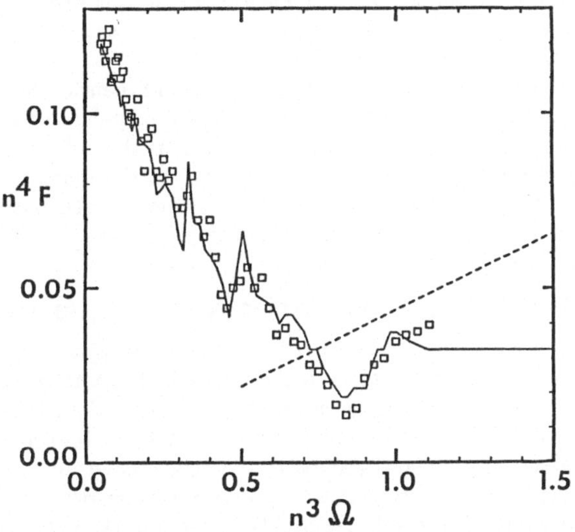

Fig. 1: A comparison of the theoretical predictions for the threshold field strengths for the onset of chaos (solid line) with the experimental values for 10 % ionization (squares) is shown along with the Casati et al.'s estimate of the quantum de-localization border (dashed line).

Besides raising a number of important questions regarding the controversial issue of "quantum chaos", the success of the classical theory poses a formidable challenge to quantum theory to account for the quantum mechanisms which mimic the onset of classical chaos and which are responsible for the experimental results. Motivated by the success of the one-dimensional classical theory a number of research groups have investigated the numerical solution of the corresponding one-dimensional Schrödinger equation.[4,10–12] These numerical studies provide a means of generating non-perturbative solutions for the quantum dynamics which can be used to suggest new analytical approximations and physical interpretations for understanding the behavior of atoms in intense electromagnetic fields.

In particular, in the last two years a new physical picture has begun to emerge which also shows good agreement with the experimental measurements for the microwave ionization of highly excited hydrogen atoms.[4,10–12] For small perturbations the initial state couples weakly to nearby states with both larger and smaller n during the pulse and returns to the initial state at the end of the pulse with almost 100% probability. (Of course some excitation and ionization always occurs, as predicted by perturbation theory, but on the time-scale of the experiments, $\sim$ 300 field oscillations, these effects are very small.) Then when the perturbation exceeds a critical value, the evolution of the perturbed quantum state exhibits a sudden explosion or delocalization in which it couples strongly to many states, including those which ionize rapidly. For the range of parameters studied in the current experiments the critical field for this transition from a localized to delocalized wavepacket has been found to coincide very closely with the onset of classical chaos and the onset of ionization in the experiments. In fact, a recent paper by Blümel and Smilansky[4] provides a comparison of their quantum calculation for the threshold field with the experimental measurements which shows comparable agreement to that for the classical theory shown in Fig. 1.

As a consequence, the agreement of the classical, quantum, and experimental results appear to confirm the remarkable vigor of Bohr's correspondence principle which might be expected to apply to these highly excited hydrogen atoms. However, the one-dimensional quantum calculations performed by Casati, Chirikov, Shepelanski, and Guarneri[10] also indicate that for scaled frequencies, $n^3\Omega > 1$, which is just outside the current range of experiments, the quantum delocalization will require significantly larger perturbations than the onset of classical chaos. Their estimate for the quantum delocalization border (drawn as a dashed line in Fig. 1), predicts a clear divergence between the classical and quantum theories as the experiments move to *higher quantum numbers* or higher frequencies. These theoretical predictions suggest that even more surprises are to be expected as the experiments enter this new regime in the response of atoms to intense electro-magnetic fields.

Before turning to the problem of laser ionization, I should mention that there are several other interesting features of the microwave ionization of highly excited atoms which are currently subjects of intense experimental and theoretical interest such as the behavior of non-hydrogenic Rydberg atoms, the appearance of non-monotonic "peaks" in the ionization probability as a function of field, and the effects of a two frequency or "two color" perturbation. Unfortunately, space does not permit the discussion of these additional topics; however, the reader is urged to consult the references for more details.

IONIZATION OF GROUND STATE ATOMS WITH INTENSE LASER PULSES

The most striking feature of the experiments on the ionization of ground state atoms in intense laser fields is the observation of multiple peaks in the photoelectron energy spectrum corresponding to energies requiring the absorption of more than the minimum number of photons to ionize the atom.[5–7] This so called Above Threshold Ionization (ATI) has been observed by several different experimental groups by exposing the ground states of Xe and Kr to laser pulses with intensities of 10^{13}W/cm^2 or more and wavelengths between 193 and 1064 nm. Moreover, the number, width and relative hights of these peaks exhibit a very interesting dependence on the intensity which has challenged theory. In particular, the observation that the first few peaks diminish in relative size as the perturbation intensity is increased and eventually disappear into the experimental background has stimulated the greatest interest. (The reader will surely find a number of examples of these experimental photoelectron spectra in the contributions of Bucksbaum and Agostini to this volume.)

At present the status of the theory is such that the presence of multiple peaks can be understood as a consequence of the nonperturbative nature of these intense fields which, as pointed out earlier, correspond to a good fraction of the Coulomb binding field. In addition a very plausible, heuristic explanation for the suppression and dissappearance of the first peaks has been proposed, based on the ponderomotive potential shift of the continuum states and the near cancelation of this shift in second-order perturbation theory for the ground state.[19] However, the complete theoretical account of the experimental data remains an open problem because of the difficulty in solving the Schrödinger equation in this nonperturbative regime. Nevertheless, a variety of attempts have been made using and a variety of non-perturbative approximations.[8–9,20] In general, these different theories have only succeded in explaining a few features of the ATI peaks but not all and in stimulating a very lively debate over the validity of different approximations and physical interpretations.

Motivated by the success of numerical studies of the one-dimensional model for the ionization of hydrogen with intense microwaves, we have suggested[21] that a numerical study of a similar one-dimensional model of a ground state atom in intense laser field could provide accurate nonperturbative solutions of the corresponding Schrödinger equation which would clarify some of the controversial issues surrounding ATI. In particular, since the numerical algorithm developed by Susskind and myself[12] to describe the microwave problem provides a detailed representation of the Coulomb continuum, this numerical "laboratory" can be used not only to study the dependence of the total ionization on the perturbing field but also to examine the physical mechanisms which are responsible for the curious properties of the ATI peaks in the continuum. In previous work we briefly described the properties of ATI peaks which were observed in numerical studies of the ionization of Rydberg atoms.[3,21] (Unfortunately, these peaks would be very difficult to observe in the microwave ionization experiments because of the low photon energy.) Here we will apply this model to a problem of more direct interest to the laser ionization community.

To illustrate the capabilities of this numerical laboratory, we considered the 2-photon ionization of the $n = 1$ state of a one-dimensional hydrogen atom in an intense laser pulse with a scaled frequency of $\Omega = 0.3$ (which corresponds to a wavelength of 152 nm) and with an scaled field of $F = 0.1$ (which corresponds to an intensity of 3.5×10^{14} W/cm^2). Two different pulse shapes were used to study the effects of the temporal behavior of the laser pulse - a smooth pulse which increases linearly to the peak field after 5 laser periods, is constant for 20 periods, and then decreases to zero over 5 periods and a square pulse which rises suddenly and remains constant for 20 periods.

The calculations were performed by numerically evolving the wave function for a one-dimensional hydrogen atom perturbed by an oscillating electric field. Our numerical algorithm expands the wavefunction in basis consisting of a finite number of bound and discretized continuum eigenstates of the unperturbed Hamiltonian. (A detailed description of our numerical algorithm can be found in Ref. 12) Because of the high laser frequency very few bound states are coupled and the bound basis could be truncated at $n = 4$, while 100 discretized continuum states were required to resolve the time evolution of the first few resonant peaks in the continuum. (The number of both the bound and continuum states were doubled in subsequent calculations to check for the convergence of the numerical results.) In addition, to explore the physical processes which are responsible for the behavior of these peaks we performed calculations with and without continuum-continuum interactions, in both the $x \cdot E$ and $p \cdot A$ repesentations of the perturbation, and with both the slow and sudden laser pulses.

The probability of ionization was determined by ploting the natural logarithm of the total probability of remaining in a bound state, P_B, versus time. The ionization rate is then determined by the slope of this curve. Typical results are shown in Figure 2 for a calculation with 4 bound states and 100 continuum states perturbed by a smooth laser pulse in the $x \cdot E$ representation. As the laser field rises the ionization rate increases then remains constant on average during the flat portion of the pulse and then decreases to zero as the pulse ends. (The oscillations in P_B over each cycle indicate a population flopping between the bound states and the continuum.) At the end of 30 cycles the total ionization probability in this very intense field is approximately 90%.

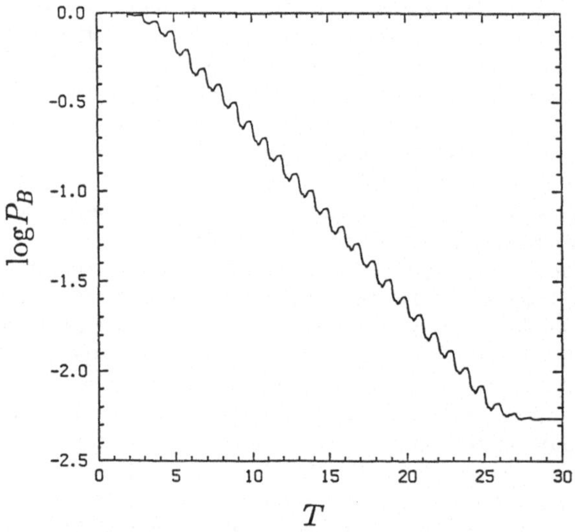

Fig. 2: The logarithm of the total probability of remaining bound is plotted as a function of time for a smooth laser pulse with $\Omega = 0.3$ and $F = 0.1$.

Effects of Pulse Shapes and Intermediate States

Since the calculation explicitly determines the excitation of the continuum states, we can plot the probabilty, P_k of exciting the contunuum states with "momentum" label k as a function of k at different times to look for ATI peaks. In Figure 3 we have graphed the $\text{Log}_{10}P_k$ versus k after 1, 5, 10, 15, 20, 25, and 30 periods of the smooth pulse which resulted in the ionization displayed in Fig.2. Since the energy of the continuum electrons is simply determined by $E_k = k^2/2$ it is straightforward to verify that the prominent peaks which grow resonantly with time near $k = 0.447$ and $k = 0.894$ correspond to the energies for 2 and 3 photon absorption from the ground state. In addition a small persistent structure is also apparent in the family of curves at $k = 0.592$ which corresponds to a 1 photon transition from the $n = 2$ state. Although this transition does not conserve energy such excitations are nevertheless permitted because of the finite pulse length.

If we use the sudden pulse, the peaks corresponding to the 1 and 2 photon transitions from the $n = 2$ state are much more prominent as indicated in the corresponding curves after 1, 5, 10, 15, and 20 cycles of the sudden pulse shown in Fig. 4. Real excitation of the intermediate $n = 2$ state occurs (it does not simply act as virtual step for the two photon transition frmo the $n = 1$ state) which can then be resonantly excited by one photon to $k \sim 0.592$ and 2 photons to $k \sim 0.975$. (The reader should note that recent experimental results presented in the papers by Bucksbaum and Agostini at this workshop, also show evidence for fine structure in the ATI peaks for very short pulses which appear to be associated with the excitation of intermediate states.)

In fact, excitation of the intermediate states occurs in both cases, as shown by the graphs of probabilities of occupying each of the 4 bound states as a function of time in Figs. 5a and 5b for the smooth and the sudden pulses. These figures clearly show the effect of the temporal behavior of the pulse on the excitation of intermediate states. With the smooth pulse (Fig. 5a) the slow turn on prepares an adiabatic superposition of the $n = 1$ and $n = 2$ statesThen

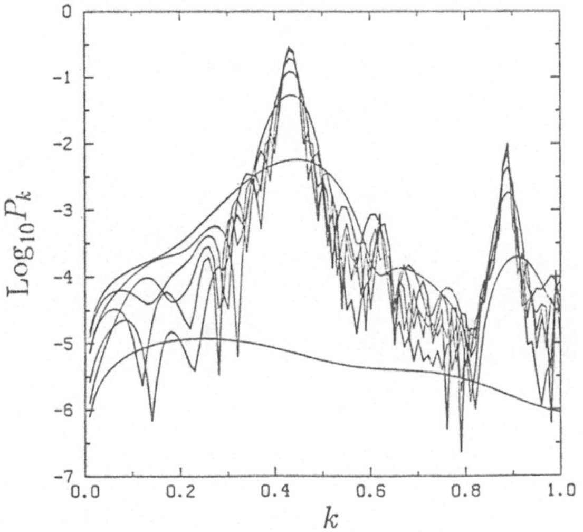

Fig. 3: The Logarithm of the probability of exciting states in the discretized continuum basis as a function of the label k (where $E_k = k^2/2$) after 1, 5, 10, 15, 20, 25, 30 periods of the smooth pulse.

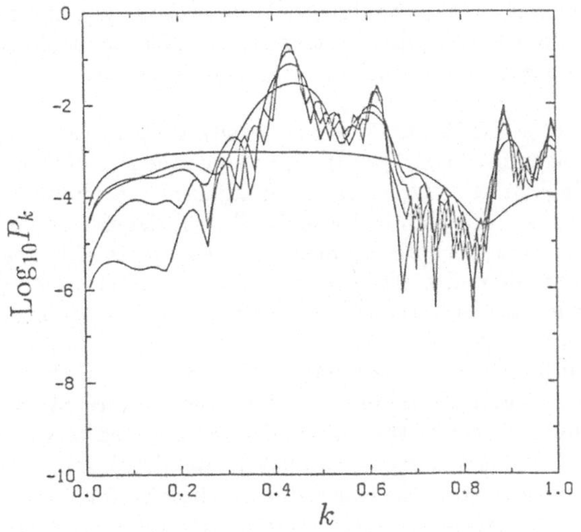

Fig. 4: Same as Fig. 3 but for the sudden pulse.

when the perturbation is slowly turned off, the adiabatic supperposition of $n = 1$ and $n = 2$ returns to the $n = 1$ state while the small nonadiabatic excitation of the $n = 3$ and 4 states remains. In contrast, for the sudden pulse (Fig. 5b) the $n = 1$ and 2 states exhibit distinct Rabi oscillations[22] and a distribution over all of the excited states remains after the pulse is turned off.

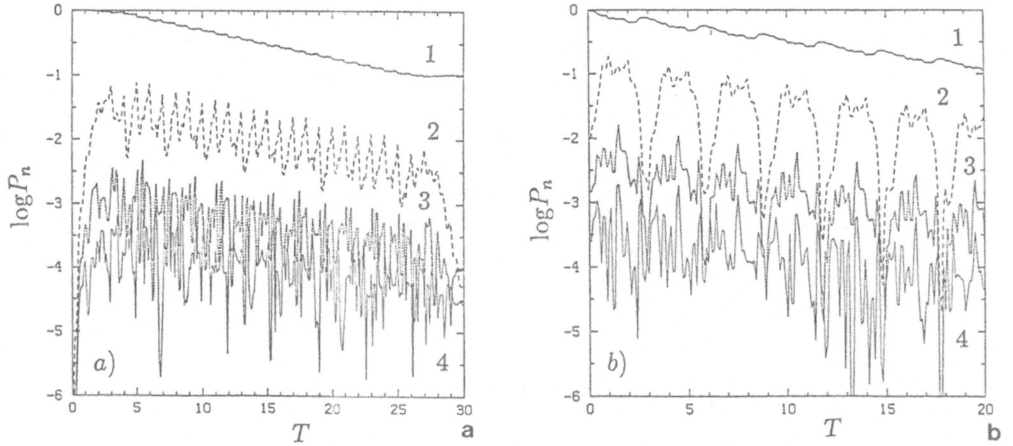

Fig. 5: The Logarithm of the bound state probabilities are plotted as a function of time for $n = 1 - 4$ withe a) the smooth pulse and b) the sudden pulse.

Role of the Continuum-Continuum Transitions

It is important to note that the calculations described above and in Ref. 21 were performed without including any coupling between the continuum states since the calculation of the continuum-continuum (c-c) transitions between the 100 or more continuum states greatly increases the computation time. As a consequence, it is clear that the presence of multiple peaks in the continuum does not require c-c transitions. The multiple ATI peaks observed in these early calculation are a consequence of direct multi-photon excitation from the bound states.

Nevertheless, by using a CRAY/XMP we can easily turn on the continuum coupling to assess the effects of the c-c transitions. We can also reduce the size of the c-c matrix elements, which exhibit a strong singularity $\sim 1/(E - E')^2$ for adjacent continuum states,[23] by using the $p \cdot A$ representation for the perturbation which multiplies all the matrix elements by a factor of $(E - E')/\Omega$. This reduces the c-c matrix elements at the expense of increasing the size of the bound-continuum and bound-bound matrix elements for transitions involving several photons. (A check of the $p \cdot A$ calculations without c-c coupling gave almost identical results to the previous ones with $x \cdot E$.)

Calculations with 10 bound states and 100 continuum states with the full bound-bound, bound-continuum, and continuum-continuum coupling were performed with the same perturbation parameters as before. Although the total ionization changed very little (the total ionization at the end of the smooth pulse increased to 93%), the distribution over continuum states changed dramatically. Figure 6 shows the results for the distribution over the continuum states after 5, 10, 15, 20, 25, and 30 periods of the slow pulse with the same perturbation parameters as before. The most striking effect of the c-c transitions is the shift of resonant peaks for 2 and 3 photon transitions from the n=1 state to lower energies. In fact, both levels are shifted down in energy by $\sim .04$ (a.u.) which is comparable to (but somewhat larger than) the ponderomotive potential shift associated with the oscillating electric field of $F^2/4\Omega^2 = .028$ that is predicted by theory.[7,19] (These shifts were also observed in similar one-dimensional model calculations by Javanainen and Eberly.[24]) Note, however, that the small structure associated with the one-photon ionization from the $n = 2$ state hardly shifted at all. Thereby confirming that the excited states show a much smaller shift relative to the continuum states than the ground state. Moreover, Figs. 3 and 6 shows that the 3 photon ATI peak is enhanced by an order of magnitude with the inclusion of c-c coupling. Although the c-c tansitions are not required for the appearance of ATI peaks, the c-c coupling does appear to play an important role in determining the relative heights of these peaks.

To assess the accuracy of these numerical "experiments", the same calculations were performed with the $x \cdot E$ representation using a basis consisting of 100 Sturmian functions which provides an alternative means of representing the c-c interactions.[12] By projecting the Sturmian wave packet onto the continuum states,[12] we can recover the distribution over continuum states at each instant of time for comparison with the previous results. In this case the same shifts of the peaks were clearly observed. Moreover, an examination of the distribution during a single cycle revealed the "sloshing" of probability in the continuum due to the field oscillation which was previously remarked upon by Javanainen and Eberly.[24] (Note that our choice of a $\cos \Omega t$ as opposed to their $\sin \Omega t$ time dependence allows us to avoid the nuisance of having to monitor the results at 1/4 cycles of the field to observe the stationary shifts of the peaks.)

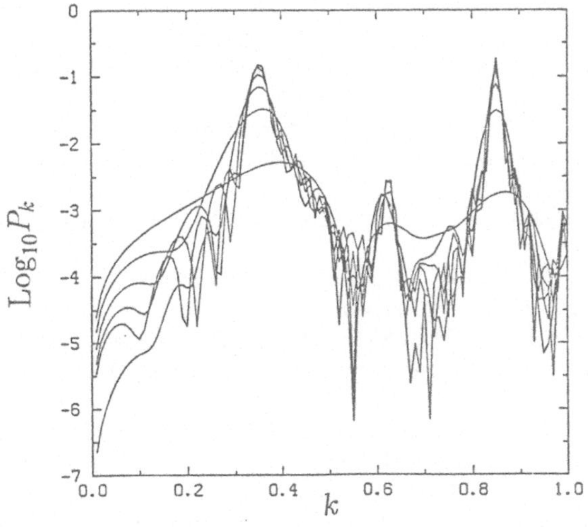

Fig. 6: Same as Fig. 3 but with the inclusion of continuum-continuum interactions.

DISCUSION

We have attempted to show how theoretical techniques developed in the successful analysis of the chaotic, microwave ionization of highly excited hydrogen atoms could be applied to the related problem of Above Threshold Ionization of ground state atoms with intense laser pulses. In particular, our numerical study of the 2 photon ionization of one-dimensional hydrogen atoms provides a textbook illustration of the quantum dynamics of an atom in an intense oscillating field which clearly shows the appearance of multiple ATI peaks in the energy spectrum of the continuum electrons and their dependence on the temporal behavior of the laser pulse, the presence of intermediate bound states, and the inclusion of continuum-continuum interactions. These results suggest that this numerical "laboratory" could be used to develop a theoretical interpretation of the real experiments which might account for the curious dependence of the ATI peaks on the laser fields as well as the recently discovered fine structure due to intermediate states observed with very short sub-picosecond laser pulses.[13]

Certainly, many open questions remain regarding the applicability of these model calculations to the ionization of real atoms, such as the relevance of a one-dimensional model to three-dimensional atoms and the differences in the atomic structure of one-dimensional hydrogen and Xe or Kr. Unfortunately, due to space limitations, these issues as well as more extensive calculations involving lower laser frequencies and higher numbers of ATI photons must be relegated to future work.[25]

The authors thank their experimental and theoretical colleagues P.M. Koch, K.A.H. van Leeuwen, J.E. Bayfield, J.N Bardsley, B. Sundaram, D. Richards, G. Casati, I. Guarneri, R. Blümel, U. Smilansky, D. Shepelansky, and B. Chirikov for many stimulating discussions of this problem. In addition, this work was supported by theAlfred P. Sloan Foundation, the National Science Foundation and the Pittsburgh Supercomputing Center.

1. J.E. Bayfield and P.M. Koch, Phys. Rev. Lett. *33*, 258 (1974).
2. K.A.H. van Leeuen, G.v. Oppen, S. Renwick, J.B. Bowlin, P.M. Koch, R.V. Jensen, O. Rath, D. Richards, and J.G. Leopold, Phys. Rev. Lett. *55*, 2231 1985); P.M. Koch, K.A.H. van Leeuwen, O. Rath, D. Richards, and R.V. Jensen, in *The Physics of Phase Space*, Lecture Notes in Physics 278, eds. Y.S. Kim and W.W. Zachary, (Springer, New York, 1987), p. 106.
3. R.V. Jensen, in *Atomic Physics 10*, eds. H. Narumi and I. Shimamura, (North-Holland, New York, 1987), p.319.
4. R. Blümel and U. Smilansky, Z. Phys. D *6*, 83 (1987).
5. P. Kruit, J. Kimman, H.G. Muller, and M.J. van der Wiel, Phys. Rev. A *28*, 248 (1983) and references therein.
6. F. Yergeau, G. Petite, and P. Agostini, in *Photons and Continuum States of Atoms and Molecules*, eds. N.K. Rahman, C. Guidotti, and M. Allegrini, (Springer, New York, 1987), p. 78.
7. T.J. McIlrath, P.H. Bucksbaum, R.R. Freeman, and M. Bashansky, Phys. Rev. A *35*, 4611 (1987).
8. Z. Deng and J.H. Eberly, J. Opt. Soc. Am. B *2*, 486 (1985).
9. H.R. Reiss, Phys. Rev. A *22*, 1786 (1980).
10. G. Casati, B.V. Chirikov, and D.L. Shepelyansky, Phys. Rev. Lett. *53*, 2525 (1984); G. Casati, B.V. Chirikov, D.L. Shepelyansky, and I. Guarneri, Phys. Rev. Letters, *57*, 823 (1986).
11. J.N. Bardsley and B. Sundaram, Phys. Rev. A *32*, 689 (1985); J.N. Bardsley, B. Sundaram, L.A. Pinnaduwage, and J.E. Bayfield, Phys. Rev. Lett. *56*, 1007 (1986). J.N. Bardsley and M.J. Comella, J. Phys. B *19*, L565 (1986).
12. S.M. Susskind and R.V. Jensen, submitted to Phys. Rev. A.
13. See contributions by Bucksbaum and Agostini to this volume.
14. J.G. Leopold and I.C. Percival, J. Phys. B *12*, 709 (1979).
15. R.V. Jensen, Phys. Rev. *30A*, 386 (1984).
16. M. Sanders and R.V. Jensen, in preparation.
17. O. Rath and D. Richards, private communication.
18. R.V. Jensen, Physica Scripta, *35*, 668 (1987).
19. L. Pan, L. Armstrong, Jr., and J.H. Eberly, J. Opt. Soc. B *3*, 1320 (1986).
20. See for example the papers in *Photons and Continuum States of Atoms and Molecules*, eds. N.K. Rahman, C. Guidotti, and M. Allegrini, (Springer, New York, 1987) and references therein.
21. R.V. Jensen and S.M. Susskind, in *Photons and Continuum States of Atoms and Molecules*, eds. N.K. Rahman, C. Guidotti, and M. Allegrini, (Springer, New York, 1987), p. 13.
22. N.B. Delone and V.P. Krainov, *Atoms in Strong Light Fields*, (Springer, New York, 1985).
23. Y. Gontier, private communication.
24. J. Javanainen and J.H. Eberly, preprint (1986).
25. R.V. Jensen and S.M. Susskind, in preparation.

ABOVE THRESHOLD PHOTODISSOCIATION WITH INTENSE IR LASERS - A COUPLED EQUATIONS STUDY

André D. Bandrauk and Nadia Gélinas

Départment de chimie, Faculté des Sciences
Université de Sherbrooke
Sherbrooke, Que, Canada, J1K 2R1

INTRODUCTION

Above threshold ionization, ATI, is a phenomenon first observed in atomic physics[1-3], whereby an atom in an intense laser field is ionized by absorption of a number of photons larger than the ionization potential E_g (E_g = ground state energy), resulting in electrons with high kinetic energies. At high intensities ($I > 10^{12}$ W/cm^2), this phenomenon is no longer describable by perturbation theories which predict an I^{N+S} intensity law for the ionization probability, where N is the minimum number of photons for ionization ($N\hbar\omega \simeq E_g$), and S is the order of the transition in the continuum, so that $S\hbar\omega$ is the minimum excess kinetic energy. At intensities $I > 10^{12}$ W/cm^2, a new phenomenon occurs, i.e., suppression of the low energy peaks and shifting of the peaks as the pulse length is shortened below the subpicosecond time regime[4]. Of note is that the absorption intensities in the continuum are quite high in spite of low "Franck-Condon" factors (continuum-continuum wave function overlaps). Clearly the large probabilities observed must be a consequence of the linearly increasing dipole moment of the electron-ion system as the particles separate, leading to an atom-field interaction $\vec{\mu}\cdot\vec{E} = e\vec{r}\cdot\vec{E}$ which can become quite large in the complete dissociation limit.

We wish to address here the question whether similar effects could happen in molecular systems. In particular we will examine here a special class of diatomics AB which are called ionic, because at the equilibrium distance R_e, these are represented by an ionic electronic configuration A^+B^-, eg. LiF[5], MgH[6]. At infinite separation, these dissociate into the neutral fragments as a result of an avoided crossing between the covalent configuration A-B and the ionic state A^+B^-, see figure 1. The case of the alkali halides[5,7] offer the interesting situation that the ground state remains ionic out to very large distances, so that IR multiphoton absorption in these simple systems should mimick the large absorptions seen in the atomic case using visible or near IR photons. Summarizing, the molecule-field coupling for ionic molecules will be $e\vec{R}\cdot\vec{E}$, where R is the separation between the A^+ and B^- moities of the molecule. This radiative interaction is calculable from the relation

$$V_{rad} = 1.17 \times 10^{-3}\ \mu(\text{a.u})\ \sqrt{I(\text{W/cm}^2)}\ \text{cm}^{-1} \quad . \tag{1}$$

For LiF, at its equilibrium $R_e \simeq 4$ a.u., and an intensity of 10^{10} W/cm^2, one obtains $V_{rad} \sim 500$ cm^{-1}, or a radiative transition rate of 1.5×10^{13} s^{-1}, which is of the order of vibrational frequencies. At the crossing point ($R_c \sim 8$ Å), the coupling is four times larger. Thus the coupling increases linearly as in the $X^+ + e^-$ atomic case, giving rise to very large radiative couplings as the system dissociates in the presence of fields of intensity greater than $I = 10^{10}$ W/cm^2. We shall examine in particular fields between 10^{12} and 10^{13} W/cm^2 by a coupled equation method which we have developed previously for electronic transitions in diatomics[8-12]. Our main conclusion is that above threshold photodissociation (ATPD) should be an observable process as a result of efficient climbing of the vibrational ladder in ionic molecules in the presence of intense IR lasers fields.

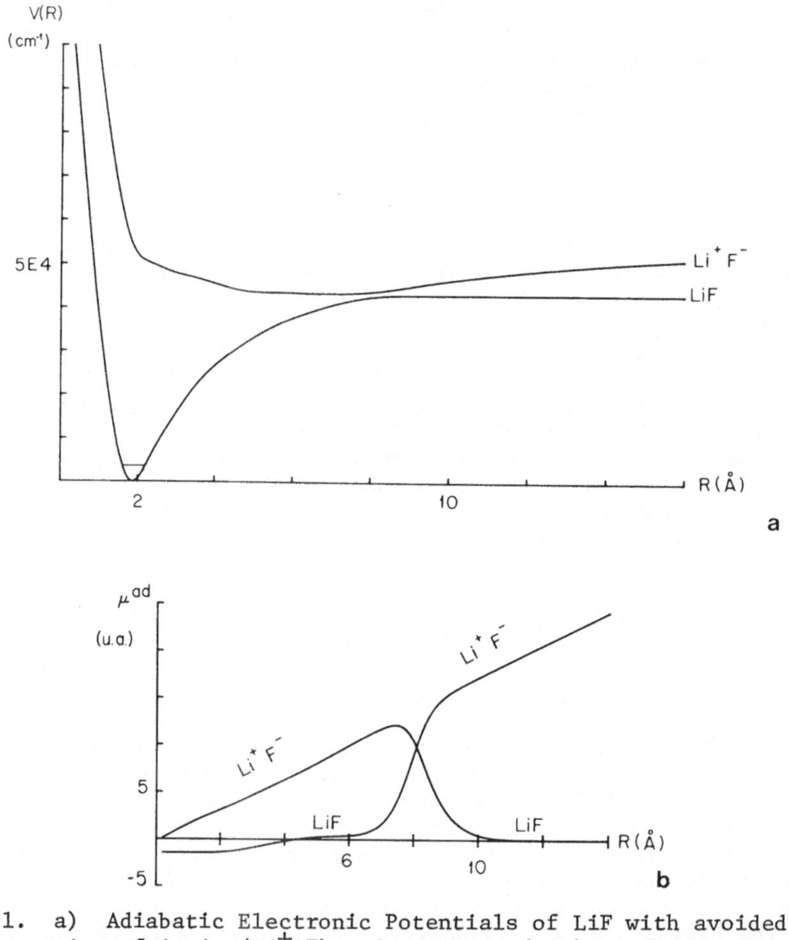

Figure 1. a) Adiabatic Electronic Potentials of LiF with avoided crossing of ionic (Li^+F^-) and covalent (LiF) configurations at $R_c = 8$ Å; b) Adiabatic Dipole Moments.

COUPLED EQUATIONS

The necessity of using coupled equations to treat molecular multiphoton processes in the regime of intense fields stems from the fact that molecules are intrinsically multilevel systems so that two or more level descriptions such as used in the dressed atom picture are no longer applicable. The appearance of continuum nuclear states corresponding to dissociative molecular states requiresthe proper use of scattering techniques to incorporate finite lifetime effects into multiphoton processes. Thus as emphasized by us previously[8,12,13], direct photodissociation is the analogue of predissociation[14], where in the former, radiative couplings replace formally the nonradiative couplings in the coupled equations formalism for the latter.

Since we wish to include continuum states as well as bound states and thus be able to extend numerical simulations of multiphoton processes into the nonperturbative regime, we exploit the technique of quantum scattering theory which enables one to handle explicitly continua, bound, and resonant states, the latter being bound states of finite lifetime as a result of their decay into a continuum state via a dissociative mechanism. In the event that both initial and final states are bound states, one would like to calculate transition probabilities between such states explicitly by including intermediate transitions and photodissociations as well. One can transform all transitions, including bound to bound transitions into a scattering problem by introducing additional continua as entrance and exit channels. This is an extension of an artifice originally employed by Shapiro[15] to calculate Franck-Condon factors in direct photodissociation by a single entrance channel. We have employed the two channel scattering technique in previous work on Resonance Raman Scattering[16] and other nonlinear optical cross sections[12,17] in order to examine the importance of nonadiabatic effects in these.

The introduction of entrance and exit channels $|C1\rangle$ and $|C2\rangle$ allows us to exploit the various relations between transition matrices[18-19] in order to extract relevant photophysical processes (see figure 2). Thus the transition amplitude or transition matrix element between entrance and exit channels is given by

$$T_{C1,C2} = \exp(i\phi_1)\, V_{1a} G_{ab} V_{b2} \exp(i\phi_2) \quad , \tag{2}$$

where ϕ_1 and ϕ_2 are elastic phase shifts for the uncoupled entrance and exit channels. V_{1a} and V_{1b} are the artificial couplings chosen to be weak so that their perturbation on the states $|a\rangle$ and $|b\rangle$ is negligible. Finally G_{ab} is the total Green's function matrix element defined now in a molecule-photon (dressed) space, $|a,n\rangle = |a\rangle \otimes |n\rangle$.

$$G_{ab} = \langle a,n|\, (E-H)^{-1}\, |b,n-m\rangle \quad . \tag{3}$$

For bound to bound transitions, these matrix elements have been previously calculated[20-21],

$$G_{ab} = T_{ab} \left[(E-E_a-T_{aa})(E-E_b-T_{bb}) - T_{ab}T_{ba}\right]^{-1} \tag{4}$$

This is a highly nonlinear expression with respect to T_{ab}, the desired transition amplitude between initial state $|a\rangle$ and final state $|b\rangle$. Setting $T_{ba} = 0$, i.e., not allowing backscattering from the final to the initial state (this produces energy shifts of these states[16]), is obtained by using unsymmetric couplings in the coupled equations, $V_{ad} \neq 0$, $V_{da} = 0$, etc. This will generate perturbative multiphoton transition amplitudes since no renormalization of the coupled states is permitted. Thus by insuring that the states are unperturbed by the radiation field, one can use the relation between G and T,[19]

$$T = V + VG_0T = V + TG_0V \quad , \tag{5}$$

$$G = G_0 + G_0TG_0 \quad , \tag{6}$$

to give the perturbative multiphoton bound-bound transition amplitude T_{ab},

$$T_{ab} = T_{C1,C2} \exp\left[-i\,(\phi_1 + \phi_2)\right] (G_o^a V_{1a} V_{b2} G_o^b)^{-1} \quad . \tag{7}$$

All quantities on the right-hand side of equation (7) are amenable to exact numerical calculations as shown below. $G_o = (E-H_o)^{-1}$ is the zeroth order resolvent corresponding to a radiation free molecular Hamiltonian H_o. We will use equation (7) to calculate the fluorescence emitted by the system in figure 1 as photons climb the vibrational ladder towards the continuum above the threshold for dissociation of LiF (figure 2).

Using the relation (5) for the transition operator one can also calculate the transition amplitude between the enhance channel |C1> and the physical channel |C>

$$T_{C1,C} = \exp\,(i\phi_1)\; V_{1a} G_o^a T_{ac} \quad , \tag{8}$$

Therefore one can extract the photodissociation amplitude T_{ac} from the initial bound state |a,n> to the final continuum state |C,n-m> after absorption of m photons, as

$$T_{ac} = T_{C1,C2} \exp\,(-i\phi_1)\; (V_{1a} G_o^a)^{-1} \quad . \tag{9}$$

Again all quantities on the right-hand side can be calculated numerically to give the desired photodissociation amplitude T_{ac} (figure 3).

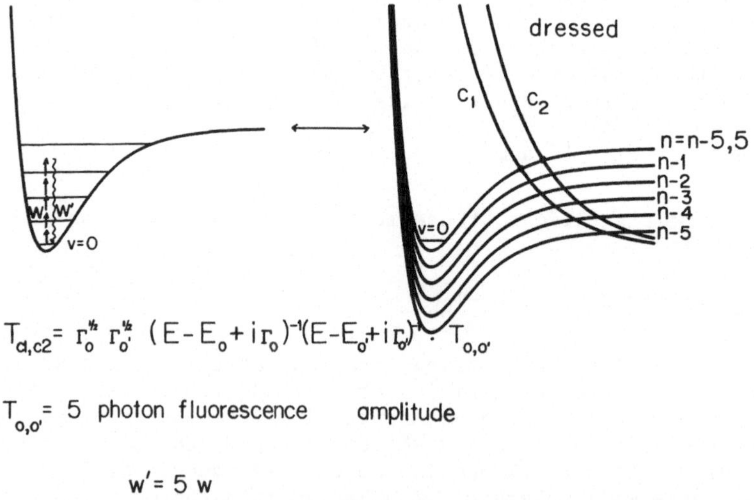

Figure 2. Perturbative and dressed (nonperturbative) representations for calculations of fluorescence amplitudes $T_{0,0'}$. C1 and C2 are entrance and exit artificial channels for the initial and final bound states in the 5 photon fluorescence.

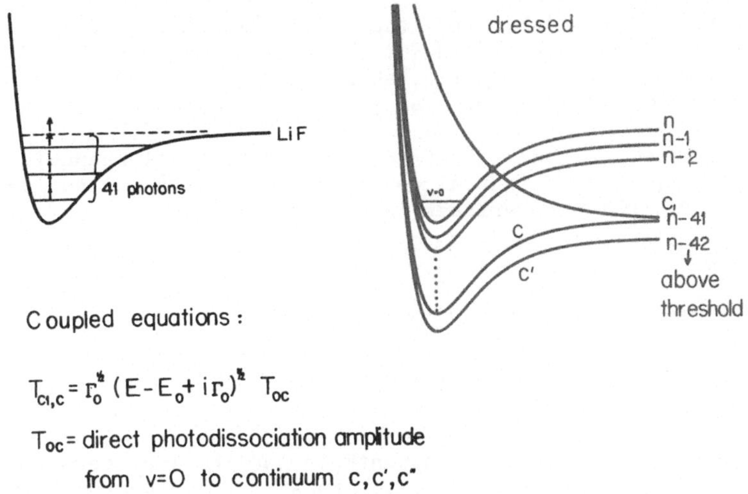

Figure 3. Perturbative and dressed representation for calculation of the Photodissociation amplitude T_{oc}. Cl is the entrance artificial channel for the initial bound state v = 0.

The amplitudes T_{ab} (equation 7) and T_{ac} (equation 9) are both perturbative, since by the artifice of unsymmetrical radiative couplings, one avoids renormalization of the molecular states by the photon field of frequency ω. One can obtain nonperturbative photodissociation probabilities $|T_{ac}|^2$ by exploiting the already mentioned analogy between radiative transitions and nonradiative transitions such as predissociation[8,12-13]. Thus in figure 3, after absorption of 41 photons, the ground vibrational state $|v = 0>$ finds itself in the continuum states $|c>$ of the dressed potential $V_{LiF}(R) - 41\ \hbar\omega$ and is coupled radiatively to the latter via all the intermediate states of the 41 potentials, $V_{LiF} - m\hbar\omega$ where m = 1, ..., 41. Putting all radiative couplings symmetric ($V^r_{ad} = V^r_{da}$), one can calculate elastic transition amplitudes T_{cc} directly, from which one can obtain resonance phase shifts η_r by the relation $T_{cc} = e^{i(\eta_c+\eta_r)}$. Using the isolated resonance expression

$$\eta_r = \text{artan}\ \frac{\Gamma_r}{(E - E_r)} \quad , \tag{10}$$

one can calculate exactly the linewidth Γ_r of the initial state in the presence of the laser field. This linewidth gives the exact photodissociation probability $|T_{ac}|^2$ from the relation

$$\Gamma_r = 2\pi\ |T_{ac}|^2 \quad . \tag{11}$$

This can then be compared to the perturbative amplitude given by equation (9). Furthermore in the case of continuum-continuum transitions, i.e., above threshold absorptions, we have shown in previous work[22-24] that numerical calculations of all elastic S matrix elements S_{cc} gives the relative linewidths Γ_c from each continuum, i.e.

$$\frac{\Gamma_c}{\Gamma} = \frac{1}{2}\ [1 - (|S_{aa}|^2)^{\frac{1}{2}}] \quad , \tag{12}$$

where Γ is the total linewidth, i.e., the total dissociation probability. We reemphasize that equations (11) and (12) are exact in the sense that all radiative couplings are symmetric, thus resulting in radiative renormalizations of the molecular states.

The total system, molecule plus field plus artificial channels yields a set of coupled differential equations for the radial nuclear wave functions F(R), which is written in matrix form[12],

$$F'' + WF = 0 \quad . \tag{13}$$

The diagonal elements of the energy matrix W are

$$W_{ii}^{nn} = \frac{2n}{\hbar^2}\,[E - V_i(R) - n\hbar\omega] - J(J+1)/R^2 \quad , \tag{14}$$

where $V_i(R)$ are the bare electronic potentials and the last term is the rotational energy. The off diagonal elements $W_{ij}^{n,n'}(R)$ are of two types:

$$\text{radiative:} \qquad W_{ii}^{n,n'}(R) = (\vec{\mu}_{ii}(R) \cdot \vec{E})\; 2M/\hbar^2 \quad , \tag{15}$$

$$\text{nonradiative:} \quad W_{ij}^{nn}(R) = \langle\phi_i/\partial/\partial R\,|\phi_j\rangle\; \partial/\partial R \quad , \tag{16}$$

In the LiF case, figure 1, only two electronic states, i.e., i = 1, 2 are of importance and W_{12}^{nn} is the nonadiabatic coupling between the two states. $\mu_{11}(R)$ and $\mu_{22}(R)$ are illustrated in figure 1b. The system (13) is solved numerically with the appropriate boundary conditions: $F_j(R) = 0$; $F_j(R \to \infty) \to 0$ for bound vibrational states; $F_j(R \to \infty) \to k_j^{-\frac{1}{2}} \sin(k_j R - J\pi/2) + k_j^{-\frac{1}{2}} R_{ij} \cos(k_j R - J\pi/2)$ for scattering states[12-24]. The S matrix elements are obtained from the relation $S = (1-iR)^{-1}(1+iR)$[25].

RESULTS

<u>Fluorescence</u> - We present in figure 4 the square of the fluorescence amplitudes $T_{00'}^{n}$, correspond to a transition v = 0 to v' = 0 with emission of n photons of frequency $\omega = 1000\ \text{cm}^{-1}$. This frequency is nonresonant with the molecular frequency $\omega_{LiF} = 800\ \text{cm}^{-1}$. The calculations were performed by the procedure explained in the previous section involving the use of two artificial channels, |C1> and |C2> as entrance and exit channels in order to convert the numerical calculation into a scattering problem (equation 7). The units of the amplitude $T_{00'}$ are energies for bound to bound transitions, since $T_{00'}$ is an effective potential created by the absorptions-emissions of successive photons on the initial state. All results are <u>perturbative</u> since all radiative couplings are assymmetric, in the actual calculation. The theoretical expression for the n photon absorption followed by emission of one photon at frequency $\omega' = n\omega$ becomes

$$T_{00'} = \gamma^n I^{n/2} \sum_{i_1, i_2, .. i_n} \frac{\langle 0/\mu/i_1\rangle\langle i_1/\mu/i_2\rangle \ldots \langle i_{n-1}/\mu/i_n\rangle\langle i_n/\mu/0'\rangle}{(E-E_{i_1})\,(E-E_{i_2})\,\ldots\,(E-E_{i_n})} \tag{17}$$

where $\gamma = 1.17 \times 10^{-3}$ is a conversion factor, equation (1), to convert $T_{00'}$ into units of cm^{-1}, I is the intensity in watts/cm^2. E is the initial energy $E = E_o + n\hbar\omega$, E_{ij} is the intermediate state energy $E_{ij} = E_o + v_j\hbar\omega_v - j\hbar\omega$. The intermediate state $i_j\rangle$ is the j'th vibrational level contributing to the transition if the selection rule $\Delta v = +1$ is applied as expected for a linear dipole $\mu = eR$.

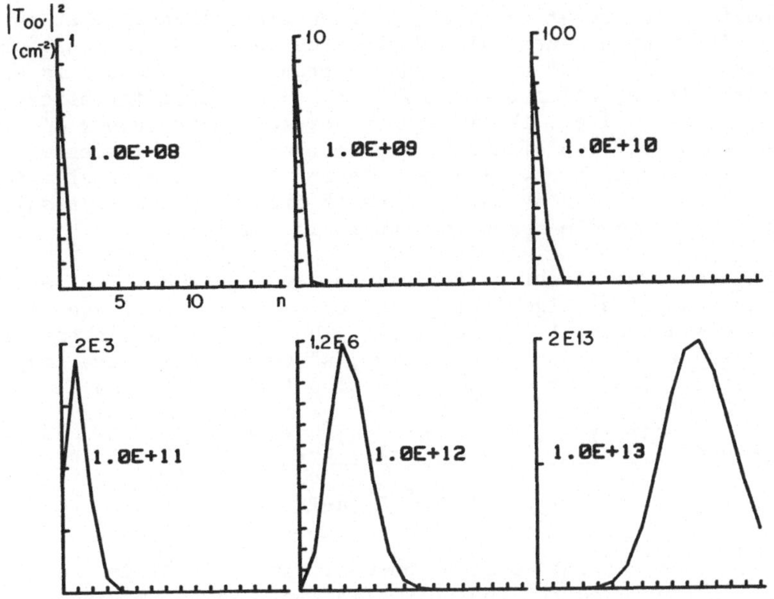

Figure 4. Fluorescence probabilities $|T_{00'}|^2$ (cm^{-2}) as a function of laser intensity and number of photons n absorbed (ω(fluorescence) = $n\omega$(laser); ω(laser) = 1000 cm^{-1}).

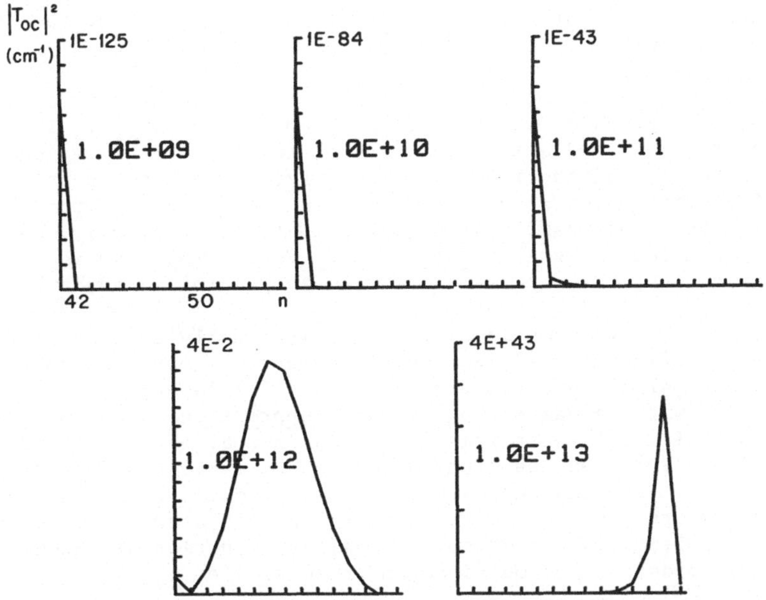

Figure 5. Photodissociation probabilities $|T_{0c}|^2$ (cm^{-1}) as a function of laser intensity and number n of photons absorbed. n=41=S=number of photons absorbed above threshold (ω(laser)=1000 cm^{-1}).

The results illustrated in figure 4 demonstrate clearly that deviations from a linear or exponential n dependence (n = # of photons) occur at intensities of 10^{11} W/cm^2 and beyond. A pronounced peak occurs at these high intensities, which peak shifts with increasing intensity to higher photon numbers, i.e., the maximum fluorescence frequency $\omega' = n\omega$ corresponding to absorption of n incident photons and reemission of a single photon at frequency ω'. The results are indicative of efficient vibrational ladder climbing by the photons of frequency ω as a result of the linearly growing dipole moment of the ionic molecule.

Photodissociation - The direct photodissociation amplitudes T_{oc} were obtained from the numerical algorithm by the consistent use of equation (9) involving one artificial channel $|C1\rangle$. The results, as in the case of fluorescence are therefore perturbative, since all radiative couplings were nonsymmetric. The appropriate expression for T_{oc} is thus

$$T^n_{oc} = \gamma^n I^{n/2} \sum_i \frac{\langle 0/\mu/i_1\rangle \ldots \langle i_{n-2}/\mu/i_{n-1}\rangle\langle i_{n-1}/\mu/C\rangle}{(E-E_{i_1}) \ldots\ldots (E-E_{i,n-1})} , \qquad (18)$$

which corresponds to absorption of n photons, where n = 41 is the minimum number of photons required for photodissociation. Above threshold absorption of S photons due to S continuum-continuum transitions would follow from the amplitude

$$T^{n+S}_{OC_S} = \underset{(\lim \varepsilon \to 0)}{\gamma^{(n+S)} I^{(n+S)/2}} \sum_i \int \frac{\langle 0/\mu/i_1\rangle \ldots \langle i_{n-1}/\mu/C_1\rangle \ldots \langle C_{S-1}/\mu/C_S\rangle}{(E-E_{i_1}) \ldots\ldots (E-E_{C_1} + i\varepsilon)} dE_{C_1} \ldots dE_{C_{S-1}} \qquad (19)$$

where the integral sign corresponds to (S-1) integrations over the (S-1) intermediate continua.

The units of T_{oc} are now cm$^{-\frac{1}{2}}$, since the linewidth of the initial state $|0\rangle$ would be given by equation (11) $\Gamma_o(\text{cm}^{-1}) = 2\pi|T^n_{oc}|^2$. Thus $|T^n_{oc}|^2$ is an effective radiative linewidth of the initial vibrational level v = 0 which photodissociates into the various continua above threshold after absorption of n photon, with n = 41 bieng the threshold photodissociation. The results illustrated in figure 5 clearly show nonlinear n dependence above I = 10^{11} W/cm^2. Furthermore, peaks occur above that intensity, which peaks shift with increasing intensity.

The accuracy of the numerical procedure is clearly evident in figure 5 where a scaling law is rigorously respected. Thus the probability of absorbing 41 photons follows an I^{41} scaling law at 10^9, 10^{10} and 10^{11} Watts (cm^2) with no maxima appearing; Suppression of the peak at n = 42 occurs at I = 10^{12} Watts/cm^2 with a maximum at n = 47. This corresponds to S = 1 and 6 excess electrons absorbed above the photodissociation threshold. Clearly, nonlinear effects predominate beyond intensities of 10^{11} Watts/cm^2. However, contrary to ATI observed in atoms[1-4], the present phenomenon is an extremely sensitive function of intensity due to the high order of the multiphoton process, i.e. n > 41.

DISCUSSION

The numerical results illustrated in figures 4 and 5 on the multiphoton IR absorption in an ionic molecule such as LiF indicate that a phonemenon analogous to ATI in atoms is to be expected in ionic molecules. The mechanism common to both physical systems is the linearly increasing dipole moment which efficiently couples high lying states and continua. The numerical results reported here are perturbative since they depend on calculations of the bound-bound amplitudes T_{ab} and bound-continuum amplitudes T_{ac} using nonsymmetric radiative couplings. A full calculation with completely symmetric couplings, thus allowing renormalization of the molecular states by the radiation field is in progress[26]. This will enable us to compare perturbative and nonperturbative probabilities for the multiphoton transitions. The present results, nevertheless do indicate the extreme sensitivity of above threshold photodissociation (ATPD) to intensity. Thus the region of intensities between 10^{11} and 10^{13} W/cm^2, as in the atomic ATI process, indicates important nonlinear effects in the photophysical processes.

Acknowledgments - We thank the National Sciences and Engineering Research Council of Canada for financial support of this project under the auspices of a special Strategic Grant programme.

REFERENCES

1. P. Kruit, J. Kimman, M.J. van der Wiel; J. Phys. B14, L597 (1981).
2. P.H. Bucksbaum, M. Bashkansky, R.R. Freeman, T.J. McIlrath, L.F. Dimauro; Phys. Rev. Lett., 56, 2590 (1986).
3. G. Petite, P. Agostini, F. Yergeau; J. Opt. Soc. Am. B4, 765 (1987).
4. P. Agostini, "Above Threshold Ionization in Strong Fields", in this proceedings.
5. R.S. Berry; J. Chem. Phys. 27, 1288 (1957).
6. M.L. Sink, A.D. Bandrauk; Can. J. Phys. 57, 1178 (1979).
7. L.R. Kahn, P.J. Hay, I. Shavitt; J. Chem. Phys. 61, 3530 (1974).
8. A.D. Bandrauk, M.L. Sink; J. Chem. Phys. 74, 1110 (1981).
9. A.D. Bandrauk, G. Turcotte; J. Chem. Phys. 77, 3867 (1982); Chem. Phys. Lett. 94, 175 (1983).
10. A.D. Bandrauk, G. Turcotte; J. Phys. Chem. 87, 5098 (1983).
11. A.D. Bandrauk, N. Gélinas; J. Comput. Chem. 8, 313 (1987).
12. A.D. Bandrauk, N. Gélinas; J. Chem. Phys. 86, 5257 (1987).
13. A.D. Bandrauk, O. Atabek, to appear in J. Phys. Chem., Nov. 1987.
14. A.D. Bandrauk, M.S. Child; Molec. Phys. 19, 95 (1970).
15. M. Shapiro; J. Chem. Phys. 56, 2582 (1972).
16. A.D. Bandrauk, G. Turcotte, R. Lefebvre; J. Chem. Phys. 76, 225 (1982).
17. A.D. Bandrauk, N. Gélinas; Chem. Phys. Lett. 129, 362 (1986).
18. M.L. Goldberger, K.M. Watson; Collision Theory (Wiley, N.Y., 1964), Chap. 8.
19. K.M. Watson, J. Nuttall; Topics in Several Particle Dynamics, (Holden-Day, San Francisco, 1967), Chap. 1.
20. L. Mower; Phys. Rev., 142, 799 (1966).
21. A.D. Bandrauk, J.P. Laplante; Can. J. Phys. 55, 1 (1977).
22. M.L. Sink, A.D. Bandrauk; Chem. Phys. Lett. 65, 246 (1979).
23. M.L. Sink, A.D. Bandrauk; J. Chem. Phys. 73, 4451 (1980).
24. O. Atabek, X. He; "H_2^+ in Intense Fields" - this volume.
25. D.W. Norcross, M.J. Seaton; J. Phys. B6, 614 (1973).
26. A.D. Bandrauk, N. Gélinas; in preparation.

LASER ENERGY ABSORPTION DUE TO LOCAL PLASMON EXCITATION IN COMPLEX ATOMS

Farhad H.M. Faisal

Fakultät für Physik
Universität Bielefeld
4800 Bielefeld 1
Federal Republic of Germany

A many-body random-phase analysis of the N-electron atomic Hamiltonian coupled to a laser pulse is made using a variation of the non-perturbative methods of Tomonaga and of Bohm and Pines. Existence of quantum oscillations of the atomic dipole operator in the form of local plasmon excitation is theoretically investigated. Laser energy absorption cross sections due to local plasmon excitation in Xe, Kr, Ar and Ne are obtained. It is shown that at high laser intensities the local plasmon excitation mechanism provides a strong <u>and</u> rapid process for transfer of a large amount of laser energy to noble gas atoms. Estimates of the corresponding ionization-saturation intensities in noblegas atoms for two CO_2 laser frequencies show remarkably close agreement with the experimental data.

My purpose in this paper is to give a many-body non-perturbative analysis of the Hamiltonian of an N-electron atom coupled to a laser pulse and to show that an intense laser can force quantum dipole oscillations in a complex atom in the form of local plasmon excitation. Numerical values of energy absorption cross sections due to this process are also calculated for a number of noble gas atoms. It is found that the mechanism of local plasmon excitation provides a strong <u>and</u> rapid process of energy transfer from an intense laser to noble gas atoms. Estimations of the ionization-saturation intensities in noble gases for two CO_2 laser frequencies show remarkably close agreement with the experimental data.

The N electrons of a complex atom in the ground state may be conceived of as a stationary inhomogeneous charge cloud confined in a microscopic volume V characterised by the atomic radius R. Instead of following the individual motion of each electron, Bloch [1] was the first to consider the excited states of the atom in which the electron cloud as a whole oscillates. This was further developed by Tomonaga [2] and by Bohm and Pines [3,4] who treated a system of fermions by the method of sound waves or of plasma oscillations. I shall use a variation of the methods of these authors to consider energy transfer by forced excitation of atomic charge cloud by an intense laser pulse. To this end I define the charge density fluctuation of a many electron atom through the operator

$$\rho(\bar{r}) = \sum_{i=1}^{N} e\delta(\bar{r}-\bar{r}_i) - Ze\delta(\bar{r}) \tag{1}$$

$$= \frac{1}{V}\sum_{\bar{K}} e^{-i\bar{K}\cdot\bar{r}}\rho_{\bar{K}} \tag{2}$$

where

$$\rho_{\bar{K}} = e\sum_{i=1}^{N}(e^{i\bar{K}\cdot\bar{r}_i} - 1) \tag{3}$$

The first term of (1) arises from the electronic charges and the second term from the nuclear charge (Z = N for a neutral atom) located at the origin. The Hamiltonian of the atomic charge cloud coupled to a laser pulse can be written as

$$H(t) = H_{at} + e\sum_{i=1}^{N} \bar{r}_i\cdot\bar{\varepsilon}_L\, E(o,t) \tag{4}$$

where

$$H_{at} = -\frac{\hbar^2}{2m}\sum_{i=1}^{N}\nabla_i^2 - Z\sum_{i=1}^{N}\frac{e^2}{r_i} + \frac{e^2}{2}\sum_{i\neq}^{N}\sum_{j}^{N}\frac{1}{|\bar{r}_i - \bar{r}_j|} \tag{5}$$

is the N-electron Hamiltonian of the atom. The electric field of the laser is assumed to have the form

$$\bar{E}\,(o,t) = \bar{\varepsilon}_L\, E'(o,t)\, \cos \omega t \qquad (6)$$

where $\bar{\varepsilon}_L$ is the unit polarization vector, ω is the frequency and $E'(o,t)$ is the pulse envelope.

Using the Fourier components $\rho_{\bar{K}}$, given by (3), and the expansion

$$\frac{1}{|\bar{x}|} = \frac{4\pi}{V} \sum_{\bar{K}} \frac{e^{i\bar{K}\cdot\bar{x}}}{K^2}$$

one obtains H_{at}, after some simple algebra, in the form

$$H_{at} = -\frac{\hbar^2}{2m} \sum_{i=1}^{N} \nabla_i^2 + \frac{2\pi}{V} \sum_{\bar{K}} \frac{1}{K^2} \rho_{\bar{K}} \rho_{\bar{K}}^* - \frac{2\pi}{V} N(N+1) e^2 \sum_{\bar{K}} \frac{1}{K^2} \qquad (7)$$

I now define a set of generalized coordinates $q_{\bar{K}}$ and momenta $p_{\bar{K}}$ by

$$\begin{aligned} q_{\bar{K}} &= -\frac{i}{e} \rho_{\bar{K}} / K \\ p_{\bar{K}} &= -\frac{i\hbar}{N} \sum_{i=1}^{N} e^{-i\bar{K}\cdot\bar{r}_i} \bar{\nabla}_i \end{aligned} \qquad (8)$$

It follows from direct calculation that

$$q_{\bar{K}}\, p_{\bar{K}'} - p_{\bar{K}'}\, q_{\bar{K}} = i\,\hbar\, \frac{\bar{K}'}{\bar{K}} \left[\frac{1}{N} \sum_{i=1}^{N} e^{i(\bar{K}' - \bar{K})\cdot\bar{r}_i} \right] \qquad (9)$$

For large N the phase-sum in the square brackets oscillates rapidly for $\bar{K}' \neq \bar{K}$, giving negligible contribution compared to the case $\bar{K}' = \bar{K}$. Hence, in this random phase approximation (RPA), equation (9) reduces to

$$q_{\bar{K}}\, p_{\bar{K}'} - p_{\bar{K}'}\, q_{\bar{K}} = i\,\hbar\, \bar{\varepsilon}_{\bar{K}}\, \delta_{\bar{K}\bar{K}'} \qquad (10)$$

where $\bar{\varepsilon}_{\bar{K}} = \frac{\bar{K}}{|\bar{K}|}$ is an unit vector along $\bar{K}$. Thus, the $q_{\bar{K}}$ and the $p_{\bar{K}}$ introduced above constitute pairs of canonically conjugate dynamical variables of the system. By direct calculation one also finds

$$\sum_{\bar{K}} p_{\bar{K}} p^*_{\bar{K}} = \frac{\hbar^2}{N} \sum_{i=1}^{N} \sum_{j=1}^{N} \left[\frac{2}{N} \sum_{\bar{K}} e^{-i\bar{K}\cdot(\bar{r}_i-\bar{r}_j)} \right] \frac{d^2}{d\bar{r}_i d\bar{r}_j}$$

$$+ i\frac{\hbar^2}{N^2} \sum_{\bar{K}} \bar{K}\cdot \sum_{i=1}^{N} \bar{\nabla}_i \qquad (11)$$

Within the RPA for large N one may replace the quantity in the square brackets by $\delta_{\bar{r}_i,\bar{r}_j}$, which gives the kinetic energy operator in terms of the canonical momenta

$$-\frac{\hbar^2}{2m} \sum_{i=1}^{N} \frac{d^2}{d_{\bar{r}_i}^2} = -\frac{N}{2m} \sum_{\bar{K}} p_{\bar{K}} p^*_{\bar{K}} - \frac{\hbar}{2m} \sum_{\bar{K}} \bar{K}\cdot p_{\bar{o}} \qquad (12)$$

Combining equations (8) and (12) with (7) I find the following useful form for the atomic Hamiltonian:

$$H_{at} = -\frac{N}{2m} \sum_{\bar{K}} p_{\bar{K}} p^*_{\bar{K}} + \frac{2\pi e^2}{V} \sum_{\bar{K}} q_{\bar{K}} q^*_{\bar{K}} - \frac{\hbar}{2m} \sum_{\bar{K}} \bar{K}\cdot p_{\bar{o}} + W^o \qquad (13)$$

where $W^o = -N(N+1)\frac{2\pi e^2}{V} \sum_{\bar{K}} \frac{1}{K^2}$ (which is independent of the dynamical variables) corresponds to a constant energy which may be absorbed in the scale of total energy of the system.

The laser-atom interaction energy may also be expressed in term of the generalized coordinates:

$$e \sum_{i=1}^{N} \bar{r}_i\cdot\bar{\varepsilon}_L\, E(o,t) = e \sum_{\bar{K}} \delta_{\bar{K},\bar{o}}\, \delta_{\bar{\varepsilon}_{\bar{K}},\bar{\varepsilon}_L}\, q_{\bar{K}}\, E(o,t) \qquad (14)$$

Combining equations (13) and (14) with (4) the total Hamiltonian can be reduced to the form

$$H(t) - W^o = \sum_{\bar{K}} \left[-\frac{N}{2m} p_{\bar{K}} p^*_{\bar{K}} + \frac{2\pi e^2}{V} q_{\bar{K}} q^*_{\bar{K}}\, \delta_{\bar{K},\bar{o}}\, \delta_{\bar{\varepsilon}_{\bar{K}},\bar{\varepsilon}_L}\, e q_{\bar{K}} E(o,t) - \frac{\hbar}{2m} \bar{K}\cdot p_{\bar{o}} \right] \qquad (15)$$

Note that the last term identically vanishes for $\bar{K} = \bar{0}$ and the corresponding sum is also zero due to the odd parity of $\bar{K}\cdot p_{\bar{o}}$. The Hamiltonian (15) consists in fact of a sum of independent Hamiltonians in the $\bar{K}$-modes. What is of greatest interest in the present context is that the dipole laser field couples only to the $\bar{K} = \bar{0}$ mode (in the field direction) of the system. Clearly, therefore, the transfer of laser energy to the complex atom in the RPA is governed by the $\bar{K} = \bar{0}$ mode Hamiltonian

$$h_o(t) = \frac{N}{2m} p_{\bar{o}}^2 + \frac{2\pi e^2}{V} q_{\bar{o}}^2 - e q_{\bar{o}} E(o,t) \tag{16}$$

To get a deeper insight into this result I return to the ordinary coordinate system. To be specific let the laser be polarized along the z-axis (the case of circular polarization can be treated in a similar way without difficulty). From equation (8) one finds

$$e q_{\bar{o}} = e \sum_{i=i}^{N} z_i \equiv D_z$$

$$N p_{\bar{o}} = - i \hbar \sum_{i=1}^{N} \frac{d}{dz_i} \equiv P_z \tag{17}$$

Where D_z is the atomic dipole operator and P_z is the momentum operator of the charge cloud (projected in the direction of the field). Thus $h_o(t)$ takes the form

$$h_o(t) = - \frac{\hbar^2}{2\mu_o} \frac{d^2}{dD_z^2} + \frac{1}{2} \mu_o \omega_o^2 (R) D_z^2 - D_z E(o,t) \tag{18}$$

where, as is usual for the inhomogeneous Fermi-gas model of the atomic charge cloud, N/V is interpreted locally [5] . Thus, $N/V \rightarrow n_o(R)$, the stationary density of the charge cloud in the ground state. We have also defined

$$\mu_o \equiv \frac{m}{Ne^2}$$

and

$$\omega_o(R) \equiv \left[\frac{4\pi n_o(R) e^2}{m}\right]^{\frac{1}{2}} \tag{19}$$

Eq. (18) shows that in the absence of the field, h_o describes quantum oscillation of the atomic dipole operator D_z at the local plasma frequency $\omega_o(R)$; in other words the $\bar{K} = \bar{0}$ mode of the density fluctuation of an N-electron atom considered as an inhomogeneous Fermi-gas corresponds to local plasmon excitations in this gas. The forced response of the inhomogeneous Fermi-gas atom to a dipolar laser pulse is governed by the corresponding schrödinger equation

$$i\hbar \frac{\delta}{\delta t} | \Psi(t)\rangle = \left[- \frac{\hbar^2}{2\mu_o} \frac{d^2}{dD_z^2} + \frac{1}{2}\mu_o \omega_o^2(R) D_z^2 - D_z E(o,t) \right] | \Psi(t)\rangle \tag{20}$$

Eq. (20) can be solved exactly for any given laser pulse. For the present purpose the quantity of greatest interest is the probability of photon absorption. Assuming that the atom is in the ground state initially (which would correspond to all the oscillators, including the $\bar{K} = \bar{0}$ mode dipole oscillator, in their 0th levels) the amplitude of excitation of the nth quantum is given by

$$A_{o\to n}(t) = \frac{(-i)^n}{(n!)^{1/2}} \exp\left[-\int_{-\infty}^{t} \dot{x}^*(t')\, x(t')dt'\right] \left[x(t)\right]^n \tag{21}$$

where

$$x(t) = \frac{i}{\hbar}\, \xi \int_{-\infty}^{t} e^{i\omega_o(R)t} E(o,t)dt \tag{22}$$

with

$$\xi = \langle o \,|D_z|\, 1 \rangle$$

The corresponding probability of excitation is then obtained as the well-known Poisson distribution

$$P_{o\to n}(\omega;R) = \frac{[p(R)]^n}{n!}\, e^{-p(R)} \tag{23}$$

where

$$p(R) \equiv |x\,(t\to\infty)|^2$$
$$= \frac{Ne^2}{2m\hbar\omega_o(R)} \left| S\,(\omega;R)\right|^2 \tag{24}$$

and

$$S(\omega;R) = \frac{E_o}{2}\sqrt{2\pi}\,\tau \left\{\exp\left[-\frac{\tau^2}{2}\,(\omega_o(R)-\omega)^2\right] + \exp\left[-\frac{\tau^2}{2}\,(\omega_o(R)+\omega)^2\right]\right\} \tag{25}$$

For the sake of definiteness, $S\,(\omega;R)$ in (25) is given for a Gaussian pulse

$$E(o,t) = E_o\, e^{-\frac{1}{2}(t/\tau)^2} \cos \omega t \tag{26}$$

The mean energy absorbed by the atom is now easily calculated using (23):

$$\langle \Delta E \rangle = \frac{4\pi}{N} \int_0^\infty n0\,(R)\; R^2\; \Delta E(\omega;R)\; dR \tag{27}$$

where

$$\Delta E\;(\omega;R) = \sum_{n=1}^{\infty} n\hbar\;\omega_0(R)\;P_{0\to n}(\omega;R)$$

$$= \hbar\omega_0\;(R)\;p\;(R) \tag{28}$$

and no(R) is the probability density of finding the atom in the ground state with radius R. One may also define a mean energy absorption cross section

$$\sigma_{ab}(\omega) \equiv \frac{1}{I\tau} \langle \Delta E \rangle \tag{29}$$

where $I = \frac{Eo^2}{8\pi} c$ is the peak intensity and τ is the duration of the pulse.

We may simplify (27) and (29) further by noting that for $\tau >> \frac{1}{\omega}$ the second exponential in (25) becomes negligible and the dominant contribution to the integral arises from the first exponential term in the vicinity of $\omega_p(R_0) = \omega$ and one obtains the useful formula [6].

$$\sigma_{ab}\;(\omega) = 4\pi^2\;\frac{\omega}{c}\;\frac{R_0^{\,2}(\omega)}{L(R_0(\omega))} \tag{30}$$

where $R_0(\omega)$ is the "resonance radius" given implicitly by the relation

$$\omega = \omega_p\;(R_0) = \left[\frac{4\pi e^2 n_0(R_0)}{m}\right]^{1/2} \tag{31}$$

and $L(R_0(\omega))$ is the logarithmic derivative of $n_0(R)$ at $R=R_0(\omega)$.

Note that the usual I^n-dependence of n-photon absorption probability at low intensities is reproduced by eq. (23). Eq. (23) also predicts a rapid saturation of the n-photon absorption probability as $\sim (I^n)e^{-\alpha I}$, similar to a form often assumed empirically [7], for high intensities. Another interesting point is to note that despite the strong non-linear dependence on intensity of individual n-photon absorption probability, the mean energy transfered, eq. (27), shows a linear dependence on the intensity which is common to all energy absorption cross sections for an effective single-photon process; this is a direct consequence of the Poisson-distribution of the n-photon absorption probability predicted by the inhomogeneous Fermi-gas model within the RPA. Recently, basing themselves on very differnt heuristic considerations, Boyer et al. [8] conjectured that the mean laser-energy absorption cross section may be estimated using again a single-photon absorption cross-section.

We now turn to numerical calculations based on the cross section formula derived above (eq. (29)).

Table 1. Photon energy absorption cross section $\sigma_{ab}(\omega)$ due to forced dipole excitation in noble gas atoms at CO_2 laser frequencies. $R_o(\omega)$ are the corresponding resonance radii.

	$\hbar\omega=0.118$eV		$\hbar\omega=0.130$eV	
Atom	$\sigma_{ab}(\omega)$ (cm^2)	$R_o(\omega)$ (a_o)	$\sigma_{ab}(\omega)$ (cm^2)	$R_o(\omega)$ (a_o)
Xe	$6.6\text{x}10^{-19}$	6.9	$7.1\text{x}10^{-19}$	6.8
Kr	$5.7\text{x}10^{-19}$	6.5	$6.1\text{x}10^{-19}$	6.4
Ar	$4.9\text{x}10^{-19}$	6.1	$5.3\text{x}10^{-19}$	6.1
Ne	$2.0\text{x}10^{-19}$	4.6	$2.2\text{x}10^{-19}$	4.5

Table 2. Comparison of the mean energy(ΔE) absorbed by noble gas atoms in CO_2 laser pulses at $I=3\text{x}10^{13}W\ cm^2$ with the ionisation energy for $\tau_{eff}=2\text{x}10^{-13}s$.

	(ΔE) (eV)		
Atom	$\hbar\omega=0.118$eV	$\hbar\omega=0.130$eV	E_{ion}(eV)
Xe	24.7	26.6	12.13
Kr	21.3	22.9	14.00
Ar	18.4	19.9	15.76
Ne	7.5	8.2	21.57

Table 3. Comparison of theoretical and experimental saturation intensities I_s for noble gas atoms at two CO_2 laser frequencies for an estimated effective interaction time $\tau_{eff}=2x10^{-13}$ s.

	$\hbar\omega=0.118eV$		$\hbar\omega=0.130eV$	
Atom	I_s(Expt) ($W\ cm^{-2}$)	I_s(Theory) ($W\ cm^{-2}$)	I_s(Expt) ($W\ cm^{-2}$)	I_s(Theory) ($W\ cm^{-2}$)
Xe	$1.9x10^{13}$	$1.5x10^{13}$	$1.2x10^{13}$	$1.4x10^{13}$
Kr	$1.7x10^{13}$	$2.0x10^{13}$	$1.9x10^{13}$	$1.8x10^{13}$
Ar	$2.5x10^{13}$	$2.6x10^{13}$	$2.4x10^{13}$	$2.4x10^{13}$
Ne	$7.0x10^{13}$	$8.8x10^{13}$	$6.5x10^{13}$	$8.0x10^{13}$

In table 1 we present the first quantitative estimates of the photon absorption cross sections due to dipole excitation in Xe, Kr, Ar and Ne atoms at the experimental CO_2 laser frequencies $\hbar\omega=0.118$ and 0.130eV. The table summarises the numerical results obtained using a simplified Hartree-Fock density [9]. Also shown are the resonance radii $R_o(\omega)$ for the noble gas atoms at the two laser frequencies. At these frequencies virtually all the contributions to the absorption cross sections arise (for all the atoms considered) from the outermost p shell only. The mean energy (ΔE) absorbed by an atom at a given CO_2 laser intensity, I, can be calculated from the cross sections $\sigma_{ab}(\omega)$ provided the effective interaction time τ_{eff} is known. In order to obtain a qualitative estimate of τ_{eff} we equate the mean energy absorbed by the atom to the ionisation energy E_{ion} at the (ionisation) saturation intensity I_s defined by [10]

$$E_{ion} = \sigma_{ab}(\omega)\ I_s\tau_{eff}. \tag{32}$$

The experimental data [11] on ionisation of Xe at $h\omega=0.118$ and 0.130eV indicate a saturation intensity somewhere between the ion appearance intensity, $10^{+13}W\ cm^{-2}$, and $2x10^{+13}W\ cm^{-2}$. This rather narrow saturation domain may be understood by considering the extremely high order of the photon absorption process involved in the ionisation. Taking the mean experimental value $I_s \simeq 1.5x10^{+13}W\ cm^{-2}$ and using E_{ion} for Xe to be

12.13eV [12], from (32) one obtains $\tau_{eff} \approx 0.2$ ps. Note that this time is much smaller than the pulse duration (approximately 1.1ns). Whether this time is determined by the possible life-time of the excitations [13] remains an open question.

The mean energies absorbed by the noble gas atoms at a typical experimental intensity $I=3x10^{13}$ W cm^{-2} for $\hbar\omega=0.118$ and 0.130eV are compared with the respective ionisation energies in table 2. The table clearly indicates that at $3x10^{13}$ W cm^{-2} not only Xe but also Kr and Ar absorb enough energy to be singly ionised but Ne does not.

In table 3 the saturation intensities obtained from experimental measurements [11] with noble gas atoms at two different CO_2 laser frequencies are compared with the prediction of saturation intensities obtained from (32). In view of the usual uncertainty in reading the saturation intensity from the experimental data, the prediction of the simple inhomogeneous Fermi-gas model is in remarkably good agreement with the experimental data.

References

1. F. Bloch, Z. für Physik, 81: 363 (1933).
2. S. Tomonaga, Prog. Theor. Phys., 5: 544 (1950).
3. D. Bohm and D. Pines, Phys. Rev., 82: 625 (1951).
4. D. Pines and D. Bohm, Phys. Rev., 85: 339 (1952); Ibid. 92: 609 and 626 (1953).
5. N.H. March, "Self-Consistent Field in Atoms", Pergamon Press, Oxford (1975) chap.4.
6. F.H.M. Faisal, J. Phys. B: At. Mol. Phys. 20: L299 (1987).
7. M. Crance, J. Phys.B: At. Mol. Phys. 17: 3505, L355(1974).
8. K. Boyer, J. Jara, J. S. Luk, I.A. McIntyre, A. McPherson, R. Rosman and C. K. Rhodes, Preprint (1987).
9. M. Horbatsch, J. W. Darewych and R. P. McEachran, J.Phys.B: At.Mol.Phys., 16: 4451 (1983).
10. M. Crance, Phys. Report. 144: 117 (1987).
11. F. Yergeau, S.L. Chin and P. Lavigne, J.Phys.B: At.Mol. Phys. 20: 733 (1987).
12. R. D. Cowan, "The Teory of Atomic Structure and Spectra", Univ. Calif. Press, Berkeley (1981).

13. C. K. Rhodes, in "Multiphoton Ionization of Atoms" ed. S. L. Chin and P. Lambropoulos, Academic Press, New York, (1984) p.31.

ATOMIC HYDROGEN IN INTENSE, HIGH-FREQUENCY LASER FIELDS

M. Gavrila and M. Pont

FOM-Institute for Atomic and Molecular Physics
Kruislaan 407, 1098 SJ Amsterdam
The Netherlands

I. FORMALISM

Under the impact of impressive advances in the development of superintense lasers, yielding many atomic units of intensity at frequencies in the VUV and beyond[1], attention has been focused on atomic behavior in this new kind of environment. Because of the high intensities at hand, the traditional theoretical approach, based on lowest order perturbation theory formulas, ceases to be valid. However, a high-frequency nonperturbative theory was recently developed by Gavrila and Kaminski[2,3], capable of dealing with atomic interactions under these circumstances. So far only one-electron atom systems have been considered. The theory was first applied to electron-potential scattering[4-6]. It was then extended to atomic structure and multiphoton ionization in the intense, high-frequency regime[3,7,8]. In the following we shall review the formalism for the bound state problem and discuss some of its recent applications.

The laser field is represented by a monochromatic plane wave, assumed here to be linearly polarized, in the dipole approximation. The electrodynamic potentials were taken as $\mathbf{A} = a\mathbf{e} \cos \omega t$, $\phi = 0$, where $\mathbf{e}$ is the real polarization vector. (The case of circular polarization is considered in ref. 8.)

The atom in the laser field is described by the time-dependent Schrödinger equation

$$\left[\frac{1}{2m}\left(\mathbf{P} - \frac{e}{c}\mathbf{A}\right)^2 + V(\mathbf{r})\right]\Psi = i\hbar\frac{\partial \Psi}{\partial t}\ , \tag{1}$$

written here in the momentum gauge. It is more convenient, however, to use its "space-translated" version, originally introduced by Kramers[9], see also ref. 10 and 11. This is obtained by making a time dependent translation of the coordinate $\mathbf{r}$ to $\mathbf{r} + \boldsymbol{\alpha}(t)$, with

$$\boldsymbol{\alpha}(t) = -\frac{e}{mc} \int^{t} \mathbf{A}(t')\, dt' = \alpha_0\, \mathbf{e} \sin \omega t \ , \tag{2}$$

and

$$\alpha_0 = -\frac{ea}{mc\omega} \tag{3}$$

For an electron, in atomic units ($e = \hbar = m = 1$) Eq.(3) becomes

$$\alpha_0 = I^{1/2}\, \omega^{-2} \ \text{a.u.} \ , \tag{4}$$

where I is the time-averaged intensity of the plane wave. The atomic unit of time-averaged intensity is $I_0 \doteq 3.51 \times 10^{16}$ W/cm^2. By also isolating a time-dependent phase factor, we introduce the new wave function $\psi(\mathbf{r},t)$:

$$\psi(\mathbf{r},t) = \Psi(\mathbf{r} + \boldsymbol{\alpha}(t)) \ \exp\left[\frac{i}{\hbar}\frac{e^2}{2\,mc^2} \int^{t} A^2(t')\, dt'\right] . \tag{5}$$

This satisfies the ("space-translated") Schrödinger equation

$$\left[\frac{1}{2m} P^2 + V(\mathbf{r} + \boldsymbol{\alpha}(t))\right]\psi = i\hbar \frac{\partial \psi}{\partial t} . \tag{6}$$

As apparent the field dependence is now contained in the vector $\alpha(t)$. Obviously, Eq.(6) is entirely equivalent to Eq.(1), if the observables are appropriately transformed.

Since Eq.(6) has periodic time-dependent coefficients, we may seek, as usual, a "quasi-energy solution" of the Floquet form

$$\psi(\mathbf{r},t) = e^{-iEt} \sum_{n=-\infty}^{+\infty} \psi_n(\mathbf{r})\ e^{-in\omega t} , \tag{7}$$

where the quasi-energy E remains to be determined. (For the background of this type of approach in the radiationless case, see ref. 12, 13, 14.) Then, we Fourier analyze the potential:

$$V(\mathbf{r} + \boldsymbol{\alpha}(t)) = \sum_{n=-\infty}^{+\infty} V_n(\alpha_0;\mathbf{r})\, e^{-in\omega t} . \tag{8}$$

By some algebraic manipulations the coefficients can be written as

$$V_n(\alpha_0;\mathbf{r}) = (i^n/\pi) \int_{-1}^{+1} V(\mathbf{r} + \alpha_0\, \mathbf{e}u)\, T_n(u)\, (1-u^2)^{-1/2}\, du \ , \tag{9}$$

where $T_n(u)$ are Chebyshev polynomials. Insertion of Eqs.(7) and (8) into Eq.(6) leads to a system of coupled differential equations for the components $\psi_n(\mathbf{r})$, which we write

$$\left[\frac{1}{2m}\mathbf{P}^2 + V_0 - (E + n\hbar\omega)\right]\psi_n = -\sum_{\substack{m=-\infty \\ (m\neq n)}}^{+\infty} V_{n-m}\,\psi_m \; . \tag{10}$$

At this point we have to specify the boundary conditions for the components $\psi_n(\mathbf{r})$, and therefore for the wave function Eq.(7), in order to insure that the solution describes correctly the decay of the atom in the radiation field. These we shall take of the Siegert form[14]:

$$\begin{aligned} \psi_n(\alpha_0,\omega;\mathbf{r}) &\rightarrow \frac{1}{r} f_n(\alpha_0,\omega;\hat{\mathbf{r}}) \exp[i(k_n r - \gamma_n \ln 2k_n r)] \;, && \text{for open channels;} \\ \psi_n(\alpha_0,\omega;\mathbf{r}) &\rightarrow 0 \text{ (exponentially)} \;, && \text{for closed channels.} \end{aligned} \tag{11}$$

Here $f_n(\alpha_0,\omega;\hat{\mathbf{r}})$ is the n-photon ionization amplitude, $\gamma_n = -Ze^2 m/\hbar^2 k_n$ with Z the asymptotic charge seen by the electron (for a short range potential $Z = 0$); for the asymptotic momenta we have $p_n = \hbar k_n$. These are related to the quasi-energy E of the initial state (associated to the channel $n = 0$) by the energy conservation equation

$$\frac{p_n^2}{2m} = E_n = E + n\hbar\omega \; . \tag{12}$$

If we assume that $\hbar\omega$ is larger than the binding energy for the ground state (see below), the open channels are characterized by $n > 0$, and the closed ones by $n \leq 0$.

As we shall immediately see, imposition of the boundary conditions Eq.(11) for the Floquet system Eq.(10) will force the quasi-energy E to be complex. This merely reflects the fact that the initial state is not a physically stable state, but an intermediate (resonant) state in a collision process for which we want to ignore the first stage (i.e. the formation of the resonance). Because E is complex, the p_n given by Eq.(12) are complex too, so that care must be exercized in choosing the determination of their square roots.

We now turn to the solution of the Floquet system, Eq.(10). The left hand side contains the Hamiltonian

$$H = \frac{1}{2m}\mathbf{P}^2 + V_0(\alpha_0,\mathbf{r}) \;, \tag{13}$$

in which the role of the potential is played by the zeroth order Fourier component of Eq.(8). By using the Green's operator $G(\Omega)$ associated to H (on the first or second Riemann sheet of the energy Ω, as needed), Eq.(10) may be formally solved as

$$\psi_n = u_\lambda \delta_{no} - G(E_n) \sum_{\substack{m \\ (m\neq n)}} V_{n-m}\psi_m \; . \tag{14}$$

Here u_λ is an eigenfunction of the (Hermitian) Hamiltonian Eq.(13) with the (real) eigenvalue W_λ:

$$Hu_\lambda = W_\lambda u_\lambda \ . \tag{15}$$

An analysis of Eq.(14) shows that, at sufficiently high ω, the terms containing the Green's operators $G(E_n)$ become small in comparison to $\psi_0 \equiv u_\lambda$, which does not depend on ω, if α_0 is considered as given (see ref. 2, 3). This shows that, *to lowest order* in $1/\omega$, the Floquet system Eq.(10) reduces to one single equation for ψ_0, Eq.(15). More specifically, the *frequency condition* under which this holds was shown to be: $\omega >> |W_\lambda{}^0(\alpha_0)|$, where $W_\lambda{}^0(\alpha_0)$ is the eigenvalue of Eq.(15) for the lowest lying state within the same symmetry class as the initial state u_λ (see below). On the other hand the value of α_0 may be arbitrary. It may be shown that for $\alpha_0 << 1$ the results obtained from the present theory agree with those obtained from standard perturbation theory (at high frequencies). The case $\alpha_0 \geq 1$ is highly nonperturbative, however.

Since Eq.(15) has real eigenvalues, *the atom is stable in the high-frequency limit* ($\omega \rightarrow \infty$), i.e. it cannot decay by multiphoton ionization. This becomes possible only in the *next order* approximation in $1/\omega$, when by iteration we obtain from Eq.(15) for n = 0:

$$\Big[\frac{1}{2m} P^2 + V_0 - \sum_{\substack{m \\ (m\neq 0)}} V_{-m}\, G(E_m)\, V_m + \cdots\Big]\, \psi_0 = E\, \psi_0 \ . \tag{16}$$

This equation determines a complex quasi-energy E.

Since the term containing $G(E_m)$ in Eq.(16) is anyway small at large ω, it can be treated as a perturbation. Then, by denoting

$$E = W_\lambda + E_\lambda{}^{(1)} + \cdots \ , \tag{17}$$

one finds for $E_\lambda{}^{(1)}$ the complex value:

$$E_\lambda{}^{(1)} = \Delta_\lambda - i\,\frac{\Gamma_\lambda}{2} \ ; \tag{18}$$

$$\Delta_\lambda = -\sum_{\substack{m \\ (m\neq 0)}} \langle u_\lambda | V_{-m}\, \mathbb{P}\, (H - W_\lambda - m\hbar\omega)^{-1}\, V_m | u_\lambda \rangle \ , \tag{19}$$

$$\Gamma_\lambda = 2p \sum_{\substack{m \\ (m\neq 0)}} \langle u_\lambda | V_{-m}\, \delta\,(H - W_\lambda - m\hbar\omega)\, V_m | u_\lambda \rangle \ . \tag{20}$$

Δ_λ and Γ_λ are real numbers; $\Gamma_\lambda > 0$, whereas Δ_λ may have either sign. Δ_λ represents a shift of the energy W_λ of the atom in the field (which itself is shifted from the unperturbed energy outside the field) due to the improved approximation in $1/\omega$ used. As usual, Γ_λ is related to the total decay rate w of the atom: $\Gamma = \hbar w$. It is easily seen that Γ_λ vanishes rapidly with ω, as $\omega \to \infty$.

At the same level of approximation in 1/ω one may derive from the asymptotic form of Eq.(14) the following expression for the multiphoton ionization amplitudes (n > 0):

$$f_n(\alpha_0,\omega;\hat{\mathbf{r}}) = -\frac{1}{2\pi} < u^{(-)}_{\mathbf{p}_n} |V_n| u_\lambda > . \tag{21}$$

Here $u^{(-)}_{\mathbf{p}_n}$ is a scattering solution of Eq.(15) for energy $E_n = W_\lambda + n\hbar\omega$, normalized to unit asymptotic amplitude, and having incoming spherical waves. It may be shown that the corrective terms are of higher order in 1/ω and therefore relatively small at high frequencies. Note that the amplitudes f_n are in fact ω-dependent because of the energy conservation equation relating the initial and final states.

The angular decay rate for n-photon ionization is[3]

$$\frac{d\Gamma^{(n)}_\lambda}{d\Omega} = p_n |f_n(\alpha_0\omega;\mathbf{r})|^2 . \tag{22}$$

It may be easily checked that Eqs.(20) and (22) are consistent, that is:

$$\Gamma_\lambda = \sum_{n=1}^{\infty} \Gamma^{(n)}_\lambda . \tag{23}$$

Let us return to the eigenvalue equation which determines the structure of the one-electron atom in the high-frequency limit, Eqs.(13) and (15). This contains the dressed potential $V_0(\alpha_0,\mathbf{r})$, Eq.(9), which is simply the time average of the oscillating potential Eq.(8). By writing it as

$$V_0(\alpha_0,\mathbf{r}) = \int_{-\alpha_0}^{+\alpha_0} \sigma(u)\, V\,(\mathbf{r}-\mathbf{e}u)\, du \tag{24}$$

with

$$\sigma(u) = \frac{1}{\pi\alpha_0} \left[1 - \left(\frac{u}{\alpha_0}\right)^2\right]^{-1/2} , \tag{25}$$

it can be looked upon as created by a linear distribution of "charges" extending from $-\alpha_0$ to $+\alpha_0$ along **e**, with density σ(u), the unit of "charge" generating the potential V(**r**). This behavior appears natural due to the rapid oscillations of the center of force in Eq.(8). If V(r) is spherically symmetric, $V_0(\alpha_0,\mathbf{r})$ is axially symmetric around the line of "charges" and has even parity with respect to its center.

The eigenvalue equation (15) has been considered earlier as a possible description for atomic structure in a radiation field. It was derived either heuristically from Eq.(6) (Henneberger[10], see also ref. 15; Lima and Miranda[16,17], see also ref. 18), or by an approach different from ours (Gersten and Mittleman[19], see also ref. 20). In the beginning the high-frequency character of the approximation was not realized[10,15,16,18]. Gersten and

Mittleman were the first to do this although they formulated its validity condition too restrictively, requiring that ω be large with respect to the *field free* value (for $\alpha_0 = 0$) of the atomic ionization potential $|W_0(0)|$, instead of its value in the field $|W_0(\alpha_0)|$. As will become apparent in the following, the true condition greatly extends the validity of the theory. Also Eqs.(17)-(20) were derived by Mittleman.[20]

II. LEVEL STRUCTURE

As an example of application of the previous theory we consider the calculation of the level structure of the hydrogen atom in a very high frequency field, according to Eq.(15). The unperturbed potential is now $V(r) = -1/r$.

We first note that because of the reduced symmetry of the dressed potential $V_0(\alpha_0,\mathbf{r})$, Eq.(24), only the magnetic quantum number m and the parity (g or u) remain good quantum numbers, whereas ℓ ceases to be so. Moreover, as easily shown, the eigenvalues W do not depend on the sign of m (twofold degeneracy for $m \neq 0$). In fact, because of the symmetry of our problem ($D_{\infty h}$), it is natural to adopt the standard classification of levels prevalent for homonuclear diatomic molecules. Within a given manifold, defined by $|m|$ and the parity, we shall continue to use the labels (n,ℓ) to designate the state which in the weak field limit is characterized by these quantum numbers (e.g. $(n\ell)\pi_u$ will designate the state with $|m| = 1$ and odd parity, deriving continuously from the unperturbed state n,ℓ).

To solve Eq.(15) we have introduced[7] a spherical harmonics basis, with the polar axis along α_0. Eq.(15) then splits into sets of radial differential equations, one for each $|m|$. Within a given set the equations are coupled by the off-diagonal elements of the potential matrix $U^{mm}_{\ell\ell'}(\alpha_0,r) = \langle \ell m|V_0(\alpha_0,\mathbf{r})|\ell' m\rangle$. In the present work we have adopted the "decoupled ℓ-channels approximation", in which we neglect the off-diagonal elements ($\ell \neq \ell'$) for the following reasons. It can be shown that for small α_0 (i.e. through order α_0^2) the off-diagonal matrix elements do not contribute at all r (see below). Further, a numerical comparison shows that the off-diagonal matrix elements are about one order of magnitude smaller than the diagonal ones for all r. In our approximation, the set of radial equations corresponding to a given $|m|$ decouples, yielding for each ℓ:

$$\left[\frac{1}{2} - \left(\frac{d^2}{dr^2} - \frac{\ell(\ell+1)}{r^2}\right) + U^{mm}_{\ell\ell}(\alpha_0,r)\right]\chi = W\chi \tag{26}$$

Now ℓ and n become good quantum numbers together with $|m|$. (The principal quantum number is defined, as usual, by $n = n' + \ell + 1$, where n' is the number of nodes of $\chi(r)$.) The effective potentials $U^{mm}_{\ell\ell}(\alpha_0,r)$ entering Eq.(3) can be calculated from the expansion of $V_0(\alpha_0,\mathbf{r})$ in a multipole series.

The α_0 dependence of the energy eigenvalues can be predicted analytically for small values of this parameter. One may treat the difference $U^{mm}_{\ell\ell}(\alpha_0,r) - V(r)$ to first order in perturbation theory. The result for the displacement of a level ("dynamic Stark-shift", "light-shift"), to lowest order in α_0 is:

$$\Delta W_{n\ell|m|} = \alpha_0^2 \left[\frac{\pi}{3} |\psi^c_{n\ell m}(0)|^2 - \frac{1}{2} \left\langle \frac{1}{r^3} \right\rangle^c_{n\ell} \langle \ell m|P_2|\ell m\rangle \right] , \tag{27}$$

where the last term should be taken zero for $\ell = 0$ and $|\psi^c(0)|^2$ and $\langle r^{-3}\rangle$ are the probability density at the origin and the average value for the Coulomb potential. It can be shown that the level-shift formula Eq.(27) coincides with the one derived from $V_0(\alpha_0,\mathbf{r})$ without making the decoupled ℓ-channels approximation. It leads to an overall n-dependence $\Delta E \propto n^{-3}$. From

$$\langle \ell m|P_2|\ell m\rangle = [\ell(\ell+1) - 3m^2] / (2\ell+3)(2\ell-1) , \tag{28}$$

and Eq.(27) we see that for each $|m|$ value, the first values of ℓ up to some $\ell_0(|m| \leq \ell < \ell_0)$ yield a $\Delta W > 0$ (i.e. the perturbed levels are more weakly bound), whereas from ℓ_0 onward, $\Delta W < 0$. For $|m| = 0,1,2$, the values of ℓ_0 are 1,2,4, respectively.

We have computed numerically the low lying energy eigenvalues of hydrogen for α_0 ranging from 0 to 100. The results are presented in Figs. 1-4, and are grouped according to their symmetry. Fig. 1 shows the α_0-dependence of the energy of the ground state $(1s)\sigma_g$. Its characteristic feature is the rapid decrease of the ionization potential over the range $0 < \alpha_0 < 20$. This greatly facilitates the possibility of satisfying the frequency condition $\omega >> |W_0{}^m(\alpha_0)|$, for m = 0, stated in Sec. I. In fact, this is fulfilled for some of the excimer lasers now in operation, e.g. $\omega = 6.3$ eV, $I = 10^{17}$ W/cm^2 see ref. 1, since α_0 can be more than 30, which corresponds to a reduction of the ionization potential by about a factor of ten. The large decrease of the ionization potential indicates strong atomic distortion and corresponds to a substantial increase in the size of the atom.

Figs. 2-4 show the behavior of the excited state levels. As they split from the unperturbed hydrogen level ($\alpha_0 = 0$), some of them increase monotonically, others, in contrast, first decrease and then increase to zero at sufficiently large α_0. As a consequence of our approximate approach, there are some curve-crossings in Figs. 2-4. These will turn into avoided crossings in an exact calculation. The small α_0 behavior in Figs. 1-4 (e.g. the sign of the light-shifts ΔW) can be entirely understood based on Eq.(27). (For very small α_0 there is excellent agreement between the numerical computation and Eq.(27).)

The ground state energy dependence on α_0 has been computed earlier for the linearly polarized case at hand by Choi et al.[15], using a different method. Results were given for $\alpha_0 < 1$, in graphical form. Over this range our results agree with theirs to better than 2%.

Very recently we have performed a highly accurate computation of the structure of hydrogen according to Eq.(15).[21] This was done by diagonalizing the Hamiltonian Eq.(13) in a multicenter (Cartesian) Gaussian basis. The location of the centers, their number, as well as the exponents of the Gaussians were optimized for each value of α_0 separately. Many of the features discussed above have been confirmed, but new remarkable aspects have emerged.

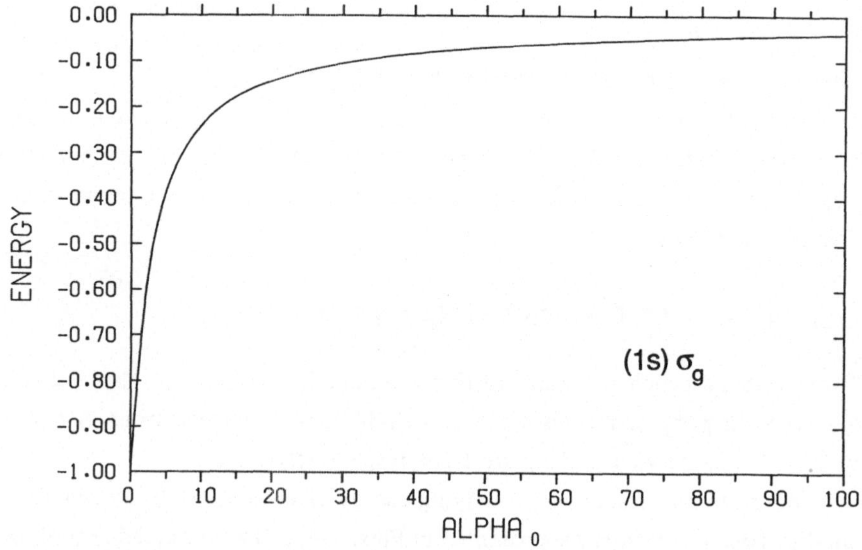

Fig. 1. Dependence of ground state $(1s)\sigma_g$ energy level of a hydrogen-like atom (in Ry units) on α_0 ($\alpha_0 = I^{1/2}\,\omega^{-2}$ a.u.).

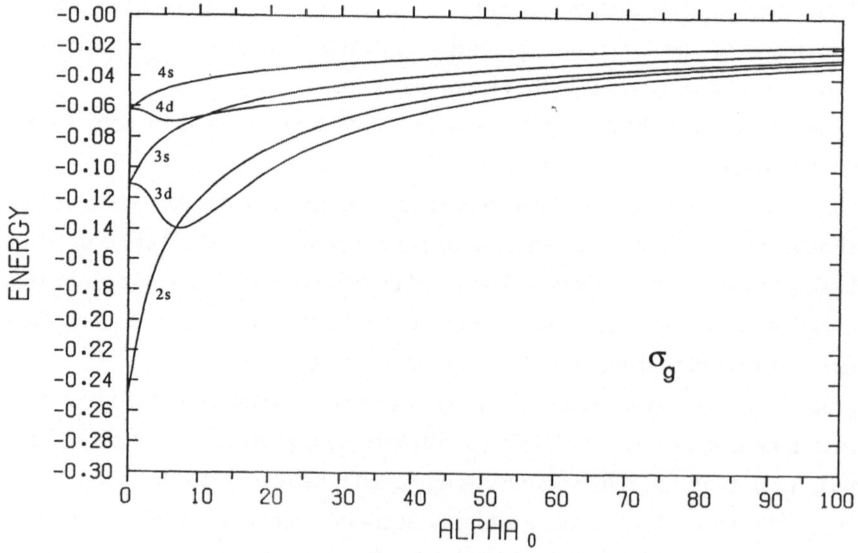

Fig. 2. Same as in Fig. 1, but for σ_g states.

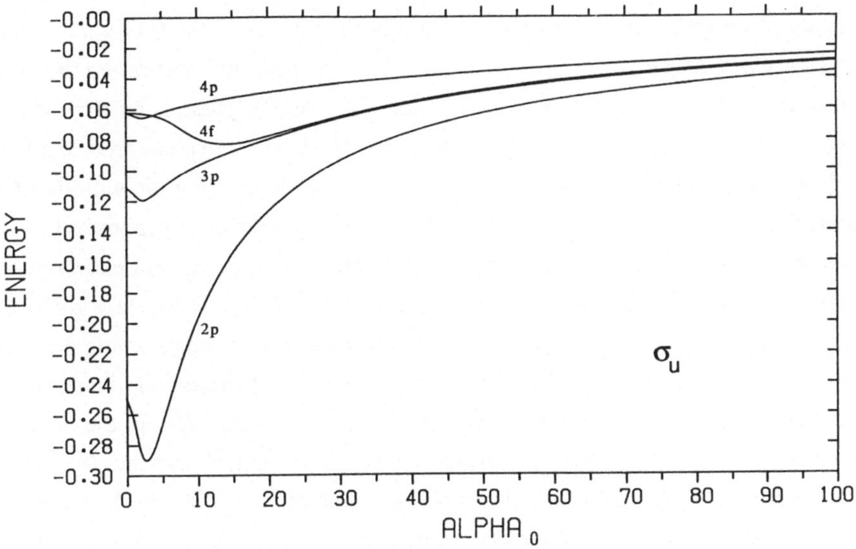

Fig. 3. Same as in Fig. 1, but for σ_u states.

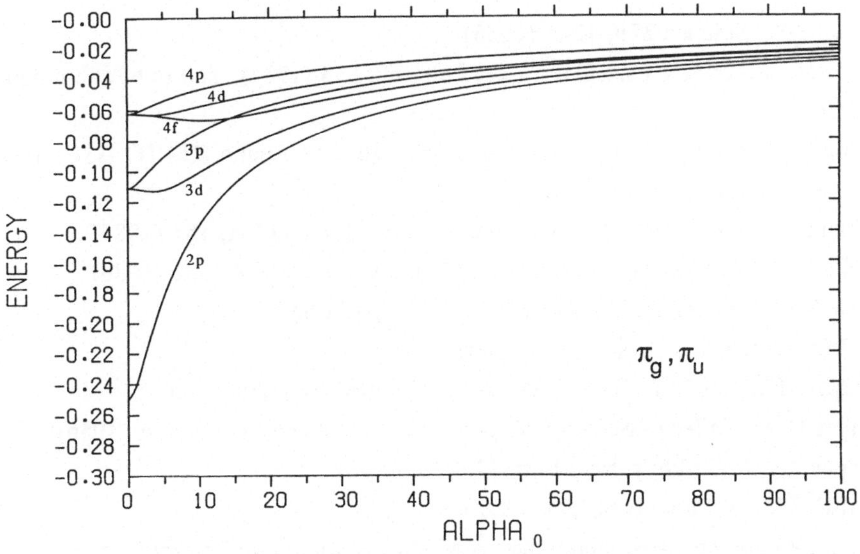

Fig. 4. Same as in Fig. 1, but for π_g and π_u states.

The decrease of the ionization potential $|W_0|$ found for the ground state has relevant experimental consequences. To begin, it contrasts with the behavior of noble gas atoms in multiphoton experiments performed in the infrared or the visible at intensities up to 10^{-1} a.u., e.g. see Kruit et al.[22] and Agostini et al.[23]. (The fact that the effective potential for the noble gas atoms is not Coulombic, as in our case, should not play a role in the following.) Under those circumstances an *increase* of the ground state ionization potential occurs and the lowest energy ionization peak may be suppressed (see Muller and Tip[24]). There is no contradiction with our results, however, since the increase or decrease of $|W_0|$ will depend on ω and I, and our theory does not apply in their case (the frequency condition limiting the validity of the theory is not satisfied). Moreover, in our regime no ionization peak can be suppressed.

The energy spectrum of the electrons leaving the atom by multiphoton or excess-photon ionization depends on the laser shifted value of $|W_0|$ at the time of emission (peaks are located at $W_\nu = -|W_0| + \nu\omega$, with $\nu \geq \nu_{min}$). On the other hand, by the time the escaping electrons reach the detector (placed outside the field) their energy may have substantially increased due to the ponderomotive force of the intense laser field (see refs. 24 and 25). In the noble gas experiments mentioned, the two effects practically cancel each other,[19] and the electrons are detected with energies W_ν as if $|W_0|$ had the unperturbed atom value (like in the weak field case). In our high-frequency, high-intensity regime the two effects are additive and this should result in a substantial increase of the electron energies at the detector in comparison to their weak field values.

REFERENCES

1. Rhodes C.K., Science **229**, 1345 (1985).
2. Gavrila M., Kaminski, J.Z., Phys. Rev. Lett. **52**, 613 (1984), and to be submitted for publication.
3. Gavrila M., in: *Atoms in Unusual Situations* , ed. J.P. Briand (NATO ASI series B, vol. 143, Plenum Press, 1987), p.225.
4. Offerhaus M.J., Kaminski J.Z., Gavrila M., Phys. Lett. **112** A, 151 (1985).
5. Gavrila M., Offerhaus M.J., Kaminski J.Z., Phys. Lett. **118** A, 331 (1986).
6. Van de Ree J., Kaminski J.Z., Gavrila M., to be published.
7. Pont M., Gavrila M., Phys. Lett., in press.
8. Pont M., Offerhaus M.J., Gavrila M., to be submitted for publication.
9. Kramers H.A., *Collected Scientific Papers*, (North Holland, Amsterdam, 1956) p. 866.
10. Henneberger W.C., Phys. Rev. Lett. **21**, 838 (1968).
11. Faisal F.H.M., J. Phys. B **6**, L89 (1973).
12. Baz A.I., Zeldovich Ya.B., Perelomov A.M., *Scattering, Reactions and Decay in Non-relativistic Quantum Mechanics* (translated from the Russian by the Israel Program for Scientific Translations, Jerusalem 1969), Chap. 5.
13. Zeldovich Ya.B., Sov. Phys. Usp. **16**, 427 (1974).
14. Siegert A.J., Phys. Rev. **56**, 750 (1939).

15. Chan K. Choi, Henneberger W., Sanders F.S., Phys. Rev. A **9**, 1985 (1974).
16. Lima C.A.S., Miranda L.C.M., Phys. Rev. **23**, 3335 (1981).
17. Lima C.A.S., Miranda L.C.M., in *Essays in Theoretical Physics*, Ed. W.E. Parry (Pergamon Press, 1984), p. 129.
18. Landgraf T.C., Leite J.R., Almeida N.S., Lima C.A.S., Miranda L.C.M., Phys. Lett. **92** A, 131 (1982).
19. Gersten J.I., Mittleman M.H., J. Phys. B **9**, 2561 (1976).
20. Mittleman M.H., *Theory of Laser-Atom Interactions*, (Plenum, 1982), Chap. 3.
21. Pont M., Walet N., McCurdy C.W., Gavrila M., contributed abstract XV ICPEAC, Brighton 1987, and to be published.
22. Kruit P., Kimman J., Muller H., van der Wiel M., Phys. Rev. A **28**, 248 (1983).
23. Agostini P., Fabre F., Petite G., in: *Multiphoton Ionization of Atoms*, Eds. S.L. Chin and P. Lambropoulos (Academic Press, 1984), p. 133.
24. Muller H., Tip A., Phys. Rev. A **30**, 3039 (1984).
25. Freeman R.R., McIlrath T.J., Bucksbaum P.H., Bashkansky M., Phys. Rev. Lett. **57**, 3156 (1986).

A CONTRIBUTION TOWARDS THE THEORY OF MULTICHANNEL MULTIPHOTON IONIZATION OF ATOMS

Riccardo Burlon and Claudio Leone

Dipartimento di Energetica e di Applicazioni di Fisica
Viale delle Scienze
90128 Palermo, Italy

and

Gaetano Ferrante

Istituto di Fisica Teorica dell'Università
Via dei Verdi
98100 Messina, Italy

INTRODUCTION

This Lecture presents a theoretical treatment based on the S-matrix to study ionization of atoms by strong laser fields, when ionization channels of different photon multiplicity are simultaneously open. Presently this atomic elementary process is intensively investigated both experimentally and theoretically, and an account of whatsoever extent of the large existing literature on the subject is out of question. By the way, representative and significant contributions to the subject may be found in several Lectures forming this Volume. Here specifically we present a treatment concerning the ionization of the hydrogen atoms. The salient features and scopes of our treatment may be summarized as follows.

(1) It is based on the S-matrix formalism, and accordingly exploits the notion of transition probability per unit time. Generally speaking, it is to be expected that not always such a notion be valid when applied to ionization by very strong lasers. Thus an implicit limitation of our treatment is that the pulse duration be much smaller than the inverse of the ionization rate, or that the depletion of the atom initial state caused by ionization be small.

(2) The treatment is essentially nonperturbative, this feature being achieved through the construction of the wavefunction for the electron in the final state. At the same time, it is non Keldysh-type, in the sense that in the final continuum state wavefunction we account for the coulomb interaction with the residual ion. Keldysh-type treatments in our treatment cor-

respond to neglect the influence of the coulomb interaction. In the calculations reported below, we compare the predictions based on formulas accounting or not for the coulomb interaction (Fig.s 1 and 2), and thus are in the position to quantitatively and qualitatively appreciate the significance of the latter (of course, restricted to the problem at hand).
(3) The laser field is treated in two models, namely, in that of an ideal, purely coherent field, and in that of a multimode equal-frequency field, with a number of modes approaching the chaotic limit. The first model is used as it generally serves as reference model; in fact most of the theoretical treatments so far are based on it. Besides, it is requested for separating out and isolating effects due to other sources of the theoretical treatment, such as the consequences of including or not the coulomb interaction in the final states. In general, it is expected that real very intense lasers be rather poorly described by the ideal model. The second laser model is used as it is expected to make a closer contact with the real experiments in the following two respects: (i) in several of the multiphoton ionization experiments the laser systems are operated in a multimode configuration; (ii) in any case, due to focussing and pulse duration, besides possibly statistical fluctuations, in any real experiment the laser enters as a system with its characteristic parameters exhibiting rather large distributions. As we are considering in practice a chaotic model with zero bandwidth, we account only for the statistical distribution of the field intensity realizations. Nevertheless, we expect that the consequences of this distribution be to some extent qualitatively similar to those of other distributions, originated by space-time macroscopic inhomogeneities. The basis for this expectation is that ultimately any fluctuation or inhomogeity produces in space and/or time in the ionization chamber a distribution of the field parameters at which the various atoms are ionized. From the experimental point of view the consequence is that invariably some sort of an average signal is measured.
(4) Due consideration is given to the gauge-invariance formulation of the S-matric theory of the ionization process. Calculations show that it may turn out be a very relevant aspect of the problem (Fig.s 5 and 6).

OUTLINE OF THE THEORETICAL TREATMENT

The starting point of an S-matrix treatment of multiphoton ionization of hydrogen atom from their ground state is given by

$$S_{if} = (-i/\hbar)\int_{-\infty}^{\infty} dt \langle \tilde{\psi}_q^-(\underset{\sim}{r},t) | e\,\underset{\sim}{E}(t)\cdot\underset{\sim}{r}\,| \psi_0(\underset{\sim}{r},t)\rangle \tag{1}$$

Provided the S-matrix formalism is valid for the problem at hand, and provided we are able to find exactly $\tilde{\psi}_q^-$, the expression (1) gives the rigorous answer to the problem. In (1) ψ_0 is the ground state of the hydrogen atom, the electron radiation interaction has been written in the Electric field gauge (E-gauge), and $\tilde{\psi}_q^-$ is a continuum solution of the Schroedinger equation (in the gauge of the vector potential, A-gauge)

$$\left\{\frac{1}{2m}\left[\hat{p} + \frac{e}{c}\underset{\sim}{A}(t)\right]^2 - \frac{e^2}{r}\right\}\psi = i\hbar\dot{\psi} \tag{2}$$

The continuum solution of Eq.(2) is looked for in the form

$$\psi = \psi_q^- = a_q(t)|-,q,t\rangle \exp\left\{\frac{-i}{\hbar}\left[\varepsilon_q t + \int_{-\infty}^{t} \frac{e^2 A^2(\tau)\, d\tau}{2mc^2}\right]\right\} \tag{3}$$

where $\varepsilon_q = \hbar^2 q^2/2m$; $\hbar \underset{\sim}{q}$ is the electron asymptotic momentum, and $|-,q,t\rangle$ is an unknown function of $\underline{r}$ and t to be determined, which asymptotically must behave as the field-free incoming coulomb wave $|q\rangle$. Adiabatic switching on is assumed. The requirement that for any t the wavefunction ψ_q^- be normalized to a delta function, similarly to the coulomb wavefunction $|q\rangle$, yields the following condition to be imposed

$$\langle q'|-,q,t\rangle = \delta(\underset{\sim}{q}'-\underset{\sim}{q}) \tag{4}$$

Now we substitute Eq.(3) into the Schroedinger equation (2), multiply from the left by $\langle q'|$ and integrate over $\underset{\sim}{q}'$ and $\underset{\sim}{r}$. As a result we obtain the following two equations for the coefficient $a_q(t)$ and $|-,q,t\rangle$:

$$a_q(t)\lambda_q(t) = i\hbar \frac{\partial}{\partial t} a_q(t) \tag{5a}$$

$$\left[H_0 - \varepsilon_q - i\hbar\frac{\partial}{\partial t}\right]|-,q,t\rangle = \left[\lambda_q(t) - W(t)\right]|-,q,t\rangle \tag{5b}$$

In Eq.s (5)

$$\lambda_q(t) = \int d^3q' \langle q'|W(t)|-,q,t\rangle \tag{6}$$

$$H_0 = \frac{p^2}{2m} - \frac{e^2}{r} \quad ; \qquad W(t) = \left(\frac{e}{mc}\right)\underset{\sim}{A}(t)\cdot\underset{\sim}{p} \tag{7}$$

We now solve Eq.s(5) by a successive approximation procedure, by letting

$$|-,q,t\rangle = |q\rangle + \sum_{j=1}^{\infty} |j,q,t\rangle \tag{8}$$

$$\lambda_q(t) = \sum_{j=1}^{\infty} \lambda_q^{(j)}(t) \tag{9}$$

Substitution of Eq.s(8),(9) into (5) gives a sequences of equations, which using $|q\rangle$ first allow to calculate $|1,q,t\rangle$ and $\lambda_q^{(1)}(t)$ through

$$\lambda_q^{(j)}(t) = \int d^3q' \langle q'|W(t)|j-1,q,t\rangle \tag{10}$$

using next $|q\rangle$, $\lambda_q^{(1)}(t)$ and $|1,q,t\rangle$ allow to calculate $|2,q,t\rangle$ and $\lambda_q^{(2)}(t)$, and so on[1].

Specifying to

$$\underline{A}(t) = A_0 \cos\omega t \tag{11}$$

the resulting wavefunction within a second-order procedure is given by

$$\Psi_q^- = \exp\left\{\frac{-i}{\hbar}\left[\varepsilon_q t + F_q^{(2)}(t)\right]\right\} \times \left[|q\rangle + |1,q,+1\rangle e^{i\omega t} + |1,q,-1\rangle e^{-i\omega t} + |2,q,0\rangle + |2,q,2\rangle e^{i2\omega t} + |2,q,-2\rangle e^{-i2\omega t}\right] \tag{12}$$

with $|j,q,m\rangle$, (j=1,2; m=0, $\pm$ 1, $\pm$ 2) corrections, due to the influence of the radiation field, distorting the coulomb wave $|q\rangle$ (see, for details, Ref.1).

When (12) (times the appropriate gauge phase factor) is used in the basic S-matrix element (1), the following picture arises. The terms containing $|q\rangle$ account for N-photon ionization via one-step transition from the ground state to the given continuum; the terms containing the first-order correction $|1,q,\pm 1\rangle$ account for the same N-photon ionization via a two-step transition, the intermediate step involving either a discrete or a continuum state of the hydrogen atom. The second-order corrections account for three-step N-photon ionization processes and so on.

The idea exploited below is that when very powerful lasers are used for ionization, the latter takes place very rapidly essentially via a one-step transition, which thus correspond to some sort of an high-intensity limit. For relatively weak fields instead multistep multiphoton ionization should take place.

$$F_q^{(2)}(t) = \int_{-\infty}^{t} d\tau \left[\frac{e^2 A^2(\tau)}{2mc^2} + \lambda_q^{(1)}(\tau) + \lambda_q^{(2)}(\tau)\right], \tag{13}$$

$$\lambda_q^{(1)}(\tau) = \int d^3q' \langle q' | \left(\frac{e}{mc}\right) \underset{\sim}{A_0} \cdot \underset{\sim}{p} | q \rangle \cos\omega\tau \tag{14}$$

and $\lambda_q^{(2)}$ a rather involved expression, which may found in Ref.1. The evaluation of $\lambda_q^{(1)}$ and $\lambda_q^{(2)}$ requires in particular the calculation of spatial matrix elements containing coulomb functions $|q\rangle$; here, for simplicity, they are evaluated using plane waves. Then $\lambda_q^{(1)}$ becames exactly the term appearing in the Volkov plane waves (in particular, it is real), and $\lambda_q^{(2)}$ vanishes exactly:

$$F_q^{(2)}(t) \longrightarrow F_q^{V}(t) = \int_{-\infty}^{t} d\tau \left[\frac{e^2 A^2}{2mc^2} + \left(e\, \underset{\sim}{q} \cdot \underset{\sim}{A_0} / mc\omega\right) \cos\omega\tau\right] \tag{15}$$

Thus, the structure of the wavefunction (12) appears to be qualitatively the following: the coulomb wavefunction (the spatial part) is distorted by the presence of the radiation field, while the time-dependent part may be considered to be that of a Volkov plane, distorted by the presence of the coulomb interaction.

In this pilot calculation we have restricted ourselves to the leading term of the constructed wavefunction plus $F_q^{(2)}$ substituted by F_q^{V} In other words in Eq.(1) we have inserted

$$\tilde{\Psi}_q^- = \mathcal{T}\exp\left\{\frac{-i}{\hbar}\left[\varepsilon_q t + F_q^V(t)\right]\right\}|q\rangle \tag{16}$$

$$\mathcal{T} = \exp\left\{i\left(\frac{e}{\hbar c}\right)\underset{\sim}{A}_0\cdot\underset{\sim}{r}\cos\omega t\right\} \tag{17}$$

being the appropriate gauge unity transformation operator transferring our approximate wavefunction from the A- to the E-gauge.

The remaining procedure to arrive at transition probabilities and cross-sections is the usual one, and accordingly details are omitted. We wish to point out only the following. 1) The required derivations are carried out without other simplifications besides those already stated; and the calculations too are carried out by means of rigorous numerical methods. 2) The assumption of a ponderomotive force acting on the ejected electrons and compensating for the shift of the ionization level is adopted.

WORKING FORMULAS

A. Purely Coherent Field

A basic starting quantity is the doubly differential ionization rate for the s-th multiphoton channel (s is the number of photons absorbed above threshold),

$$\left(d^2W/d\Omega\, d\varepsilon\right)_s$$

$$= Q_0(I/I_a)\left[1-\exp(-2\pi\nu)\right]^{-1}\left|T_{n_0+s}(\underline{q},\underline{E}_0)\right|^2\delta(\varepsilon+\Delta-\varepsilon_s) \tag{18}$$

$$T_\ell(\underline{q},\underline{E}_0) = -\int_{-\pi}^{\pi} d\alpha\,\cos\alpha\, B(\underline{q},\underline{E}_g\sin\alpha)\, f_\ell(\alpha) \tag{19a}$$

$$f_\ell(\alpha) = \exp\left[i\ell\alpha - i(\Delta/2\hbar\omega)\sin\alpha - i\lambda\cos\alpha\right] \tag{19b}$$

$$B(\underline{q},\underline{E}_g\sin\alpha) = -\left(16\pi a_0^4\right)^{-1}\times$$

$$\times\int d^3r\,\exp\left[-r/a_0 + i(\underset{\sim}{q}+\underset{\sim}{E}_g\sin\alpha)\cdot\underset{\sim}{r}\right](\hat{e}\cdot\underset{\sim}{r})\,F\left(i\nu,1,i(qr+\underset{\sim}{q}\cdot\underset{\sim}{r})\right) \tag{19c}$$

$$Q_0 = (2^5e^2/\hbar a_0\pi^2);\quad \Delta = e^2E_0^2/4m\omega^2;\quad \lambda = e\underset{\sim}{q}\cdot\underset{\sim}{E}_0/m\omega^2;$$

$$\underset{\sim}{E}_g = e\underset{\sim}{E}_0/\hbar\omega;\quad \nu = 1/qa_0;\quad \varepsilon_s = (n_0+s)\hbar\omega - I_0;$$

$I_0 = 13.6$ eV; $I_a = 3.51\times10^{16}$ W/cm^2; I- field intensity;
n_0- minimum ionization photon number; s=0,1,2,...

Starting from (18) a number of characteristic quantities of the ionization process may be defined and calculated, namely

1. $(dW/d\Omega)_s$ - the differential ionization rate for the s-th channel

2. $W_s = \int \left(\frac{dW}{d\Omega}\right)_s d\Omega$ - the angle-integrated s-th channel ionization rate

3. $W = \sum_s W_s$ - the ionization rate summed over all the available multiphoton channels

4. $(d\sigma/d\Omega)_s$ - the s-th channel differential cross section

5. σ_s - the angle-integrated s-th channel cross section

6. $\sigma = \sum_s \sigma_s$ - the total cross-section summed over the multiphoton channels.

B. Zero Bandwidth Chaotic Field

For this particular case of chaotic field, the corresponding quantities may be simply obtained by averaging the purely coherent field formulas over the threshold shift distribution

$$\mathcal{P}(\Delta) = \langle\Delta\rangle^{-1} \exp\left\{-\Delta/\langle\Delta\rangle\right\} \tag{20}$$

related to the intensity one by

$$\langle\Delta\rangle = (2\pi e^2/mc\omega^2)\langle I\rangle$$

$\langle I\rangle = \langle|\mathcal{E}(t)|^2\rangle$ being the variance of the field amplitude $\mathcal{E}(t)$ with $\langle\mathcal{E}(t)\rangle = 0$.

In particular, the averaging of the doubly differential ionization rate (18) gives[2]

$$\langle(d^2W/d\Omega d\varepsilon)_s\rangle = \int_0^\infty d\Delta \, (d^2W/d\Omega d\varepsilon)_s \, \mathcal{P}(\Delta)$$

$$= (e^2 2^5/\hbar a_0 \pi^2 \Delta_a)\left[1-\exp(-2\pi\nu)\right]^{-1} \eta \exp(-\eta) \left|T_{n_0+s}(q, \tilde{E}_0)\right|^2 \tag{21}$$

$$\Delta_a = e^3/4m\omega^2 a_0 ; \tag{22a}$$

$$\bar{E}^2 = (4m\omega^2/e^2)(\varepsilon_s - \varepsilon) ; \tag{22b}$$

$$\eta = (\varepsilon_s - \varepsilon)/\langle\Delta\rangle . \tag{22c}$$

Using this result all the other quantities immediately follow.

CALCULATIONS

We remind that the calculations reported below in the Fig.s 1-6 are for ionization of hydrogen atoms from their ground state by a linearly polarized laser field with photon energy $\hbar\omega$ =1.17 eV.

Fig. 1 shows total cross sections (in a.u.) for different numbers s of excess photons above threshold vs the laser intensity for the case when the purely coherent field model and coulomb waves are used. s goes from 0 to 4.

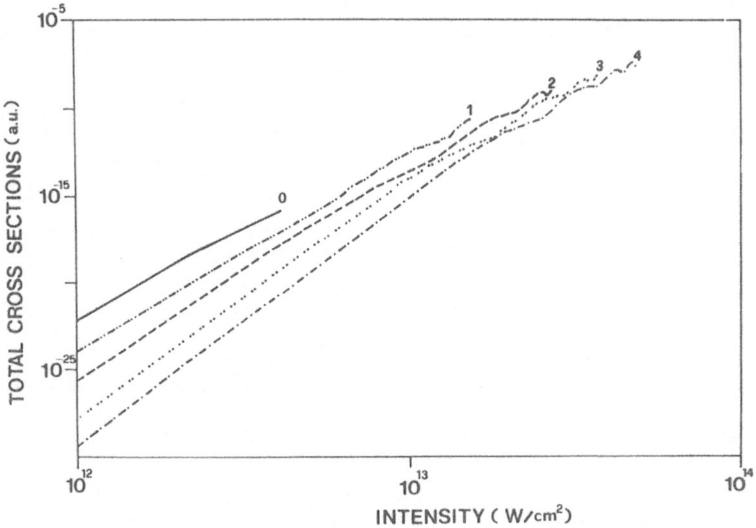

Fig. 1 Total cross sections vs the laser intensity (coulomb wave)

Fig.2 shows the same total cross sections as Fig.1 with the only difference that plane waves are used instead of coulomb ones.

Fig.3 shows ionization rates (in a.u.) vs the mean field intensity (W/cm^2) for the first five channels of lower energy, when coulomb wave and the chaotic field model are used. $1E\pm n = 10^{\pm n}$.

Fig. 4 reports photoelectron energy spectra (normalized to the first peak) for various mean field intensities (in W/cm^2): a) 5.0×10^{11}; b) 2.5×10^{12}; c) 5.0×10^{12}; d) 7.5×10^{12}. Coulomb waves and the chaotic field model are used.

We observe that while the use of the purely coherent field model gives quite different results concerning respectively the coulomb (Fig.1) and the plane waves (Fig.2), nothing similar takes place when the chaotic field model is used. Calculations not reported here for the lack of space show that, when used jointly with the chaotic field model, the plane waves predict largely the same behaviours as the coulomb waves, with however a sort of systematic overestimate.

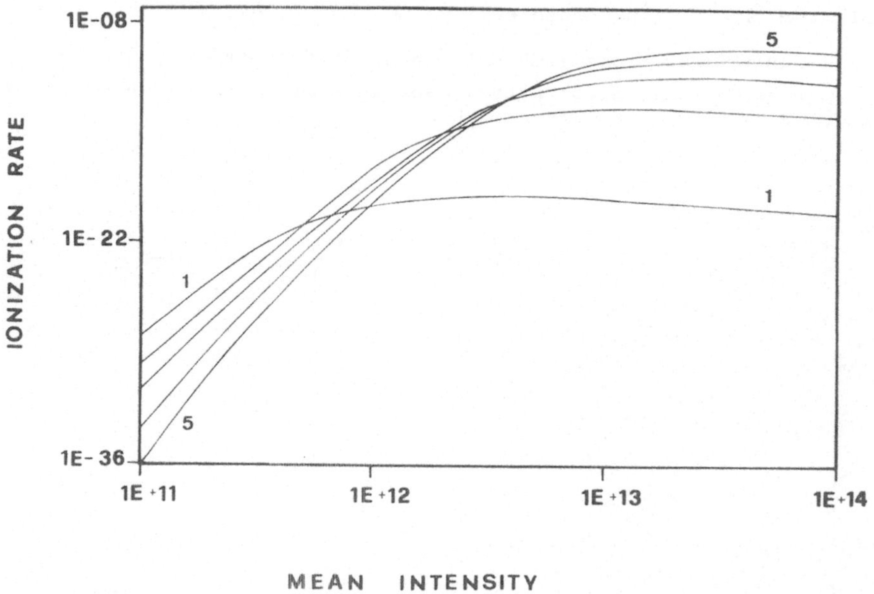

Fig. 2 Total cross sections vs the laser intensity (plane wave)

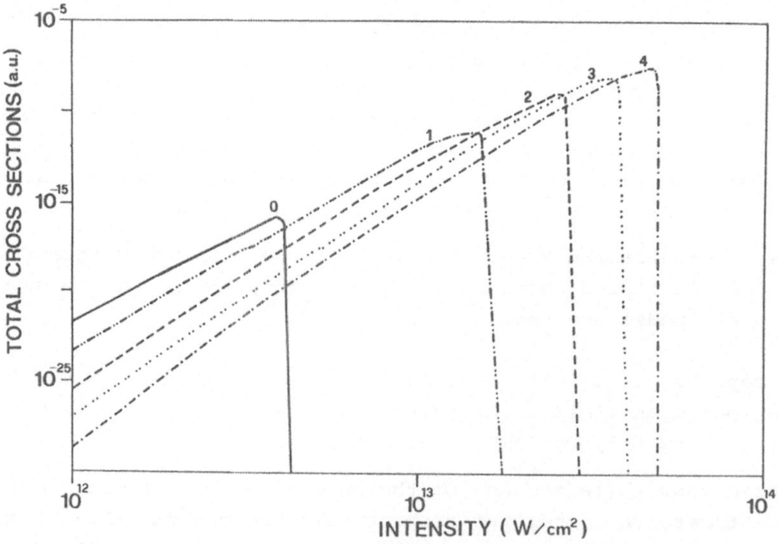

Fig. 3 Ionization rates (a.u.) vs the laser mean intensity

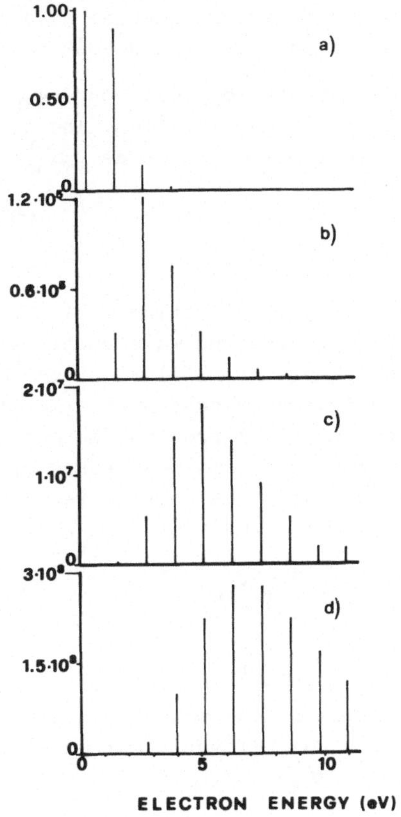

Fig. 4 Photoelectron energy spectra, normalized to the first peak

GAUGE ASPECTS

Recently, in connection with the several measurements on multiphoton ionization, there has been a revival of interest in Keldysh-type models as prototypes of nonperturbative treatments. In particular, other formulations have been proposed[3,4], which are slightly different as compared to the original Keldysh'one[5]. These new formulations, however, are likely to be not better than the Keldysh's treatment as they are not completely satisfactory, among others, from the gauge invariance standpoint. In the reference 6 we have addressed in general terms the problem of gauge-invariance of the pertinent physical quantities of ionization formulated in different gauges. Here we very concisely quote the results of our derivation, which are of direct pertinence to the above mentioned Keldysh-type treatments. Those results are characterized by four spatial matrix elements[6].

1. $$M_K^E = \langle T'\chi_A^V \,|\, e\,\underset{\sim}{E}(t)\cdot\underset{\sim}{r} \,|\, \Psi_0 \rangle \tag{23}$$

appears in the original Keldysh's treatment[5], and is written in the E-gauge, where all the wavefunctions and operators are gauge consistent.

$$T'\chi_A^V = \chi_E^V = \exp\left\{\frac{i}{\hbar}\frac{e}{c}\underset{\sim}{A}(t)\cdot\underset{\sim}{r}\right\}\exp\left\{i\underset{\sim}{q}\cdot\underset{\sim}{r} - \frac{i}{2\hbar m}\int_{-\infty}^{t}\left[\hbar q + \frac{e}{c}A\right]^2 d\tau\right\} \tag{24}$$

is the Volkov plane wave in the E-gauge, and the first exponential gives the unitary operator T' connecting the E- to A-gauge (and viceversa).

Having in mind to perform some selected numerical calculations, ψ_0 is taken to be again the hydrogen atom ground state wavefunction. The angle-integrated cross section obtained using (23) is labelled below as σ_K^E .

2. $$M_H^E = \langle \chi_A^V | e\, \underset{\sim}{E}(t) \cdot \underset{\sim}{r} | \psi_0 \rangle \tag{25}$$

is a 'hybrid' matrix element, which is obtained by (23) letting T'=1. Thus (25) correspond to a matrix element in which operators and wavefunctions are not simultaneously transformed. In the E-gauge, this simplification has been in practice adopted by Becker et al[4] Calculations show that such simplifications may have a drastic consequence on the numerical results. In particular, it gives an incorrect behaviour of the cross sections near threshold, where peak inversion take place[6]. The corresponding cross section is labelled as σ_H^E .

3. $$M_K^A = \langle \chi^V_A | \left(\frac{e}{mc}\right) \underset{\sim}{A}(t) \cdot \underset{\sim}{p} + \frac{e^2 A^2}{2mc^2} | T'^{+} \psi_0 \rangle \tag{26}$$

is a matrix element in the A-gauge, which can be shown[6] to be in this gauge the closest analogue of the Keldysh's matrix element. We point out that where is no exact counterpart in the A-gauge to the Keldysh's matrix element. The corresponding cross section is labelled as σ_K^A .

4. $$M_H^A = \langle \chi_A^V | \left(\frac{e}{mc}\right) \underset{\sim}{A} \cdot \underset{\sim}{p} + \frac{e^2 A^2}{2mc^2} | \psi_0 \rangle \tag{27}$$

is a 'hybrid' matrix element in the A-gauge, which is obtained from (26) letting $T'^{+} = 1$. It has been used by several authors[3,7], lately by Reiss in particular[3]. The corresponding cross-section is labelled as σ_H^A .

Fig.s 5 and 6 show representative calculations showing the behaviour vs intensity of the ratios

$$R_{AK}^{EK} = \sigma_K^E / \sigma_K^A \quad ; \quad \text{and} \quad R_{AH}^{EK} = \sigma_K^E / \sigma_H^A$$

The results of Fig.s 5 and 6 refers to the first five ionization channels. The purely coherent field model is used. The reported calculations together with those of other ratios[6] show that gauge-invariance considerations are far from being of little significance in the present problem.

SUMMARIZING REMARKS

The reported calculations suggest the following summarizing remarks.

1. The inclusion of the coulomb interaction in the final state within a treatment in which the laser is taken as a purely coherent field yeld only sequential channel closing (peak disappearence) without inversion. This result has been predicted by other authors as well within different treatments[8] but it does not appear to be out of question (see, for instance, other con-

tributions to this Volume). In our case, the point to be checked further is the consequence of our using a Volkov-like factor in our working wavefunction, instead of the exact one. Work is in progress on this point.

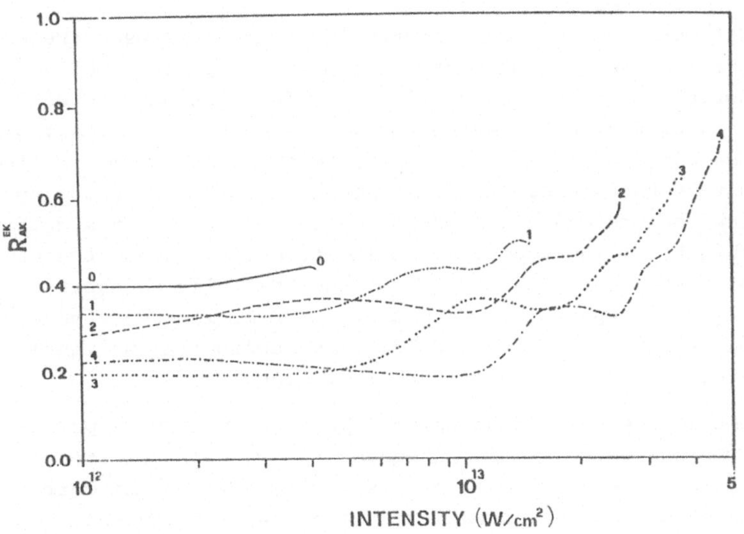

Fig. 5 Variation of the ratio R^{EK}_{AK} as a function of the laser intensity for different numbers s of excess photons above threshold

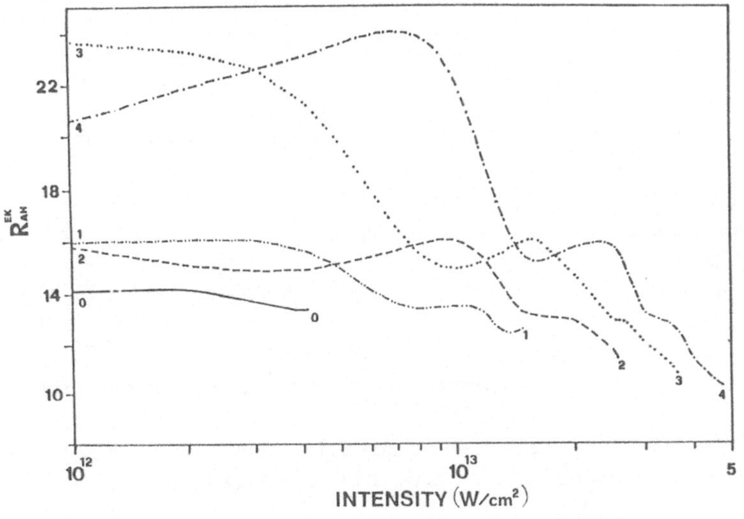

Fig. 6 As in Fig. 5 for the ratio R^{EK}_{AH}

2. Neglecting of the coulomb interaction jointly to the use of a purely coherent field model gives only apparently a better agreement with experimental phenomenology, at least as far as peak inversion is concerned. Actually, near threshold the use of plane waves for the ejected electrons gives simply a wrong behaviour of the pertinent physical quantities. Besides, relatively far from threshold, use of plane waves gives overestimates compared with the value obtained when the coulomb interaction is retained.

3. The results obtained when the chaotic field model is used are worth of attention. Qualitatively, the theory is found to reproduce most of experimentally observed features. Moreover, the differences between coulomb and plane waves are significantly washed out, the latter giving only some overestimates. The fluctuation of the field intensity imply similar fluctuations of the ionization threshold. It in turn contributes to wash out the differences between coulomb and plane waves, where they are most significant, i.e., near threshold. The reported results strongly suggest the need to incorporate into any theoretical treatment the averaging mechanisms always present in the real experiments, and that, when it is done, the controversy of "inclusion vs neglect" of the coulomb interaction in the final continuum states is likely to loose some of its significance.

4. Gauge aspects have been found to be important in many respects. Hybrid procedures, where wavefunctions and operators are not consistently transformed may give quite different results as compared to the correct ones, concerning both the absolute numerical values and the way they behave near threshold, which is particularly critical for the evolution of peak inversion. Thus caution is called for in dealing with Keldish-type treatments.

Work is presently in progress to improve and to explore further the possibilities of the formal treatment presented above.

ACKNOWLEDGEMENTS

The authors wish to thank Dr.s Salvatore Basile and Fabio Trombetta, with whom part of the reported results has been obtained. This work is supported in part by the Italian Ministry of Education and the National Group of Structure of Matter.

REFERENCES

1. C. Leone, R. Burlon, F. Trombetta, S. Basile and G. Ferrante, Nuovo Cimento D (in press)
2. S. Basile, F. Trombetta, G. Ferrante, R. Burlon and C. Leone, "Multiphoton Ionization of Hydrogen by a Strong Multimode Field", (to be published)
3. H. R. Reiss, Phys. Rev. A22, 1786 (1980); J. Phys. B20, L79 (1987)
4. W. Becker, R. R. Schlicher and M. O. Scully, J. Phys. B19, L785 (1986)
5. L. V. Keldysh, Sov. Phys. JETP 20, 1307 (1965)
6. R. Burlon, C. Leone, F. Trombetta and G. Ferrante, Nuovo Cimento D (in press)
7. I. V. Karapetyan, IZV. VUZ. Radiofizika 17, 236 (1975) in russian
8. S. I. Chu and J. Cooper, Phys. Rev. A32, 2769 (1985)

H_2^+ IN INTENSE FIELDS. A COUPLED EQUATIONS STUDY

O. Atabek and X. He

Laboratoire de Photophysique Moléculaire -Université de Paris Sud - Orsay - France

1. Introduction

The advent of intense laser sources leads to increasing interest, both experimental and theoretical, in photodissociation problems.

At weak fields, perturbative treatments involve application of Fermi's golden rule to a bound-to-continuum transition for the photon absorption step which is the basis of the Franck-Condon principle. A proper theoretical description going beyond the previous limit by taking on an equal footing photon absorption, emission and molecular dissociation should, however, be of non perturbative nature and include the field in appropriate models. In these treatments the Fock (photon occupation number) representation is used which leads to dressed molecule or electronic field surfaces[1]. A direct consequence is the prediction of field induced resonances which of course cannot be described by usual perturbation techniques. Photodissociation can be viewed as a predissociation with discrete states embedded in continua.

Very important non linear effects arise at intermediate and strong fields. In this paper we study the gradual changes occuring in the optical spectrum of H_2^+ molecule with the increase of light intensity. For doing so we refer both to time dependent and time independent versions of quantum scattering theory.

2.Time dependent formalism for transition amplitudes

The total hamiltonian for a molecular system submitted to an electromagnetic field can be written as:

$$H = H_m + H_r + V \tag{1}$$

where H_m and H_r are the isolated molecule and the free radiation hamiltonians respectively. V stands for the radiation matter interaction:

$$V(cm^{-1}) = 1.17 \times 10^{-3} \quad \mu(a.u.) \quad [I(W/cm^2)]^{\frac{1}{2}} \tag{2}$$

μ being the electronic transition moment and I the field intensity. The unperturbed states of the system are eigenstates of $H_m + H_r$ and are described by direct products of molecular ($|a\rangle$) and field ($|n\rangle$) states in the

Fock representation[2]

$$(H_m + H_r)|a,n\rangle = (E_a + n\hbar\omega)|a,n\rangle \tag{3}$$

n and ω being the number and frequency of photons.

We are interested in calculating the temporal evolution of some initial state $|\Psi(0)\rangle$ under the influence of the total hamiltonian H, which is formally given by:

$$|\Psi(t)\rangle = \exp(-iHt/\hbar) \quad |\Psi(0)\rangle \tag{4}$$

Expanding in terms of the eigenstates of $H_m + H_r$ we can write (Eq.4) as:

$$|\Psi(t)\rangle = \sum_c \int dE_c|c\rangle \left(\sum_a I_{ca}(t)\langle a|\Psi(0)\rangle \right) \tag{5}$$

$I_{ca}(t)$ is the time dependent transition amplitude between unperturbed bound $|a\rangle$ and continuum $|c\rangle$ states:

$$I_{ca}(t) = (2i\pi)^{-1} \int dE \quad \exp(-iHt/\hbar) \quad G_{ca}(E^+) \tag{6}$$

where:

$$G(E^+) = \lim_{\epsilon \to 0}(E + i\epsilon - H)^{-1} \tag{7}$$

Using the well known relation between resolvent (G) and transition (T) operators, $I_{ca}(t)$ can be expressed in terms of energy dependent transition amplitudes $T_{ca}(E^+)$. It can ultimately be shown[3] that:

$$\lim_{t \to \infty} |\Psi(t)\rangle = \sum_c \int dE_c \quad T_{ca}(E_c) \quad \exp\left[-i(E_c + n_c\hbar\omega)t/\hbar\right] \quad |c\rangle \tag{8}$$

provided the initial state be $|\Psi(0)\rangle = |a\rangle$. Thus the photodissociation probability P_{ca} for a transition from the initial bound state $|a\rangle$ to continuum state $|c\rangle$ becomes:

$$P_{ca} = \int dE_c \quad |\langle c|\Psi(t_\infty)\rangle|^2 = C \int dE_c \quad |T_{ca}(E_c)|^2 \tag{9}$$

where C is a constant. Finally T_{ca} is amenable to numerical calculations via coupled equation techniques for the S matrix.

As far as the weak field limit is considered this can successfully be achieved by introducing an artificial open channel $|c_1\rangle$ as was first suggested by Shapiro[4] which transforms the photodissociation process into a collision problem. When in addition the coupling between $|a\rangle$ and $|c\rangle$ is taken as asymmetric ($V_{ca} \neq V_{ac} = 0$) no radiative shifts and widths affect the state $|a\rangle$ and only the second order term of the Born expansion contributes to the S matrix element:

$$S_{cc_1}(E) \propto \sum_a \frac{\langle c|V|a\rangle\langle a|V|c_1\rangle}{E - E_a - \hbar\omega} \tag{10}$$

The quantity which is reached as the residue of S_{cc_1}:

$$\sigma_a(\omega) \propto |\langle c|V|a\rangle|^2_{E_a+\hbar\omega} \tag{11}$$

is a first order approximation to $|T_{ca}|^2$ calculated at the energy $E_a + n\hbar\omega$ and has the interpretation of a cross section. Implicit in this derivation is that the initial state $|a\rangle$ is a well defined isolated and unperturbed state .Clearly this model may be extended to multiphoton transitions to a final continuum $|\Psi_c^{(-)}\rangle$ provided the first radiative transition from the state $|a\rangle$ is weak,without any constraint concerning all other couplings involved in the exact calculation of the incoming wavefunction $|\Psi_c^{(-)}\rangle$ conducted with physical symmetric radiative couplings .Such a model has been used by Shapiro[5a] and Bandrauk[5b] to calculate the two-photon photodissociation cross section in IBr. We note that if physical symmetric couplings are to be taken into account for the first radiative transition from the state $|a\rangle$,higher order terms would contribute in the Born expansion of S_{cc_1}-from which the cross section we are looking for,could no longer be extracted.

Intense incident fields will produce coherent mixing of the eigenstates $H_m + H_r$ via the radiative couplings V.Superposition of these stationary states as the initially prepared state must be incorporated into the calculation of appropriate amplitudes.Bandrauk and Turcotte[3a] suggest the introduction of a second artificial bound channel $|d\rangle$ which plays the role of the true initial unperturbed (zero field) molecular state weakly coupled to the total molecular-field manifolds. In practice $|d\rangle$ is some state $|a,n\rangle$,a condition which is achieved using the electronic potentiel V_d identical to V_a and displacing V_d in energy with respect to V_a ensuring $\langle d|a\rangle = \delta_{ac}$ and $V_{ad} = \langle a|V|d\rangle = 1$, $V_{da}=0$, whereas $V_{ac} = V_{ca} \neq 0$ The scattering matrix element between $|c_1\rangle$ and $|c\rangle$ leads to the determination of T_{ca} valid at any order of radiation-molecule coupling:

$$T_{ca}(E) \propto \frac{(E - E_d + i\Gamma_d/2)}{(2\pi\Gamma_d)^{1/2}} \quad S_{cc_1}(E) \tag{12}$$

Γ_d being a small linewidth resulting from the weak coupling of the two artificial channels ($|d\rangle$ to $|c_1\rangle$). As for the formal definition of T_{ca}:

$$T_{ca}(E) = \sum_L \frac{\langle c|t|L\rangle\langle L|a\rangle}{E - E_L + i\frac{\Gamma}{2}} \tag{13}$$

where $t = V + VP(E^+ - PHP)^{-1}PV$, P being the projection over all continuum states,it has the following interpretation.[3b] $|L\rangle$ are complex molecular states dressed by the incident field with finite lifetimes due to the field induced widths Γ_L. The overlaps $\langle L|a\rangle$ describe the preparation of the initial state into the dressed $|L\rangle$ states.Finally the system transits into the physical continuum $|c\rangle$ via the transition operator t connecting these dressed states with the physical continuum. Invoking the weak field limit,one has $|L\rangle \rightarrow |a\rangle$,$\Gamma_L \rightarrow 0, E_L \rightarrow E_a$ and $t = V$ which result in Fermi's golden rule for (Eq.13).

An application of this method to the case of H_2^+ photodissociation in strong fields ($I > 10^{14} W/cm^2$) is currently under investigation in our group[3c]

3. Time independent formalism for resonances

Photodissociation is mediated by laser induced molecular resonances,i.e. bound states which acquire finite lifetimes due to radiative decays[1]. As has been previously emphasized by Bandrauk[1b] direct photodissociation is the analogue of predissociation, with radiative couplings instead of non radiative ones between the discrete and continuum states. Time independent close coupled equations derived from scattering techniques are well adapted to calculate the characteristics of these resonances: position and width.

Resonances are complex poles of the scattering matrix and in the limit of isolated poles they can be obtained as solutions of an implicit equation [6] :

$$E - (E_a + n_a\hbar\omega) - F_a(E) + i\pi|\langle c|V|a\rangle|^2 = 0 \tag{14}$$

where $F_a(E)$ is some real energy-shift function. The width Γ_a which is the imaginary part of the solution is approximatively given in the weak field limit as:

$$\Gamma_a = \pi|\langle c|V|a\rangle|^2_{E_a+\hbar\omega} \tag{15}$$

As far as this limit is considered the photodissociation cross section σ_a is proportional to Γ_a.

A direct evaluation of the resonances may be attempted, in a single photon description,by solving a close coupled system subjected to some appropriate boundary conditions known as Siegert wave behaviour[7]. The coupled system follows from the expansion of the total wavefunction on some adiabatic basis set which ,in the case of a diatomic molecule with two B.O. electronic states,takes the form:

$$\Psi_E(r,R) = \psi_a(r,R)\chi_a(R)|k,e\rangle + \psi_c(r,R)\chi_{C.E}(R)|vac\rangle \tag{16}$$

where the R-dependent χ's are unknown nuclear wavefunctions. In (Eq.16) r and R stand for electronic and nuclear coordinates respectively, k and e being the wave number and the polarization of the incident photon. It can be shown that complex energy solutions of the resulting close coupled system with regular behaviour at the origin ($R = 0$) and presenting only outgoing waves at $R \to \infty$ are quantized and correspond to the scattering amplitude poles. The imaginary parts of the corresponding energies have a first order interpretation given by (Eq.15) leading to a linear behaviour as a function of field intensity (fig.1).

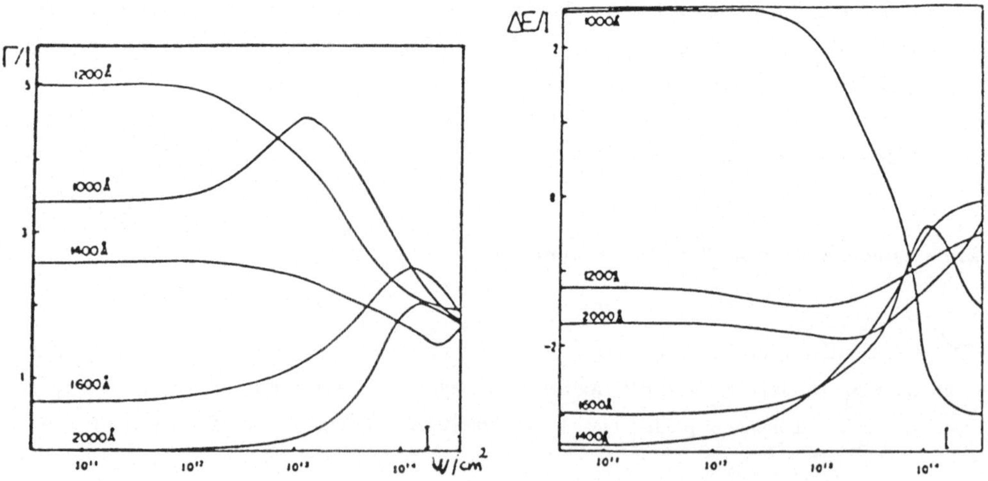

Fig. 1 Resonance widths and shifts as a function of the field strength for different wavelengths.

When the electronic Rabi frequency $\omega_{ca}(\hbar\omega_{ca} = \langle c|V|a\rangle)$ is of the same order as a vibrational frequency ω_v, we are in an intermediate field regime. In the case of H_2^+ photodissociation we are studying as an example, this roughly requires an intensity of $10^{11} W/cm^2$. A perturbative treatment for σ_a performed by summing up successive contributions (VG_0VG_0V + higher order terms) clearly leads to divergence, whereas the coupled equations technique for the calculation of resonances remains a useful tool for understanding the non linear effects in the plot of Γ/I or of $\Delta E/I$ as a function of I (fig.1). It is also noteworthy to observe that the lowest adiabatic (avoided-crossing) electronic field channel obtained by diagonalising the radiative coupling is a fairly good appoximation producing the non linear effects (fig.2).

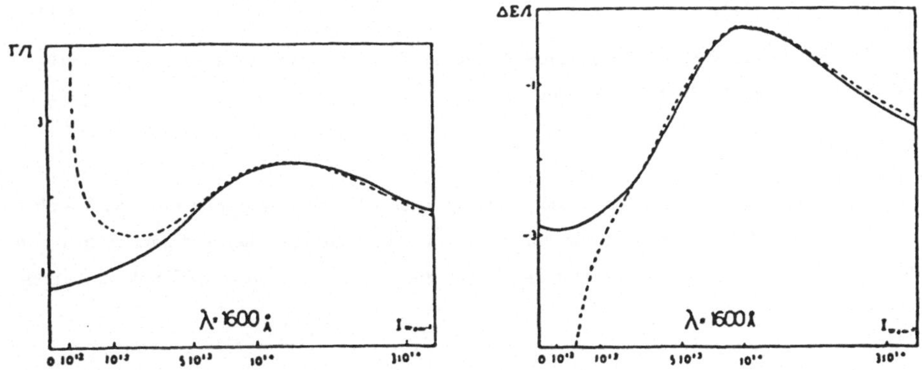

Fig. 2 Resonance widths and shifts as a function of the field strength for λ = 1600A. Solid line: single photon exact calculation. Dotts: adiabatic approximation.

For strong fields ($\omega_{ca} >> \omega_v$; a situation which occurs in our example for $I > 10^{13} W/cm^2$),the previous single photon description is no more valid .All relevent absorption and emission processes so far neglected on the basis of the rotated wave approximation, are to be taken into account. A set of dressed channels correponding to different photon numbers is defined which leads to the time independent Floquet hamiltonian[8].The coupled equations, after integration over field variables can be given the close form:

$$[H_a(R) + (n+1)\hbar kc - E]\,\chi_{a.n+1}(R) + V_{ac}(R)\,[\chi_{c.n}(R) + \chi_{c.n+2}(R)] = 0$$

$$[H_c(R) + n\hbar kc - E]\,\chi_{c.n}(R) + V_{ca}(R)\,[\chi_{a.n+1}(R) + \chi_{a.n-1}(R)] = 0 \qquad (17)$$

where each electron field channel function is labelled by a pair of indices (a or c for the electronic part together with the specification of the photon number n).As is pointed out by Chu[8],the quasi energy eigenvalues(resonances) of this problem have imaginary parts which are related to photodissociation rates.In particular for non overlapping resonances,the relation is direct and a generalized cross section may be defined by:

$$\sigma_a(\omega) = \frac{4\pi^2\omega}{3c}\frac{\Gamma_a}{I} \qquad (18)$$

Resonances are calculated, as in the single photon description, by the well documented complex coordinate technique incorporated into a Fox-Goodwin propagation scheme [9].The number of Floquet blocks to be retained (resonant or non resonant) depends on the field strength and the accuracy which is looked for. Specific examples concerning H_2^+ photodissociation are indicated in fig.3.

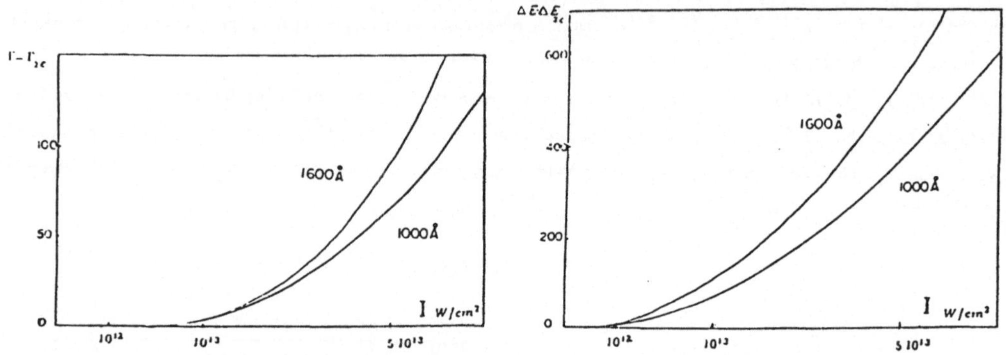

Fig. 3 Differences between 8 Floquet blocks converged results and 1 Floquet block calculations for the widths and the shifts as a function of the field strength.

Fig.4 displays a complete lineshape from $(1s\sigma_g, v = 0, j = 1)$ with increasing laser intensity.We observed,in agreement with Chu [8],that photodissociation spectrum is blue shifted when the field intensity increases and, which is more interesting its bell structure flattens due to important mixing of molecular levels.

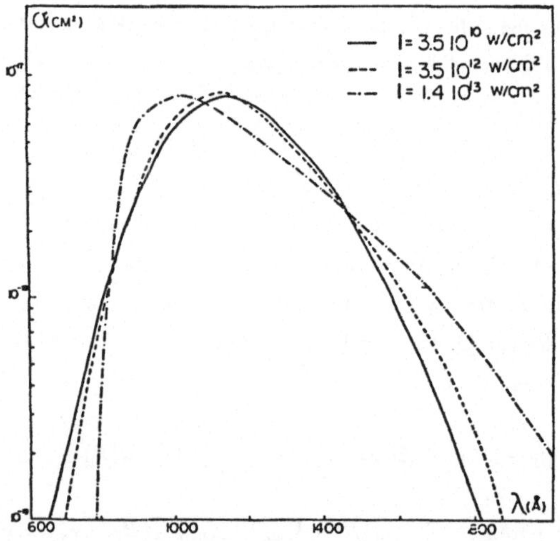

Fig. 4 Complete photodissociation lineshape for field strengths covering the range between weak to strong regimes.

In summary we have emphasized in this paper the role played by field induced resonances in the interpretation of non linear effect. Their direct calculation through close coupled equations provides an alternative to time dependent formalisms for the evaluation of transition amplitudes. This work will be pursued in the context of multiphoton transitions in larger molcules.

Acknowledgments

We are indebted to Dr. Annick Giusti-Suzor for many valuable discussions.

References

1 – a) J. M. Yuan and T. F. George,*J. Chem. Phys.*,**68**,3040 (1978);

b) A. D. Bandrauk and M. L. Sink, *J. Chem. Phys.*,**74**,1110 (1981).

2 – R. Loudon,The Quantum Theory of Light, Oxford press,Oxford (1973)

3 – a) A. D. Bandrauk and G. Turcotte,*J. Phys. Chem.*,**89**,3039 (1985);

b) A. D. Bandrauk and O. Atabek, *Adv. Chem. Phys.*(in press);

c) O. Atabek ,A. D. Bandrauk and S. Miret-Artes, (in preparation).

4 – M. Shapiro,*J. Chem. Phys.*,**56**,2582 (1972).

5 – a) M. Shapiro and H. Bony,*J. Chem. Phys.*,**83**,1588 (1985);

5 – b) A. D. Bandrauk ,G. Turcotte and R. Lefebvre,*J. Chem. Phys.*,**76**,225 (1982).

6 – U. Fano,*Phys. Rev.* ,**124**,1866 (1961).

7 – A. F. J. Siegert,*Phys. Rev.* ,**56**,750 (1939).

8 – S. I. Chu,*J. Chem. Phys.*,**75**,2215 (1981).

9 – O. Atabek and R. Lefebvre, *Phys. Rev.* ,**A22**,1817 (1980).

INTENSE-FIELD RADIATIVE PROCESSES IN THE COULOMB CONTINUUM

Alfred Maquet and Valérie Véniard

Laboratoire de Chimie Physique†, Université Pierre et Marie Curie

11, Rue Pierre et Marie Curie, F.75231 Paris Cedex 05. France

It has been known since almost twenty years now that atomic or molecular species submitted to an intense laser field are ionized when the radiation intensity $I \geq 10^{-5}\, I_0$, where $I_0 = 3.5\,10^{16}$ W/cm^2 is the (time averaged) atomic unit of field intensity. It is only in the late 70's however, it has been realized that,at such intensities, the system can absorb in fact more photons than strictly necessary for being ionized.[1] This process has been termed, somewhat unproperly, as Above Threshold Ionization (ATI). Our current understanding of ATI can be schematized as follows : [2] the first stage of the process takes place when the atom experiences the intensity rise of the laser pulse.[3] The atom can then absorb the minimum number N_0 of photons required to be ionized. We note that a sensible approximate theoretical description of this stage can be deduced from a lowest order perturbative approach, which can be thought to be valid so long as the energy of the system remains appreciably negative.

In the second stage of the process, as the atom absorbs more photons from the field and, accordingly, its total energy becomes slightly larger than zero, a (semi-) classical description of the dynamics becomes plausible. It appears in fact that the system must acquire some additional energy, i.e. absorb more photons, in order to possess an energy larger than the so-called ponderomotive energy Δ :

$$\Delta = \left\langle \frac{e^2 A^2}{2\, m\, c^2} \right\rangle = \frac{I}{4\, \omega^2} \qquad (1)$$

† Unité associée au CNRS, UA 176

so that the photoelectron has enough kinetic energy to leave the interaction region and reach the detector. Here **A** is the vector potential of the field with frequency ω (atomic units will be used in the following).

This stage of the process, in the course of which the electron is submitted to both the laser field and the Coulomb potential from its parent ion, plays a determining role in the dynamics of ATI. As a matter of fact, the system can then exchange energy and angular momentum with the laser field, since each interaction with the field results in a simultaneous exchange of one photon (absorption or stimulated emission) and of one unit of angular momentum (as a result of the dipole selection rule $\Delta l=\pm1$). In the following, we discuss two different aspects of the dynamics of such radiative processes, occurring into the Coulomb continuum in the presence of a (strong) external laser field.

A. A random walk description of ATI

We have very recently modelled this stage of the process as a random walk of the system onto a two dimensional lattice whose nodes can be conveniently labelled with the angular momentum l and the net number N of photons absorbed. The model provides a fair account of the relative intensities of the ATI peaks as well as of the angular momentum distributions in each peak.[4] Within this framework, it is assumed that, at each step, four channels are in general open, corresponding to the transitions $\Delta N=\pm1$, $\Delta l=\pm1$. It then appears that the key quantities determining the evolution of the system are the relative probabilities attached to each open channel and it is natural to impose that these relative weights are proportional to the corresponding one-photon transition probabilities.[5] Given the relative weigths, the walk of each "photoelectron" on the (l,N) lattice is performed at random, i.e. the choice of the way at each step is made via a random number generator. As we will show next, this very simple model provides us with interesting views on the ATI process.

First, from the actual values of the dipole free-free transition probabilities, one observes that the transitions $l \rightarrow l+1$ ($l \rightarrow l-1$) are dominant for absorption (emission). This trend is general and is verified whatever the laser frequencies are, at least in the optical and infrared ranges we have considered and corresponding to most experimental situations.[6] Interestingly enough one observes also that the emission steps ($N \rightarrow N-1, l \rightarrow l-1$) are noticeably favored in the low energy region of the ATI spectrum, i.e. in the continuum range lying close to the ionization limit. As a consequence we can predict that a notable fraction of the photoelectron population will be, to some extent, "trapped" into the lowest energy accessible peaks, even at high intensity. However, at higher photoelectron energies, emission and absorption become almost equiprobable and the dominance of emission is less marked.

Another interesting result of our analysis is that the distributions of the angular momenta in the photoelectron peaks depends on the total number of steps N_T the system is allowed to take. This number N_T is the only adjustable parameter of the model and it permits to account for the effects of the laser intensity on the dynamics of the process.More precisely, we observe that,since when the intensity I increases, so do the (stimulated) transition probabilities, which indicates that it is natural to link the intensity variations to the value of N_T used in the calculation.Within this framework, we have shown that, at lower intensities/ small values of $N_{T,}$ the angular momentum distribution strongly depends on the one resulting from the first stage of the process, i.e. is very sensitive to atomic structure effects. On the contrary, as I (and N_T) increases, this dependance is less marked and the angular momentum distributions become more even, as actually observed in eperiments.[2]

One of the limitations of the above discussed model is that it implicitly implies that radiative transitions into the continuum are saturated[1,2,7] and that accordingly, multiphoton free-free transitions (FFT) can be considered as stepwise. This amounts to neglect the contributions of the principal values in the multiphoton amplitudes and to retain only the resonant, energy-conserving contributions arising from the presence of $i\pi\delta$ terms into the spectral expansion of the resolvent. This makes the predictions of this model (and similar ones) rather crude and renders clear the need for other more quantitative approaches. We want to stress however that this very simple model (all the computations have been carried out on a Macintosh microcomputer) provides a nice interpretation of several salient features of the observed ATI spectra.

B. Soft-photon vs. 1st Born approximation in multiphoton FFT

Among the strong field models recently proposed, many of them describe the last stage of ATI processes in a way much similar to the ones used to treat the potential scattering of dressed electron (Volkov-) waves in the presence of a strong external radiation field.[2] Within this framework, one implicitly uses either a soft-photon,[8] or a 1st Born approximation treatment of the radiative collision processes. It is thus interesting to discuss in which extent such approaches, including scattering of the photoelectron in the Coulomb potential of its parent ion, can be used to account for ATI spectra.

We have recently completed an exact perturbative calculation of various two-photon transition amplitudes in a Coulomb potential.[9] In order to check our results, obtained from a rather intricate analytic calculation, we have compared them to corresponding ones derived in the soft-photon and 1st Born limits. The main outcomes of this comparison can be summarized as follows :[10] it appears that in the typical energy ranges studied so far in ATI experiments, such approximations lead to significant errors, ranging from 100% and more if $\omega = 0.1$ Ry and E_e=0.1Ry (here E_e is the energy of the incoming electron) to about 10% at the same frequency if E_e=1.0 Ry . Though our results are obtained within a perturbative framework we believe these error ranges are characteristic of such approximations when used in the low electron energy domains, typical of ATI spectra, and would be of the same order of magnitude in the strong field regime. This shows that great care must be exercised when using such approximate treatments in trying to model ATI data.

References

[1] Recent reviews include : P. Agostini, F. Fabre and G. Petite in "Multiphoton Ionisation of Atoms", edited by S. L. Chin and P. Lambropoulos (Academic Press, New York, 1984) p. 133; M. Crance, Physics Reports **144**, 117 (1987).

[2] See, for instance, the special issue of J. Opt. Soc. Am. B **4**, pp 705 et seq. (1987).

[3] P. Lambropoulos, Phys. Rev. Lett. **55**, 2141 (1985).

[4] A. Maquet and V. Véniard, post-deadline poster presented at the IVth International Conference on Multiphoton Processes (Boulder, Colorado, July 1987).

[5] W. Gordon, Ann. Phys. (Leipzig) **2**, 1031 (1929).

[6] Recent experimental results on atomic Hydrogen were reported in : H. G. Muller, H. B. van Linden van den Heuwell and M. J. van der Wiel, Phys. Rev. A **34**, 236 (1986) and D. Feldmann, B. Wolff, M. Wemhöner, H. Rottke and K. H. Welge, in IVth International Conference on Multiphoton Processes (Boulder, Colorado, July 1987).

[7] Recent references include : M. Crance and M. Aymar, J. Phys. B **13**, L421 (1980); Z. Bialynicka-Birula, J. Phys B **17**, 3091 (1984); Z. Deng and J. H. Eberly, Phys. Rev. Lett. **53**, 1810 (1984); M. Edwards, L. Pan and L. Armstrong Jr., J. Phys. B **17**, L515 (1984); K. Rzazewski and R. Grobe, Phys. Rev. A **33**, 1855 (1986).

[8] N. M. Kroll and K. M. Watson, Phys. Rev. A **8**, 804 (1973).

[9] M. Gavrila, A. Maquet and V. Véniard, Phys. Rev. A **32**, 2537 (1985); see also: Phys. Rev. A **35**, 448 (1987).

[10] V. Véniard, A. Maquet and M. Gavrila, to be published.

ADIABATIC SEPARABILITY OF DRESSED

MOLECULAR SYSTEMS

T. Tung Nguyen-Dang

Departement de Chimie
Universite de Sherbrooke
Sherbrooke, Quebec, J1K 2R1, Canada

INTRODUCTION

In the Bloch-Nordsieck representation of dressed molecules[1-4], which is the quantal counterpart of the space-translation semiclassical method[5], the interaction between the radiation field and the molecular system is transformed into field-dependent displacements in the molecular potential energy. An adiabatic separation between the field and the molecule had then been proposed, as an appropriate representation for the description of intense field molecular dynamics[1,2]. Radiative couplings are then identified as new, non-adiabatic interactions, consistently with the view that the dressing of the molecule amounts to the addition of new degrees of freedom, the field modes. Although the adiabatic states associated with the Bloch-Nordsieck hamiltonian already incorporate a part of the radiative interactions, the resulting non-adiabatic couplings are still of first order in the original coupling constant. In the present work, we apply a recently developped[6÷8] method for the analytic resummation of non-adiabatic couplings to obtain, from the Bloch-Nordsieck representation, a second order adiabatic representation for dressed molecules, in which adiabatic basis states are coupled only by second order residual interactions.

BLOCH-NORDSIECK REPRESENTATION

Consider a molecular system in a single mode radiation field with a Coulomb-gauge vector potential

$$\vec{A}(\vec{r},t) = 2c\left(\frac{\pi}{V}\right)^{\frac{1}{2}} \vec{\varepsilon}\{\ \sin(\vec{k}.\vec{r})Q(t) + \omega^{-1}\cos(\vec{k}.\vec{r})P(t)\} \qquad (1)$$

where c = velocity of light, V = field quantization volume, $\omega = c|\vec{k}|$, $\vec{k}$ = wavevector and $\vec{\varepsilon}$ = polarization of the field. The variables Q(t) and P(t) become operators $\hat{Q}$ and $\hat{P} = (\hbar/i)\partial_Q$, resp., upon quantization, and enter the definition of the field energy through

$$\hat{H}_R = \tfrac{1}{2}(\ \hat{P}^2 + \omega^2 Q^2\ ). \qquad (2)$$

In the electric dipole approximation, (E.D.A.), the dressed molecule hamiltonian is given by

$$\hat{H} = \sum_i \frac{\hat{P}_i^2}{2m} + \sum_\alpha \frac{\hat{P}_\alpha^2}{2M_\alpha} + \tfrac{1}{2}\left(\frac{\hat{P}^2}{m_r} + \omega^2 Q^2\right) + V_{coul}(\vec{r}_i,\vec{R}_\alpha) - \theta\{\sum_i \vec{P}_i - \sum_\alpha (m/M_\alpha) Z_\alpha \vec{P}_\alpha\}.\vec{\varepsilon}P \qquad (3)$$

where $i = 1,2,\ldots,N_e$ are electronic labels, $\alpha = 1,2,\ldots,N_n$, are nuclear ones,

$$\theta = (2e/m\omega)(\pi/V)^{\frac{1}{2}}, \qquad (4)$$

and m_r denotes a new photon mass which incorporates the effect of the usually neglected A^2 terms,

$$m_r = \{ 1 + \theta^2 m(N_e + \sum_\alpha Z_\alpha^2 (m/M_\alpha)) \}^{-1} . \qquad (5)$$

It has previously been shown[1,2] that, under the action of

$$\hat{U}_{BN} = \exp\{ (i/\hbar)\theta Q(\sum_i \vec{p}_i - \sum_\alpha (m/M_\alpha) Z_\alpha \vec{P}_\alpha).\vec{\varepsilon} \}, \qquad (6)$$

the dressed hamiltonian of eq. (3) can be transformed into

$$\begin{aligned}\hat{H}_{BN} &= \hat{U}_{BN} \hat{H} \hat{U}_{BN}^{\dagger} \\ &= \sum_i \frac{\hat{p}_i^2}{2m} + \sum_\alpha \frac{\hat{P}_\alpha^2}{2M_\alpha} + \tfrac{1}{2}(\frac{\hat{P}^2}{m_r} + \omega^2 Q^2) \pm V_{coul}(\vec{r}_i + \theta Q\vec{\varepsilon}, \vec{R}_\alpha - \frac{m}{M_\alpha} Z_\alpha \theta Q\vec{\varepsilon}) \\ &= \hat{H}_{mol}(Q) + \hat{H}_R . \end{aligned} \qquad (7)$$

In this new, Bloch-Nordsieck hamiltonian, the appearance of Q as displacements in the molecular potential has previously motivated the use of an adiabatic, Born-Oppenheimer like representation for the dressed molecule eigenstates[1,2]. Let us recall the results: If we write

$$\Psi(\vec{r}_i, \vec{R}_\alpha, Q) = \sum_I \phi_I(\vec{r}_i, \vec{R}_\alpha; Q)\, \chi_I(Q) , \qquad (8)$$

where the ϕ_I's are eigenfunctions of $\hat{H}_{mol}(Q)$, then the field states χ_I satisfy a set of coupled differential equations with non-adiabatic (radiative) couplings governed by the matrix elements

$$< \phi_L;Q \mid \hat{P} \mid \phi_I;Q > = \theta\vec{\varepsilon}.< \phi_L;Q \mid \sum_i \vec{p}_i - \sum_\alpha \frac{m}{M_\alpha} Z_\alpha \vec{P}_\alpha \mid \phi_I;Q >. \qquad (9)$$

Although the states $\mid \phi_I;Q >$ include a part of the radiative interactions, these coupling matrix elements are still of first order in the coupling constant θ. We now apply the recently developed " phase corrected adiabatic approximation " method[6-8] to the Bloch-Nordsieck hamiltonian. In this method, one looks for a further transformation of the form

$$\hat{U} = \exp\{ -(i/2\hbar) f(\vec{r}_i, \vec{R}_\alpha, Q)\hat{P} + \text{h.c.} \}, \qquad (10)$$

such that the ensuing f-dependent displacements in the electronic and nuclear momenta- displacements which are linear in P- would give rise to new momenta coupling terms, which will cancel the first order non-adiabatic interactions exactly, leaving only residual second order coupling terms. Note that, with $\vec{A} \propto P.\vec{\varepsilon}$, as resulting from eq. (1) and the E.D.A., eq. (10) defines a proper gauge transformation. It is however to be applied to H_{BN} and not to the original hamiltonian H of eq. (3).

SECOND ORDER ADIABATIC REPRESENTATION

We now show that an appropriate form for f is the following Q-independent quadratic expression

$$f(\vec{r}_i,\vec{R}_\alpha,Q) = \sum_i (\vec{a}_i.\vec{r}_i + \gamma r_i^2) + \sum_\alpha (\vec{a}_\alpha.\vec{R}_\alpha + \gamma_\alpha R_\alpha^2) + \delta \qquad (11)$$

where $\vec{a}_{i(\alpha)}$, $\gamma(\alpha)$ and δ are parameters to be found such as to ensure cancellation of first order non-adiabatic couplings. Under the action of U defined by eqs. (10) and (11), the Bloch-Nordsieck hamiltonian is exactly transformed into

$$\hat{U}\hat{H}_{BN}\hat{U}^\dagger = \sum_i \frac{\hat{p}_i^2}{2m} + \sum_\alpha \frac{\hat{P}_\alpha^2}{2M_\alpha} + G(\vec{r}_i,\vec{R}_\alpha)\frac{\hat{P}^2}{2m_r} + V_{tot}(\vec{r}_i,\vec{R}_\alpha,Q) + F(\vec{r}_i,\vec{R}_\alpha,\vec{p}_i,\vec{P}_\alpha)\hat{P} \qquad (12)$$

where, for notational convenience, we have defined

$$V_{tot}(\vec{r}_i,\vec{R}_\alpha,Q) = \frac{\omega^2}{2}(Q-f)^2 + V_{coul}(\vec{r}_i+\theta\vec{\varepsilon}\{Q-f\},\vec{R}_\alpha - \frac{m}{M_\alpha}Z_\alpha\theta\vec{\varepsilon}\{Q-f\}) \qquad (13a)$$

$$G(\vec{r}_i,\vec{R}_\alpha) = 1 + \sum_i \frac{m_r}{m}(\vec{a}_i + 2\gamma\vec{r}_i)^2 + \sum_\alpha \frac{m_r}{M_\alpha}(\vec{a}_\alpha + 2\gamma_\alpha\vec{R}_\alpha)^2 \qquad (13b)$$

$$\hat{F}(\vec{r}_i,\vec{R}_\alpha,\vec{p}_i,\vec{P}_\alpha) = \tfrac{1}{2}\{ \frac{1}{m}\sum_i (\vec{a}_i + 2\gamma\vec{r}_i).\vec{p}_i + \sum_\alpha \frac{1}{M_\alpha}(\vec{a}_\alpha + 2\gamma_\alpha\vec{R}_\alpha).\vec{P}_\alpha + h.c.\} \qquad (13c)$$

the new momenta coupling term F.P and the photon mass factor G originating from the anticipated $\nabla f.\varepsilon P$ shifts in the electronic and nuclear momenta.

Define now the new Q-parametrized molecular hamiltonian by

$$\hat{h}_{mol}(Q) = \hat{U}\hat{H}_{BN}\hat{U}^\dagger - \{ \hat{F}.\hat{P} + G.\frac{\hat{P}^2}{2m_r} \}, \qquad (14)$$

and write the eigenstates of $\hat{U}\hat{H}_{BN}\hat{U}^\dagger$ in the form of a Born-Huang series as in eq. (8), but with the states $| \phi_I;Q >$ defined as eigenstates of $\hat{h}_{mol}(Q)$, with eigenvalue $\varepsilon_I(Q)$. The following coupled equations are then obtained:

$$\{ <G>_{II}\frac{\hat{P}^2}{2m_r} + <(\hat{F} + G\frac{\hat{P}}{m_r})>_{II}\hat{P} + <G\frac{\hat{P}^2}{2m_r}>_{II} + \varepsilon_I(Q) \}\chi_I(Q)$$
$$+ \sum_{I'\neq I}\{ <G>_{II'}\frac{\hat{P}^2}{2m_r} + <G\frac{\hat{P}^2}{2m_r}>_{II'} + <(\hat{F} + G\frac{\hat{P}}{m_r})>_{II'}\hat{P} \}\chi_{I'}(Q) = E\chi_I(Q). \qquad (15)$$

So far, no approximation has been made, and an exact adiabatic separation would result if one could fix the parameters of f, such as to cancel exactly all coupling terms (with $I' \neq I$), of eq. (15). We are aiming however at obtaining residual couplings of second order. We thus note that the first coupling term on the l.h.s. of eq. (15) would certainly be of order θ^2, if $\vec{a}_{i(\alpha)}$, $\gamma(\alpha)$ are all of order θ^1. Let us make this assumption, and examine the condition for the cancellation of the last coupling term, up to second order in θ, i.e. the condition for obtaining

$$< \phi_I;Q \mid \hat{F} + \frac{\hat{P}}{m_r} \mid \phi_{I'};Q > = O(\theta^2), \qquad (I \neq I') \qquad (16)$$

since $G = 1 \pm O(\theta^2)$, with the above assumption. We may also write this condition in the form

$$< \phi_I;Q \mid [\hat{F},\hat{h}_{mol}] + (\hbar/i)\partial_Q\hat{h}_{mol} \mid \phi_{I'};Q > = O(\theta^2) \qquad (17)$$

using the definition of $| \phi_I;Q >$ as eigenfunctions of $\hat{h}_{mol}(Q)$. Evaluating the commutator and the derivative in eq. (17), we find

$$< \phi_I \,|-(\frac{\gamma}{m^2}\sum_i \hat{p}_i^2 + \sum_\alpha \frac{\gamma_\alpha}{M_\alpha^2} \hat{P}_\alpha^2) + \{ \hat{F}(\vec{r}_i,\vec{R}_\alpha,\vec{\nabla}_i,\vec{\nabla}_\alpha) + \partial_Q \}V_{tot} \,|\, \phi_{I'} > = O(\theta^2). \qquad (18)$$

If we now choose $\gamma_\alpha =(M_\alpha/m)\gamma$, for all α, then the first term on the l.h.s of eq. (18) can be written as

$$- \frac{\gamma}{m}(\frac{1}{m}\sum_i \hat{p}_i^2 + \sum_\alpha \frac{1}{M_\alpha} \hat{P}_\alpha^2) = -2\frac{\gamma}{m} \{ \hat{h}_{mol}(Q) - V_{tot} \}, \qquad (19)$$

so that, using the orthogonality of the eigenstates $|\phi_I;Q >$ of $\hat{h}_{mol}(Q)$, eq. (18) reduces to

$$< \phi_I \,|\, \{ 2\frac{\gamma}{m} V_{tot} + (\hat{F}(\vec{r}_i,\vec{R}_\alpha,\vec{\nabla}_i,\vec{\nabla}_\alpha) + \partial_Q)V_{tot} \} \,|\, \phi_{I'} > = O(\theta^2). \qquad (20)$$

Equation (20) states that the unknown values of the parameters $\vec{a}_{i(\alpha)}$, γ and δ of f are governed by the properties of the potential V_{tot}. Thus, recalling the definition of V_{tot}, eq. (13a), exploiting the homogeneity of the coulomb potential, which we write as

$$V_{coul}(\vec{r}_i+\theta\vec{\varepsilon}\{Q - f\},\vec{R}_\alpha- \frac{m}{M_\alpha} Z_\alpha\theta\vec{\varepsilon}\{Q - f\}) =$$

$$-\{ \sum_i(\vec{r}_i - \theta\delta\vec{\varepsilon}).\vec{\nabla}_i + \sum_\alpha(\vec{R}_\alpha + \frac{m}{M_\alpha} Z_\alpha\theta\delta\vec{\varepsilon}).\vec{\nabla}_\alpha \}V_{coul} + O(\theta), \qquad (21)$$

and using the explicit expression for F, eq. (13c), we find that the curly brackets on the l.h.s. of eq. (20) can be expressed as

$$\omega^2\{ \sum_i(\vec{a}_i.\vec{r}_i + \gamma r_i^2) + \sum_\alpha(\vec{a}_\alpha.\vec{R}_\alpha + \gamma \frac{M_\alpha}{m} R_\alpha^2) \}(6\frac{\gamma\delta}{m} - 1) + \{ Q \}$$

$$+ \{ \frac{1}{m}\sum_i(\vec{a}_i + 2(\gamma\delta + m)\theta\vec{\varepsilon}).\vec{\nabla}_i + \sum_\alpha(\frac{\vec{a}_\alpha}{M_\alpha} - \frac{Z_\alpha}{M_\alpha}(2\gamma\delta + m)\theta\vec{\varepsilon}).\vec{\nabla}_\alpha \}V_{coul} + O(\theta^2)$$

$$(22)$$

where $\{ Q \}$ denotes a function of Q only, which obviously contributes zero to the coupling matrix element of eq. (20). This function, as well as the first two terms of eq. (22) originate from the quadratic component $(Q - f)^2$ of V_{tot}. The contribution of this component to eq. (20) would thus vanish, to order θ^2, if we set

$$\delta = (m/6\gamma). \qquad (23)$$

Substituting eq. (23) into eq. (22), we finally find that eq. (20) would be satisfied, implying the fulfilment of eqs. (17) and (16), if we choose

$$\vec{a}_i = -(4/3)m\theta\vec{\varepsilon}, \qquad i = 1,2,\ldots,N_e \qquad (24a)$$

$$\vec{a}_\alpha = (4/3)Z_\alpha m\theta\vec{\varepsilon}, \qquad \alpha = 1,2,\ldots,N_n. \qquad (24b)$$

Notice that γ is left undetermined by the above analysis. Its value can in fact be chosen arbitrary, provided that it is of order θ, so that our previous assumptions are well satisfied. Incidentally, we have also ensured that the remaining coupling matrix elements of eq. (15) are also of order θ^2, as can be seen by inserting the completeness relation for the states $|\phi_I >$ into $< P^2 >_{II'}$, and by using eq. (16) to express matrix elements of P.

With the above remark, we have shown that the unitary transformation $\hat{U}$ defined by eq. (10) and eq. (11), yields a new dressed hamiltonian, which admits an adiabatic separation of second order between the molecule and the field, in the sense that the associated adiabatic basis states $\phi_I\chi_I$ are coupled to each other by residual coupling terms that are strictly of order θ^2. These states define a second order adiabatic representation for the dressed system.

DISCUSSIONS

In the second order representation found above, the molecular adiabatic states ϕ_I are governed by the hamiltonian $h_{mol}(Q)$, defined by the total transformed potential V_{tot}, eq. (13a). Since $Q\varepsilon \propto E$, the electric field, and f involves the molecular dipole operator, h_{mol} thus includes the usual dipolar interaction $\mu.E$, as a part of the dressed molecular potential, and not as couplings. The molecular potential also contains the term $(\omega^2/2)Q^2$, which carries over , via $\varepsilon_I(Q)$, to the decoupled equation (15) for the field states χ_I. To zeroth order in θ, this equation is seen to reduce to the usual harmonic oscillator Schroedinger equation, so that the characterization of the field states by photon number n, or by intensity $I \propto n$, remains valid. Corrections due to the diagonal matrix elements in the decoupled equation (15) enter only at order θ^2, while all residual $O(\theta^2)$ couplings, (non-diagonal), contribute corrections to the field energy only at order θ^4. We thus expect that the presently derived second order adiabatic representation will be particularly useful for the description of processes occuring at high field-intensity, as perturbation series expressed in this basis will converge more quickly, due to the reduction of the couplings, as seen above.

Acknowledgements: The author wishes to thank the Natural Sciences and Engineering Research Council of Canada, (NSERC), for grants supporting this research.

REFERENCES

1. T. T. Nguyen-Dang and A. D. Bandrauk, J. Chem. Phys. 79, 3256 (1983).
2. T. T. Nguyen-Dang and A. D. Bandrauk, J. Chem. Phys. 80, 4926 (1984).
3. A. D. Bandrauk and T. T. Nguyen-Dang, J. Chem. Phys. 83, 2840 (1985).
4. A. D. Bandrauk, O. F. Kalman and T. T. Nguyen-Dang, J. Chem. Phys. 84, 6761 (1986).
5. G. Ferrante and C. Leone, Phys. Rev. A26, 3101 (1982).
6. Y. Maréchal, J. Chem. Phys. 83, 247 (1985).
7. T. T. Nguyen-Dang and A. D. Bandrauk, J. Chem. Phys. 85, 7224 (1986).
8. T. T. Nguyen-Dang, J. Chem. Phys. 86, to appear (1987).

THE INTERACTION OF INTENSE PICOSECOND INFRARED PULSES WITH ISOLATED MOLECULES

Eric Mazur

Division of Applied Sciences and Department of Physics
Harvard University
Cambridge, MA 02138, USA

Introduction

In the past decade there has been much interest in the dynamics of highly vibrationally excited and dissociating molecules. Selectivity at high levels of excitation may eventually lead to the realization of laser-controlled photochemistry, with broad applications in such diverse areas as laser-assisted chemical vapor deposition, isotope separation, and photosynthesis. Polyatomic molecules in the ground electronic state can reach levels of excitation up to the dissociation threshold by absorbing a large number of photons from a resonant high-power infrared laser. Despite the selectivity of infrared excitation at low energy, however, at high excitation the excitation energy is no longer confined to one 'mode'. It has been shown experimentally that for molecules excited close to or above the dissociation threshold equilibration of energy occurs, in agreement with theoretical predictions. There is no agreement, however, as to the validity of theoretical models that presuppose equipartitioning of energy in the region *below* the dissociation threshold. Recent spontaneous Raman spectroscopy experiments on infrared multiphoton excited molecules in our laboratory provide information on the intramolecular vibrational energy distributions of highly vibrationally excited molecules in this region. The experimental results show that an excess of energy can remain in the pumped mode up to levels of excitation close to the dissociation threshold. This paper provides a review of the results that were obtained in the past three years, part of which were published previously.

Background

In 1973 it was discovered that isolated molecules in the ground electronic state can be dissociated by a short, intense pulse from a CO_2 laser.[1] Since then the absorption of large numbers of monochromatic infrared photons by isolated molecules has been studied extensively.[2–8] The early work in this field was motivated by the hope of driving chemical reactions in either a bond-specific or isotopically selective fashion by 'localized' deposition of energy in a small subset of modes.

In the past ten years many experimental techniques have been applied to study infrared multiphoton excitation. Photoacoustic measurements were applied to determine the energy absorbed by the molecules,[9] and to study the excitation as a function of various parameters, such as pumping fluence, intensity, and wavelength, pressure, etc. Photoacoustic spectroscopy was also used at high intensities to study dissociation yields as a function of absorbed energy. More detailed information on infrared multiphoton dissociation, such as the species of the dissociation fragments, branching ratios of different dissociation channels, and the translational energy distribution of the fragments, was obtained by mass and time-of-flight spectrometry.[10] Pump-and-probe experiments have also provided more detailed knowledge of the infrared multiphoton excitation and dissociation. For example, laser induced fluorescence[11,12] was used to measure the vibrational energy distribution of the infrared multiphoton dissociation fragments. Infrared double-resonance experiments[13–15] were done to determine the rotational relaxation rate and the population depletion of the vibrational ground state. Spontaneous and coherent anti-Stokes Raman scattering were used to probe the distribution of vibrational energy over the different modes of infrared multiphoton excited molecules.[16–24]

The following qualitative picture has emerged from the experimental results. Basically, one can distinguish between three different regions in the molecular vibrational spectrum depending on the level of excitation. At low excitation the energy is essentially confined to the pumping mode, just as in ordinary one-photon spectroscopy: the first few photons absorbed by a 'cold' molecule produce transitions between separate discrete vibrational states located in the resonant mode (region I). At higher levels of excitation, the spacing between individual vibrational states becomes increasingly smaller due to molecular anharmonicities, and other nonresonant modes also acquire energy during the excitation (region II, often referred to as the 'quasicontinuum'). Molecular excitation in this region is thought to occur through stepwise incoherent transitions between homogeneously broadened states that are superpositions of various normal mode states. Once in region II, many polyatomic molecules easily absorb large numbers of infrared photons and reach the continuum above the dissociation threshold (region III). Clearly the excitation process is very different in each of these three regions, and experimental results often reflect a combination of the spectroscopies of different regions.

Most of the experiments carried out to date have centered around characterizing the gross features of infrared multiphoton excitation by relatively large molecules. The parameters that have been measured, such as average number of photons absorbed per molecule, dissociation rates and branching ratios, are the product of a number of mechanical and kinetic processes and, hence, are incapable of probing the detailed dynamics of the excitation process. It has been established, however, that the infrared multiphoton dissociation branching ratios and the energy distributions of the dissociation fragments are generally consistent with statistical theories, such as the RRKM theory. This means that when molecules are excited into region III, equilibration of the intramolecular vibrational energy distribution occurs, and dissociation takes place along a thermodynamically favored path, resulting in a loss of the initial 'selectivity'. Whether equilibration occurs for highly excited molecules *below* the dissociation threshold (region II) remains an open question.

Direct information on the intramolecular energy distribution in highly excited molecules was obtained experimentally with pump-probe type experiments, in particular by Raman probing. Raman spectroscopy was first employed by Bagratashvili and coworkers[16] and later by our group[19] as a tool for studying infrared multiple photon excitation. In the Raman experiments, the population in various vibrational modes is probed after excitation of the molecules into region II with an intense infrared pulse. Since the Raman signal intensities are a measure for the amount of energy in each mode,[19] this type of experiment provides direct experimental information on the intramolecular energy distribution.

Intramolecular energy distributions

The experimental technique and apparatus have been described previously.[20,21] Four different molecules, CF_2HCl, CF_2Cl_2, SF_6 and CH_3CHF_2, varying in size from five to eight atoms, were studied with the present apparatus.[20,22–24] An overview of experimental results is presented in Table I. All measurements were carried out at room temperature, with gas pressures ranging from 14 to 500 Pa and with infrared fluences up to 8 $\times 10^4$ J/m^2. The commercially obtained gases have a reported purity better than 99.99%.

The first molecule studied, SF_6, has only one accessible Raman active mode, ν_1, with a Raman shift of 775 cm^{-1}. Data were obtained for CO_2-laser frequencies between the P(12) and the P(28) lines of the 10.6 μm branch, which are resonant with the triply degenerate infrared active ν_3-mode (944 cm^{-1}). Two different infrared pulse durations were employed: 0.5 and 15 ns full-width at half-maximum pulses.

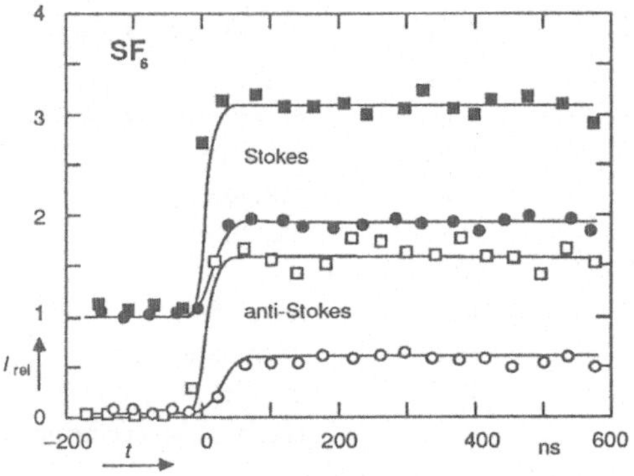

Fig. 1. Intensity of Stokes (closed symbols) and anti-Stokes (open symbols) signal as a function of the time delay between pump and probe pulses at a pressure of 67 Pa for SF_6. Infrared excitation with 0.5 ns (squares) and 15 ns (circles) pulses at the 10.6 μm P(20) line. Average fluence: 0.8×10^4 J/m^2. Data from Ref. 20.

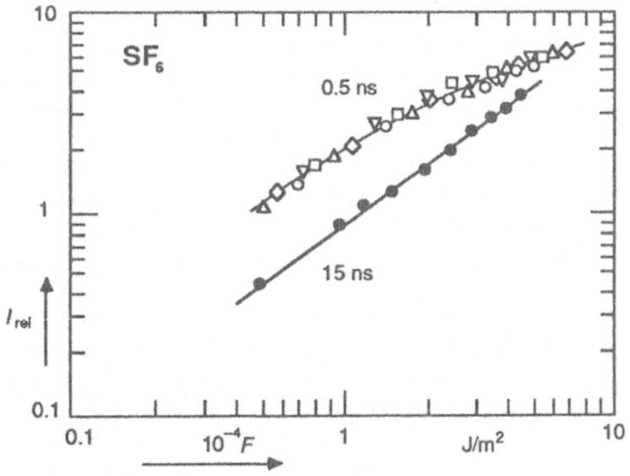

Fig. 2. Relative anti-Stokes signal of SF_6 as a function of the infrared pumping fluence for various pressures. Excitation at the 10.6 μm P(20) line with two pulse durations: 0.5 ns (open symbols) and 15 ns (closed symbols). Data from Ref. 20.

□ : 33 Pa; ○: 67 Pa; △: 133 Pa; ▽: 200 Pa; ◇: 267 Pa; ● : 133 Pa

Fig. 1 shows the increase in Stokes and anti-Stokes signals, measured at 356.7 and 338 nm respectively, as a function of the time delay between the pump and the probe pulse for two infrared pulse durations. The signals are normalized with the room temperature Stokes signal (for $t < 0$, room temperature equilibrium data are automatically obtained). At $t = 0$ infrared excitation takes place and both Stokes and anti-Stokes signals increase. The rise time of the signals is determined by the 20 ns pulse duration of the second harmonic of the probe laser. However, although not resolved in these measurements, the increase in signal clearly occurs on a time-scale that is much shorter than the mean free time between collisions (about 200 ns at a pressure of 67 Pa). The pressure dependence of the signals further shows that the increase in signal is not due to collisions, but is truly a *collisionless* phenomenon.[20] Interestingly enough the signals remain constant, even on a time scale on which collisional vibrational energy relaxation occurs.[20] For longer delay times ($t > 2$ μs), diffusion of the excited molecules out of the probing region causes the signals to revert to their original values.[20]

Table I. Spectroscopic data for the molecules studied in this paper. The vibrational data for SF_6 and CF_2Cl_2 are from literature. All data are in cm^{-1}, vs = very strong, s = strong, m = medium, and w = weak.

Molecule	CO_2 line	Wavenumber	Mode	Activity	Remarks[20,22–24]
SF_6[25]			$\nu_1 = 775$	R (s)	changes after excitation
			$\nu_2 = 644$	R (w)	not probed
	10.6 μm P(20)	944	$\nu_3 = 965$	IR	pumped
			$\nu_4 = 617$	IR	
			$\nu_5 = 524$	R (w)	not probed
			$\nu_6 = 363$	inactive	
CF_2Cl_2[26,27]			$\nu_1 = 1101$	IR(s)	
			$\nu_1 = 1098$	R (m)	changes after excitation
			$\nu_2 = 667$	IR (s)	
			$\nu_2 = 667.2$	R (s)	changes after excitation
			$\nu_3 = 457.5$	R (s)	not probed
			$\nu_4 = 261.5$	R (s)	not probed
			$\nu_5 = 322$	R (w)	not probed
			$\nu_6 = 1159$	IR (s)	
			$\nu_6 = 1167$	R (w)	not probed
			$\nu_7 = 446$	IR (w)	not probed
	10.6 μm P(32)	933	$\nu_8 = 922$	IR (vs)	pumped
			$\nu_8 = 923$	R (w)	changes after excitation
			$\nu_9 = 437$	IR (w)	
			$\nu_9 = 433$	R (m)	not probed
CH_3CHF_2			870	R	changes after excitation
	10.6 μm P(20)		944	IR	pumped
			1140	R	no change
			1460	R	no change
			2980	R	no change
CF_2HCl			590	R	no change
			800	R	no change
	9.4 μm R(32)		1086	IR	pumped
			1130	R	no change
			1330	R	no change
			3030	R	no change

The dependence of the anti-Stokes signal intensity on the infrared laser fluence (energy per unit area) is shown in Fig. 2 for different pressures and pulse durations. The data obtained for the two pulse durations show that at low fluence the signals depend on the exciting laser pulse intensity: a larger increase in Raman signal occurs at the shorter, higher intensity, pulses. At low excitation one needs a high intensity for *coherent* multiphoton excitation through the lower part of the vibrational ladder. At the higher fluences, once the

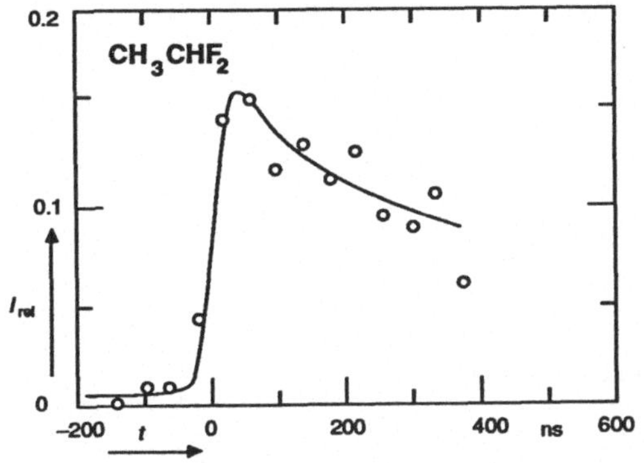

Fig. 3. Intensity of the anti-Stokes signals versus time delay between pump and probe pulse for CH_3CHF_2 at 660 Pa. Infrared excitation: 10.6 μm P(20) line, 0.5 ns pulse with average fluence 1.5×10^4 J/m^2. Data from Ref. 22.

molecules are highly excited, the curves for the 0.5 and 15 ns pulse durations approach each other, and the dependence of the signal intensity on laser pulse intensity vanishes in agreement with the behavior observed in photoacoustic measurements.[9]

The observed collisionless changes in Raman signals provide clear and direct evidence that some of the nonresonant modes do indeed participate in the excitation process. The main purpose of this research is to obtain information on the role of nonresonant modes in the multiphoton excitation of polyatomic molecules. Since the intensity of the signals is proportional to the average energy in the mode, E_R, one can determine E_R from the ratio of the anti-Stokes intensity to the thermal room temperature value of the Stokes signal, I_{rel}. Unfortunately SF_6 has only one accessible Raman active mode, so that it is not possible to compare the values of E_R for different Raman active modes. This limits us therefore to a comparison of energy in the ν_1 mode with the average total energy absorbed per molecule, $\langle E \rangle$, known from photoacoustic measurements. If one assumes an equilibrium distribution of the excitation energy $\langle E \rangle$, the amount of energy in the ν_1 mode agrees remarkably well with the value for E_R that one obtains from the Raman measurements,[19,20] suggesting that for SF_6 the intramolecular energy distribution indeed equilibrates. The absence of a decay of the Raman signals in Fig. 1 further supports this suggestion. Even though the initially nonequilibrium *intermolecular* distribution of energy equilibrates,[18] E_R remains constant once intramolecular equilibrium is achieved. In the absence of intramolecular equilibrium, one would expect E_R, and consequently the signal intensities, to change on a much shorter time scale because of a rearrangement of energy over the various vibrational modes.

The asymmetric CF_2HCl molecule has five accessible Raman active modes of widely different energy (600–3000 cm^{-1}). The peak absorption of this molecule coincides with the 9.4 μm R(32) CO_2 laser line at 1086 cm^{-1}. Even at the maximum fluence at this line (2×10^4 J/m^2), *none* of the five Raman lines show a detectable change in intensity.[22] Photoacoustic studies[23] of the infrared multiphoton excitation of this molecule have shown that at such a fluence the molecules absorb about ten infrared photons (10,000 cm^{-1}). The absence of anti-Stokes scattering from low lying levels, such as the Raman active mode at 587 cm^{-1}, suggests that not all modes participate in the excitation process, and that the energy distribution for this molecule does not equilibrate without collisions.

The asymmetric isomer CH_3CHF_2 has four accessible Raman modes. Data were obtained for 0.5 ns long pulses at the P(20) line of the 10.6 μm branch, which is resonant with the infrared active C—F stretch mode at 942 cm^{-1}. Only one of the Raman active modes, at 870 cm^{-1}, shows an increase in signal after excitation.[22] Fig. 3 shows the time-

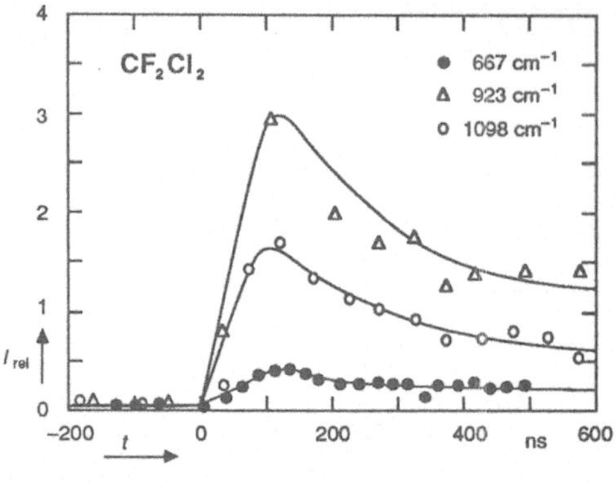

Fig. 4. Intensity of the anti-Stokes signals as a function of the time delay between pump and probe pulse for CF_2Cl_2 at 400 Pa. Infrared excitation: 10.6 μm P(32) line, 15 ns pulse with average fluence 1.8×10^4 J/m^2. Data from Ref. 23.

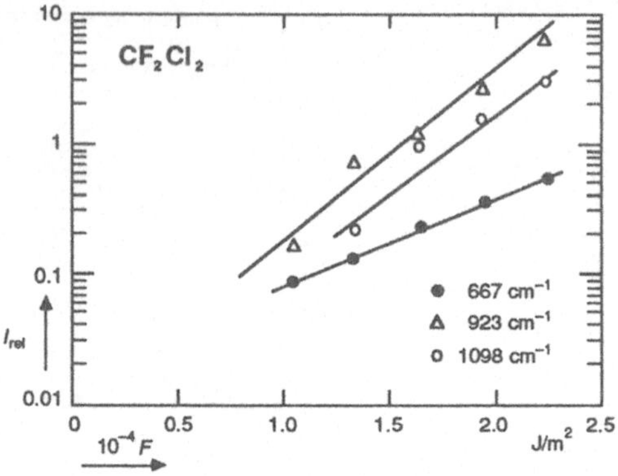

Fig. 5. Semilogarithmic plot of the fluence dependence of the anti-Stokes signals of CF_2Cl_2 at 400 Pa. The dependence is exponential for all three modes. Excitation with 15 ns pulses at the 10.6 μm P(32) line. Data from Ref. 23.

dependence of this signal at a pressure of 660 Pa and a fluence of 1.5×10^4 J/m^2. Again a short collisionless increase in signal occurs, but in contrast to SF_6 the signal now decays on a time scale of the order of collisional relaxation times. This, combined with the fact that no other Raman active mode exhibits any change, leads to the conclusion that for this particular molecule too, the excitation energy does not equilibrate.

The most complete set of data was obtained for CF_2Cl_2.[23,24] This five atom molecule has four accessible Raman active modes, three of which (at 667, 923, and 1098 cm^{-1} respectively) were measured after infrared multiphoton excitation. The C—Cl stretch mode at 923 cm^{-1} is both infrared and Raman active and can be pumped with the P(32) line of the 10.6 μm branch of the CO_2 laser. This allows one to directly observe the energy in the pump mode and compare it with the energy in other modes. The measurements presented here were all carried out at a gas pressure of 400 Pa.

Fig. 4 shows the time-dependence of the anti-Stokes signals, each normalized with its corresponding room temperature Stokes signals. The signals rise in 20 ns and show a clear decay, especially for the two highly excited modes (923 cm^{-1} and 1098 cm^{-1}). This decay is most likely attributed to collisional transfer of energy to other, initially 'cold' vibrational modes. The fluence dependence of the anti-Stokes signals of CF_2Cl_2 is nearly exponential (Fig. 5): above 1×10^4 J/m^2, the signals double roughly every 0.3×10^4 J/m^2 increment. Since CF_2Cl_2 is smaller than SF_6, fewer vibrational modes are available and one expects a stronger bottleneck effect in CF_2Cl_2. The observed slow rise of the signals at low fluence, which is in sharp contrast with the linear fluence dependence of SF_6 (*cf.* Figs. 2 and 5), indeed suggests that this is the case. Measurements of the Raman signal with shorter infrared pulses would provide a better understanding of the role of intensity effects.

Since more than one Raman active mode was measured for this molecule, one can directly compare the intensities from the various modes. In equilibrium, the intensities of the normalized signals are given by a Maxwell-Boltzmann distribution. The results in Figs. 4, and 5, however, show that the signal intensities after infrared multiphoton excitation cannot be described by such a distribution. Especially the normalized intensity of the pumped vibrational mode at 923 cm^{-1} is considerably higher than the corresponding intensities of the other two modes: at all fluences most of the energy remains in the pumped mode. In addition, as is clear from Fig. 5 the rate of increase is different for the three modes. It appears that there is a stronger coupling of the pump mode with the 1098 cm^{-1} mode than with the less energetic 667 cm^{-1} mode, notably at the high fluence end. Note also that although the intensities of the anti-Stokes signals increase significantly between 1.5 and 2.4×10^4 J/m^2, the intensity *ratio* does not change much. This rules out the possibility that the observed nonequilibrium distribution is a result of averaging a 'hot' equilibrium ensemble and a 'cold' bottlenecked ensemble, since the ratio would change as the fraction of molecules in the hot ensemble becomes larger with increasing fluence. Adding up the energy content of the three modes for CF_2Cl_2 calculated from the signal intensities in Fig. 5, it follows that a complete equilibration of energy does not occur below 10,000 cm^{-1} of excitation. Preliminary measurements show that after pumping the 1098 cm^{-1} mode an excess of energy is found in both the 923 and the 1089 cm^{-1} cm mode.

Conclusion

This paper presents an overview of the results of measurements on various collisionless infrared multiphoton excited molecules ranging in size from 5 to 8 atoms. The amount of energy in various modes of these molecules is determined from the spontaneous Raman scattering signals from each of these modes. Most of these molecules have more than one Raman active mode and thus allow *direct* observation of the intramolecular distribution of vibrational energy among these modes after the infrared multiphoton excitation. The experiments unambiguously show: (1) that collisionless intramolecular transfer of energy to Raman active modes takes place within the 20 ns time resolution, and (2) that for highly excited molecules below the dissociation threshold the final distribution of energy—after excitation, before collisional relaxation—*is not necessarily in equilibrium.* For CF_2Cl_2 in particular it was found that the pumped mode contains an excess of energy up to *at least* 10,000 cm^{-1} of excitation energy. This implies a certain degree of 'localization' of excitation energy in the pump mode up to fairly high levels of excitation. Although this is at variance with observations made in the Soviet Union, that claim complete equilibration at about 7,000 cm^{-1}, it agrees with recent theoretical studies of the intramolecular dynamics of model systems, that show that for some molecules equilibration occurs only for energies very close to the dissociation limit.[25] To the best of our knowledge this is the first direct experimental evidence that region II may indeed extend quite close to the dissociation threshold.

Acknowledgments

We are pleased to acknowledge financial support by the Army Research Office, the Joint Services Electronics Program and by the Hamamatsu Corporation.

References

1 N.R. Isenor, V. Merchant, R.S. Hallsworth and M.C. Richardson, *Can. J. Phys.* **51**, 1281 (1973)

2 V.N. Bagratashvili, V.S. Letokhov, A.A. Makarov, E.A. Ryabov, *Multiple Photon Infrared Laser Photophysics and Photochemistry* (Harwood Academic Publishers, New York, 1985)

3 N. Bloembergen and E. Yablonovitch, *Physics Today* **5**, 23 (1978)

4 C.D. Cantrell, S.M. Freund, J.L. Lyman, *Laser Handbook*, Vol. **3**, Ed. M.L. Stitch (North-Holland, Amsterdam, 1979)

5 P.A. Schultz, Aa. S. Sudbø, D.J. Krajnovitch, H.S. Kwok, Y.R. Shen, and Y.T. Lee, *Ann. Rev. Phys. Chem.* **30**, 379 (1979)

6 V.S. Letokhov, *Physics Today* **11**, 34 (1980)

7 W. Fuss and K.L. Kompa, *Prog. Quant. Electr.* **7**, 117 (1981)

8 D.S. King, *Dynamics of the Excited State*, Ed. K.P. Lawley (Wiley, New York, 1982)

9 J.G. Black, P. Kolodner, M.J. Schultz, E. Yablonovitch, N. Bloembergen, *Phys. Rev.* **A19**, 704 (1979)

10 Y.T. Lee and Y.R. Shen, *Physics Today* **33**, 52 (1980)

11 J.D. Campbell, G. Hancock, J.B. Halpern, and K.H. Welge, *Chem. Phys. Lett.* **44**, 404 (1976)

12 D.S. King and J.C. Stephenson, *Chem. Phys. Lett.* **51**, 48 (1977)

13 D.S. Frankel and T.J. Manuccia, *Chem. Phys. Lett.* **54**, 451 (1978)

14 R.C. Sharp, E. Yablonovitch and N. Bloembergen, *J. Chem. Phys.* **74**, 5357 (1981)

15 P. Mukherjee and H.S. Kwok, *J. Chem. Phys.* **84**, 1285 (1986)

16 V.N. Bagratashvili, Yu.G. Vainer, V.S. Dolzhikov, S.F. Kol'yakov, A.A. Makarov, L.P. Malyavkin, E.A. Ryabov, E.G. Silkis, and V.D. Titov, *Appl. Phys.* **22**, 101 (1980)

17 V.N. Bagratashvili, Yu.G. Vainer, V.S. Dolzhikov, S.F. Kol'yakov, V.S. Letokhov, A.A. Makarov, L.P. Malyavkin, E.A. Ryabov, E.G. Sil'kis, and V.D. Titov, *Sov. Phys. JETP* **53**, 512 (1981)

18 V.N. Bagratashvili, Yu.G. Vainer, V.S. Doljikov, V.S. Letokhov, A.A. Makarov, L.P. Malyavkin, E.A. Ryabov, and E.G. Sil'kis, *Opt. Lett.* **6**, 148 (1981)

19 E. Mazur, I. Burak, and N. Bloembergen, *Chem. Phys. Lett.* **105**, 258 (1984)

20 Jyhpyng Wang, Kuei-Hsien Chen, and Eric Mazur, *Phys. Rev. A* **34**, 3892 (1986)

21 Eric Mazur, *Rev. Sci. Instrum.* **57**, 2507 (1986)

22 Eric Mazur, Kuei-Hsien Chen, Eric Mazur, *Proc. Int. Conf. on Lasers '86*, 359 (1986)

23 Jyhpyng Wang, Kuei-Hsien Chen and Eric Mazur, *Laser Chemistry*, in press

24 Kuei-Hsien Chen , Jyhpyng Wang and Eric Mazur, to be published

25 G. Herzberg, *Molecular spectra and molecular structure*, Vol. 2 (Van Nostrand Reinhold, New York, 1979)

26 Charles A. Bradley, Jr., *Phys. Rev.* **40**, 908 (1932)

27 T. Shimanouchi, *J. Phys. Chem. Ref. Data* **6**, 993 (1977)

28 B.G. Sumpter and D.L. Thompson, J. Chem. Phys. **86**, 2805 (1987)

INTRAMOLECULAR VIBRATIONAL RELAXATION AND THE DYNAMICS OF THE HIGH-POWER TWO-PHOTON EXCITATION OF NO_2

Laurence Bigio* and Edward R. Grant**

Department of Applied Physics, Department of Chemistry
Cornell University
Ithaca, NY 14853

INTRODUCTION

Dramatic progress over recent years in the development of advanced narrow-bandwidth and short-pulse-width laser sources has significantly accelerated the experimental investigation and understanding of intramolecular vibrational energy redistirbution (IVR)[1]. By the properties chosen for the excitation source, such experiments can now be seen to cast the investigator in the role either of observer or active participant. Under high resolution in the frequency domain, one observes the consequences of intramolecular coupling in the resolved pattern of spectroscopic level positions and intensities. Line-widths, densities of states and the pattern of their relative spacings all combine to paint a picture of global radiationless decay, detailed by specific rovibronic interactions[2]. In the time domain, pump-probe experiments employing transform-limited picosecond and sub-picosecond lasers launch and then probe time-evolving coherent superposition states[3]. Thus, the distinction emerges: In the first case, one finds the intrinsic or static coupling patterns associated with intramolecular state mixing, whereas, in the second case, one participates and even partially controls the progress of intramolecular dynamical evolution according to the choice of parameters which include the initial state from which the system is excited, the coherence properties of the pump and probe optical pulses, and the time delay between them.

This distinction is reflected in the points of view of two complementary descriptions of coupled molecular excited states. For example, consider the (stationary) eigenstates, $|\psi_i\rangle$, of a complete (vibronic, anharmonic, coriolis-coupled) Hamiltonian, H. Let us assume that these states can be appropriately described by superpositions of solutions, $|a_j\rangle$, to a simpler zeroth-order Hamiltonian, H_o:

$$|\psi_i\rangle = \sum_j \alpha_{ij} \; |a_j\rangle \qquad (1).$$

* Present Address: General Electric
Corporate Research and Development Center
PO Box 8 Bldg. K1-4C34
Schenectady, NY 12301

** Present Address: Purdue University
Department of Chemistry
West Lafayette, Indiana 47907

High resolution excitation of any state $|\psi_i\rangle$ from some lower state produces no meaningful dynamical process. However, the absorption or emission spectrum from $|\psi_i\rangle$ will exhibit transitions evidencing the character of the basis states $|a_j\rangle$, and thus, in the language reserved for the easily describable zeroth-order states one says the system exhibits static IVR. The stationary states $|\psi_i\rangle$ are said to be mixed; the expansion coefficients in Equation (1) are not functions of time.

On the other hand, excitation from a lower state, $|g\rangle$, with a short pulse whose transform-bandwidth spans more than one state $|\psi_i\rangle$ will coherently prepare a superposition state (of those eigenstates spanned) which propagates in time:

$$|\phi(t)\rangle = \sum_i c_i(t)\ |\psi_i\rangle \qquad (2).$$

The initial character of $|\phi(t)\rangle$ at zero time following the pump pulse is determined by the photoprojection of $|g\rangle$ onto the states $|\psi_i\rangle$. The subsequent propagation is coordinated by the complex coefficients $c_i(t)$, containing the phase factors $\omega_i t$, which control the evolution of interference between the various states $|\psi_i\rangle$. Thus, the overall time-dependent character of the wave-packet $|\phi(t)\rangle$ will depend on several factors including the extent of static mixing given in Equation (1). Some of these factors may be fixed experimentally to achieve a measure of control over the dynamic intramolecular motion. Among the factors available to the experimentalist are the choice of initial state which establishes the t=0 projection, the coherence width of the excitation source which determines the number of eigenstates spanned, and the time from pump to probe allotted for the wave-packet propagation. In the language of IVR, energy is said to be dynamically redistributed between the various modes $|a_j\rangle$ in the expansion of Equation (1).

Thus, the direct initiation and control of dynamic vibrational energy redistribution requires a coherence condition between various energy eigenstates of the excited molecular system. The use of ultra-short pulses has produced elegant examples of such dynamical processes[3]. However, other means exist to transform broaden an intermediate superposition, which do not depend on the coherence properties of the laser itself. One such method makes use of the collisional quenching properties of a buffer gas to shorten the lifetime, and thereby broaden the coherence width of a fluorescent molecule[4]. In this manner, dramatic changes in the emission spectrum have been seen in complicated molecules as a function of the pressure of the quenching gas. Another approach is to rely on the transition time in a multiphoton excitation using, in essence, power broadening to produce a coherence width[5-7]. This effect has been demonstrated in atoms, as manifested by power dependent photoelectron angular distributions[5,6]. In each of these methods a degree of control over the intermediate superposition is achieved by varying the coherence properties of its photopreparation, as established respectively by the pulse duration, quenching gas pressure, or laser intensity.

In this paper we describe the observation of an intensity-induced coherence over the vibronic levels of a molecule. The subject of this investigation is NO_2. We begin by demonstrating the highly mixed nature of energy eigenstates in the visible region of this molecule, thus establishing, in a manner analogous to previous Raman and fluorescence experiments, the presence of static IVR. These measurements rely on two-laser Optical-UV-Double-Resonance (OUDR) where the mixed character of the intermediate state is determined with reference to the vibrational levels of a higher excited electronic state. We then probe the dynamic character of optically prepared states above the first dissociation threshold by one-color two-photon resonant absorption spectroscopy (TPA). The intermediates in this case are both highly mixed and dissociative. We find that vibronic intensity distributions for this real-intermediate two-photon process depends on the laser in a way that tracks the power-dependent coherence width of the first-photon absorption step. In this way, control over the dynamic propagation of a coherently excited intermediate superposition state is achieved according to the choice of laser intensity.

EXPERIMENTAL METHOD

A) Optical-UV-Double-Resonance (OUDR)

The experimental arrangement is shown in Figure 1. We employ a Lambda Physik EMG-150 ES excimer laser which operates with two discharge cavities triggered by a common thyratron. Pulse durations are about 10 ns with cavity-one firing a jitter-free 15 ns before cavity-two. The output of cavity-one pumps a Lambda-Physik FL-2002 dye laser whose output (labeled λ_1) is used to photoprepare a mixed eigenstate of NO_2 in the 454 nm region. Cavity-two pumps a second FL-2002 whose doubled output at about 291 nm (labeled λ_2) subsequently excites the intermediate to selected vibrational levels of the 3pσ $^2\Sigma_u^+$ Rydberg state. Ionization from a Rydberg level is achieved by the absorption of a second λ_2 photon. This third step is known to be facile, and saturates easily insuring that signal intensities depend only on the transition strength from the intermediate to the Rydberg state. The unfocused beams counterpropate and cross the output of an NRC BV-100 molecular beam value (seeded 1:1:20, NO_2: O_2: Ar, at 1-atm. stagnation pressure). Mass 46 ions are collected using a quadrupole mass spectrometer, and the resulting current is amplified and averaged to give an ionization signal. Figure 2 gives an energy level diagram (left side) showing the initial preparation by λ_1 followed by further excitation by λ_2.

B) One-Laser Two-Photon Resonant Absorption (TPA)

The arrangement is similar to that in Figure 1 except that only cavity-two is employed to pump a single dye laser whose output from λ_2 = 360 to 335 nm is focussed across the axis of the pulsed jet. Once again, mass 46 ions are collected and an ionization signal generated. Figure 2 gives the energy level diagram (right side) for this two-photon resonant excitation to the Rydberg state followed by one-photon ionization. Laser output energy is monitored by means of an Epply Thermopile power meter placed after the exit window. Flux intensities are in the range of 1-20 GW/cm^2.

BACKGROUND

The spectroscopic system of interest is that of the 3pσ $^2\Sigma_u^+$ Rydberg state as diagrammed in Figure 2. This state has been thoroughly studied by conventional VUV absorption spectroscopy[8]. A distinctive feature of the one-photon absorption spectrum is the absence of an observable origin band. This is readily explained by the fact that the NO_2^+ core of this Rydberg state is isoelectronic with CO_2 and therefore predictably linear, while the ground state is bent (134°). Thus, it is not surprising that direct one-photon Franck-Condon intensity only develops upon reaching rather high levels of vibrational excitation in the upper state. However, when the Rydberg state is

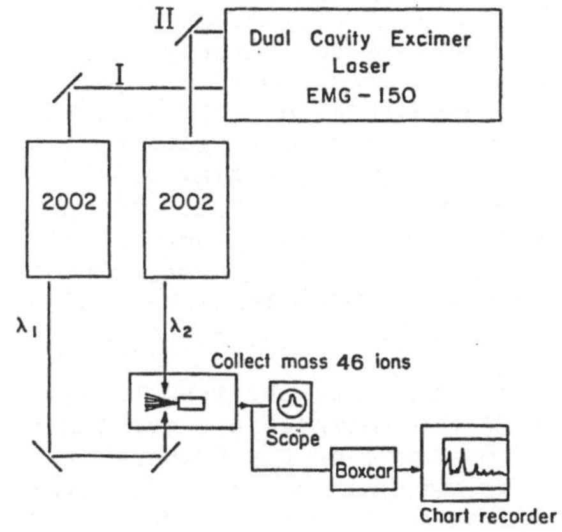

Fig. 1 - Experimental arrangement for OUDR: Counter-propagation λ_1 and λ_2 laser beams overlap on the axis of a free-jet expansion of NO_2 seeded in O_2: Ar (1:1:20). The mass spectrometer collects m/e = 46 NO_2^+ ions. For TPA, only cavity II is used, and the λ_2 beam is focussed across the jet.

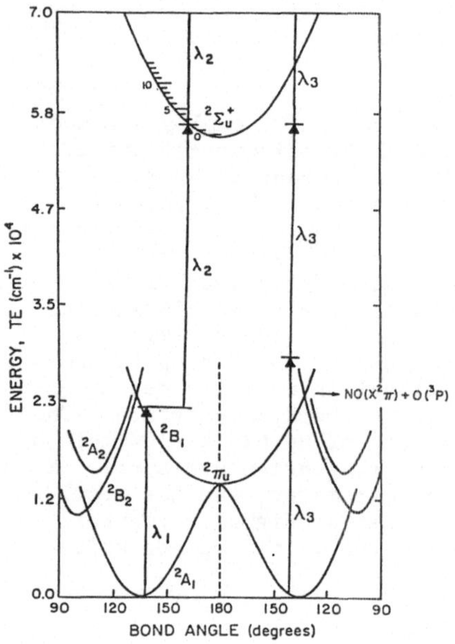

Fig. 2 - (Left; OUDR) Energy level diagram showing the two-photon ionization by λ_2 of the state which was photoprepared by λ_1. The separation of the two arrows denotes the temporal separation of the two pulses by ~ 15 nsec. (Right; TPA) Resonant two-photon excitation to the Rydberg levels by λ_3 followed by facile ionization from the same laser pulse.

accessed via a two-photon excitation, the simple Franck-Condon picture no longer applies.

The bent ground state of NO_2, $^2A_1$, is well approximated at low energies by a zeroth-order separable adiabatic (bent) harmonic oscillator Hamiltonian[9]. Likewise, the $^2\Sigma_u^+$ Rydberg state is also well approximated in terms of a similar, linear zeroth-order basis[10]. The visible region, on the other hand, is not well characterized. It exhibits a complex and dense system of spectral features which arise from vibronic mixing of four low-lying electronic states[11,12]. The vibrational nature of these states is highly mixed when expanded in terms of any appropriate zeroth-order basis[12]. Therefore, either the ground or Rydberg states serve equally well as adequate, but far from diagonal, vibrational basis sets. In both Optical-UV-Double-Resonance (OUDR) and one-laser two-photon absorption (TPA), the experimental expansion is carried out over the vibrational levels of the Rydberg state, since detection is via sequential 2 + 1 photon ionization spectroscopy.

RESULTS

A) Optical-UV-Double Resonance (OUDR)

Figure 3 shows a laser-induced-fluorescence (LIF) spectrum of NO_2 in the 454.7 nm region. The,lines carrying most of the oscillator strength have been assigned as $^2B_1(080)K=0$ levels[13]. We tune λ_1 to each of the three positions marked in the Figure and compare the OUDR signals for transitions to several of the Rydberg vibrational levels. The relative intensities of the OUDR signals are listed in Table 1 for the positions I=a,b, and c. These intensities (labeled $|\beta_i^I|^2$ in the Table) are related to the relative weighting of the Rydberg bands making up the character of the mixed intermediate state being prepared by λ_1, as will be discussed in the next section. The numbers given are derived by averaging over the J rotational contours of the Rydberg vibrational bands. Intensities for different bands are all normalized for equal λ_2 laser powers.

Typical of our survey of lines in the 450 nm region are the results seen In Table 1 for λ_1 set on peaks (a) and (b). Here, the intermediate overlaps best with the origin and low lying bending overtone bands of the Rydberg state. The ΔK selection rules for this sequential two-photon absorption process are crucial in determining which vibrational bending overtones are

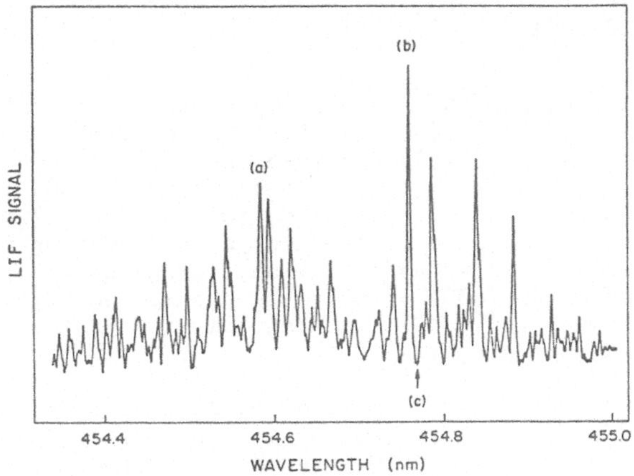

Fig. 3 - Laser-induced-fluorescence spectrum of NO_2 in the 454.7 nm region

accessible in the Rydberg state. The upper state K' for linear 3pσ NO_2 is determined by the vibrational angular momentum quantum number, $l' = v_2, v_2 - 2, \ldots, 0$. Given an isolated intermediate state with well defined total parity, photoexcitation will access either all-odd or all-even bending quantum numbers of the Rydberg state. We find, for I=(a) in Table 1, that the even quanta of v_2 are preferred, indicating that λ_1 must have originated in a K"=1 level, with a weak degenerate transition from K"=2 also active. This tends to substantiate the assignment by Sugimoto, et al.,[13] of the intermediate as $^2B_1(080)K'=0$, since this band, with total parity positive, can only optically couple to states of total parity negative, which are the even bending overtones in the $^2\Sigma_u^+$ Rydberg state. For I=(b), it is clear that two degenerate transitions from K"=0 and K"=1 in the ground state are excited by λ_1, since strong intensity is found for both (000) and (010) bands from the λ_2 transition. Such prominence in a rotational hot-band transition must be ascribed to the finite rotational temperature produced by the pulsed supersonic jet (~ 20°K) and the high density of states at the first photon (λ_1) level. Perhaps the most dramatic feature evident in Table 1 is the strong variation in OUDR intensity patterns with slight changes in first photon frequency. This can be seen in the results for the valley marked (c) which is immediately to the red of the peak (b) in the LIF spectrum of Figure 3. In contrast to both peaks (a) and (b), we find that most of the OUDR intensity is confined to one band alone, (010).

Table 1 - OUDR intensities (arb. units) to various low bending vibrational bands of the $^2\Sigma_u^+$ Rydberg state for three different λ_1 settings.

$\lvert\beta_j^I\rvert^2$	(a)	(b)	(c)
(000)	11	11	.8
(010)	4.5	11	25
(020)	10	6	.3
(030)	.2	-	.02
(100)	3.4	-	.1
(110)	1.2	-	.8
(120)	2.5	-	-

B) Two-Photon Resonant Absorption (TPA)

The overall two-photon X $^2A_1$----->3pσ $^2\Sigma_u^+$ transition exhibits selection rules $\Delta K = \pm 1$. The bands (υ_1, υ_2, υ_3) with even bending quantum number υ_2 and their various ΔJ branches are accessible only from J" = 1,2,3,...,K"=1,3,5,... The K rotational constant of $^2A_1$ NO_2 is 7.58 cm^{-1}, so, at the rotational temperatures of our pulsed jet, a significant Boltzmann correction is required to compare intensities of neighboring bands[7,14]. This temperature (approximately 20°K) as established by rotational analysis of the high-resolution spectrum of (000), places the population primarily in K"=0 and 1 with the ratio given by 1.0:0.56. Intensities of transitions to states with even υ_2 are corrected accordingly.

Figure 4 shows a scan over a number of bands of this system. These bands are readily assigned to a vibrationless origin and associated progressions and combinations in bending and symmetric stretching modes of the linear 3pσ $^2\Sigma_u^+$ Rydberg state. The linearity and hence Rydberg character of this state together with its specific term symmetry are confirmed by the high resolution structure of these bands[10,14]. To higher energy in Figure 4 we see in addition the first elements of the 3pπ $^2\Pi_u$ Rydberg state, showing again a strong origin band.

The power dependence of the relative band intensities is interesting. Figure 5 shows a set of scans across the (100) band as a function of laser intensity. As the laser pulse energy, shown by the bold horizontal trace, is raised, we see the off-resonant background uniformly increase. The resonant signal also increases at low powers, but at high power, this peak signal clearly decreases with increasing power. Thus we can conclude that the apparent cross section for resonant enhancement through this bending origin is strongly power dependent, decreasing dramatically with increasing high power.

This behavior is evident in a number of other low bending quantum number states. Figure 6(A) compares measured power dependences for the (100), (110), and (130) bands. Interestingly, note that the threshold for the apparent turnover in the cross section increases with increasing bending quantum number.

Continuing this latter trend, Figure 6(B) compares (000) with (050), the latter of which is a band that should start to show some overlap in a one-photon vertical transition. Here, at the point where (000) is starting to lose intensity, the cross section for the higher bend actually appears to increase. This antisaturation behavior characterizes the higher bending transitions.

DISCUSSION

Figure 2 presents a bending-coordinate potential energy diagram for these transitions. The ground and three lower state surfaces are taken from calculations by Gillespie and Khan[15], while the Rydberg state harmonic potential and levels are derived from the present spectra. Clearly, the transitions diagrammed to low-lying bending states are well outside the vertical one-photon Franck-Condon band.

The two-laser OUDR experiment provides us with an opportunity to discern the character of highly mixed intermediate states, or static IVR. By optically promoting an intermediate eigenstate to the Rydberg state, we observe an experimental expansion of the dipole operator acting on the selected highly mixed vibronic state, with reference to the zeroth-order vibrational bands of the upper Rydberg state. To see this, first consider the photoprepared intermediate as an expansion in terms of Rydberg vibrational levels. This is given by Equation (1) with the states $|a_j\rangle$ representing the various vibrational origin, overtone, and combination bands of the 3pσ $^2\Sigma_u^+$ Rydberg electronic state. Measurement of intensities for photoexcitation of $|\psi_i\rangle$ to the actual set of $|a_j\rangle$ states yields the relative cross-sections, or equivalently, matrix

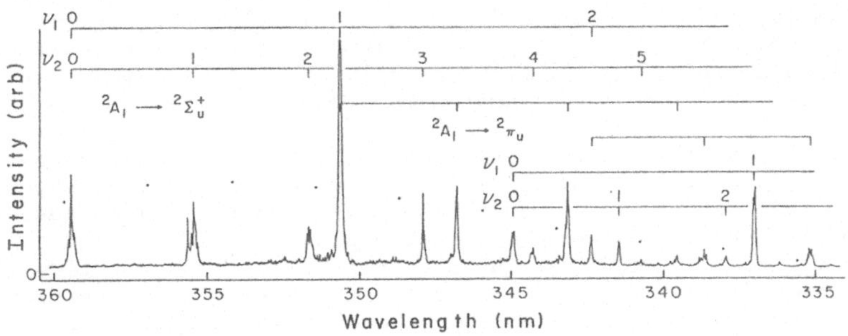

Fig. 4. Two-photon absorption spectrum of NO_2 in the energy region from 55,000 to 59,000 cm^{-1}. Ladders indicate vibronic assignments to elements of the 3pσ $^2\Sigma_u^+$ and 3pπ $^2\Pi_u^+$ Rydberg states.

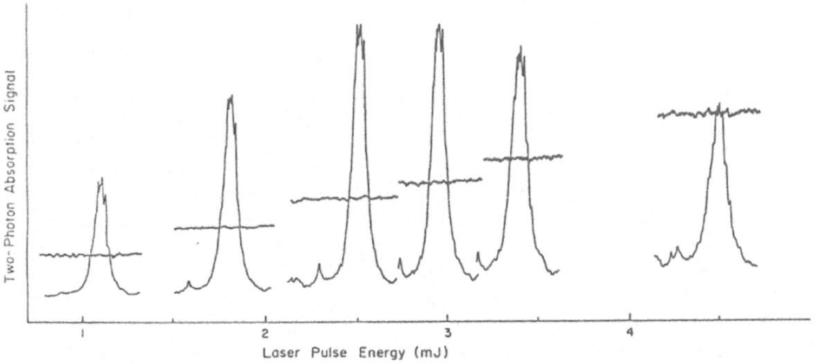

Fig. 5. Set of ionization signals, due to the two-photon absorption across the (100) $^2\Sigma_u^+$ band of NO_2, for various values of the laser intensity. Rotational lines are averaged over using appropriate boxcar time constants. The laser pulse energy is shown by the horizontal trace, for each scan. Note the eventual decrease in resonant signal for sufficiently increasing laser pulse energy. Nonresonant background signal, however, continues to rise monotonically with power.

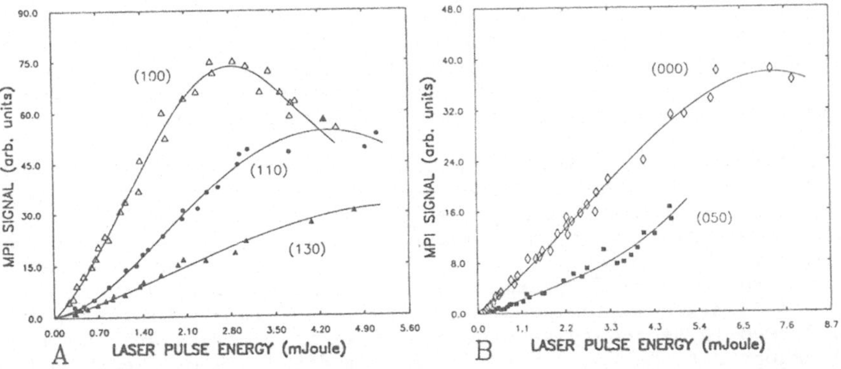

Fig. 6. A) Comparison of the measured power dependences for the (100), (110), and (130) bands of the $^2\Sigma_u^+$ state. Data points are derived by integrating the area under the curves of the sort shown in Fig. 5. B) Comparison of the power dependences for the (000) and (050) bands.

elements $|\langle\psi_i|\mu|a_j\rangle|^2$. In the Condon approximation this becomes the overlap integral $\mu_e^2|\langle\psi_i|a_j\rangle|^2 = \mu_e^2|\alpha_{ij}|^2$. Thus, apart from a constant, which is the same for all bands, one obtains directly the relative weighting of the different $|a_j\rangle$ states in the expansion for $|\psi_i\rangle$.

Unfortunately, the Condon approximation may not be as good here as for certain diatomic cases. More correctly, in a photoexcitation experiment, one never truly observes expansion of the sort in Equation (1) because the system in question is really that of the molecule plus the light field. This is made evident in the above example by the presence of the constant term μ_e^2. To treat the composite system more generally, we rewrite the intermediate and zeroth-order basis states in a more complete rovibronic form, denoting explicitly the composite quantum numbers evr. Now writing the expansion of the dipole operator acting on the intermediate eigenstate we have:

$$\vec{\mu}.\vec{E}\;|\;\psi^{I}_{evr}\rangle = \sum_j \beta^{I}_{j}\;|\;\psi^{j}_{evr}\rangle\,. \tag{3}$$

Acting on the left by $\langle\psi^{j}_{evr}|$, using $\vec{\mu} = \vec{\mu}_e + \vec{\mu}_N$, and assuming the separability of ψ_{evr} into $\psi_e\;\psi_v\;\psi_r$ gives:

$$\beta^{I}_{j} = \langle\psi^{j}_{evr}\;|\;\mu\;|\;\psi^{I}_{evr}\rangle = \int\psi^{j}_{v}\psi^{I}_{v}(\psi^{j}_{r}\cos\phi\psi^{I}_{r})\left[\int\psi^{j}_{e}\mu_e\psi^{I}_{e}d\tau_e\right]d\tau_N \tag{4}$$

A photoexcitation of $|\psi^{I}_{evr}\rangle$ to the various vibrational states of a single Rydberg electronic state yields directly the relative coefficients $|\beta^{I}_{j}|^2$ for those vibrational bands in the expansion, according to the relative intensities of the spectroscopic signals to each band. Note that these coefficients are different than the α_{ij} found in Equation (1), since the β^{I}_{j} explicitly include the effects of photoselection. For a single Rydberg state, the electronic integral in brackets remains fixed from band to band, while the rotational and vibrational terms yield the ΔK selection rules. Thus, for allowed transitions, the overall relative intensities are still governed by vibrational wavefunction overlap.

As mentioned earlier, given an isolated intermediate state with well defined total parity, photoexcitation will potentially excite only all even or all odd bending quantum numbers of the Rydberg state. This was seen in Table I for the intermediate state marked (a) (of Figure 3) in which the $|\beta^{I}_{j}|^2$ coefficients were dominant mainly for the even values of υ_2. The other results given in Table I as well as other OUDR data not presented here suggest an interesting character for vibronic mixing in sections of the visible spectrum of NO_2. Different states, even those nearly degenerate seem to be mixed in completely arbitrary ways. That is, the coefficients $|\beta^{I}_{j}|^2$ show no recognizable pattern as one proceeds through different intermediate states, $|\psi^{I}_{evr}\rangle$. Clearly, a much more detailed OUDR examination, employing a colder supersonic beam, is necessary to discern what larger patterns may exist.

Distinct from the two-laser OUDR results in which static IVR is observed, the one-laser TPA experiment evidences some variation in the dynamical aspects of intramolecular vibrational energy redistribution according to the choice of laser power. We explain this in qualitative terms as follows. At low powers the pumping process is relatively slow. Under such conditions, a scan of the laser through the system of intermediate states sweeps a relatively narrow superposition of mixed molecular eigenstates. Thus long vibrational progressions are formed in secondary absorption. This is somewhat analogous to the OUDR case except for the fact that many intermediate states are passed through in the scan. Thus individual eigenstates are not characterized, but the non-Franck-Condon transitions exhibited do globally indicate strong intermediate state mixing. As laser intensity is increased the time scale for excitation will fall into the sub-picosecond regime. At higher powers, still faster excitation of the upper state with associated Rabi cycling of probability between lower and intermediate states and between intermediate and upper states continues to shorten intermediate state lifetime, with the coherence width tracking the power-broadened bandwidth. Precisely such a phenomenon has been seen in atoms, as evidenced by a power-dependent multiphoton photoelectron angular distribution.[5,6] Thus we can anticipate, by extending this idea to molecular systems, that higher powers with broader coherence widths will give vibrational overlaps more in accord with zeroth-order Franck-Condon expectations. This indeed is the case. To see how this follows and to understand the effect of this stimulated cycling, we write down the wavepacket generated by the light field <u>during</u> the duration of the pulse.

$$| \Phi(t)\rangle = d_1(t) \; | g\rangle + \sum_I d_I(t) \; | \psi^I_{evr}\rangle + d_j(t) \; | \psi^j_{evr}\rangle \tag{5}$$

This differs from Equation (2) by the inclusion of the initial state $|g\rangle$ and the final Rydberg vibrational band, $|\psi^j_{evr}\rangle$. Equation (2) is appropriate for pump-probe type experiments in which one examines the propagation of a wavepacket previously excited by a pump pulse, whereas in the TPA experiment, excitation and subsequent ionization occurs only for the duration of the laser pulse. The character of the ground and upper excited state are mixed in the intermediate state superposition by the field.[16] The power broadened bandwidth ultimately determines the number of intermediate states $|\psi^I_{evr}\rangle$ that are spanned in the superposition. The photoprojection of $|g\rangle$ onto a large enough superposition of intermediates will generate an expansion of $|g\rangle$ in terms of those intermediates. Thus, for high powers, the effectiveness of two-photon resonant enhancement should decline for low bending quantum states which have poor overlap with $|g\rangle$, while increasing for high bends which have good overlap with $|g\rangle$.

This effect of declining cross section with increasing power for low bends is so strong that the band intensity (a measure of the competition of second photon absorption with one-photon photodissociation) actually diminishes on an absolute scale at higher powers, while the band intensities of higher bends increase. Presumably at some sufficiently high power-broadened bandwidth, the band intensity envelope will approach that of the one-photon vertical transition. At this limit a wavelength tuned to resonance with a low bend should be no more effective than an off-resonant background wavelength in promoting ionization over dissociation.

Recognizing that we are short of this limiting bandwidth in the present experiments, we can use the incident laser power together with estimated transition dipole matrix elements to put an interesting lower limit on the strength of vibronic coupling in NO_2. The power density at the focus for the highest pulse energy in Figure 5 is about 30 GW cm^{-2}. From the measured oscillator strength of the first-photon transition we calculate a representative transition dipole matrix elements of 0.2 cm^{-1}.[9] For these parameters the power-broadened bandwidth of the transition (taken simply as a two-level system) is 4 cm^{-1}.[7] This should be contrasted with a typical coupling strength for excited electronic state IVR in larger polyatomics, as observed for example in anthracene (0.1 cm^{-1}).[3]

Thus, the data appear consistent with the idea that the time scale of an MPI process, as manifested in its power broadened bandwidth, can serve as a probe of coupling strengths and intramolecular dynamics in dissociative intermediate states.

ACKNOWLEDGMENTS

This work was supported by the U.S. Army Research Office. Acknowledgment is gratefully made to the Department of Defense for a DOD-University Research Instrumentation Grant.

REFERENCES

1. P.M. Felker and A.H. Zewail, J. Chem. Phys. 82, 2961 (1985); N. Bloembergen and A.H. Zewail, J. Phys. Chem. 88, 5459 (1984); S. Mukamel, J. Chem. Phys. 82, 2867 (1985); C.S. Parmenter, Faraday Discuss. Chem. Soc. 75, 7 (1983); D.D. Smith, S.A. Rice and W. Struve, ibid. 75, 173 (1983).
2. E. Haller, H. Köppel and L.S. Cederbaum, Chem. Phys. Lett. 101, 215 (1983); K.K. Lehmann and G.J. Scherer, in Advances in Laser Spectroscopy, Vol. 3, edited by B.A. Garetz and J.R. Lombardi (John Wiley & Sons Ltd, New York, 1986): P.J. deLange, K.E. Drabe, and J Kommandeur, J. Chem. Phys. 84, 538 (1986); K.W. Holtzclaw and D. W. Pratt, ibid. 84, 4713 (1986).
3. P.M. Felker and A.H. Zewail, Chem. Phys. Lett. 102, 113 (1983); P. M. Felker and A.H. Zewail, ibid. 108, 303 (1984); M. Shapiro and P. Brumer, J. Chem. Phys. 84, 4103 (1986); D.J. Tanner, R. Kosloff, and S.A. Rice, ibid. 85, 5805 (1986).
4. K.W. Holtzclaw and C.S. Parmenter, J. Phys. Chem. 88, 3182 (1984); R.A. Coveleskie, D.A. Colson, and C.S. Parmenter, ibid. 89, 654 (1985).
5. S. Geltman and G. Leuchs, Phys. Rev. A. 31. 1463 (1985); S. Geltman, J. Phys. B: Atom. Molec. Phys. 13, 115 (1980).
6. R. Parzyński, Opt. Common. 51, 151 (1984).
7. L. Bigio and E.R. Grant, J. Chem. Phys. 83, 5361 (1985); L. Bigio and E.R. Grant, ibid. 83, 5369 (1985).
8. R.K. Ritchie and A.D. Walsh, Proc. Soc. London Ser. A 267, 395 (1962).
9. D.K. Hsu, D.L. Monts, and R. N. Zare, Spectral Atlas of Nitrogen Dioxide (Academic, New York, 1978).
10. R.S. Tapper, R.L. Whetten, and E.R. Grant, J. Phys. Chem. 88, 1273 (1984).
11. R.E. Smalley, L. Wharton, and D.H. Levy, J. Chem. Phys. 63, 4977 (1975); K.K. Lehmann and S.L. Coy, ibid. 83, 3290 (1985).
12. H. Köppel, W. Domcke, and L.S. Cederbaum, in Advances in Chemical Physics, Vol. 57, edited by I. Prigogine and S.A. Rice (John Wiley & Sons, New York, 1984); E. Haller, H. Köppel, and L.S. Cederbaum, J. Mol. Spectrosc. 111, 377 (1985).
13. N. Sugimeto, S. Takezawa, and N. Takeuchi, J. Mol. spec. 102, 372 (1983).
14. M.B. Knickelbein, K.S. Haber, L. Bigio, and E.R. Grant, Chem. Phys. Lett. 131, 51 (1986).
15. G.D. Gillespie and A.V. Khan, J. Chem. Phys. 65, 1624 (1976).
16. For a discussion of dressed molecular states, see: A.D. Bandrauk and G. Turcotte, J. Phys. Chem. 87, 5098 (1983); T.T. Nguyen-Dang and A.D. Bandrauk, J. Chem. Phys. 79, 3256 (1983); A.D. Bandrauk, O.F. Kalman, and T.T, Nguyen-Dang, ibid. 84, 6761 (1986).

COHERENT EXCITATION OF POLYATOMIC MOLECULES BY

ULTRASHORT LASER PULSES - A MODEL CALCULATION

H.S. Kwok and P. Mukherjee*

Department of Electrical & Computer Engineering
State University of New York at Buffalo
Bonner Hall, Amherst, NY 14260

* Present address: CLS-5, Los Alamos National Laboratory
Los Alamos, NM 87545

INTRODUCTION

In a recent series of experiments,[1-4] we have found that the interaction of polyatomic molecules excited into the quasicontinuum (QC) with ultrashort CO_2 laser pulses can be characterized by the following observations:[2] (1) the spectral lineshape is well-defined for short pulses, but becomes broader and more diffused as the pulse duration increases, and (2) the line center absorption cross section decreases with the interaction time. A qualitative explanation for these observations in terms of the multi-tier energy level structure[5] and the development of the homogeneous ensemble of energy levels in the QC[6] has been given and was found to be quite satisfactory. In the present article, we wish to present a model calculation, based on the coherent Bloch equation[7], on the change in small signal absorption cross section as a function of the laser pulse duration.

MODEL

In this model calculation, the optically active states are represented by many two-level systems. For each 2-level structure, the subsequent tiers of states are lumped together into a "heat bath" which is coupled to the level being excited. Fig. 1 shows the schematic diagram. For each level that is excited by the laser, there is a corresponding heat bath, and a corresponding two-level structure. The two-level scheme is physically justifiable because of the stipulation that the exciting laser is weak, and cannot induce more than 10^{-3} quanta of excitation in the system. Otherwise a multi-level scheme has to be employed.[7,8] Using the above simplifications, the Bloch equation can be written as

$$d\alpha/dt = \Delta\omega\beta - \alpha/T_2 \tag{1}$$

$$d\beta/dt = -\Delta\omega\alpha - (\rho_{22} - \rho_{11})\,\Omega/2 - \beta/T_2 \tag{2}$$

$$d\rho_{22}/dt = \Omega\beta - \rho_{22}/T_1 - \rho_{22}/T_1^* \tag{3}$$

$$d\eta/dt = \rho_{22}/T_1^* \tag{4}$$

In eqs. (1) - (4), α and β are the real and imaginary parts of the off-diagonal density matrix after the rotating wave approximation, T_1, T_2^* are the population and dephase relaxation times of the two-level system and T_1* describes the transfer of population to the heat bath η. Ω is the Rabi frequency of the transition given by

$$\Omega = \mu E/\hbar \tag{5}$$

where μ is the dipole matrix element and E is the peak electric field of the laser, and $\Delta\omega$ is the frequency mismatch between the laser (ω_o) and the energy separation between the two levels (ω).

$$\Delta\omega = \omega - \omega_o \tag{6}$$

In the calculation, ω was allowed to vary to cover the entire inhomogeneous absorption line profile of the molecule. In view of the 1 cm^{-1} linewidth of the laser, the number of two-level systems considered in the inhomogeneous averaging can be kept to $\sim$ 20. This simplified the numerical calculation considerably.

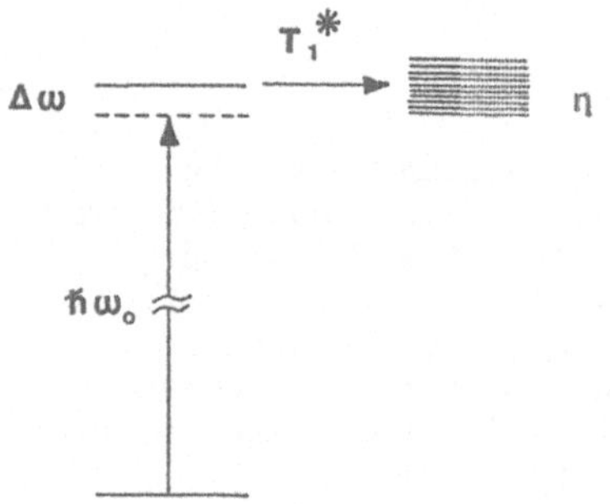

Fig. 1. Schematic diagram of the energy level systems used in the model calculation. Approximately twenty of these quasi-two-level systems were used in the calculation, for various values of $\Delta\omega$.

RESULTS AND DISCUSSIONS

In the numerical calculation, E(t) was taken to be a triangular pulse which was a good approximation of the optical free induction decay pulse shape. The values of μ, ω_o are taken to be those of the SF_6 ν_3 band. The laser intensity was varied and decreased until it no longer had any effect on the calculated absorption cross section σ. Moreover, to reduce the number of free parameters in the calculation, T_1 was taken to be equal to T_2.

Figures 2 to 4 represent the results of the model calculation for the cases of SF_6, C_2F_5Cl and C_3F_7I. The circles in those figures are the experimental results taken from references 1 to 4. The cross sections are normalized to the 120 ps values for the sake of comparison. The solid lines are the best theoretical fits by adjusting T_1 and T_1^*. The final results for T_1 and T_1* are listed in Table I. It can be seen that T_1* is generally in the range of 100-200 ps while T_1 is comparable to the shortest CO_2 laser pulse duration used in the experiments (20-30 ps).

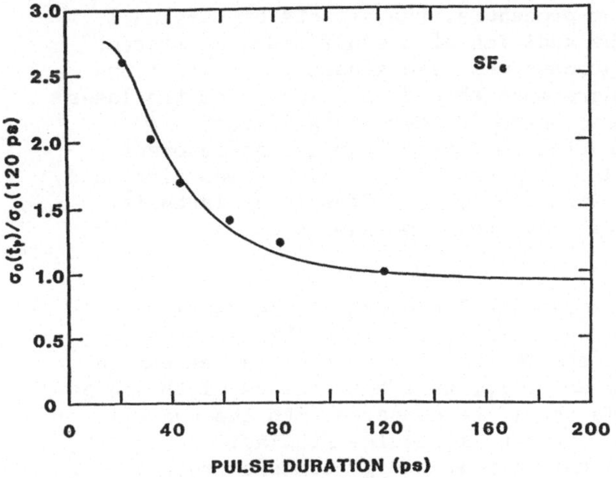

Fig. 2. Experimental data and best fit theoretical curve for SF_6.

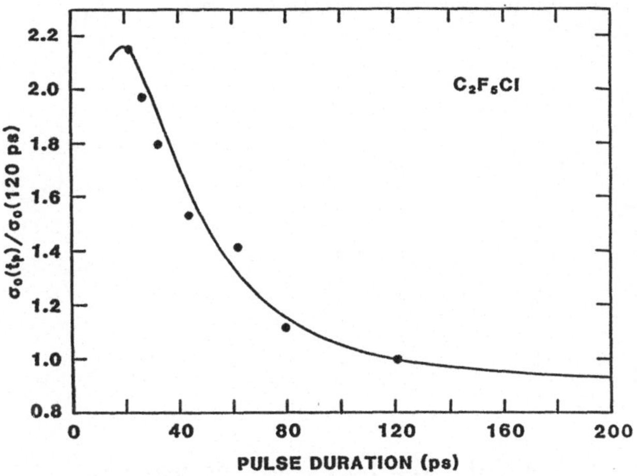

Fig. 3. Experimental data and best fit theoretical curve for C_2F_5Cl.

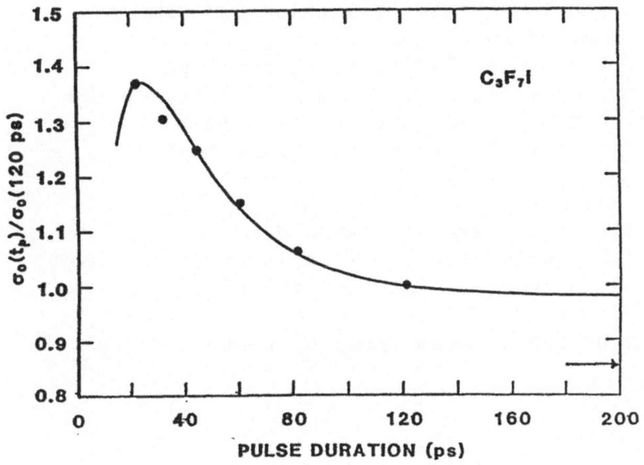

Fig. 4. Experimental data and best fit theoretical curve for C_3F_7I.

In the theoretically fitting procedure, the effects of varying T_1, T_1* were examined carefully. The addition of the off-resonant systems in the inhomogeneous averaging process was also studied. It was found that for 2-levels systems that were more than 3 cm^{-1} away from the laser frequency, their contribution to σ became rather insignificant. As anticipated, most of the contribution to σ came from the on-resonant transition. A method of systematically varying T_1 and T_1^* was also used to obtain the best theoretical fits. In general, the laser intensity was low enough such that there was less than one Rabi oscillation within the laser pulse duration.

T_1 and T_2 represents the homogeneous linewidth of the two-level structure. It is related to the spectral width of the homogeneous ensemble of energy states within the QC that are coupled to the pumped level. Using the calculated values of T_1, this homogeneous width can be estimated to be ∿ 0.5 cm-1. This should be compared with the overall inhomogeneous widths of ∿ 12 cm^{-1} for the molecules studied. On the other hand, T_1^* describes the intramolecular energy transfer rate out of the excited mode. It is quite surprising to find that they are much longer than T_1. A general trend can also be observed from Table I, that T_1* tends to be longer for the smaller molecules. This is intuitively obvious since the density of states is higher for the larger molecules, resulting in a faster energy transfer.

Table I: Summary of best fit results

Molecule	T_1 (ps)	T_1^* (ps)
SF_6	20	188
C_2F_5Cl	24	171
C_3F_7I	28	104

In summary we have carried out a model calculation of the observed change in absorption cross section for various excited polyatomic molecules as a function of the laser pulse duration. These changes are coherent in nature and is due to the incomplete relaxation and damping of the upper states within the laser pulse duration. It should be emphasized that the observations are caused by the short laser pulse duration or the interaction time, and is not due to the intensity of the laser. The intensity effect which results in "super-excitation" of these molecules, is different in nature, but also results in similar observations[7].

The calculated energy transfer times are in the hundreds of picoseconds regime. This points to the distinct possibility of inducing mode-selective chemical reactions in these excited molecules.

This research was supported by the US Department of Energy, Grant Number DEFG 0285 ER 13405.

REFERENCES

1. P. Mukherjee and H.S. Kwok, J. Chem. Phys. 84, 1285 (1986).
2. P. Mukherjee and H.S. Kwok, J. Chem. Phys. 85, 4912 (1986).
3. P. Mukherjee and H.S. Kwok, J. chem. Phys. 85, 5041 (1986).
4. P. Mukherjee and H.S. Kwok, J. Chem. Phys. 87, 128 (1987).
5. E.L. Sibert, W.P. Reinhardt and J.T. Hynes, J. Chem. Phys. 81, 1115 (1984).
6. H. Dubal and M. Quack, Chem. Phys. Lett. 72, 342 (1980).
7. P. Mukherjee and H.S. Kwok, Chem. Phys. Lett. 111, 33 (1984).
8. J. Stone and M.F. Goodman, J. Chem. Phys. 71, 408 (1979).

PICOSECOND DYNAMICS OF INTRAMOLECULAR VIBRATIONAL

REDISTRIBUTION IN JET-COOLED MOLECULES

Gary W. Leach, David R. Demmer, James W. Hager,
Grant A. Bickel and Stephen C. Wallace

Department of Chemistry
University of Toronto
80 St. George Street
Toronto, Ontario. M5S 1A1
Canada

ABSTRACT

We report the observation of quantum interference effects in the energy-resolved fluorescence decays of the second excited singlet state of jet-cooled azulene. The results are analyzed and discussed in terms of a real-time view of intramolecular vibrational relaxation. Preliminary results regarding the role of the methyl rotor in the IVR process of N-methylindole are also presented.

INTRODUCTION

The investigation of quantum interference phenomena (quantum beats) have provided extremely detailed information regarding singlet-singlet state coupling[1], singlet-triplet coupling[2], and more recently, intramolecular vibrational redistribution (IVR) within a single electronic state.[3-8]

In this publication, we present results of investigations of the S_2 excited state dynamics of jet-cooled azulene following picosecond excitation. Quantum interference effects have been observed in the energy-resolved fluorescence decays following excitation of specific vibrational levels of the second excited electronic singlet state. Preliminary investigations of N-methylindole would indicate that the methyl rotor enhances the extent of IVR in this system.

RESULTS AND DISCUSSION

Details of the experimental arrangement have been published elsewhere.[9] Briefly, the output of a cavity dumped, frequency doubled dye laser crossed a continuous free jet expansion of azulene in He. The azulene fluorescence was dispersed by a grating monochromator and detected by a microchannel plate photomultiplier using the time-correlated single photon counting method. The overall time response of this system is approximately 100 ps (fwhm).

Excitation of azulene at relatively low energies above its S_2

origin, to 0^0+662 cm^{-1}, gives rise to single exponential fluorescence decay with a lifetime of 3.2 ns. The lifetime is seen to decrease smoothly with increasing excitation energy, until it is 2.6 ns at 1299 cm^{-1} excess vibrational energy.

In this energy region the fluorescence emission spectra are rather complex, each exhibiting activity in a number of vibrational normal modes. Figure 1 shows emission spectra obtained when azulene is excited to 1315 cm^{-1} and 1299 cm^{-1} above its S_2 origin, respectively. These two spectra are quite similar, both showing activity of fundamental frequencies ν_6, ν_7, ν_8, and ν_9 as reported by Fujji et al.[10] The only major difference between these two spectra is the activity of ν_{27} in the spectrum of the 1315 cm^{-1} band, and its absence in that of the 1299 cm^{-1} band.

Monitoring individual (energy resolved) emission bands following excitation to 1299 cm^{-1} gives rise to simple fluorescence decays of 2.6 ns lifetime. The fluorescence decays of most of the bands in the emission spectrum of the 1315 cm^{-1} excitation exhibit well resolved quantum beats modulated at 2.75 ± 0.02 GHz. The presence of only a single beat frequency would indicate that the laser prepares a coherent superposition of two molecular eigenstates.

Figure 2 shows that while the beats display no detection wavelength dependent frequency change, they do show a detection wavelength dependent phase change. We use the convention that beats which are at a maximum at t=0 (as determined by the peak of the instrument response profile) will be said to be "in-phase", while those that are at a minimum are denoted "out-of-phase" or "phase-shifted".

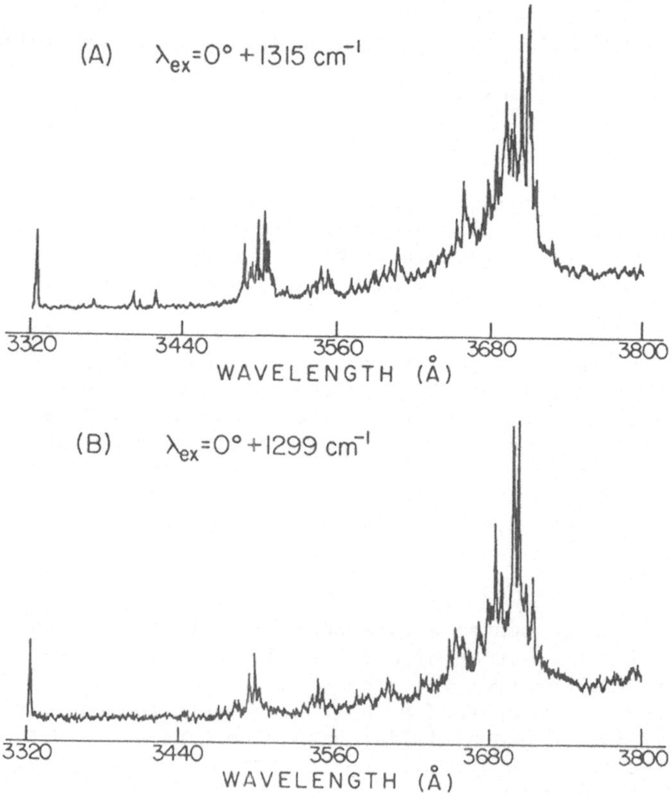

Fig. 1. Dispersed emission spectra of jet-cooled azulene. The spectral resolution of the monochromator is approximately 2Å.

The pioneering work of Felker and Zewail[3-7] and Mukamel[8], has shown that the observation of phase shifted quantum beats provides a direct view of IVR. In-phase emission decays arise from "unredistributed" fluorescence, while out-of-phase emission results from states populated via IVR, that is, "redistributed" fluorescence. The unredistributed fluorescence arises from optically active vibrational modes, i.e., those level directly accessible from v=0 of the ground electronic state. The redistributed fluorescence, however, arises from the population of optically inactive modes via IVR.

Analysis of the beat decays following excitation to 1315 cm^{-1} show that all bands which are out-of-phase are those which involve a fluorescence transition which terminates at the 27_1 level in S_0. Since this fluorescence results from optically inactive levels and follows the $\Delta v=0$ selection rule, this allows us to identify a specific vibronic level populated via IVR, namely 27^1. Also interesting is the fact that ν_{27} is inactive in the emission spectrum of the 1299 cm^{-1} band, where all fluorescence transitions show simple exponential decay.

Excitation to higher energies, roughly 1950 cm^{-1} above the $S_2 0^0$ level results in fluorescence decays of the (in-phase) unrelaxed emission which are at least bi-exponential, consisting of a very rapid (30 ps) major component and a slow (ca. 2 ns) component as displayed in Fig. 3. The slow component is very nonexponential and appears to consist of a large number of Fourier components. Emission detected from (out-of-phase) relaxed bands consists only of the slow, irregularly modulated component.

These results would indicate that in the high energy region, one coherently prepares a large number of molecular eigenstates. The very rapid component corresponds to the dephasing of a large number of Fourier components and cannot meaningfully be referred to as a lifetime. In this case, coupling between the zero-order optically active state and the manifold of optically inactive states is strong leading to essentially irreversible energy flow reflected in the fast decay component of Fig. 3.

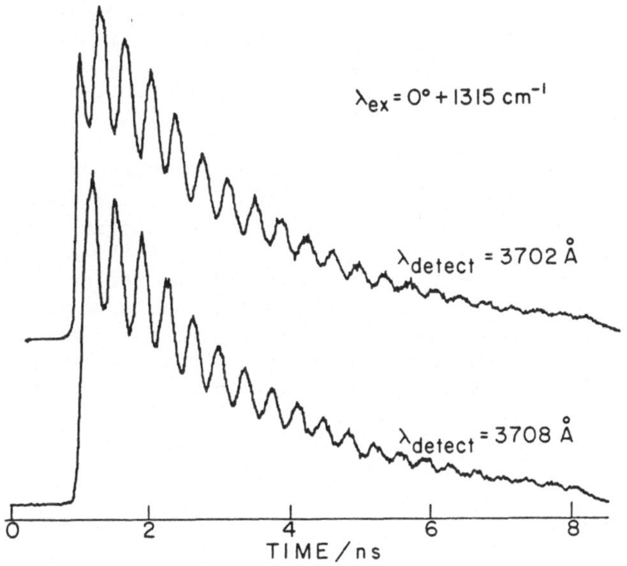

Fig. 2. The energy-resolved fluorescence decays following excitation to the S_2;0^0+1315 cm^{-1} level. The spectral resolution of the monochromator was approximately 2Å

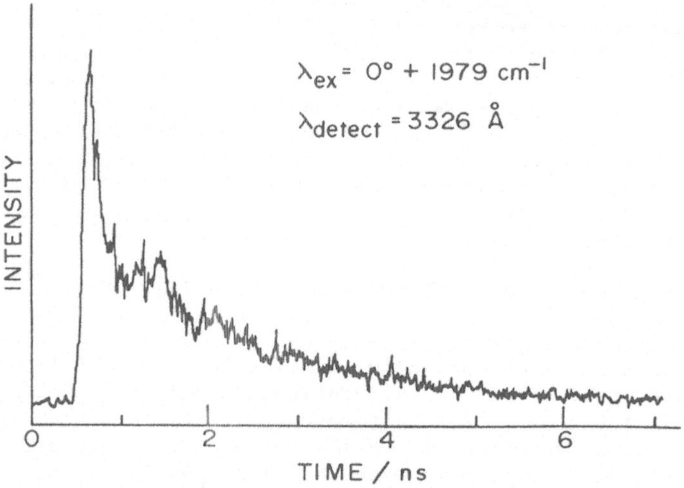

Fig. 3. The energy-resolved fluorescence decay following excitation to the $S_2;0^0+1979\ cm^{-1}$ level. The spectral resolution of the monochromator is approximately 3Å.

IVR processes have been observed in N-methylindole as well. The 650 cm^{-1} band of its first excited singlet state is well isolated, appearing in the low energy, uncongested region of the excitation spectrum. Fig. 4 shows that the dispersed fluorescence spectrum obtained by exciting this mode is quite complex. Each peak in the spectrum has built on it a band like structure extending to lower energy. This structure has been attributed to redistributed fluorescence. Fig. 5 shows that there are significant differences in rise time when monitoring the 717 cm^{-1} ("direct") fluorescence and the nearby redistributed structure ("relaxed") fluorescence. Although, we do not have time resolved data for the other excitation bands, the congestion in the spectrum increases as the excited state energy is increased and shows no mode specificity.

Redistributed fluorescence is not observed for the parent indole molecule at these excess vibrational energies,[11] implying that the methyl rotor promotes the IVR. One possible explanation for this dynamical behavior, is that the oscillation or free rotation of the methyl rotor modulates or scrambles the normal mode motion of the parent moiety through electrostatic interactions. Parmenter and Stone[12] have recently suggested that methyl substitution enhances the extent of IVR for the case of p-fluorotoluene. Further studies to determine the specific role of the methyl rotor in the IVR process are currently underway in this laboratory.

ACKNOWLEDGMENTS

We acknowledge the support of the National Science and Engineering Research Council of Canada and the Killam Foundation.

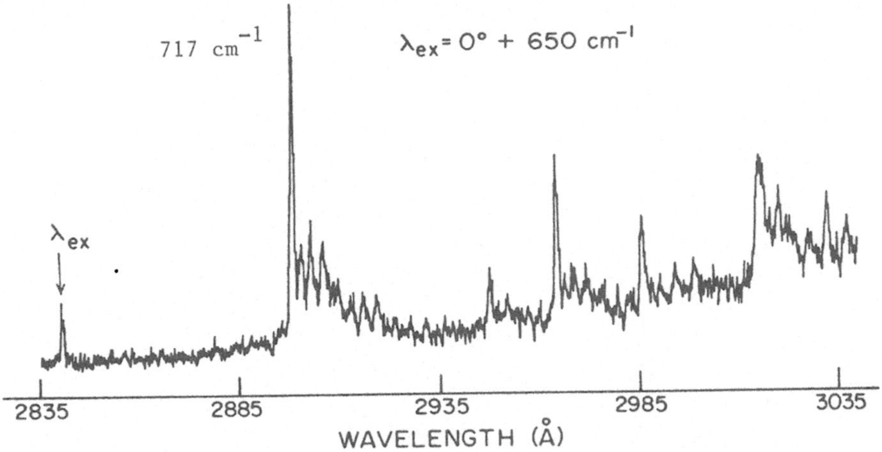

Fig. 4. Dispersed emission spectrum of jet-cooled N-methylindole. The strongest peak is at 717 cm^{-1} above the S_0 origin.

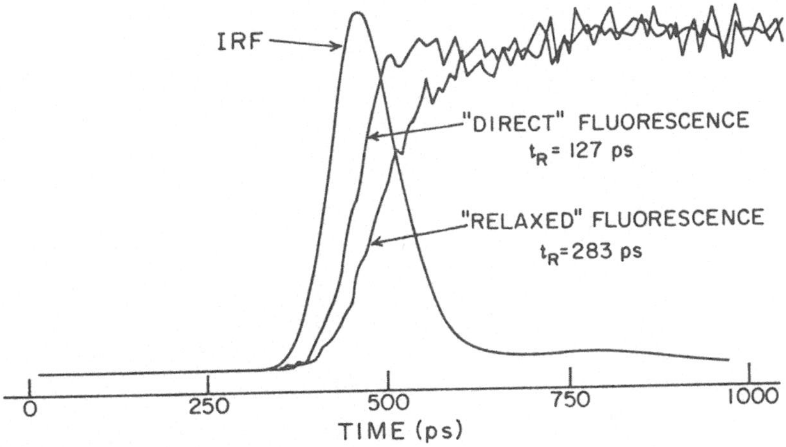

Fig. 5. The early portion of the fluorescence decay curves monitoring the peak at 717 cm^{-1} (direct) and the peaks just to the red (relaxed).

REFERENCES

1. M. Ivanco, J. Hager, W. Sharfin, and S.C. Wallace, J. Chem. Phys. 78(1983) 6531.
2. (a) J. Chaiken, M. Gurnick, and J.C. McDonald, Chem. Phys. Lett. 61(1979) 195.
   (b) J. Chaiken, M. Gurnick, and J.C. McDonald, J. Chem. Phys. 74(1981) 106.
3. P.M. Felker and A.H. Zewail, Chem. Phys. Lett. 102(1983)113.
4. P.M. Felker and A.H. Zewail, J. Chem. Phys. 82(1985) 2961.
5. P.M. Felker and A.H. Zewail, J. Chem. Phys. 82(1985) 2975.
6. P.M. Felker and A.H. Zewail, J. Chem. Phys. 82(1985) 2994.
7. P.M. Felker and A.H. Zewail, J. Chem. Phys. 82(1985) 3003.
8. S. Mukamel, J. Chem. Phys. 82(1985) 2867.
9. D.R. Demmer, J.W. Hager, G.W. Leach, and S.C. Wallace, Chem. Phys. Lett. 136(1987) 329.
10. M. Fujii, T. Ebata, N. Mikami, and M. Ito, Chem. Phys. 77 (1983) 191.
11. Y. Nibu, H. Abe, N. Mikami and M. Ito, J. Phys. Chem. 87(1983) 3898.
12. C.S. Parmenter and B. Stone, J. Chem. Phys. 84(1986) 4710.

SUBPICOSECOND UV/IR ABSORPTION SPECTROSCOPY

J.H. Glownia, J. Misewich, and P.P. Sorokin

IBM Thomas J. Watson Research Center

Yorktown Heights, N.Y.

ABSTRACT

An apparatus based on subpicosecond pulse amplification in excimer gain modules for simultaneously generating both ultrafast, energetic UV excitation pulses and ultrafast UV/IR probe continua is described. Subpicosecond pulses at both 308 and 248 nm, with energies of several millijoules at each wavelength, are available from this apparatus for purposes of excitation. A simultaneously generated UV probe continuum spans the range ~450 to ~235 nm. Both the UV probe continuum and the seed pulses for amplification at 248 nm are produced by focussing the 308-nm pulses into high-pressure gases (Ar and H_2, respectively). Ultrafast IR continuum pulses, extending from 2.2-2.7μm, result from focussing the 308-nm pulses into Ba vapor. The IR continuum can be upconverted to the visible for ease of detection. Application of the apparatus is made to the measurement of the $\tilde{B}$ to $\tilde{A}$ internal conversion rate in DABCO vapor.

INTRODUCTION

Eight years ago the first subnanosecond, broadband absorption measurements in a dilute vapor were reported [1]. In that experiment, the visible supercontinuum [2] generated by focussing an intense mode-locked (~10 psec) 1060-nm laser pulse into liquids was used to probe a sample of *trans* stilbene vapor excited by UV pulses of a few picoseconds duration. A strong absorption band peaking at ~525 nm was detected following photoexcitation. This band has widely been interpreted as a characteristic $S_N \leftarrow S_1$ absorption of *trans* stilbene [1,3]. Its decay is thought to mirror the irreversible evolvement of the initially prepared S_1

state into a denser manifold of states associated with a twisted form of the molecule. However, no transient absorptions due to the latter have ever been reported, and the exact pathways by which vibrationally hot ground state *cis* or *trans* molecules are ultimately formed still remains unknown.

In the belief that further progress in understanding the dynamical behavior following photoexcitation of gas phase molecules can be stimulated by the availability of improved time-resolved spectrographic apparatus (i.e., shorter, more intense, UV excitation pulses; probe continua in spectral ranges other than what is provided by standard condensed matter supercontinua [2]), we have recently made advances toward developing such equipment. In recent papers [4-6] we have described apparatus capable of simultaneously generating both intense subpicosecond UV (308, 248.5-nm) excitation pulses and subpicosecond continua for probing photoexcited molecules via broadband absorption spectroscopy. Both UV (230-450 nm) [4] and IR (2.2-2.7 μm) [5] continua have thus far been produced. We have also demonstrated a method of upconverting the IR to the visible for ease of detection [5].

Other groups have also recently developed new subpicosecond continua in various spectral regions. Production of a visible continuum through the focussing of amplified subpicosecond pulses into high pressure gases was independently reported by P. Corkum's group [7]. In Ref.[7], the primary beam was an intense ultrafast pulse at ~ 630 nm, while the work of Ref.[4] featured a subpicosecond pump beam at ~308 nm. The characteristics of the continua produced are generally the same, except that the wavelength ranges are biased about the respective primary beam wavelengths.

Self-phase modulation (SPM) in various semiconductors is thought to be the mechanism through which intense, subpicosecond pulses in the ten micron region can be spectrally broadened to cover a wide range in the mid-IR [8].

Although the original publication [9] describing highly efficient ~1.64 μm stimulated electronic Raman scattering (SERS) in Tl vapor induced by 15-psec, 248-nm laser pulses did not highlight this effect as an ultrashort IR continuum source, in retrospect it should probably be listed as such. This is because the observed [10] low threshold for SERS in this system should allow the use of spectrally broadened 248.5-nm pump beams (vide infra).

Even more progress has been reported in the generation of intense subpicosecond UV excitation pulses. As an example, Szatmari et. al. [11] generated ~80-fsec, ~ 10^{12}-W pulses at ~248.5 nm through amplification in KrF excimer gain modules.

All the developments sketched above should help promote advances in the area of ultrafast molecular science. Recently we utilized some of the developments to perform the first subpicosecond broadband IR absorption spectroscopy experiment [12]. This particular experiment combines subpicosecond 248-nm excitation with subpicosecond IR continuum probing to measure the $\tilde{B} \rightarrow \tilde{A}$ internal conversion rate in 1,4-diazabicyclo[2.2.2]octane (DABCO) vapor. A diagram of the photophysical processes involved is shown in Fig. 1. The idea that the $\tilde{B} \rightarrow \tilde{A}$ internal conversion rate in DABCO might be fast enough to require ultrafast techniques for its measurement is contained in an earlier study [13], in which the population of the $\tilde{A}$ state was monitored following the application of a 30-nsec, 248.5-nm, KrF laser excitation pulse. A rapid $\tilde{B} \rightarrow \tilde{A}$ internal conversion rate for DABCO was also implied in a recent two-color laser photoionization spectroscopy study [14].

The above-mentioned DABCO experiment well illustrates the general technique of using a subpicosecond continuum to measure an ultrafast process. Accordingly, we now present a brief description of this experiment.

EXPERIMENTAL ASPECTS OF THE SUBPICOSECOND DABCO EXPERIMENT

Collimated 2-mJ, 250-fsec, 248.5-nm pulses were sent unfocused (beam dimensions: 2 cm x 1 cm) into a 60-cm long cell containing DABCO at its ambient vapor pressure (~0.3 Torr) together with 100 Torr of H_2. The linear absorption of the DABCO at 248.5 nm (40229 cm^{-1}) was more than fifty percent, even though this wavelength lies near the point of minimum absorbance between the $v' = 0 \leftarrow v'' = 0$ (39807 cm^{-1}) and next highest vibronic peaks of the lowest energy, dipole allowed, band [15,16]. This band system has been assigned [17,18] as $\tilde{B}(^1E'[3p_{x,y}(+)]) \leftarrow \tilde{X}(^1A_1'\ [n(+)])$. Optical transitions from the ground state $\tilde{X}$ to the first excited state, the $\tilde{A}\,^1A_1'\ [3s(+)]$ (origin at 35785 cm^{-1}) are one-photon forbidden, two-photon allowed [17,18].

The 160-fsec IR continuum pulses which probe the $\tilde{A}$ state population were directed through the vapor colinearly with the UV photoexcitation pulses, upconverted to the visible, and then dispersed in a spectrograph equipped with an unintensified OMA detection system. The pump-probe delay could be varied up to ± 1 nsec by means of an optical delay arm.

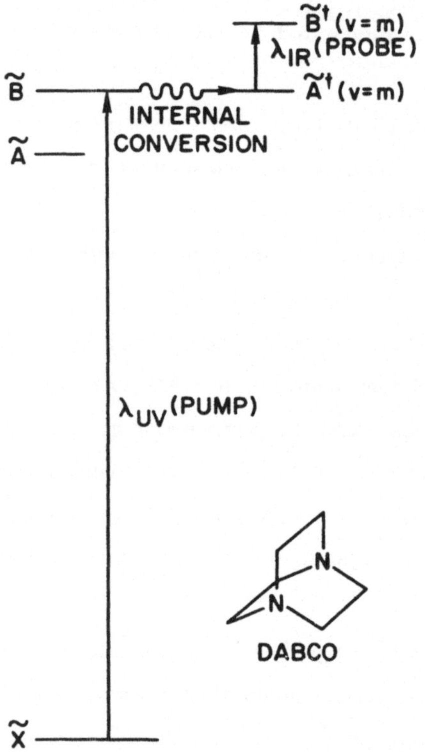

Fig. 1 Diagram of photophysical processes involved in the DABCO experiment.

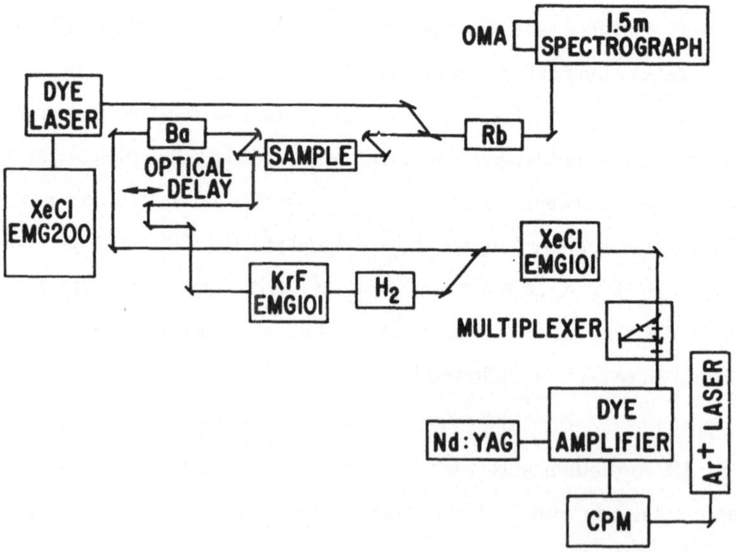

Fig. 2 Diagram of the experimental apparatus used in the DABCO experiment.

Absorbances were computed by comparison of upconverted intensities recorded with and without the UV pump blocked.

A block diagram of the experiment is shown in Fig. 2. Subpicosecond pulses at ~616 nm are formed in a CPM laser, amplified, then frequency doubled, forming seed pulses at ~308 nm for further amplification in the XeCl excimer gain module [6]. Amplification of the UV pulses in the latter occurs in the form of pairs of orthogonally polarized pulses, spaced 2-3 nsec apart, formed in the multiplexer [4]. The 160-fsec amplified UV pulse pairs are separated by a polarization sensitive coupler into pump and probe channels. The pump channel 308-nm pulses are Raman shifted in H_2 gas to form seed pulses for amplification at 248.5 nm in a KrF module [4]. The probe channel pulses are Raman shifted in Ba vapor to form IR probe continuum pulses [5]. The narrow band pulsed dye laser drives the Rb upconverter [5].

RESULTS AND DISCUSSION OF THE DABCO EXPERIMENT

Figure 3a shows the absorbance recorded when the probe is delayed ~4 psec with respect to the pump (point (a) in Fig. 4), while Fig. 3b displays the absorbance with the probe arriving just before the pump (point (b) in Fig. 4). The absorbance recorded at 2.494 μm, as a function of probe delay, is shown in Fig. 4. The large absorption band that develops represents transitions $\tilde{B}^{\dagger} \leftarrow \tilde{A}^{\dagger}$ of vibrationally excited $\tilde{A}$ state molecules, containing up to 4400 cm^{-1} of vibrational energy. Since the $\tilde{B} \leftarrow \tilde{A}$ transition is one that occurs between Rydberg states, vertical ($\Delta v = 0$) transitions are expected. Thus it is not surprising that the peak of the band in Fig. 3a appears very close to the $\tilde{B} \leftarrow \tilde{A}$ peak for vibrationally equilibrated $\tilde{A}$ state molecules (see [12] and [13] for a spectrum of the latter).

The computer generated curve in Fig. 4 is a nonlinear least squares fit to the data. The fit indicates a risetime of ~500 fsec. Although the infrared and ultraviolet pulsewidths were determined by autocorrelation to be ~160 fsec and ~250 fsec, respectively, the cross-correlation between these pulses has not been measured. Thus, the risetime in our experiment could be limited by the laser system cross-correlation. In any case, the process converting DABCO states accessed by the subpicosecond 248.5-nm pump beam into vibrationally excited $\tilde{A}$ states is observed to occur on a time scale that is at least as fast as ~500 fsec. That internal conversion to vibrationally excited $\tilde{A}$ state molecules is the dominant process for photoexcited DABCO molecules, even at UV pump intensities ~ 4GW/cm^2 , is also underscored by the fact that there is no apparent diminution in the

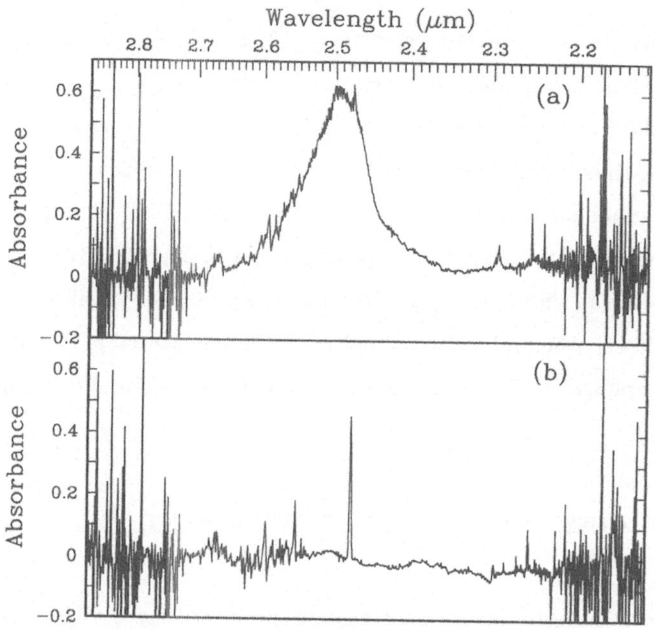

Fig. 3 (a) Absorbance (base 10) with probe delayed ~4 psec with respect to pump. (b) Absorbance with probe pulse preceding pump pulse.

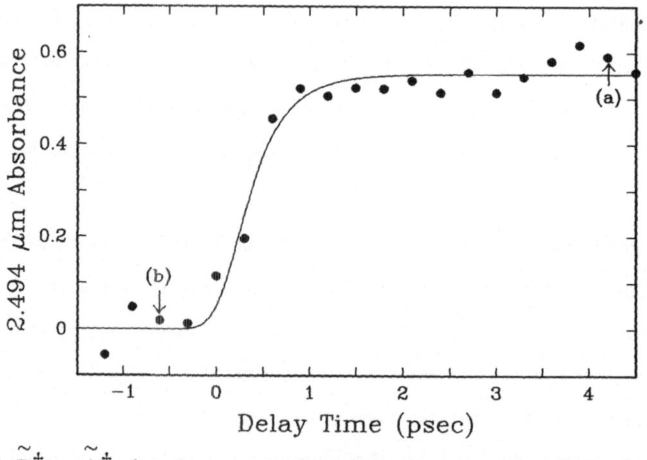

Fig. 4 Peak $\tilde{B}^{\dagger} \leftarrow \tilde{A}^{\dagger}$ absorbance as a function of probe pulse delay time with respect to pump pulse.

integrated intensity of the 2.5-μm absorption band induced at these pump intensities, as compared with the 2.5-μm integrated intensity induced by 25-nsec UV pulses at comparable fluence levels.

As mentioned above, the spectrum of our subpicosecond excitation pulse happens to fall roughly halfway between the $v' = 0 \leftarrow v'' = 0$ and next highest vibronic peaks of the $\tilde{B} \leftarrow \tilde{X}$ band. Thus a question exists as to whether it is really the $\tilde{B}$ state that is accessed by the 248.5-nm excitation pulse. The underlying continuum apparent between the $\tilde{B} \leftarrow \tilde{X}$ vibronic peaks may be the result of a transition to another state or states. In an attempt to gain further insight into the excited state dynamics, we spectrally broadened and shifted the 248.5-nm subpicosecond laser pulse to overlap the $\tilde{B} \leftarrow \tilde{X}$ ($v' = 0 \leftarrow v'' = 0$) transition. This was accomplished by focussing the 3.5-mJ, 250-fsec, 248.5-nm amplified pulses into a 80-cm long cell containing N_2 gas at ~600 lbs/in^2 pressure. A 47.5-cm f.l. lens was used to focus the beam into the high pressure cell, the windows of which were CaF_2. The spectrum of the output beam from the cell was observed to be ~1000 cm^{-1} wide; its peak was shifted to ~251 nm, a wavelength almost coincident with the DABCO $v' = 0 \leftarrow v'' = 0$ peak. Applying this spectrally shifted UV beam to DABCO vapor in a 10-cm-long cell, we repeated the time-resolved experiment described above. Although the absorbance of the induced 2.5-μm band was reduced to ~0.15, the signal-to-noise ratio remained sufficiently high to ascertain that this band again appeared with a comparable subpicosecond risetime. No evidence of an additional transient absorption feature was observed. In view of the measured ~3 cm^{-1} $\tilde{B} \leftarrow \tilde{X}$ ($v' = 0 \leftarrow v'' = 0$) width for jet-cooled DABCO molecules [14], the intrinsic internal conversion time for $\tilde{B}(v' = 0)$ state molecules should be no shorter than ~3 picoseconds. (The room temperature $\tilde{B} \leftarrow \tilde{X}$ ($v' = 0 \leftarrow v'' = 0$) width is ~80 cm^{-1}, but this most likely is due to inhomogeneous broadening arising from thermally excited ground state molecules.) Clearly, the $\tilde{B} \rightarrow \tilde{A}$ internal conversion could occur on a longer time scale. However, we did not observe the $\tilde{B} \rightarrow \tilde{A}$ gain that one would expect if this were the case. Presently we have no explanation for this discrepancy.

CONCLUSIONS AND PLANS FOR FUTURE WORK

We have developed a new system for subpicosecond pump-probe absorption experiments, with a UV (248.5, 308 nm) pump and an IR (2.2-2.7 μm) probe. This system has been applied to study internal conversion of DABCO vapor photoexcited in the region of the $\tilde{B} \leftarrow \tilde{X}$ ($v' = 0 \leftarrow v'' = 0$) transition. A subpicosecond ($\lesssim$500 fsec) rate has been determined for this process. There are plans to extend these techniques to cover other IR probe wavelength ranges, to provide better signal-to-noise ratios via the incorporation of a reference channel, and to shorten the pulsewidth of the 248.5-nm excitation pulse to ~50 fsec, the bandwidth limit for amplification in KrF excimer modules [11]. Such an improved system should be useful in real time spectral studies of photoexcited molecules.

ACKNOWLEDGEMENTS

We acknowledge the technical support of L.H. Manganaro. This work is supported in part by the U.S. Army Research Office.

REFERENCES

1. B.I. Greene, R.M. Hochstrasser, and R.B. Weisman, J. Chem. Phys. 71 (1979) 1544.
2. R.R. Alfano and S.L Shapiro, Phys. Rev. Lett. 24 (1970) 584, 592.
3. K. Yoshihara, A. Namiki, M. Sumitani, and K. Nakashima, J. Chem. Phys. 71 (1979) 2892.
4. J.H. Glownia, J. Misewich, and P.P. Sorokin, J. Opt. Soc. Am. B 3 (1986) 1573.
5. J.H. Glownia, J. Misewich, and P.P. Sorokin, Opt. Lett. 11 (1987) 19.
6. J.H. Glownia, J. Misewich, and P.P. Sorokin, J. Opt. Soc. Am. B, 4 (1987) 1061.
7. P.B. Corkum, C. Rolland, and T. Srinivasan-Rao in Ultrafast Phenomena V , edited by G.R. Fleming and A.E. Siegman, Springer-Verlag, 1986.
8. P.B. Corkum, P.P. Ho, R.R. Alfano, and J.T. Manassah, Opt. Lett. 10 (1985) 624.
9. P.H. Bucksbaum and J. Bokor in Excimer Lasers -1983, AIP. Conf. Proc. 100 , edited by C.K. Rhodes, H. Egger, and H. Pummer (1983).
10. P.H. Bucksbaum, private communication.
11. S. Szatmari, F.P. Shaefer, E. Mueller-Horsche, and W. Mueckenheim, Optics Communication, in press.
12. J.H. Glownia, J. Misewich, and P.P. Sorokin, Chem. Phys. Lett. (in press).
13. J.H. Glownia, G. Arjavalingam, and P.P. Sorokin, J. Chem. Phys. 82 (1985) 4086.
14. M.A. Smith, J.W. Hager, and S.C. Wallace, J. Phys. Chem. 88 (1984) 2250.
15. A.M. Halpern, J.L. Roebber, and K. Weiss, J. Chem. Phys. 49 (1968) 1348.
16. Y. Hamada, A.Y. Hirakawa, and M. Tsuboi, J. Mol. Spectr. 47 (1973) 440.
17. D.H. Parker and P. Avouris, Chem. Phys. Lett. 53 (1978) 515.
18. D.H. Parker and P. Avouris, J. Chem. Phys. 71 (1979) 1241.

VIBRATIONAL PREDISSOCIATION DYNAMICS OF THE NITRIC OXIDE DIMER

Michael P. Casassa, John C. Stephenson, and David S. King

Molecular Spectroscopy Division
National Bureau of Standards
Gaithersburg, MD 20899 USA

ABSTRACT. Details of experimental measurements of the total energy distribution and time dependence of the vibrational predissociation of the nitric oxide dimer are presented. Energy disposal measurements indicated the fragments to exhibit low average rotational energy, full equilibration of the lambda doublet species, approximately equal populations in both spin-orbit states, no significant degree of alignment, an isotropic flux distribution, and kinetic energies of 400 cm^{-1} per fragment. Although approximately 75% of the available energy goes into fragment translation, picosecond laser pump - probe experiments showed that ν_1, excited at 1870 cm^{-1} decayed exponentially with a 880 ps lifetime. Excitation of ν_4 at 1789 cm^{-1} gave a 39 ps predissociative lifetime.

I. INTRODUCTION

Vibrational predissociation is observable for van der Waals molecules (vdW) since a single quantum of vibrational excitation in a constituent of such a cluster generally exceeds the binding energy of the complex. Thus, vibrational excitation leads to the rupture of the weak vdW bond. Herein lies the appeal of these systems as models for vibrational dynamics: rich and dramatic photochemistry is caused by excitation to low lying vibrational levels. A combination of time-resolved and final-state-resolved measurements are needed to thoroughly characterize the predissociation dynamics of these clusters. Experimental observables include product kinetic, rotational and vibrational energy distributions, vector velocity and angular momentum distributions, and dissociation lifetimes. Although there have been many studies of vibrational energy flow within electronically excited states of vdW molecules, due to experimental difficulties most studies involving ground electronic states of vibrationally energized clusters have been limited to the measurement of infrared absorption and photodissociation spectra.[1]

In this paper, experiments are discussed which utilize molecular beam and pulsed laser techniques to characterize the rates and mechanisms important in the predissociation of the nitric oxide dimer excited to the v=1 level of either the symmetric (ν_1) or antisymmetric (ν_4) stretch within the electronic ground state. The photochemistry which is observed is:

$$(NO)_2 \xrightarrow{h\nu} (NO)_2^*(v=1) \xrightarrow{t_{vp}} 2\ NO(J,\Omega,\Lambda) + E_K$$

where final states are designated by their total angular momentum J, the electronic angular momentum Ω, and the lambda doublet component Λ . E_K is the relative kinetic energy of the fragments and $t_{vp} = k_{uni}^{-1}$ is the lifetime of the energized complex. Conservation of energy arguments imply a $(NO)_2$ vdW bond energy (see below) $D_0 \approx 800$ cm^{-1}. Photodissociation in the region of the stretching fundamentals therefore produces ≈ 1000 cm^{-1} of excess energy which must be disposed into product rotational, vibrational, and translational degrees of freedom. Due to the small NO rotational constant (B=1.7cm^{-1}), energy gap arguments would lead to the expectation of long lived vibrationally excited dimers.[1]

II. EXPERIMENTAL

Van der Waals complexes were formed by expansions of gas mixtures through the 0.75 mm diameter nozzle of a pulsed free jet. The valve was of the Gentry-Giese (hairpin) design and produced molecular pulses of 70 μs full-width at half-maximum (FWHM) duration. The infrared (5μm) pump laser traversed the molecular beam at a distance of 15 nozzle diameters downstream. At a 5 atm. backing pressure the on-axis molecular beam density in this region was $\approx 10^{17}$ cm^{-3}. At some time (t_d) after the excitation pulse the nitric oxide fragments formed from the vibrationally excited vdW complexes were probed by laser excited fluorescence techniques (LEF).[2-4]

The probe laser for the energy distribution measurements was a frequency doubled, Nd^{+3}:YAG- pumped dye laser with a 9 ns FWHM duration and either a 0.3 or 0.017cm^{-1} FWHM bandwidth at 44 200 cm^{-1}. When tuned to resonance with an A NO(V'=0;J') <-- X NO(V=0;J,Ω,Λ) transition, the resulting LEF intensity is proportional to the density of NO fragments in the level (J,Ω,Λ) at the time t_d during which the probe laser passed through the beam chamber. The fluorescence was collected by f/0.8 optics, passed through spectral filters to discriminate against laser scatter and a spatial filter to define the viewing region and detected by a solar blind (CsTe) photomultiplier tube. The LEF signals were recorded using a gated integrator and were normalized shot-by-shot to the probe laser energy. The state distributions of the fragments were determined by scanning the wavelength of the probe laser. By sufficient narrowing of the bandwidth of the excitation laser (i.e., to 0.017 cm^{-1}) the kinetic energy of the fragments was determined by Doppler profile measurements. The ir pump and uv probe lasers were aligned either colinear or at right angles. Independent control over both the orientation of laser electric vectors and propagation directions allowed for the assessment of fragment flux distributions and alignment.

The picosecond laser system used to measure the predissociation lifetimes employed a mode locked cw YAG laser to synchronously pump two tunable dye lasers. The outputs of which were amplified by pulsed dye amplifiers pumped by a frequency doubled YAG laser (of 2 ns duration). The amplified outputs of these red-yellow lasers were each approximately 0.5 mJ per pulse. These pulses occured at 20 Hz and were transform limited in duration and spectral bandwidth. Presently, with 3-plate birefringent tuning elements in the dye laser cavities, the amplified pulses had 3 cm^{-1} bandwidths and 7.5 ps FWHM autocorrelations and 10 ps crosscorrelations. The ps pulses can be shortened in time, if necessary, at the expense of a broader spectral output.

To generate tunable ir pulses, difference frequency mixing between the two amplified dye laser outputs was done in $LiIO_3$ (for ω_{ir} = 1780 to 3500 cm^{-1}). The ir pulse energy obtained was about 10 μJ. Vibrational predissociation was observed as fragment appearance using the NO(fragment) uv LEF. Picosecond uv probe pulses near 226 nm were generated by mixing the frequency doubled output of one amplified dye laser with the 2 ns fundamental of the amplifier YAG in KD*P. This leg of the ps system produced 5 to 10 μJ of uv energy. The time-resolved ir pump - LEF probe measurements of $(NO)_2$ fragmentation reported herein were performed using a dye laser wavelength of $\lambda_{dye\ 1}$ = 576.7 nm (ω_1=17 341 cm^{-1}) to produce uv probe pulses tuned to the P_{22} bandhead, i.e., probing fragments formed in low J-levels of the NO(Ω=3/2) spin-orbit state. For ν_1 photodissociation, the wavelength of the second dye laser was tuned to $\lambda_{dye\ 2}$ = 646.4 nm as required to generate the difference frequency $\omega_1 - \omega_2 = \omega_{ir}$ = 1870 cm^{-1}; for ν_4 excitation, $\lambda_{dye\ 2}$ = 643.0 nm, gave ω_{ir} = 1789 cm^{-1}. The time evolution of the fragment concentration was measured by optically varing the delay time between ir pump and uv probe pulses.

III. RESULTS AND DISCUSSION

III. A. Final State Distributions

Final state population distributions[3] were measured under a wide range of molecular beam expansion conditions. In particular, expansions of 1% NO in He at a backing pressure of 5.7 atms and 0.67% NO in H_2 at 8.3 atms gave rise to photolysis fragments that could be attributed predominantly to the fragmentation of the dimer. Following ν_1 excitation, the data for fragment species with rotational energy less that 450 cm^{-1} (>97% of all population lies in these levels) were well fit by the rotational temperatures T_r(Ω=1/2)=101±11 K and T_r(Ω=3/2)=112±12 K; both spin- orbit states were fit as separate ensembles. The ratio of total population in the two spin-orbit states was 0.9±0.2. The population distributions for expansions in He and H_2 were essentially indistinguishable.

The two spin-orbit states differ in energy, for a given value of J, by an amount approximately equal to the 123 cm^{-1} spin-orbit separation. The ratio of population in these two spin-orbit states was substantially different from the value expected if the spin-orbit and nuclear rotational degrees of freedom were equilibrated, as observed for NO fragments formed from a number of thermal processes.[2] If this condition were achieved, the populations in the Ω=3/2 state would have been a factor of 5 lower than observed. That the ratio of spin-orbit populations was near unity implies a strong spin correlation in the dissociation mechanism which is consistent with the reaction mechanism $(NO)_2$ --> NO(Ω=1/2) + NO(Ω=3/2).

Many direct electronic photodissociation reactions result in an alignment of the fragments.[5] For the NO A-X band system, high-J, Q branch transitions are polarized parallel to the axis of nuclear rotation. Thus a measure of the polarization dependence of the LEF signal can in principle map out the alignment of the rotating fragments, i.e., their m_J distribution. Using counter-propagating lasers and monitoring LEF signal intensities from fragments in levels J<9.5 the electric vector of the probe laser was rotated, with respect to the electric vector of the pump laser, through 180° degrees using a half-wave retardation plate. There was no measurable polarization anisotropy; i.e., no significant alignment of the fragments.

Kinetic energy distributions and fragment flux distributions were measured by Doppler spectroscopy with a uv probe laser bandwidth of 0.017 cm^{-1}. Doppler profiles of nearly rectangular shape, independent of experimental geometry and expansion conditions, and with a FWHM of 0.165 cm^{-1} was observed. Computer simulations of Doppler profiles were performed taking into account the actual laser resolution, pump laser photon energy, and final state distributions. Assuming the dimers in the expansion to be essentially equilibrated with the beam temperature (T≃1K), and allowing the internal energy of the co-fragment to be randomly chosen from the values available gave good fits to the observed FWHM=0.165 cm^{-1} near top-hat Doppler profiles. This gives an average per-fragment kinetic energy of 400±50 cm^{-1}. Subtraction of the internal and kinetic energies from the IR photon yields a vdW bond energy of 800±150 cm^{-1}.

III. B. Rates

Picosecond ir and uv pulses were used to measure the actual dissociation behavior of NO-dimers following separate excitations of the $(NO)_2$ ν_1(v=1) and ν_4(v=1) bands at 1870 and 1789 cm^{-1} respectively. If the ir laser excited an ensemble which dissociated with a single unimolecular rate constant, $k_{uni} = 1/t_{vp}$, the expected fragment appearance behavior would be: $S(t_d) = S(\infty)[1-\exp(t_d/t_{vp})]$. Of course non-exponential or multi-exponential decay could occur. Non- linear least-squares fits of single exponential decays to the data, yielded lifetimes of t_{vp} = 880±260 picoseconds for ν_1 and 39±8 picoseconds for ν_4. In the experiments these lifetimes represent average values for a distribution of NO-dimer rotational levels. Assuming an infrared bandwidth of 4-5 cm^{-1} (the convolution of two 3 cm^{-1} Gaussians) and thermalization of the dimers with the beam (T≃1K), the pump laser interacts with essentially all the NO-dimers in the molecular beam (i.e., the NO-dimer population in levels with $J''\leq 3$, $K_a''\leq 2$ accounts for 95% of the photodissociation signal).

That ν_4(1789 cm^{-1}) decays much faster than ν_1 (1870 cm^{-1}) is inconsistent with standard statistical theories of unimolecular reaction which predict unimolecular rates to increase with reactant internal energy. On the other hand, models of vibrational predissociation involving pure V-T energy transfer[6] correlate increased lifetime with increased energy release. However, predissociation from ν_1 only releases 81 cm^{-1} more energy than ν_4 and theory quantitatively predicts a much smaller difference in rate (factor of ≤2) than was observed. In the case of $(NO)_2$, the predissociation might be nonadiabatic since the $^2\Pi$ configurations of the NO fragments combine to form several low-lying electronic states in the dimer. Such nonadiabatic effects have been suggested to explain the anomalously large collisional cross-section observed for vibrational deactivation of NO(v=1) by NO(v=0).[7] The different overall symmetries of the ν_1 and ν_4 levels of the dimer $^1A_1$ ground state may constrain coupling to low lying repulsive state(s), causing the observed difference in lifetimes. Ongoing measurements of product energy distributions and lifetimes for states of different vibrational symmetries (e.g., combination bands involving the vdW modes) might elucidate the source of the difference in predissociation rates.

IV. REFERENCES

1. For a review, see K.C. Janda, Adv. Chem. Phys. 60, 201 (1985).
2. D.S. King and J.C. Stephenson, J. Chem. Phys. 82, 5826 (1985).
3. M.P. Casassa, J.C. Stephenson, and D.S. King, J. Chem. Phys. 85, 2333 (1986).
4. J.C. Stephenson and D.S. King, J. Chem. Phys. 78, 1867 (1983).
5. P. Andresen, G.S. Ondrey, B. Titze, and E.W. Rothe, J. Chem. Phys. 80, 2548 (1984).
6. J.A. Beswick and J. Jortner, Adv. Chem. Phys. 47, 363 (1981).
7. J.T. Yardley, Introduction to Molecular Energy Transfer, Academic Press (New York, 1980).

MULTIPHOTON ABSORPTION FRAGMENTATION MECHANISMS OF CH_3I BY PICOSECOND AND NANOSECOND LASER MASS SPECTROMETRY

Diane M. Szaflarski and M.A. El-Sayed

Department of Chemistry and Biochemistry
University of California
Los Angeles, California 90024-1569

INTRODUCTION

Previous attempts to determine pathways of fragment formation in multiphoton ionization studies have been made by recording and analyzing power dependent mass spectra. These measurements are not satisfying in that accurate power measurements are difficult, and the exponents extracted from log-log plots of ion signal verses laser power pertain only to the rate determining steps. This leads to some ambiguity in the assignment of ion formation mechanisms. Thus even though this method has been used to determine trends in the MPI process, more quantitative measurements need to be developed in order to definitively assign these mechanisms.

Recently, a technique involving a pulsed field linear reflectron mass spectrometer and a pulsed laser has been developed[1,2] to measure ion dynamics on the microsecond timescale. Using this technique ion dynamics such as formation times of fragment ions and kinetic energies of fragments can be measured. In this paper we report the application of this technique, in addition to wavelength and power dependence studies, to determine the mechanism responsible for the formation of I^+ from CH_3I when picosecond and nanosecond lasers are used as ionizing sources.

CH_3I was choosen because it is a molecule which can form ionic fragments such as I^+ by two mechanisms, each of which would produce I^+ with drastically different kinetic energies. Thus these mechanism can be distinguished by measuring the kinetic energy of I^+.

Iodine ion, I^+, whose appearance potential is 12.9 eV,[3] can be formed when the molecule absorbs three 266 nm photons. In this case I^+ is formed by a statistical dissociation of CH_3I. This mechanism is referred to as the ionic ladder mechanism. The kinetic energy of I^+ in this case is determined by statistically partitioning the excess energy (three 266 nm photons - appearance potential) among the degrees of freedom. The Franklin equation[4] is used to predict the kinetic energy of I^+ to be 0.04 eV.

Secondly, I^+ could be formed by a ladder switching mechanism. In this case one 266 nm photon is absorbed causing the C-I bond to break. Then I can be ionized by a two photon resonant, one photon ionization process. Here the kinetic energy of I^+ is approximated by assuming the internal energy of CH_3 is negligible and using the conservation of energy and momentum. According to this approximation the kinetic energy of I^+ is 0.25 eV.

EXPERIMENTAL

The ion dynamic experiments are performed using the fourth harmonic of the Nd^{+3}:YAG picosecond (Quantel International) or nanosecond (Quanta Ray) laser in conjunction with a reflectron mass spectrometer. The reflectron mass spectrometer is equipped with a high voltage pulser (PARR 1211) which extracts the ions out of the ionization region at variable times after they are formed by a focused laser pulse. The ion signals are monitored as a function of delay time between the laser pulse and the electric field pulse. The ion signals decay due to the fact that they are moving with some velocity which carries them out of the collection volume. This motion is due to thermal velocity and in some cases recoil velocity which results from the photodissociation.

The ion decays are fit to single exponentials. The decay constants are converted to kinetic energies following the developement of Tai and El-Sayed.[1] The parent ion is assumed to be formed with thermal kinetic energy. With this assumption the geometric parameters of the mass spectrometer such as the effective collection volume can be determined.

Another type of mass spectrometer, a time-of-flight mass spectrometer, and the picosecond laser were used to record the power and wavelength dependent mass spectra. The fourth harmonic of the Nd^{+3}:YAG is used for the power dependence. For the wavelength dependence the 266 nm light is anti-stokes Raman shifted using methane gas. The first anti-stokes Raman line (246 nm) is used.

RESULTS AND DISCUSSION

The mass spectra generated using the picosecond and nanosecond lasers at 266 nm are shown in figure 1. The nanosecond spectra (top) has mass peaks from CH_3I and C_6H_6 (benzene). In this spectra essentially no parent ion from methyl iodide is observed. For this reason benzene had to be added for calibration purposes and normalization in the dynamic experiments.

The mass spectrum generated by the picosecond laser is also shown in figure 1 (bottom). In this case p-dichlorobenzene is also present. Here the parent ion peak is 25% the size of I^+. This picosecond mass spectrum was at first surprising since it is well known that one 266 nm photon accesses a dissociative state. In order to observe the parent ion, the second photon absorption process must effectively compete with the dissociation. The observation of this parent ion led us to further investigate this molecule to determine mechanism from which the fragment ions are being formed.

The picosecond laser fragmentation pattern of CH_3I^+ at 266 nm was examined at different laser powers. This is shown in the top (9 uJ/pulse) and middle (30 uJ/pulse) spectra in figure 2. At 30 uJ the parent ion is observed in small quantity. It is being fragmented and

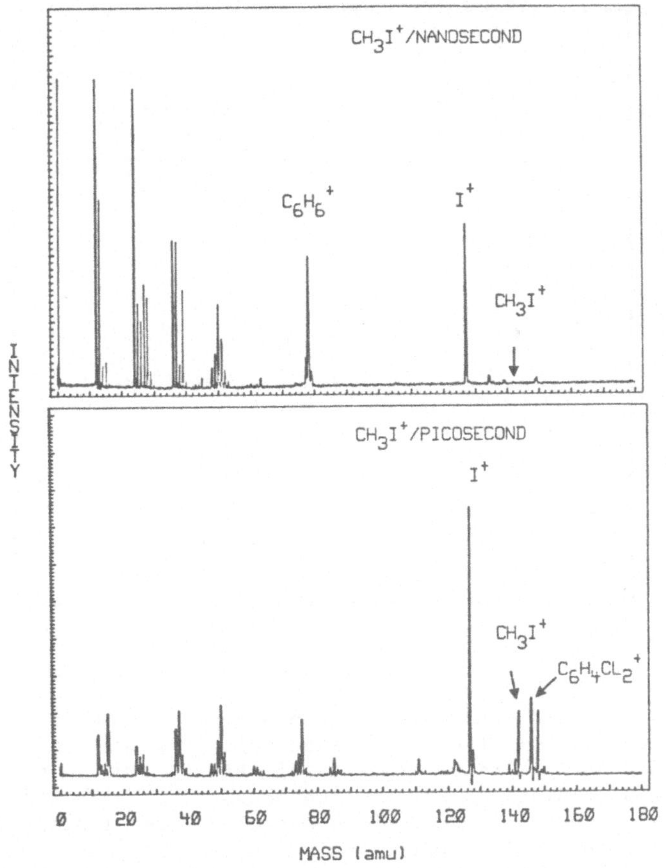

Figure 1: The mass spectrum of CH_3I recorded using a reflectron mass spectrometer and a nanosecond (top), and a picosecond (bottom) laser. The parent ion, CH_3I^+, is observed when the picosecond laser is used, but not when the nanosecond laser is used.

forming the smaller fragments I^+, CH_3^+, CH_2^+, CH^+ and C^+. When the laser power is decreased to 9 uJ the parent ion is enhanced relative to the fragment ions. The fragment ions cannot be eliminated completely. This is because the ionization potential of CH_3I is 9.5 eV,[5] thus three 266 nm (4.67 eV) photons are required to ionize. Once three photons are absorbed the molecules can either ionize or ionize and fragment resulting in parent and fragment ion formation.

If the photon energy is increased to 246 nm (5.03 eV) two photons will exceed the ionization potential forming only the parent ion. A third photon will fragment the parent ion. The mass spectrum resulting from 246 nm photon absorption (5 uJ/pulse) is shown in figure 2 (bottom). The parent ion is generated as expected. None of the fragment ions appear because the laser intensity too low. Unfortunately, this mass spectrum was generated using the maximum 246 nm intensity that we could obtain.

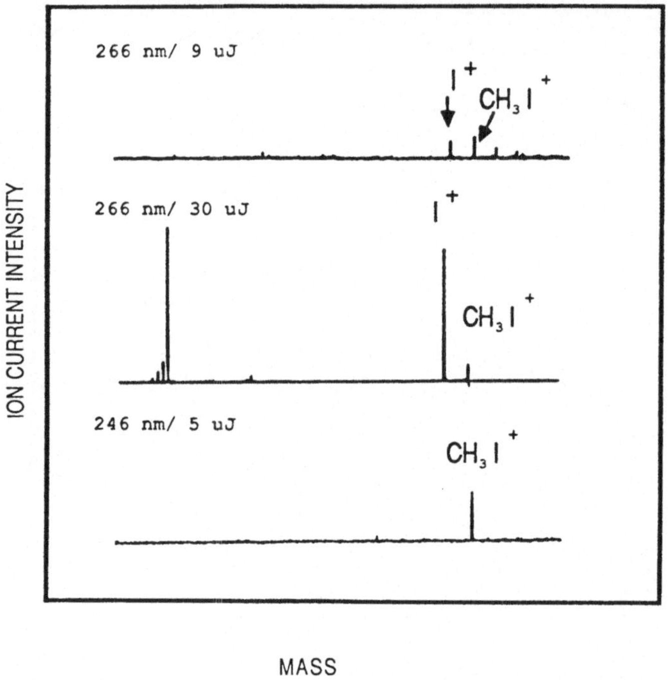

Figure 2: Picosecond power dependent mass spectra of CH_3I at 266 nm (top two spectra). As the power is decreased the relative amount of the parent ion is increased. Also shown is the mass spectra at 246 nm (bottom). In this case 2 photons exceed the ionization potential and CH_3I is efficiently ionized.

Finally the kinetic energy of I^+ formed by the picosecond and nanosecond lasers was measured. Using the reflectron mass spectrometer the ions were monitored as a function of delay time between the laser and the pulsed electric field. The results are shown in figure 3. On the left and right are the ion decays from the nanosecond and picosecond experiments, respectively. The kinetic energy for I^+ generated by the picosecond laser is thermal, 0.05 ± 0.03, whereas for the nanosecond laser the kinetic energy is 0.25 ± 0.03 eV. Comparison of these values to the predictions made previously suggests that the picosecond laser produces I^+ from the ionic-ladder mechanism whereas the nanosecond laser produces I^+ from the ladder-switching mechanism. The kinetic energy measurements along with the wavelength and power dependent mass spectra give strong support for this explanation. This observation of the two mechanisms when different pulsewidth lasers are used is discussed in a previous publication.[6]

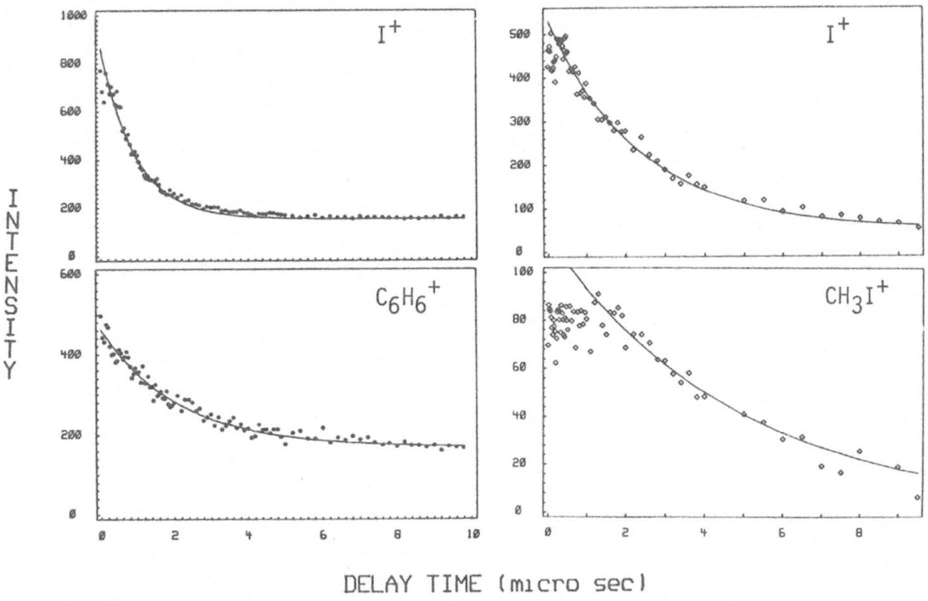

Figure 3: The ion current of I^+ (top) and parent ion (bottom) as a function of the delay time between the laser pulse and the electric field pulse. On the left and right are the results from the nanosecond and picosecond experiments, respectively. The nanosecond pulse produces I^+ with kinetic energy of 0.25 ± 0.03 eV, translationally hot, while the picosecond pulse produces thermal I^+, 0.05 ± 0.03 eV.

ACKNOWLEDGEMENT

The financial support of the National Science Foundation is gratefully acknowledged.

REFERENCES

1. Tai, T.-L.; El-Sayed, M. A. J. Phys. Chem. 1986, 90, 4477.
2. Tai, T.-L.; El-Sayed, M. A. Chem. Phys. Lett. 1986, 130, 224.
3. Levin, R. D.; Lias, S. G. Ionization Potential and Appearance Potential Measurements, 1971-1982; NBS Publication, U.S. Government Printing Office: Washington D.C., 1982.
4. Haney, M. A.; Franklin, J. L. J. Chem. Phys. 1968, 48, 4093.
5. Morrison, D. D.; Hurzeler, H.; Inghram, M. G.; Stanton, H. E. J. Chem. Phys. 1960, 33,821.
6. Szaflarski, D.M.; El-Sayed M. A. submitted to J. Phys. Chem..

LASER CATALYSIS AND CONTROL OF CHEMICAL REACTIONS

Moshe Shapiro

Dept. of Chemical Physics
The Weizmann Institute
Rehovot, Israel 76100

ABSTRACT

The feasibility of laser catalysis of bimolecular reactions is discussed. We describe a scheme based on resonant light scattering via a bound upper electronic state. The laser acts as a catalyst since no net photons are absorbed or emitted. We present computations for the $H + H_2$ reaction in which we show how optically induced reaction-barrier crossings can be achieved. Laser enhancement is found to be pronounced at thermal and sub-thermal energies and to scale linearly with laser power over a wide range of intensities. Common features of the Brumer-Shapiro two-colour unimolecular control and laser catalysis in the bimolecular case are discussed.

I. INTRODUCTION

Since the advent of tunable and high power laser technology, reaction dynamicists have sought a means of altering reaction pathways via laser irradiation[1]. In particular, the possibility of controlling dissociation (and ionization) via laser "pulse shaping" has excited the imagination of many investigators. Recently[2],[3], it has been demonstrated, however, that regardless of the pulse shape a single pulse is insufficient for controlling the selectivity of reactions. It was however shown[3] that this goal can be attained using multiple pulse configurations.

The complementary field in which lasers are meant to assist and catalyze, otherwise forbidden, bimolecular reactions, has been extensively discussed in the theoretical literature[4]-[16]. So far, most of the theoretical expectations have not been realized, although the experimental investigations[17]-[26], have been progressing steadily.

Experimental follow up of the theoretical ideas have been hindered mainly by the projected need for extremely high laser powers. Such high powers are required in many of the schemes involving free-free type transitions of one form or another suggested in the past[6d],[10] [12],[14]. Reliance on such transitions severely limits the use of narrow-band light sources, such as lasers, because the oscillator strengths are spread over a wide continuous range, i.e., the strength per unit wavenumber tends to be small.

In contrast, free-bound transitions, which involve relatively sharp lines are more suitable to excitation by narrow band sources[8]. A reactive free-bound scheme that may overcome the need for very high laser powers was recently suggested[16]. It involves using stimulated light scattering via an excited adduct made up of the reaction partners. Thus, reactants whose energy is insufficient to surmount a reaction barrier, would undergo an optical excitation to a bound barrier-free excited state. While in this state, the system may shuttle freely between the reactants' and products' configurations. Due to the bound nature of the excited state, the reaction can only come to conclusion when the products return to the ground electronic state by emitting a photon. The photon emission is most effective when stimulated, preferably at the same wavelength initially absorbed. In this way the laser merely serves as a catalyst, since the number of laser photons remains unchanged.

The $H+H_2$ exchange reaction has all the necessary ingredients for the successful application of the above scheme[16], save for the fact that the H_3 transition state would only absorb light in the VUV. Since powerful VUV lasers are, as yet, non-existent, this reaction is not ideal for demonstrating the effect. Nevertheless because of its fundamental importance, (in a sense it is the "Hydrogen atom" of reaction dynamics), we use the $H + H_2$ reaction as our primary example for laser catalysis.

In this paper we discuss both unimolecular control and bimolecular laser catalysis and bring several computational examples. We then argue that the unifying theme of both fields lies in the crucial dependence on the attainment of (laser and material) coherence.

II. THEORY OF BIMOLECULAR LASER CATALYSIS

Bimolecular laser catalysis requires, almost by definition, a non-linear theory of the interaction with the radiation field. This follows because one has only a very limited time, namely the collision time, to influence the outcome of the chemical event. This is in sharp contrast to the unimolecular case for which the molecule to be dissociated is available at all times. The free-bound scheme which was developed by us[16],[27] and described above, is no exception, although the molecular matrix elements are effectively much larger than those associated with free-free transitions.

In what follows, we consider the interaction of the $|1)|n,\omega\rangle$ and $|2)|n-1,\omega\rangle$ electronic-field states, where $|1)$ denotes the ground (scattering), and $|2)$ the excited (bound) electronic state. In order to account for the interference between the "natural" non-radiative process and the field-induced reaction, we draw upon the theory of scattering by two potentials. Accordingly, we partition the total (matter+radiation) Hamiltonian,

$$H= H_m + H_{rad} + H_{int} , \qquad (1)$$

where H_m is the matter Hamiltonian H_{rad} the radiation field Hamiltonian and H_{int} the radiation-matter interaction, into two parts,

$$H = H_1 + H_2 , \qquad (2a)$$

where,

$$H_1 = H_m , \qquad (2b)$$

and

$$H_2 = H_{rad} + H_{int} \cdot \quad (2c)$$

The non-radiative part of the transition matrix, $\underline{\underline{T}}_1$, can be singled out by partitioning the full transition matrix as[29],

$$\underline{\underline{T}} = \underline{\underline{T}}_1 + \underline{\underline{T}}_2, \quad (3)$$

$\underline{\underline{T}}_2$ contains the effects of the radiation field and the coupling between the radiative and non-radiative processes.

This partitioning may be recast in a matrix-product form for the S-matrix[29],[30],

$$\underline{\underline{S}} = \underline{\underline{S}}_1^{1/2} \cdot \underline{\underline{S}}_2 \cdot \underline{\underline{S}}_1^{1/2} \quad (4a)$$

where,

$$\underline{\underline{S}}_1 = \underline{\underline{I}} - 2\pi i \underline{\underline{T}}_1 , \quad (4b)$$

$$\underline{\underline{S}}_2 = I - 2\pi i \underline{\underline{\Gamma}}_2 \quad (4c)$$

with,

$$\underline{\underline{\Gamma}}_2 \equiv \underline{\underline{S}}_1^{-1/2} \cdot \underline{\underline{T}}_2 \cdot \underline{\underline{S}}_1^{-1/2} . \quad (4d)$$

We now turn our attention to the non-perturbative evaluation of $\underline{\underline{S}}_2$, the radiative S-matrix. It is most conveniently done by first calculating the radiative reactance matrix - $\underline{\underline{R}}_2$, whose relation to $\underline{\underline{S}}_2$ is given by the well known expression,

$$\underline{\underline{S}}_2 = (\underline{\underline{I}} - i\pi \underline{\underline{R}}_2) \cdot (\underline{\underline{I}} + i\pi \underline{\underline{R}}_2)^{-1}. \quad (5)$$

The computation of $\underline{\underline{R}}_2$ proceeds by first partitioning $|E,v,\gamma\rangle$ - the standing wave solutions of the total (radiative + non-radiative) Schroedinger equation,

$$(E - H)\ |E,v,\gamma\rangle = 0, \quad (6)$$

into its two electronic components. Due to the nature of the interaction Hamiltonian, only $(2|E,v,\gamma\rangle$ must be considered explicitly. It is then possible to write $R_{2,v'v}^{\beta\gamma}$ as[30],[29],

$$R_{2,v'v}^{\beta\gamma} = -\langle E,v',\beta,1|\ \vec{\mu}_{1,2} \cdot \vec{\varepsilon}\ (2.|E,v,\gamma\rangle. \quad (7)$$

$$\text{where, } \vec{\mu}_{1,2} = (1|\ \vec{\mu}\ |2) \ . \quad (8)$$

In Ref. 30, a partitioning technique[29] was applied to the standing-waves case, using real optical potentials. Following that work, we define two orthogonal projectors Q and P, where,

$$P = |1)|n,\omega\rangle\langle n,\omega|(1|, \quad (9a)$$

and

$$Q = |2)|n-1,\omega\rangle\langle n-1,\omega|(2|. \quad (9b)$$

Obviously $QP = 0$ and since we only consider two electronic states, $Q + P = I$.

For the ω values considered, the states spanned by Q are closed. Hence the Q-projection of the total wavefunction - $|E,v,\gamma\rangle$, can be related to $P|E,v,\gamma,1\rangle$ - the the solution of the non-radiative motion on the ground state, in the following way,

$$Q|E,v,\gamma\rangle = -(E - Q\Xi Q)^{-1}\, Q\, \vec{\mu}\cdot\vec{\varepsilon}\, P|E,v,\gamma,1\rangle, \tag{10a}$$

where the standing-waves optical potential - $Q\Xi Q$ - is given as,

$$Q\Xi Q = QHQ + Q\, \vec{\mu}\cdot\vec{\varepsilon}\, P\, P_\upsilon(E - PHP)^{-1}\, P\, \vec{\mu}\cdot\vec{\varepsilon}\, Q\ , \tag{10b}$$

with P_υ signifying the Cauchy Principal Value.

Substituting Eqs. (10) into Eq. (8), we obtain that,

$$R_{2,v'v}^{\beta\gamma} = \langle E,v',\beta,1|P\, \vec{\mu}\cdot\vec{\varepsilon}\, Q(E - Q\Xi Q)^{-1}Q\, \vec{\mu}\cdot\vec{\varepsilon}\, P|E,v,\gamma,1\rangle, \tag{11}$$

where it was assumed that $\mu_{1,1} = \mu_{2,2} = 0$.

Because Q spans a closed manifold, one can compute $Q\Xi Q$ in a discrete representation. Writing Eq. (11) in matrix notation, we have that,

$$\underline{\underline{R}}_2 = \underset{\approx}{\mu}^{\dagger}(E)\cdot(E\, \underset{\approx}{I} - \underset{\approx}{\Xi})^{-1}\cdot\underset{\approx}{\mu}(E)\ , \tag{12}$$

where † denotes a matrix-transpose, the $\approx$ underscore - a matrix having two pairs of indices. $\underset{\approx}{\Xi}$ is a matrix representation of the $Q\Xi Q$ optical potential. From Eq. (10b) we have that,

$$\underset{\approx}{\Xi} = \underset{\approx}{\mathbf{h}} + P_\upsilon \int dE'\ \ \underset{\approx}{\mu}(E')\cdot\underset{\approx}{\mu}^{\dagger}(E')/(E - E')\ . \tag{13}$$

Thus, the entire computation is based on knowing the $\underset{\approx}{\mathbf{h}}$ and $\underset{\approx}{\mu}$ matrices. $\underset{\approx}{\mathbf{h}}$ is given in terms of (two-dimensional) bound-bound integrals of the H_3 excited state Hamiltonian,

$$\{\underset{\approx}{\mathbf{h}}\}_{nm,n'm'} \equiv \langle E_{n,m}|QHQ|E_{n',m'}\rangle. \tag{14}$$

We use the SLTH[32] form for the $H + H_2$ ground-state potential and the empirical G1 surface of Maine et. al.[28] for the H_3 excited state potential. The continuum - $|E,v,\gamma,1\rangle$ and bound nuclear wavefunctions - $|E_{n,m}\rangle$ were approximated as products of <u>vibrationally-adiabatic</u> functions and, for the ground state, the uniform Airy scattering functions[27].

The $\underset{\approx}{\mu}(E)$ matrix is defined as,

$$\{\underset{\approx}{\mu}(E)\}_{nm,v\gamma} \equiv \langle E_{n,m}|\vec{\mu}_{2,1}\cdot\vec{\varepsilon}|E,v,\gamma,1\rangle\ . \tag{15}$$

It involves the same transition-dipole bound-free integrals as in first order perturbation theory. Explicit formulae for both the bound-bound and bound-free integrals are given in Ref. 27.

$\mathbf{R}_2$, obtained via Eqs. (12-15), is a real symmetric matrix having poles on the real energy axis. These poles become resonances, i.e., they acquire widths, only for the **S** matrix, (obtained via Eqs. (4) and (5)). As shown below, both the initial spread of collision energies and the laser band-widths are usually much larger than the absorption line widths. Under these circumstances the absorption line-strengths are simply proportional to the <u>area</u> under each line,

$$k_{abs}(E,v,\gamma|\nu) = \int d\omega \; |T^{\gamma\gamma}_{2,vv}(E,\omega)|^2 \tag{16a}$$

where ν denotes the center of the line, and $\underline{\underline{T}}_2$ is computed via Eqs. (4), (5). Likewise, the laser-assisted reaction rates are proportional to the area under the reactive lines, i.e., the frequency dependence of the (natural + radiative) reaction probabilities,

$$k_{reac}(E,v,\beta\leftarrow\gamma|\nu) = \int d\omega \; |T^{\beta\gamma}_{vv}(E,\omega)|^2 \tag{16b}$$

In the three dimensional world, summation over the impact-parameter-dependent $|T^{\beta\gamma}_{vv}(E,\omega)|^2$ should also be performed.

III. LASER CATALYSIS - A COMPUTATIONAL EXAMPLE

Our calculated reactive lineshapes ($|T^{\beta\gamma}_{vv}(E,\omega)|^2$), for the collinear H + H_2 reaction at three laser intensities, are shown in Fig. 1. At a first glance laser catalysis seems to be achievable at all field intensities, since the reaction probability reaches the value of 1 at some frequency, regardless of the laser intensity. However, since both the laser bandwidth and spread of the initial collision energies are larger than the reactive line-widths, the effect really depends on the integrated lineshape, namely the line-strength. This magnitude is crucially dependent on the field strength, since the line broadens (while still reaching 1 at some point), as the field intensity grows. Roughly speaking the catalytic effect is therefore determined by the line-widths.

The increase in reaction rate (and line-width) is linear over a wide range of intensities. This is shown in Fig. 2, for a number of lines. The linearity in field intensity is due to the crucial role of the excitation process in promoting the radiative reaction. Once the system is in the excited state it can react, irrespective of the rate of its down-transitions. Even if there were no external fields to de-excite the system to the ground state, the system, once excited, would still form products, (with a 50% probability - if all nuclei are identical), even if all coherence were lost.

As shown in Fig. 1 the main distinguishing feature of a coherent laser catalysis is that the lines are non-symmetric. The asymmetry is due to an interference between the (direct) non-radiative reactive route and the (complex) radiative one. Due to this interference the reaction is hindered on the red side of the line and enhanced on the blue side of the line.

IV. UNIMOLECULAR CONTROL

Contrary to the bimolecular case, unimolecular control can be achieved in the weak field limit. This is due to the fact that the reactants are available at all times and are not scattered away if not acted upon by the laser, as in the bimolecular case. One is faced however with a different problem, that of guiding a reaction which readily takes place to a desired destination. This is where one can greatly benefit from the coherence of lasers.

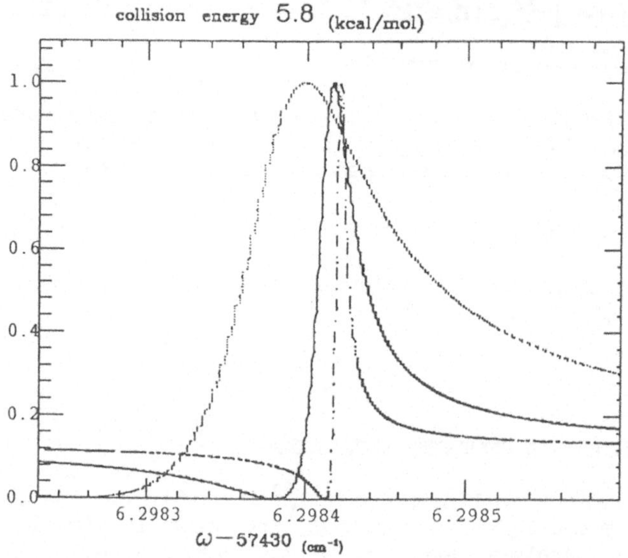

Figure 1. Calculated reactive lineshapes ($|T^{\beta\gamma}_{\gamma v}(E,\omega)|^2$), at three laser intensities. (-.-. 21 MW/cm^2); (—— 83 MW/cm^2); (... 338 MW/cm^2).

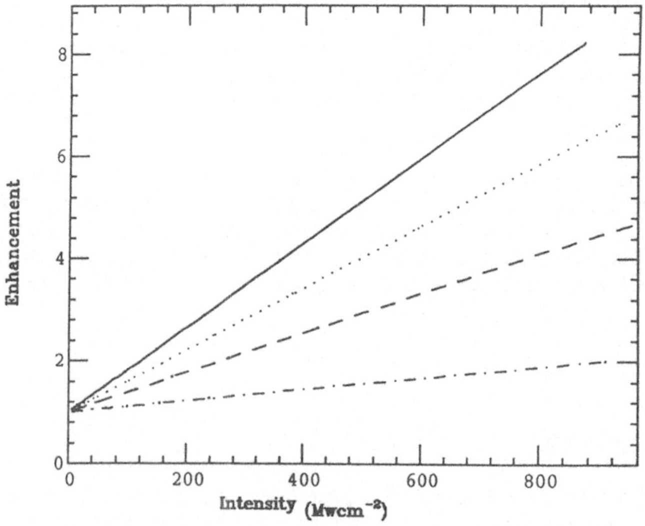

Figure 2. Enhancement factor of room-temperature H + H_2 reaction rates as function of field intensity for four different lines. (—— 55650 cm^{-1}); (-.-. 58080 cm^{-1}); (... 59345 cm^{-1}); (--- 66200 cm^{-1}).

Our approach to coherent control has been described in detail elsewhere[3]. Here we only give a brief outline of the theory. Consider an explicit example in which we wish to control the branching ratio to two product channels denoted $|E,a\rangle$ and $|E,b\rangle$, accessible at energy E. $|E,q\rangle$ are free translational states, with q=a,b signifying the chemical identity of the photofragments. The essence of the method[3] is to irradiate a molecule first prepared in a superposition of bound states, given by,

$$|\chi(t=0)\rangle = c_1|E_1\rangle + c_2|E_2\rangle \tag{17}$$

by two frequencies, ω_1 and ω_2, related as,

$$\omega_1 = \omega_2 + (E_2 - E_1)/\hbar \quad . \tag{18}$$

The action of the combined electric field,

$$\vec{\varepsilon}(t) = \vec{\varepsilon}(\omega_1)\cos(\omega_1 t + \theta_1) + \vec{\varepsilon}(\omega_2)\cos(\omega_2 t + \theta_2) \tag{19}$$

where $\vec{\varepsilon}(\omega)$ is a field amplitude at frequency ω, and θ_ω is a time independent phase associated with light at that frequency, is to give rise to either the "main" line, in which level $|E_1\rangle$ absorbs the ω_1 photon and level $|E_2\rangle$ absorbs the ω_2 photon, or to two "sattelites" in which level $|E_1\rangle$ absorbs the ω_2 photon and level $|E_2\rangle$ the ω_1 photon. In general, the sattelites cannot be avoided, unless the resulting energies fall outside the absorption band.

The main line is distinct from the sattelites in being composed of two exactly degerate continuum states of energy E, where,

$$E = \hbar\omega_1 + E_1 = \hbar\omega_2 + E_2 \;. \tag{20}$$

We choose to represent these degenerate continuum states, by the incoming scattering states, denoted $|E,a^-\rangle$ and $|E,b^-\rangle$, because each correlates, in the distant future, with a unique free state. We have that,

$$\langle E',p|E,q^-\rangle \exp(-iEt/\hbar) \underset{t\to\infty}{=} \delta(E-E')\,\delta_{p,q}\exp(-iEt/\hbar)\;,\quad p,q = a,b. \tag{21}$$

The superposition state due to the main line absorption, can be written, in the weak field limit, as[3],

$$|\Psi(t)\rangle = (\pi i/\hbar)\exp(-iEt/\hbar)\cdot$$
$$\{c_1\,\varepsilon(\omega_1)\exp(-i\theta_1)\,[\,|E,a^-\rangle \mu_1^a + |E,b^-\rangle \mu_1^b\,] +$$
$$c_2\,\varepsilon(\omega_2)\exp(-i\theta_2)\,[\,|E,a^-\rangle \mu_2^a + |E,b^-\rangle \mu_2^b\,]\}\;, \tag{22}$$

where,

$$\mu_j^q = \mu_{j,j}^q \;,\quad j=1,2;\; q=a,b. \tag{23}$$

Contrary to ordinary photodissociation[32] the photo-production rates of asymptotic states such as $|E,a\rangle$, are now a result of a coherent sum of the two routes of photo-exciting $|E,a^-\rangle$. From Eqs. (21),(22),

$$P_a(E) = (\pi/\hbar)^2\,\{\,c_1\,\varepsilon(\omega_1)\,\mu_1^a + c_2\,\varepsilon(\omega_2)\,\mu_2^a\exp[i(\theta_1-\theta_2)]\,\}^2 \quad . \tag{24}$$

In the two color scheme, the rates, hence the branching ratios, are influenced by two external parameters, the relative time-independent phase,

$$\theta_{1,2} \equiv \theta_1 - \theta_2 , \tag{25a}$$

and the relative strength,

$$x \equiv |c_1 \, \varepsilon(\omega_1)/c_2 \, \varepsilon(\omega_2)| \; . \tag{25b}$$

Specifically, the a:b branching ratio is given as,

$$R(a{:}b|E) \equiv P_a(E) \;/\; P_b(E) =$$

$$\frac{\{ \; |\mu_{1,1}^{a}| + x^2|\mu_{2,2}^{a}| + 2x \cos[\theta_{1,2} + \alpha_{1,2}^{a}]|\mu_{1,2}^{a}| \; \}}{\{ \; |\mu_{1,1}^{b}| + x^2|\mu_{2,2}^{b}| + 2x \cos[\theta_{1,2} + \alpha_{1,2}^{b}]|\mu_{2,2}^{b}| \; \}} \tag{26}$$

where $\mu_{i,j}^{q}$ and $\alpha_{i,j}^{q}$ are defined by the following expression,

$$\mu_{i,j}^{q} \equiv \mu_i^{q} \cdot \mu_j^{q} \equiv |\mu_{i,j}^{q}| \; \exp(i\alpha_{i,j}^{q}), \; i,j=1,2 \; . \tag{27}$$

It follows immediately from Eq. (27) that by tuning the external strength ratio and relative phase to,

$$x = |\mu_{1,2}^{a}|/|\mu_{2,2}^{a}| \tag{28a}$$

and,

$$\theta_{1,2} = \pi - \alpha_{1,2}^{a} \tag{28b}$$

we shut-off channel a. Hence the reaction products are funneled exclusively to channel b. In an identical manner we can tune in the field intensities and relative phase to direct all the molecules to channel a.

For polyatomic photodissociation (and that of most diatomics) there are however more than just two final states. The molecule breaks apart to two (or more) composite fragments, each having a set of internal (vibrational, rotational, etc...) quantum numbers, denoted collectively as **n**. The expression for the branching ratio (Eq. (26)) still holds, but the molecular matrix elements must be replaced by cummulative sums over the internal quantum numbers[3],

$$\mu_{i,j}^{q} = \Sigma_{\mathbf{n}} \; \langle E_i|\mu|E,\mathbf{n},q^{-}\rangle\langle E,\mathbf{n},q^{-}|\mu|E_j\rangle \; . \tag{29}$$

It follows from Eq. (29) that $\mu_{i,j}^{q}$ is no longer guaranteed to be factorizable as in the simple two channel case, (Eq. (27)). Hence by the Schwarz inequality,

$$|\mu_{i,j}^{q}|^2 \leq |\mu_{i,i}^{q}| \; |\mu_{j,j}^{q}| \; , \tag{30}$$

with the equality being the exception rather than the rule. If the strict inequality holds, there are no zeros to either numerator or denominator in the branching ratio expression, (Eq.(26), and 100% control cannot be achieved. Nevertheless, as demonstrated elsewhere[3], and in the following example, it is still possible to substantially enhance (deplete) the branching fraction R(a:b|E). VI. DISCUSSION

In this paper we have reviewed two ways of altering the natural flow of reactive events by lasers. On way, pertinent to bimolecular reactions, was to affect a reaction barrier crossing by a transient excitation to a bound excited state. We found that catalysis is, in principle, possible and that the effect scales linearly, over a wide range, with intensity. Another way, of relevance to unimolecular reactions, was to use a coherent two colour excitation to control the branching ratio in photodissociation reactions. The process was found to be very effective in altering electronic branching ratio in direct photodissociation such as that of CH_3I.

Although the two methods seem quite different they both depend on coherence in one form or another. In the unimolecular case, control is achieved by the coherent interference between two optically controlled pathways. Each pathway involves the absorption of a photon. Coherence, more easily created between a pair of bound states, is transferred to a more useful form, that between a degenerate pair of continuum states.

In the bimolecular case, the laser-induced enhancement and hindrance of the reaction are also a result of interference, this time between one path which is optically controlled and one path which is not. The optically controlled route involves a virtual two-step process, the absorption and emission of a photon. The non-radiative path is "natural" tunneling.

There are some important differences: The bimolecular effect increases linearly with laser power, hence relies, in practical applications, on the availability of powerful light sources. Control in the unimolecular case works well, its limits best understood, in the weak field limit. On the other hand the unimolecular effect completely relies on coherence. If the latter is not maintained the interference terms of Eq. (26) drop out and with it the entire effect. In the bimolecular case the asymmetry of the reactive lineshape is lost in the absence of coherence but not the entire effect. Once excited to the bound state, a process independent on coherence, the system has an even (50% in the symmetric cases) chance of reaction. To be sure reaction probability will not reach 100% at some point but the efect will still be substantial for reactions with high ground state barriers, such as the "Woodward-Hoffman forbidden" reactions.

REFERENCES

1. a) A. Ben Shaul, Y. Haas, K.L. Kompa and R.D. Levine, "Lasers and Chemical Change" (Springer-Verlag, Berlin, 1981). b) N. Bloembergen and E. Yablonovitch, Physics Today, **31**, 23 (1978). c) R.L. Woodin and A. Kaldor, Adv. Chem. Phys. **47(b)**, 3 (1981).

2. M. Shapiro and P. Brumer, J. Chem. Phys., **84**, 540 (1986)

3. a) P.Brumer and M. Shapiro, Chem. Phys. Lett., **126**, 541 (1986); b) M. Shapiro and P. Brumer, J. Chem. Phys. **84**, 4103 (1986); c) P. Brumer and M. Shapiro, Faraday Disc. Chem. Soc. **82**, xxx (1986).

4. N.M. Kroll and K.M. Watson Phys. Rev. A **8**, 804 (1973); N.M. Kroll and K.M. Watson Phys. Rev. A **13**, 1018 (1976).

5. A.M.F. Lau Phys. Rev. A **13**, 139 (1976); ibid A **18**, 172 (1978); ibid A **25**, 363 (1981).

6. a) J.M. Yuan, T.F. George and F.J. McLafferty, Chem. Phys. Lett., **40**, 163 (1976). b) J.M. Yuan, J.R. Laing and T.F. George, J. Chem. Phys., **66**, 1107 (1977). c) T.F. George, J. Yuan, I.H. Zimmermann, Faraday Disc. Chem. Soc. **62**, 246 (1977); d) P.L. deVries and T.F. George, Faraday Disc. Chem. Soc., **67**, 129 (1979). e) T.F. George J. Phys. Chem. **86**, 10 (1982), and references therein.

7. J.I. Gerstein and M.H.Mittleman, J. Phys. B **9**, 383 (1976)

8. V.S. Dubov, L.I.Gudzenko, L.V. Gurvich and S.I. Iakovlenlo, Chem. Phys. Lett., **45**, 351 (1977).

9. H.J. Foth, J.C. Polanyi, H.H. Telle, J. Phys. Chem. **86**, 5027 (1982);

10. a) A.E. Orel and W.H. Miller J. Chem. Phys., **70**, 4393 (1979). b) A.E. Orel and W.H. Miller, J. Chem. Phys., **73**, 241 (1980).

11. K.C. Kulander and A.E. Orel, J. Chem. Phys. **74**, 6529 (1981); ibid **75**, 675 (1981).

12. J.C. Light and A. Altenberger-Siczek, J. Chem. Phys. **70**, 4108 (1979).

13. M. Hutchinson and T.F. George, Mol. Phys., **46**, 81 (1982).

14. I. Last, M. Baer, I.H. Zimmerman and T.F. George, Chem. Phys. Lett., **101**, 163 (1983); M. Baer, I. Last, and Y. Shima, Chem. Phys. Lett., **110**, 163 (1984); I. Last and M. Baer, J. Chem. Phys., **82**, 4954 (1985).

15. T. Ho, C. Laughlin and S. Chu, Phys. Rev. A, **34**, 122 (1985).

16. M. Shapiro and Y. Zeiri, J. Chem. Phys. **85**, 6449 (1986).

17. S.E. Harris and J.C. White, IEEE J. Quantum Electron, **QE-13**, 972 (1977).

18. P.H. Cahuzac and P.E. Toscheck, Phys. Rev. Lett., **40**, 1087 (1978).

19. P. Hering, P.R. Brooks, R.F. Curl, R.S. Judson and R.S. Lowe, Phys. Rev. Lett., **44**, 687 (1980).

20. a) J.C. Polanyi, Faraday Disc. Chem. Soc., **67**, 129 (1979); b) P. Arrowsmith, F.E. Bartoszek, S.H.P. Bly, T. Carrington, P.E. Chaters and J.C. Polanyi, J. Chem. Phys., **73**, 5896 (1980).

21. P. Polak-Dingles, J.-F. Delpech and J. Weiner, Phys. Rev. Lett., **44**, 1663 (1980); P. Polak-Dingles, R. Bonanno, J. Keller and J. Weiner, Phys. Rev. A **25**, 2539 (1982).

22. B.E. Wilcomb and R. Burnham, J. Chem. Phys., **74**, 6784 (1981).

23. H.P. Grieneisen, K. Hohla and K.L. Kompa, Opt. Commun. **37**, 97 (1981); H.P. Grieneisen, Hu Xue-jing and K.L. Kompa, Chem. Phys. Lett., **82**, 421 (1981).

24. P.R. Brooks, R.F. Curl and T.G. Maguire, Ber. Bunsenges. Phys. Chem. **86**, 401 (1982);

25. a) G. Inoue, J.K. Ku and D.W. Setser, J. Chem. Phys., **76**, 735 (1982). b) J.K. Ku, D.W. Setser and D. Oba, Chem. Phys. Lett., **109**, 429 (1984).

26. D. Kleiber, A.M. Lyyra, K.M. Sando, S.P. Heneghan and W.C. Stwalley, Phys. Rev. Lett., **54**, 2003 (1985).

27. T. Seideman and M. Shapiro, "Laser Catalysis and Transition State Spectra of the H + H_2 Exchange Reaction" submitted for publication.

28. H.R. Mayne, R.A. Poirier, J.C. Polanyi, J. Chem. Phys. **80**, 4025 (1984).

29. R.D. Levine, "Quantum Mechanics of Molecular Rate Processes" (Clarendon Press, Oxford, 1969).

30. H. Shyldkrot and M. Shapiro, J. Chem. Phys., **79**, 5927 (1983).

31. a) B. Liu, J. Chem. Phys., **58**, 1925 (1973). b) D.G. Truhlar and C.J. Horowitz, J. Chem. Phys., **68**, 2566 (1978); **71**, 1514(E) (1979).

32. M. Shapiro and R. Bersohn, Ann. Rev. Phys. Chem. **33**, 409 (1982).

MAKING MOLECULES DANCE:

OPTIMAL CONTROL OF MOLECULAR MOTION

Herschel Rabitz

Department of Chemistry
Princeton University
Princeton, New Jersey 08544

INTRODUCTION

The aim of this research was to use optimal control theory to establish whether it is feasible in principle as well as in practice to employ optical fields for controlling molecular scale motion. Attempts to achieve control over molecular motion, particularly for site specific chemistry in polyatomic molecules, has been a long sought after goal with virtually no success. Heretofore, experiments and theory aimed at this goal have basically relied on physically inspired intuition for a choice of pumping schemes. Given this background, one may conclude that either the problem of molecular motion control is not soluble or that the solution is far more complex and defies simple guessing at the nature of the optical field. The mathematical tools of optimal control theory, largely developed in the engineering disciplines[1], provide a well defined means of settling this issue. Essentually the problem is one of determining the existence of <u>any</u> possible optical field capable of meeting (quantum) mechanical molecular objectives such as local bond excitation or selective bond rupture. Some hints at going in this direction exist in the current molecular dynamics literature[2] and approaches of this type are proving to be quite successful at solving pulse shaping problems in the field of magnetic resonance imaging[3]. The analysis and numerical calculations summarized below clearly show the feasibility of using optimal control theory to establish the answer to the basic problem being posed.

OPTIMAL CONTROL THEORY

In the establishment of a control theory approach to any problem a number of physical and mathematical issues always arise. In the context of the present molecular control problem, we shall be concerned with the physical objective of stretching a particular bond or dissociating it at a specified time T. We desire T to be sufficiently large such that the active structure of the control field is not influenced by the magnitude of T. In the present study we shall focus exclusively on the possibility of achieving such objectives by directly pumping the vibrational modes of a polyatomic molecule, although other indirect schemes might also be considered. Rather than attempting to *a priori* choose an optical electric field E(t) for this purpose, the question of concern is whether a field of any temporal structure exists capable of reaching such objectives.

An important formal step underlying this research is recent work on the controllability[4] of Schrödinger's equation.

$$i\hbar \frac{\partial\psi(t)}{\partial t} = [H_0 + U(t)]\ \psi(t) \qquad (1)$$

Here H_0 is the field free molecular Hamiltonian for a system of an arbitrary number of degrees of freedom and U(t) is the (unknown) control term. Controllability implies that it is possible, in principle, to find a control term U(t) such that one may access any region of the physical state space (or configuration space, momentum space, etc. as appropriate).

An optimal control problem is set up by defining an optimizing functional J having the following components

$$J = \begin{matrix}\text{Physical Target}\\ \text{Objective}\end{matrix} + \begin{matrix}\text{Penalty}\\ \text{Costs}\end{matrix} + \begin{matrix}\text{Constitutive}\\ \text{Relations}\end{matrix} \qquad (2)$$

The physical target objective to be satisfied at time T might be the stretching of a particular bond, the dissociation of a particular bond, the achievement of a wavepacket with an especially desirable structure, or some other physically meaningful goal. In realistic molecular control problems the specification of the objectives alone will not suffice. For example, when desiring to dissociate a particular bond, we certainly want to build in the penalty of not also dissociating other bonds on perhaps otherwise excessively disturbing the remainder of the molecular. In addition it may also be desirable to consider excessive optical field strengths as a penalty. Constitutive relations can arise for a number of reasons but the simplest case is a desire that while achieving the overall objective we also build in the proper equation of motion (e.g., equation 1). Therefore, we may include constitutive relations by the introduction of appropriate Lagrange multiplier functions $\lambda(t)$. Another possible constitutive relation might arise associated with the process of additionally demanding that the optimal electric field E(t) be minimally sensitive to laboratory field errors.

In general, each of the terms in J will either be explicitly or implicitly dependent on the unknown optical electric field E(t). The problem is physically and mathematically well posed when the optimizing functional J has at least one minimum with respect to E(t), establishing the existence of the optimum optical field. In the simple but important case of controlling the molecular wavepacket, it is possible to prove the existence and hence the well-posedness of the molecular control problem for an arbitrary polyatomic molecule[5].

As an example consider the simple objective of achieving a desired magnitude for the expectation value of the i-th bond displacement $\langle x_i(T)\rangle$ at time T. With the problem posed in this form alone, there will clearly be many wavepackets $\psi(T)$, and hence control strategies, specified by the optical field E(t), $0 \le t \le T$ capable of achieving this objective. As a first step towards a laboratory realization of this objective we might be willing to accept any one of these control strategies. However, even neglecting practical laboratory constraints, this problem will likely be highly unstable on the computer. This comment follows since the iterative process to minimize J with respect to the optical field will likely not achieve a solution without further guidance. This guidance can often be provided by the introduction of the penalty cost terms. In practice, these terms can narrow the family of possible solutions and thereby stabilize the numerical aspects of the field design. Naturally, the dual achievement of numerical stablization as well as the desired physical goals will typically also have the deleterious effect of decreasing the quality of the ultimately achieved objective. Nevertheless, the positive aspects of introducing desireable penalty costs will typically outweigh this negative feature.

NUMERICAL EXAMPLES OF OPTIMAL MOLECULAR CONTROL

The following examples were chosen to illustrate the basic techniques and practical feasibility of carrying out molecular control calculations for the design of optimum fields. The molecules are all idealizations and other cases could just as well be pursued.

Dissociation of a Diatomic Molecule[5,6]

Radiatively induced molecular dissociation has already been established for diatomic molecules, and the purpose of the present calculation is to show one way of designing an optimal optical field for that purpose. In this case the molecule is modelled by a Morse oscillator and the optical interaction term has the form $U(x,t) = \mu_0(x-x_0)E(t)$ where x_0 is the equilibrium bond distance and μ_0 is the transition dipole. The external forcing function $U(x,t)$ and the evolution of the wavepacket are shown in Figure 1. A careful analysis of the figure will show that the field has a complex initial structure corresponding to a phase adjustment period followed by one of intense pumping where the field effectively provides the role of actively guiding the molecular wavepacket to its target objective. This general theme is also apparent in the other example below.

End Bond Stretching of a Linear Molecular Chain[7]

The physical model under consideration consists of a linear chain of 20 atoms where the dipole is at bond 1 and the objective is to stretch bond number 19 at the opposite end at a selected time $T = 0.3$ps. The results of this optimal control effort are shown in Figure 2. We see that the optimal field is signaled by the molecule to be turned off at long times since after a critical period essentially no information may be transmitted from the optically pumped end to the opposite objective end of the molecule. In simplest terms, the objective is achieved by an acoustic pulse travelling to the objective end of the molecule from its localized pumping source. Finally the detailed structure in the optical field is quite important and it is possible to show that the quality of the achieved objective is highly sensitive to field errors in this case.

CONCLUSION

The research presented here should be viewed as illustrating the potential capabilities of optimal control techniques for addressing key issues concerning bond selective chemistry. Although it is early to make definative conclusions about the achievability of any classes of molecular objectives, some points and comments can be made.

The optimal optical pulses can be very intense with structure on the near femtosecond time scale. Lower intensity, longer time pulses can also be obtained by the imposition of field amplitude penalties. The quality of the achieved molecular objectives are highly sensitive to the details of the coherent structure in the temporal pumping fields. A successfully attained molecular objective requires a high degree of cooperativity to exist between the structure of the external forcing field and the dynamical capabilities of the molecule. Finally, successful laboratory optical field designs for

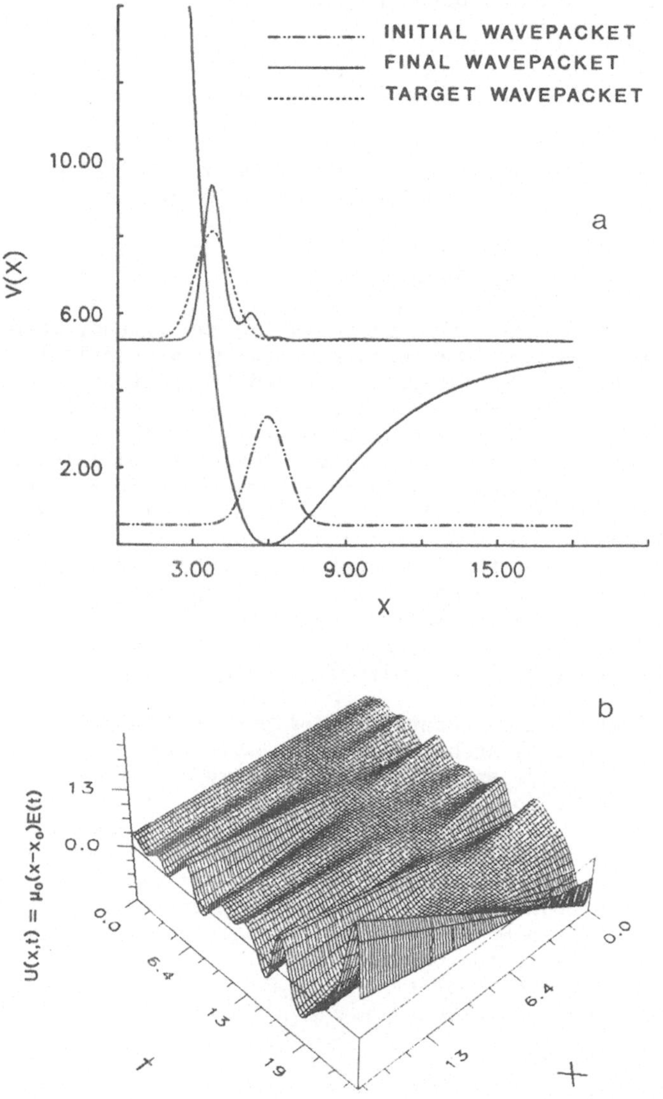

Fig. 1(a-b) Plot of the Morse potential, the initial ground state wavepacket, the target wavepacket and the finally achieved optimal wavepacket. The wavepackets are drawn at their respective energies, and it is evident that the final wavepacket corresponds to a dissociative state as targeted; b) The radiative interaction term U as a function of position and time. Since the spatial part has a fixed form, the only control is through the temporal electric field. Therefore, at any instant of time the spatial potential is a linear ramp of positive or negative slope. This swinging ramp when added to the Morse potential in Figure (a) provides the control for guiding the wavepacket to the desired final objective;

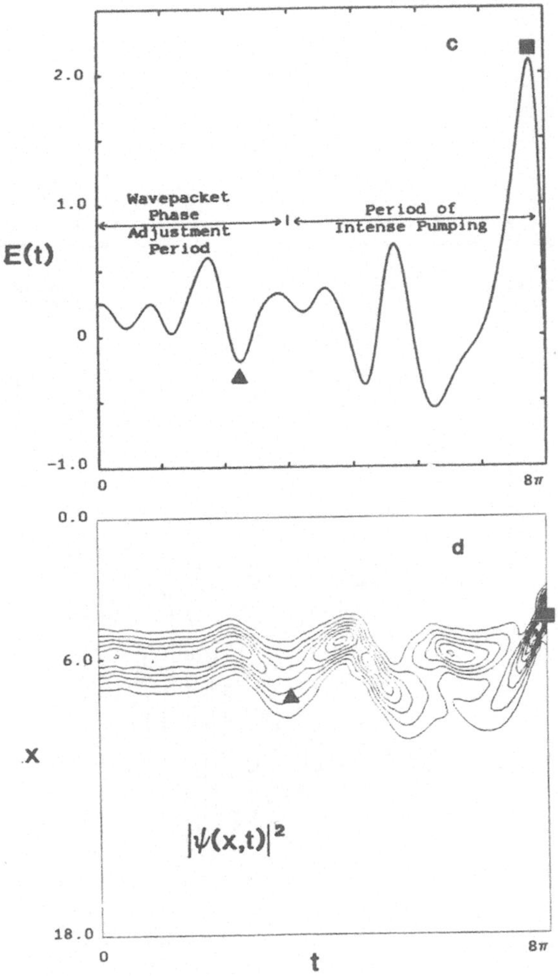

Fig. 1(c-d) Figure (c) presents the time dependent electric field corresponding to a slice through Figure (b). Figure (d) is a plot of the probability density as a function of position and time. From an examination of Figures (c) and (d) it is possible to identify an early period of time where the optical field primarily adjusts the wavepacket phase and deposits little energy followed by a period of intense pumping when most of the energy is deposited. The plots indicate a strong correlation between the behavior of the wavepacket and the temporal structure of the driving field. The field tends to lead the wavepacket motion by a time period of approximately $\pi/2$. This is evident, for example, in that the point labelled by ▲ in the field leads the local extrema in the corresponding probability density plot. A similar comment also applies to the final intense push by the field occuring at the point indicated by ■. Note also that the field near the last point provides the <u>guidance</u> to squeeze the wavepacket, and hence compress the molecule, to its desired target.

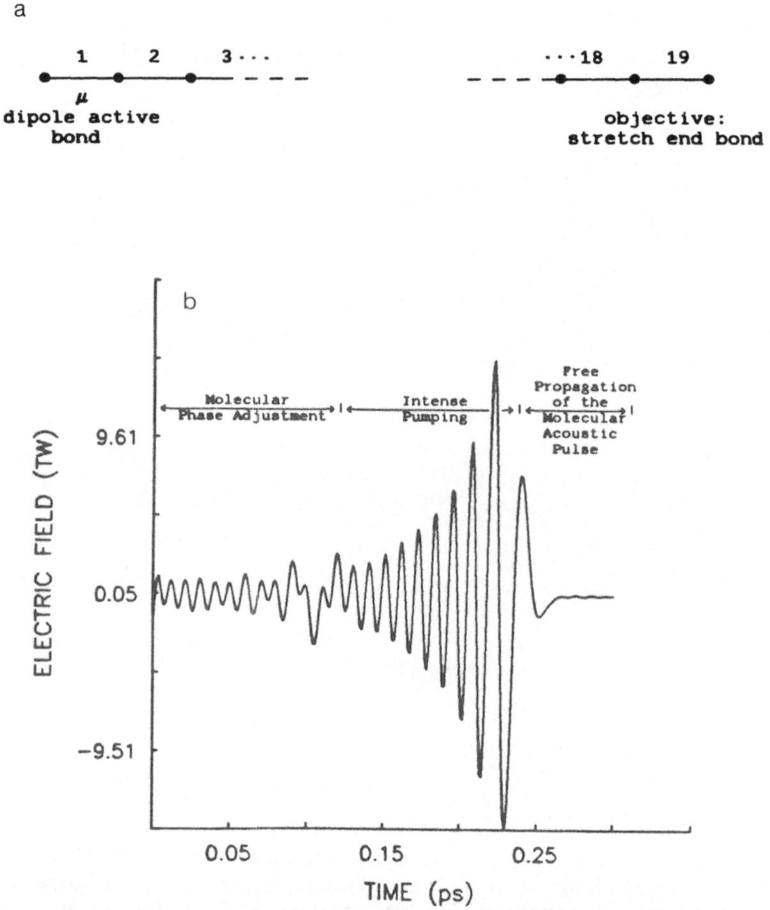

Fig. 2(a-b) Linear chain bond stretching objectives. (a) The molecule of 20 atoms with its dipole at one end and the objective of substantially stretching the bond at the other end at time t=0.3 ps while minimally disturbing the remainder of the molecule; (b) The time-dependent electric field achieving the desired objective. The phase adjustment period is central to preparing the entire molecule for subsequent intense pumping when most of the enegy is deposited. Finally, the molecule signals a turn off of the radiative field to allow for the transmission of the acoustic pulse from the energy absorbing end of the molecule to the target bond at the other end.

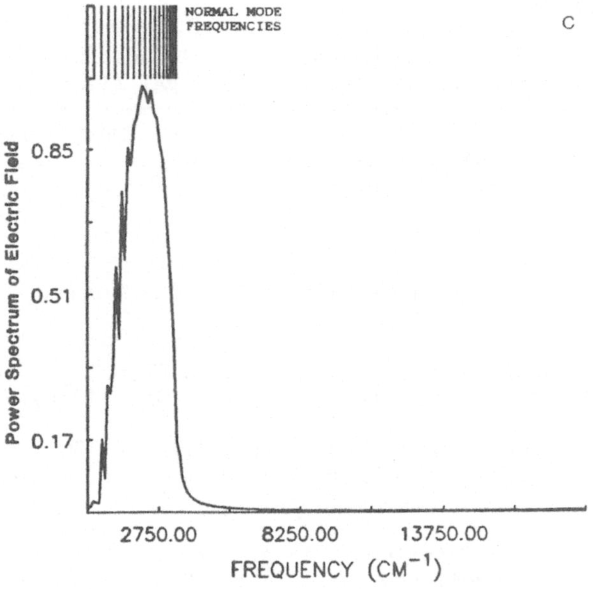

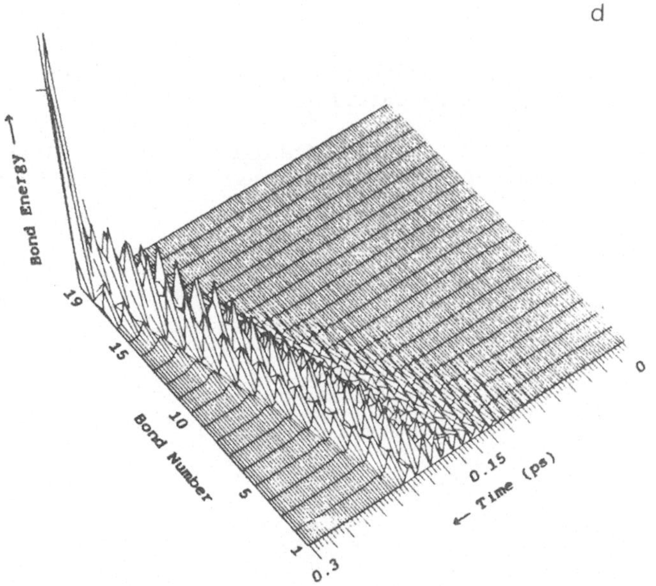

Fig. 2(c-d) Linear chain bond stretching objectives. (c) The frequency spectrum corresponding to the temporal pulse in Figure(b). The discrete normal mode frequency is presented at the top. A highly broadband excitation is involved. (d) Localized bond energy as a function of time and molecular bond number. The coherent travelling acoustic energy pulse is quite evident resulting in a high degree of excitation at the end bond at the desired time.

achieving molecular objectives must exhibit a reasonable level of insensitivity to Hamiltonian uncertainty, dipole function uncertainty and laboratory field errors.

A firm conclusion is that under almost all circumstances it will be impossible to simply guess the nature of the optical fields needed to successfully achieve molecular objectives. This comment establishes why previously attempted intuitive efforts have failed. Much further research needs to be performed to fully develop and ultimately implement optimal control techniques for chemical systems.

REFERENCES

1. D. Luenberger, 1979, "Introduction to Dynamics Systems", John Wiley and Sons, New York.

2. D. Tannor, R. Kosloff and S. Rice, J. Chem. Phys., 85, 10 (1986); M. Shapiro and B. Brumer, J. Chem. Phys., 84, 4103 (1986).

3. M. Silver, R. Joseph, and D. Hoult, Phys. Rev. A, 31, 2753 (1985); S. Conolly and A. Macovski, In Proc. 4th Meeting Soc. Magnetic Res. Med., 958 (1985).

4. A. Butkovskii and Y. Samoilenko, Auto. Rem. Control, 4, 485 (1979); A. Butkovskii and Y. Samoilenko, Dokl. Akad. Nauk. SSSR, 250, (1980); G. Huang, T. Tarn and J. Clark, J. Math. Phys., 24, 2608 (1983).

5. A. Peirce and H. Rabitz, "Optimal Control of Quantum Mechanical Systems: Existence, Numerical Approximation and Applications", manuscript in preparation.

6. S. Shi and H. Rabitz, "Observation Based Optimal Control of Molecular Motion", manuscript in preparation.

7. S. Shi, A. Woody and H. Rabitz, "Optimal Control of Selective Vibrational Excitation Inharmonic Linear Chains", manuscript in preparation.

QUANTUM DYNAMICS OF SHORT PULSE VIBRATIONAL OVERTONE SPECTROSCOPY AND PHOTOCHEMISTRY

John S. Hutchinson and Kenneth T. Marshall

Department of Chemistry and Rice Quantum Institute
Rice University
P.O. Box 1892, Houston, Texas 77251 U.S.A.

INTRODUCTION

Highly vibrationally excited states of polyatomic molecules corresponding to excitations of overtones of "local bond modes" have been studied throughout this decade as a means of revealing the intramolecular vibrational dynamics important to chemical reactivity.[1-3] Under conditions of low temperature and low pressure with highly sensitive photoacoustic detection and intense lasers, spectroscopists observe progressions of vibrational bands at energies up to 25,000 cm^{-1} in the ground electronic state, accessible via direct, one photon transitions from the ground vibrational state, which very closely fit a Morse oscillator energy relation.[4-9] These data are naturally interpreted as arising from overtone excitations of light atom/heavy atom bonds (typically HC, HN, HO, or HSi) behaving as "local" diatomic oscillators in polyatomics.[2] This implies that one can exploit these transitions to "deposit" several quanta of vibrational energy into a single bond in a molecule. Such a prospect is clearly extremely enticing to the laser photochemist, for whom localized excitations hold out the possibility of laser control of chemical reactivity in the form of "mode-selective chemistry."[1,10] As we shall discuss in some detail, the experiments performed to date do not prepare truly localized excitations. Rather more correctly, the observed high energy vibrational bands derive their intensity from a pure overtone state, "a doorway state," of a local bond mode.[11] This intensity is distributed over typically very many vibrational eigenstates due to strong coupling of the local bond mode with the normal modes of the intramolecular "bath," so that the laser excitation actually prepares a state which is substantially delocalized.

Nonetheless, local mode overtones are of great interest and significant value in the study of intramolecular vibrational energy redistribution (IVR) and unimolecular reaction. Bray and Berry's initial study[4] reported broad, nearly Lorentzian shapes of the gas phase overtone bands of high CH overtones in benzene, with widths varying from 50 to 100 cm^{-1}. These widths have been ascribed to lifetime (uncertainty) broadening, related to the time (50 to 100 fs) for relaxation of an initially excited local bond overtone state.[11,12] Classical and quantum dynamics calculations on realistic models of benzene confirm that relaxation is indeed expected on this ultrashort time scale.[12-14] Moreover, calculations by Sibert, Hynes, and Reinhardt[12] have revealed a detailed mechanism for the CH relaxation in benzene which accounts for the time scales and for the observed, but unexpected, non-monotonic dependence of the overtone band widths on excitation level. Similar light atom/heavy atom overtone spectra have been observed in alkanes,[9] alkenes,[15] alkynes,[16] amines,[17] alcohols,[18] and substituted aromatics[19] with increasing resolution,[7] at higher energies,[16] in frozen crystals,[6] and in supersonic jets.[5] Theoretical analysis of these overtone bands as a function of level of excitation and of the "environment" of the local bond has provided a nearly complete picture of the IVR pathways and couplings involving highly energized local bonds.[20-25]

Further interest in the properties of local mode overtones arises from the possibility of inducing mode-selective chemistry with these high energy states. The recent and continuing studies[1,26] of overtone-induced chemical reactivity provide perhaps the cleanest measurements of state-selected unimolecular reaction rates, with two very notable examples. Butler, Ticich, Likar, McGinley and Crim[5] have measured the high resolution jet-cooled spectra of overtone states of both hydrogen peroxide and tetramethyldioxetane at energies above unimolecular dissociation threshold. This provides direct observation of single vibrational predissociative resonances whose widths give directly state-specific unimolecular rate constants.[27] In the somewhat opposite limit, Scherer and Zewail[28] have photodissociated hydrogen peroxide by exciting OH overtones with intense sub-picosecond laser pulses. It may be assumed that these ultrashort pulses prepare initial states which are at least partially localized in the OH bond. Hence, Scherer and Zewail have come the closest yet to "depositing" energy into a local bond via overtone excitation.

These studies provide a wealth of data for theoretical analysis of the intramolecular dynamics responsible for the unimolecular reaction rates. However, we believe it is also important to address the perhaps more preliminary issue of what the initial state is that is prepared and probed in local mode overtone experiments. We have recently studied the dynamics of the excitation of high energy vibrational states in polyatomic molecules by local mode overtone excitation; as alluded above, these studies have shown that such excitations are actually not localized, despite the local nature of the doorway state which carries the transition dipole moment.[29-31]

In this study, we will discuss the quantum dynamics of overtone excitation in three cases. In the first and simplest of these, bound overtone transitions in a small molecule limit are induced by continuous laser irradiation, and the excitation highly specifically prepares only a delocalized eigenstate. However, if the overtone excitation induces predissociation, as in the studies by Crim et al.,[5] the initially prepared state is a sensitive function of the duration of the laser irradiation, and furthermore, is not eigenstate specific in general, even in the small molecule limit. Alternatively, if the excitation source is an ultrashort pulse, as in the studies by Zewail et al.,[28] or if the overtone relaxes "irreversibly" into an intramolecular bath in the typical large molecule limit, then the initially prepared state and the observed spectrum are sensitive functions of the length of the excitation pulse. We will illustrate each of these conclusions. Most importantly, we believe the results will show that one can use ultrashort pulse length dependent overtone spectroscopy to observe IVR dynamics directly.

DYNAMICS OF EXCITATION OF A BOUND LOCAL MODE OVERTONE

The overtone excitation dynamics of a diatomic oscillator, modelled by a Morse oscillator, were first studied classically by Christoffel and Bowman,[32] Gray,[33] and Medvedev,[34] semiclassically by Gray,[33] and Voth and Marcus,[35] and quantally by Dardi and Gray.[36] In particular, Gray and Medvedev demonstrated that, for reasonable laser intensities, overtone excitation is forbidden for a classical Morse oscillator. Holme and Hutchinson[29] first studied the excitation of a local mode overtone in a polyatomic molecule, finding again that classical mechanics fails to describe the process.

Of course, overtone state preparation is observed experimentally and therefore is allowed quantum mechanically. Holme and Hutchinson noted that the quantum manifestation of the classical forbiddenness is that molecular overtone excitation is very, very slow, even for very intense laser sources. This can be illustrated rather easily for a simple two-mode model. We consider a Morse oscillator for a CH bond coupled kineticly to an adjacent harmonic CCH bend. The CH bond, x, is assumed to be dipole active, with a dipole derivative chosen to match an *ab initio* calculation for formyl fluoride. Representing the laser field by a harmonic oscillator H_f, the full model Hamiltonian for the molecule-field system is

$$H = H_{mol} + H_f + H_{int} , \tag{1}$$

where

$$H_{mol} = \frac{P_x^2}{2\mu_{HC}} + D(1 - e^{-\alpha x})^2 + \frac{P_y^2}{2\mu_H (x+x_o)^2} + \frac{1}{2}ky^2 , \tag{2}$$

and the coupling H_{int} of the molecule to the field with intensity I is given by the dipole approximation as

$$H_{int} = -i \left(\frac{I}{2\epsilon_o c}\right)^{\frac{1}{2}} \left(\frac{\partial\mu}{\partial x}\right)_o x \left(|v_f\rangle\langle v_f-1| + |v_f\rangle\langle v_f+1| \right) . \tag{3}$$

The initial state in the excitation process is of the form $|0\rangle|0\rangle|v_f\rangle$, the molecular ground state dressed with a harmonic oscillator state for the field. Due to localization of the dipole in the CH mode, the initial state is coupled by the field only to states of the form $|v_x\rangle|0\rangle|v_f-1\rangle$. The time scale for the transition is defined by the on-resonance Rabi period, τ, which is calculated from

$$\tau = \pi\hbar/2\Delta , \tag{4}$$

where Δ is the coupling

$$\Delta = \langle v_f|\langle 0|\langle 0|H_{int}|v_x\rangle|0\rangle|v_f-1\rangle . \tag{5}$$

This simplifies upon substitution for H_{int} to

$$\Delta = -i \left(\frac{I}{2\epsilon_o c}\right)^{\frac{1}{2}} \left(\frac{\partial\mu}{\partial x}\right)_o x \langle 0|x|v_x\rangle . \tag{6}$$

The factor of interest is the transition matrix element $\langle 0|x|v_x\rangle$. For a harmonic oscillator, of course, this integral is zero unless v_x=1. For a Morse oscillator, the integral is nonzero for all choices of v_x; however, the zero-order harmonic nature of Morse oscillator is substantial, and the transition matrix element is very small unless v_x=1. As an approximate guide, the value of the matrix element decreases by an order of magnitude for every unit increase in quantum number v_x. Hence, Δ becomes quite small rapidly.

τ is clearly the time required for complete transfer of probability from the ground state to the pure local mode overtone state $|v_x\rangle|0\rangle$, a zeroth-order nonstationary state of the molecular Hamiltonian, and is thus one possible measure of "how slow" the excitation is. We have calculated and tabulated in Ref. 29 these times for a variety of laser intensities and overtone levels, v_x. For example, with a fairly intense 1 MW/cm^2 laser tuned to the fifth overtone of the CH bond, v_x=6, we find that $\Delta = 3.5 \cdot 10^{-6}$ cm^{-1} and $\tau = 2.4$ μs. This time decreases only as the inverse square root of the laser intensity. Hence, even with an unrealistically intense 1 TW/cm^2 laser tuned to $v_x = 6$, $\tau = 2.4$ ns.

Are these excitation times to be considered slow? We must compare them to the competing intramolecular events. In this simple case, the only competitor is state-to-state probability flow out of the pure local mode overtone state and into combination states of the form $|v_x'\rangle|v_y'\rangle$, generating mode-mode energy flow. Again, we can measure the time scale for this process from the intramolecular coupling

$$\delta = \langle 0|\langle v_x|H_{mol}|v_x'\rangle|v_y'\rangle \; . \tag{7}$$

For v_x=6, in this case, probability flow is greatest into the state $|5\rangle|2\rangle$, due both to a substantial value of δ and to near degeneracy of $|5\rangle|2\rangle$ with $|6\rangle|0\rangle$. We find that δ = 130 cm^{-1}, so that the time scale for IVR is 32 fs. This is quite reasonable in comparison to experimental bandwidth measurements and theoretical calculations for more realistic molecular systems. It is thus very clear that, in a race between local mode overtone state preparation and relaxation, the preparation is a very poor loser. Even with a 1 TW/cm^2 laser, the rate at which energy can be "deposited" into the CH bond is four to five orders of magnitude smaller than the rate at which the CH bond dissipates this energy into the molecule. Not surprisingly, then, the initially prepared state is delocalized.

In the small molecule limit illustrated here, we can easily identify the delocalized initial state. The dipole "bright" zero-order state $|6\rangle|0\rangle$ has a significant projection onto two principal eigenstates:

$$\begin{aligned} \phi_1 &= 0.7253|6,0\rangle -0.4150|5,2\rangle -0.2465|4,4\rangle +0.2057|3,6\rangle +0.3301|2,8\rangle \\ \phi_2 &= 0.6482|6,0\rangle +0.5328|5,2\rangle -0.0115|4,4\rangle -0.3167|3,6\rangle -0.3197|2,8\rangle \end{aligned} \tag{8}$$

These states are split in energy by 212 cm^{-1}. By contrast, we have already calculated the field-molecule coupling Δ = 0.0000035 cm^{-1}. Substantial excitation occurs only if the laser is tuned within Δ of an eigenstate. Obviously, therefore, the excitation in this limit is eigenstate specific: at all times during the laser excitation, the prepared state is one of the delocalized stationary states above.

The simplicity of the state preparation in this limit arises, in our view, from the fact that the energy "relaxation" processes are in reality fully reversible on the time scale of the excitation. As such, a stationary distribution of probability and energy can be established. We expect that this will not be true when the state preparation competes with a decay which is either truly irreversible (unimolecular reaction) or practically irreversible (large molecule IVR). In these cases, a stationary state is not prepared, and more subtly, the state which is prepared is a sensitive function of the frequency, duration, and shape of the laser pulse.

ULTRASHORT PULSE OVERTONE-INDUCED PHOTOCHEMISTRY

We have illustrated the quantum dynamics of vibrational predissociation stimulated by laser overtone excitation with a two-mode dissociative system, similar to the bound system discussed earlier. Full details of the model may be found in Ref. 30. A Morse oscillator CH bond, x, is coupled via kinetic and potential terms to a dissociative CC bond, y, with the Hamiltonian

$$H_{mol} = \frac{P_x^2}{2\mu_{HC}} + D(1 - e^{-\alpha x})^2 + \frac{P_y^2}{2\mu_{CC}} + \frac{1}{2}ky^2e^{-\beta y^3} + H_c \quad . \tag{9}$$

The CC mode dissociates with a reaction barrier of 6000 cm^{-1}, so that overtone states of the CH bond above this energy are only quasi-bound resonances. When an intense laser is tuned to transitions above the barrier, photodissociation occurs, involving the simultaneous dynamics of local mode state preparation, IVR, and bond fragmentation.

To calculate the dynamics of these processes, we applied the complex coordinates method[37] to find the energies and lifetimes of the predissociative resonances above the barrier. The resonance states have been projected onto a set of zero-order states $|v_x\rangle|v_y\rangle$ describing the bond modes, thus permitting a spectroscopic assignment of the resonances. Again, full details and results of the calculation are in Ref. 30, and we will focus on the most revealing of these.

Two prominent resonance states are observed near energy 11,700 cm^{-1} above the ground state. The more intense of these, which we label $4\nu_1$, has the greater projection onto the bright state, $|4\rangle|0\rangle$, energy 11,838 cm^{-1} and width 1.4 cm^{-1}. The other state, $3\nu_1+2\nu_2$, has a weaker transition moment, energy 11,768 cm^{-1}, and width 20.5 cm^{-1}. Because of the 70 cm^{-1} energy splitting, it might be expected that, as in the bound system above, only a single delocalized (quasi-bound) eigenstate would be excited by a laser. Indeed, if we tune to the $4\nu_1$ state, only this state is excited. This "reactive intermediate" state rapidly achieves a steady state at very low probability, in accord with usual kinetic theory: the first step of the photodissociation, laser excitation, is the rate-limiting step.

However, if we tune the laser to the $3\nu_1+2\nu_2$ state, off-resonance from the brighter "pure overtone" state, the dynamics are significantly more interesting, as illustrated in Fig. 1. Again, the state to which the laser is tuned rises in probability to a steady state, but there is also comparable probability in the nearby state $4\nu_1$. The contribution of $4\nu_1$ to the dynamics oscillates rapidly, due to the Rabi detuning frequency from this state, and the asymptotic steady-state population of $4\nu_1$ is nearly as great as in $3\nu_1+2\nu_2$. Two points should be clear: first, since two quasi-bound eigenstates are excited, the laser prepares a nonstationary state, so that true intramolecular dynamics are involved in the unimolecular reaction. Second, if the laser were to be shut off at any time prior to about 20 ps, the prepared state would be a sensitive function of length of the laser pulse.

These results arise from the competition between irreversible molecular decay and off-resonance excitation of the local mode overtone state. Since the IVR rate and the unimolecular reaction rate are both in the subpicosecond range, we would expect this competition to be observable with pulses of subpicosecond duration. This proves to be the case. We now apply a Gaussian

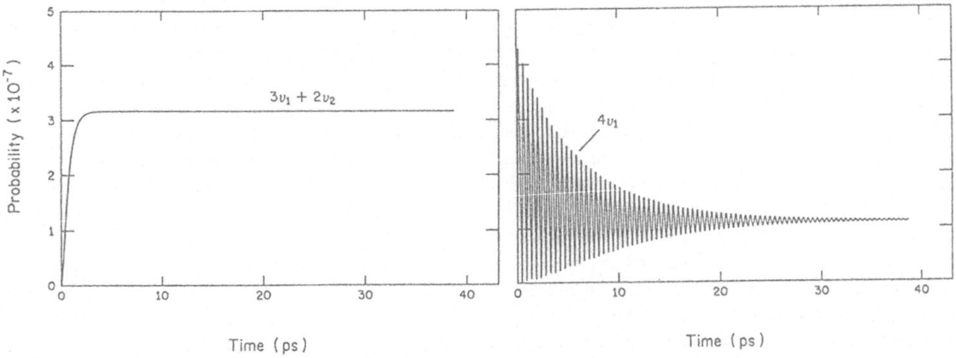

Fig. 1. Photodissociation dynamics for the model system of Eq.(9) with a 200 GW/cm^2 laser tuned to the $3\nu_1+2\nu_2$ state. Probabilities are shown for the two main predissociative resonance states.

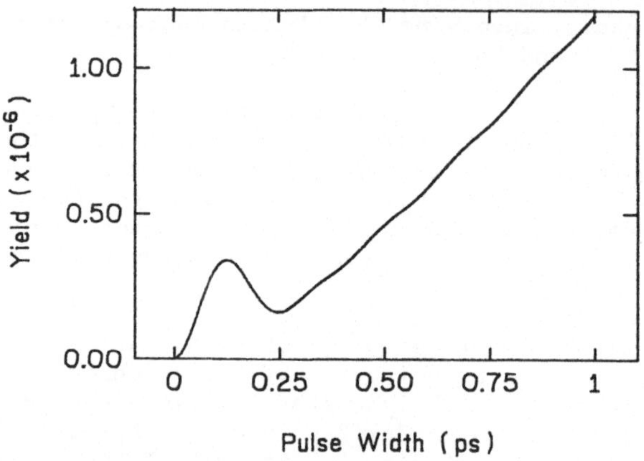

Fig. 2. Non-monotonic photochemical yield as a function of ultrashort pulse width. The intensity is Gaussian in time, peaking at 200 GW/cm^2.

shaped pulse of variable duration and with frequency tuned to the $3\nu_1+2\nu_2$ transition, and we calculate the asymptotic photochemical yield following the pulse (found from the total absorption probability after the pulse). The results, shown in Fig. 2, display a striking anomaly: the photochemical yield is a *non-monotonic* function of the length of the laser pulse. For pulse durations between 0.10 and 0.25 ps, the yield decreases with increasing pulse length! Note that this time scale coincides with the time for state-to-state probability flow, so that off-resonant pulse-length dependent photochemistry, as in Fig. 2, might provide a direct experimental observation of the intramolecular dynamics responsible for unimolecular reaction.

ULTRA-SHORT PULSE OVERTONE SPECTROSCOPY

The competition between state preparation and bound unimolecular dynamics should be most interesting in the large molecule limit, with a high density of bath states. In this case, the dissociative continuum of the previous system is replaced by a quasi-continuum, but on physical grounds, it seems unlikely that the distinction will be observable on a sub-picosecond time scale. As such, we expect to observe that the initially prepared state will be a sensitive function of the excitation pulse length, as in the dissociative system.

To devise an appropriate model for a local mode overtone relaxing into a high density of states, we refer to the work of Sibert, Hynes, and Reinhardt,[12] and Hutchinson, Reinhardt, and Hynes[22] in describing CH overtones in realistic molecular models with limited numbers of states. These studies reveal that overtone relaxation occurs sequentially, as the highly excited local bond transfers energy one quantum at a time into the molecular bath. For example, the anharmonic frequency of the CH bond may, upon excitation, fall into a 1:1 or 1:2 frequency ratio with one or more of the bath modes. This creates a near degeneracy and strong coupling between zero-order states representing these modes, resulting in state-to-state probability flow and transfer of energy. The de-excitation of the CH bond raises its frequency, creating new frequency resonances and degeneracies with different bath states and further probability flow. The result is a division of the bath states into "tiers" in which probability flow occurs between states in adjacent tiers. The tier mechanism has been demonstrated to apply to local mode overtone relaxation in benzene,[12] linear alkane chains,[22] cyanoacetylene,[23] and propargyl alcohol.[24]

We have created a model tier structure for this study.[31] The single first tier state is the only spectroscopically bright state, and represents the pure local mode overtone state. There are eight adjacent tiers, containing 8, 25, 60, 100, 150, 210, 280, and 400 states respectively, randomly distributed in a range of $\pm 100\ cm^{-1}$. The tenth and final tier contains 8767 states in the same energy range, for a total of 10,000 states and a density of 50 states/cm^{-1}. Each state is only coupled to states in neighboring tiers, and the couplings are chosen randomly in the range $\pm 25\ cm^{-1}$. We have studied the dynamics and spectroscopy of many such systems, but in the remainder of this paper, we will concentrate on only one system, for which the first five tiers are shown in Fig. 3.

To calculate the dynamics of this system, we have used the Recursive Residue Generation Method (RRGM) of Wyatt and co-workers,[38] which utilizes the Lanczos algorithm[39] to reduce the full Hamiltonian to smaller tridiagonal form. The method consists of recursively multiplying the Hamiltonian matrix onto the first tier state; the number of recursions required is determined by the time scale of the dynamics desired, or equivalently, the resolution of the spectrum desired. If N recursions are performed, then N eigenvalues are generated along with the projection of N eigenstates onto the first tier state. This is sufficient to generate the survival probability $|\langle 1|\Psi(t)\rangle|^2$ of the initial state $|1\rangle$, as plotted in Fig. 4. Clearly, on a several picosecond time scale, IVR into the quasi-continuum is irreversible, and the decay appears nearly exponential. We have also calculated, somewhat more tediously, the time-dependence of probability flow, and as expected, probability appears sequentially "left to right" through the tiers of Fig. 3.

We are interested here in the nature of the prepared state and the observable spectrum as a function of the length of the excitation pulse. The pulse-length dependent spectrum is calculated by convoluting the energies and intensities from the RRGM calculation with the Fourier transform of the pulse time profile. In Fig. 5a, the spectrum following a 240 fs pulse appears as a single featureless band. This clearly corresponds to preparation of the pure first tier state, which decays exponentially as in Fig. 4. With longer pulses of 480 and 720 fs, in Figs. 5b and 5c, two broad and overlapping features dominate the spectrum. With a still longer 1 ps pulse, only a few additional spectral features are observed.

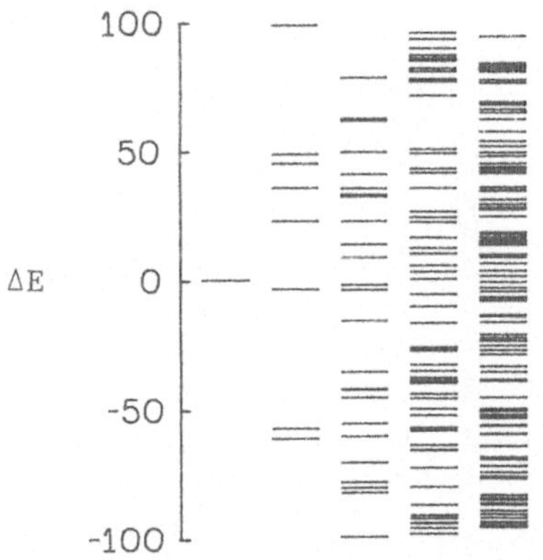

Fig. 3. Distribution of zero-order states for the model in the text. Energies are in cm^{-1}.

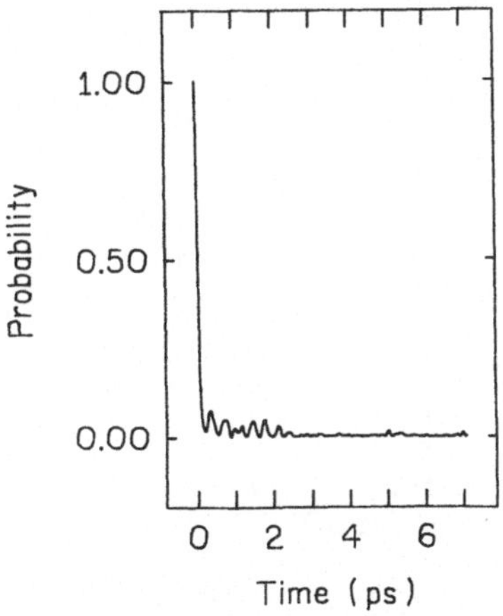

Fig. 4. Probability $|\langle 1|\Psi(t)\rangle|^2$ to remain in the first tier state of Fig. 3.

Figures 5 reveal the dependence of the prepared state on the laser pulse duration. The slow sequential appearance of new bands in the spectrum clearly corresponds to the progression of probability through the tiers of Fig. 3. Most interestingly, the persistence of a few broad bands in the spectrum implies the existence of comparatively long-lived initially prepared states. These states should be analogous to long-lived vibrational predissociative resonances: the discrete nature of the final tier as a quasi-continuum instead of a true continuum is undoubtedly not of consequence to these states. As such, we consider them "IVR resonances." They are superpositions of a small number of states from the first few tiers in the relaxation scheme, intermediate in character between a pure localized excitation and a delocalized pure eigenstate, which would be expected to decay exponentially.

The appearance of these IVR resonances is of clear experimental interest. Observation of such states clearly reveals the presence of the sequential relaxation mechanism. Furthermore, the widths of these states will be sensitive probes of the state-to-state couplings responsible for IVR. On the other hand, these IVR resonances are intriguing theoretically as well, in that they suggest possible ways to calculate long-time quasi-irreversible

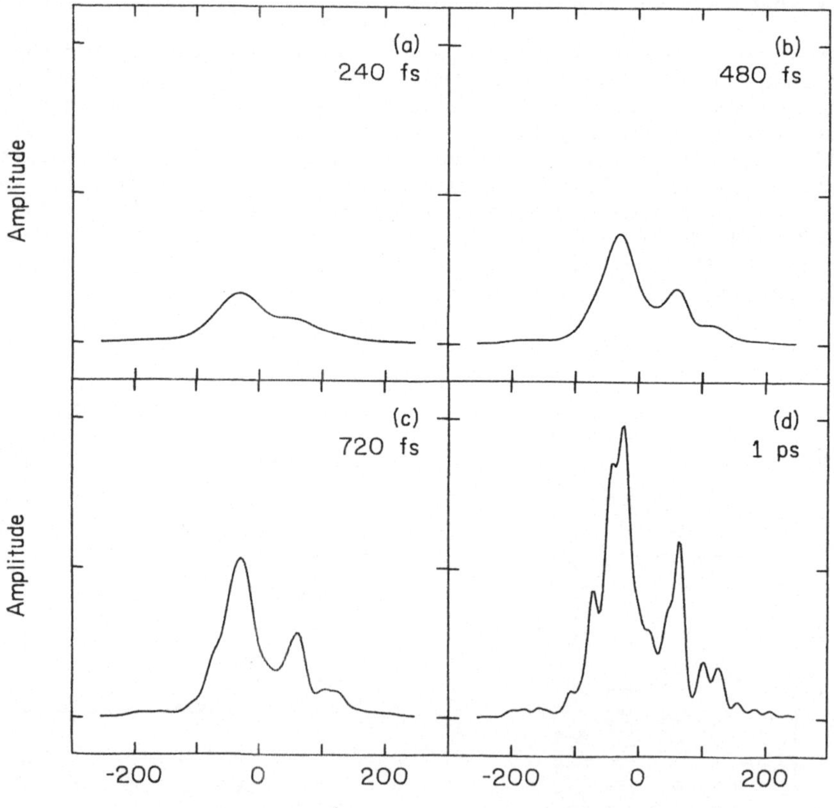

Fig. 5. Absorption spectrum for the system of Fig. 3 as a function of the width of the Gaussian excitation pulse. The ultrashort pulse spectrum is seen to be dominated by a few IVR resonances.

quantum relaxation. Specifically, the quasi-continuum might be accounted for via the complex coordinates or optical potential methods used to treat the continuum in vibrational predissociation. This would simplify calculations involving large numbers of states, which currently require supercomputing capability, and is the subject of our current research.

Of additional experimental interest is the trend of the overall band shape in the pulse length dependent spectrum. A careful comparison of the spectra in Figs. 5 shows that the overall "breadth" of the band increases with increasing pulse length. This provides an interesting comparison with the pulse length dependent fundamental absorption spectra of C_3F_7I observed experimentally by Mukharjee and Kwok.[40] They report a similar increase with pulse length of the band width, and postulate this effect to be due to a sequential IVR mechanism. Our calculations tend to confirm this postulate, in demonstrating that additional states participate in the spectrum with increasing pulse length, although the systems and conditions under study are admittedly quite different.

The appearance of new spectral features is obviously and correctly explained by the improved resolution with longer pulse times. It is nonetheless unexpected and significantly interesting that the improvement in the resolution occurs slowly and non-monotonicly. In particular, the appearance of only a few states at a time is a clear consequence of the sequential decay mechanism of Fig. 3.

We believe that the results discussed here illustrate the utility of a pulse-length dependent view of spectroscopy in revealing the details of initial state preparation in competition with intramolecular dynamics. With rapidly improving pulsed laser technology,[41] experimental observation of overtone spectra with pulses continuously tunable between 10 fs and 1 ps would now seem not only valuable but imminent.

ACKNOWLEDGMENTS

The authors wish to recognize the substantial early contributions of Dr. Thomas A. Holme to this work. Much of this work was feasible only due to a generous grant of NEC SX2 supercomputer time from the Computer Systems and Applications Research Center (CSARC) at the Houston Area Research Center (HARC). This work was supported in part by grants from the Robert A. Welch Foundation of Houston, Texas, and the National Science Foundation. Acknowledgment is made to the Donors of the Petroleum Research Fund, administered by the American Chemical Society, for partial support of this work.

REFERENCES

1. F. F. Crim, Ann. Rev. Phys. Chem. 35:657 (1984).
2. B. R. Henry, Acc. Chem. Res. 10:207 (1977); L. Halonen and M. S. Child, Adv. Chem. Phys. 57:1 (1984).
3. J. S. Hutchinson, Local Modes Overtones and Mode Selectivity, in "Lasers, Molecules, and Methods," J. O. Hirschfelder, R. E. Wyatt, and R. D. Coalson, ed., Adv. Chem. Phys., in press.
4. R. G. Bray and M. J. Berry, J. Chem. Phys. 71:4909 (1979); K. V. Reddy, D. F. Heller, and M. J. Berry, J. Chem. Phys. 76:2814 (1982).
5. L. J. Butler, T. M. Ticich, M. D. Likar, and F. F. Crim, J. Chem. Phys. 85:2331 (1986); E. S. McGinley and F. F. Crim, J. Chem. Phys. 85:5741 (1986).
6. J. W. Perry and A. H. Zewail, J. Chem. Phys. 80:5333 (1984).
7. G. J. Scherer, K. K. Lehmann, and W. Klemperer, J. Chem. Phys. 78:2817 (1983).

8. H. L. Fang, D. M. Meister, and R. L. Swofford, J. Phys. Chem. 88:410 (1984).
9. J. S. Wong and C. B. Moore, J. Chem. Phys. 77:603 (1982).
10. V. S. Letokhov, Nature 305:103 (1983).
11. D. F. Heller and S. Mukamel, J. Chem. Phys. 70:463 (1979).
12. E. L. Sibert, III, J. T. Hynes, and W. P. Reinhardt, Chem. Phys. Letters 92:455 (1982); J. Chem. Phys. 81:1115,1135 (1984)
13. K. L. Bintz, D. L. Thompson, and J. W. Brady, J. Chem. Phys. 86:4411 (1987); J. Chem. Phys. 85:1848 (1986); Chem. Phys. Letters 131:398 (1986).
14. D.-H. Lu, W. L. Hase, and R. J. Wolf, J. Chem. Phys. 85:4422 (1986).
15. J. M. Jasinski, Chem. Phys. Letters 123:121 (1986).
16. M. J. Berry, Proceedings of the Robert A. Welch Foundation Conference, XXVIII:133 (1984); R. R. Hall, PhD. Dissertation, Rice University (1984).
17. H. L. Fang, R. L. Swofford, and D. A. C. Compton, Chem. Phys. Letters 108:539 (1984).
18. H. L. Fang and R. L. Swofford, Chem. Phys. Letters 105:5 (1984); H. L. Fang, D. M. Meister, and R. L. Swofford, J. Phys. Chem. 88:405 (1984); J. M. Jasinski, Chem. Phys. Letters 109:462 (1984).
19. J. W. Perry and A. H. Zewail, J. Phys. Chem. 85:933 (1981).
20. J. S. Hutchinson, J. T. Hynes, and W. P. Reinhardt, J. Phys. Chem. 90:3528 (1986).
21. T. A. Holme and J. S. Hutchinson, J. Chem. Phys. 83:2860 (1985).
22. J. S. Hutchinson, W. P. Reinhardt, and J. T. Hynes, J. Chem. Phys. 79:4247 (1983); Chem. Phys. Letters 108:353 (1984).
23. J. S. Hutchinson, J. Chem. Phys. 82:22 (1985).
24. K. T. Marshall and J. S. Hutchinson, J. Phys. Chem. 91:3219 (1987).
25. B. G. Sumpter and D. L. Thompson, J. Chem. Phys. 86:2805 (1987).
26. K. V. Reddy and M. J. Berry, Chem. Phys. Letters 66:223 (1979); D. W. Chandler, W. E. Farneth, and R. N. Zare, J. Chem. Phys. 77:4447 (1982); J. M. Jasinski, J. K. Frisoli, and C. B. Moore, J. Chem. Phys. 79:1312 (1983); and references within these.
27. T. Uzer, J. T. Hynes, and W. P. Reinhardt, Chem. Phys. Letters 117:600 (1985); J. Chem. Phys. 85:5791 (1986).
28. N. F. Scherer and A. H. Zewail, J. Chem. Phys. 87:97 (1987); N. F. Scherer, F. E. Doany, A. H. Zewail, and J. W. Perry, J. Chem. Phys. 84:1932 (1986); L. R. Khundar, J. L. Knee, and A. H. Zewail, J. Chem. Phys. 87:77, 115 (1987).
29. T. A. Holme and J. S. Hutchinson, J. Chem. Phys. 84:5455 (1986).
30. J. S. Hutchinson, J. Chem. Phys. 85:7087 (1986).
31. J. S. Hutchinson and K. T. Marshall, Local Mode Overtones: Ultrashort Pulse Excitation, Intramolecular Relaxation, and Unimolecular Reactions, "Faraday Symposium on Molecular Vibrations", Faraday Disc. Chem. Soc., in press.
32. K. M. Christoffel and J. M. Bowman, J. Chem. Phys. 85:2159 (1981).
33. S. K. Gray, Chem. Phys. 75:67 (1983); 83:125 (1984).
34. E. S. Medvedev, J. Mol. Spectrosc. 114:1 (1985).
35. G. A. Voth and R. A. Marcus, J. Phys. Chem. 89:2208 (1985).
36. P. S. Dardi and S. K. Gray, J. Chem. Phys. 80:4738 (1984).
37. W. P. Reinhardt, Ann. Rev. Phys. Chem. 33:223 (1982), and references therein.
38. R. E. Wyatt, The Recursive Residue Generation Method, in "Lasers, Molecules, and Methods," J. O. Hirschfelder, R. E. Wyatt, and R. D. Coalson, ed., Adv. Chem. Phys., in press.
39. C. Lanczos, J. Res. Natl. Bur. Stand. 45:255 (1950); J. K. Cullum and R. A. Willoughby, "Lanczos Algorithms for Large Symmetric Eigenvalue Computations," Birkhäuser, Boston (1985).
40. P. Mukharjee and H. S. Kwok, J. Chem. Phys. 85:4912 (1986); 87:128 (1986).
41. C. V. Shank, Science 233:1276 (1986).

IR LASER EXCITATION IN MOLECULES: CHAOS AND DIFFUSIVE ENERGY GROWTH

Jay R. Ackerhalt and Peter W. Milonni

Theoretical Division T-12
Mail Stop - J569
Los Alamos National Laboratory
Los Alamos, New Mexico 87545

ABSTRACT

After a short review of a generic vibrational model of IR multiple-photon excitation, we generalize the model to include rotations. It is shown that the combination of chaotic dynamics and rotational averaging leads to fluence-dependent absorption which removes the sensitivity of the results to model-dependent parameters. The classical rotation-vibration dynamics observed in this model correlate very well with one's quantum intuition based on a molecule's P, Q and R-branch structure and on the red-shift of the vibrational absorption with excitation. The implication of these results for MPE experiments is discussed.

INTRODUCTION

Infrared multiple-photon excitation (MPE) of polyatomic molecules was a major area of research for roughly a decade beginning in 1971 with the work of Isenor and Richardson.[1] Their experiments showed that modest laser powers could easily dissociate a polyatomic molecule. The promise for the future was bond-selective laser photochemistry and laser isotope separation. As this research field grew over the next ten years, it became clear that molecules were not as discriminating with respect to laser wavelength and intensity as was first hoped. By the mid 1980's the field had been nearly abandoned. A few diehards, like ourselves, refused to give up because a solid theoretical explanation of the phenomenon had not yet been found. So here we are today, still talking about MPE.

As the MPE era was winding down in the early eighties, we were aware of the data collection efforts of Judd[2]. He had graphed experimental data on roughly fifty different molecules, all on basically the same fluence-dependent absorption curve, after scaling out the small-signal cross section and excited population fraction. Based on this result we decided that MPE must be a general consequence of the interaction of the laser with the molecule, independent to a certain extent of an individual molecule's idiosyncrasies. Therefore we formulated a generic model of MPE incorporating only the essential features of the phenomenon.[3]

VIBRATIONAL MODEL

The Hamiltonian chosen to represent this vibrational model of MPE was

$$\hat{H} = \Delta\, \hat{a}^{\dagger}\hat{a} - \chi\, (\hat{a}^{\dagger}\hat{a})^2 + \Omega\, (\hat{a}+\hat{a}^{\dagger}) + \sum_m (\Delta+\epsilon_m)\, \hat{b}^{\dagger}_m \hat{b}_m + \sum_m \beta_m\, (\hat{a}^{\dagger}\hat{b}+\hat{b}^{\dagger}\hat{a}) \quad (1)$$

where the terms in order of appearance from left to right represent the anharmonic pump mode (two terms), laser-pump mode coupling, harmonic background modes, and the pump mode-background mode coupling. This form of the Hamiltonian assumes that the laser can be treated classically, the rotating-wave approximation is valid, and the excitation number is conserved. The parameters describing the process are Δ, the frequency detuning of the laser from pump-mode resonance; χ, the pump-mode anharmonicity; Ω, the Rabi frequency; ϵ_m, the frequency offset of the mth background mode from the pump mode; and β_m, the coupling strength of the mth background mode with the pump mode. Several approximations were made in solving this model based on the general nature of the MPE process; the background modes are equally spaced ($\epsilon_m = \Delta_0 + m\rho^{-1}$, where Δ_0 is the frequency offset of the zero mode), the background-pump mode coupling is constant ($\beta_m=\beta$), and since it takes roughly 30 to 40 photons to dissociate many of the larger molecules like SF_6, the dynamics can be solved classically. In addition, the parameters were chosen to be consistent with one's knowledge of molecules: Δ specified for Q-branch excitation, $\chi = 2$ cm^{-1}, $\Omega = .3\ cm^{-1}$ consistent with roughly 10 MW/cm^2 in SF_6, $\rho = 4/cm^{-1}$ and $\beta = .2\ cm^{-1}$.

Using the Hamiltonian (1), making the approximation that the number of background modes is infinite (an approximation justified in the context of radiationless transition theory by Bixon and Jortner[4]), eliminating the background modes, and simplifying the equations using the Poisson summation formula[5], the dynamical equations of motion become

$$\dot{a}(t) = -\, i(\Delta-\chi)\, a(t) + 2i\chi\, |a(t)|^2 a(t) - i\Omega - (\gamma/2)\, a(t) - \gamma\, s(t) \quad (2a)$$

$$s(t) = e^{-i\varphi}\, [s(t-\tau_R) + a(t-\tau_R)] \quad (2b)$$

where $\gamma \equiv 2\pi\beta^2\rho$ is the Fermi Golden Rule rate, $\tau_R \equiv 2\pi\rho$ is the memory recurrence time caused by having an equally-spaced background, and $\varphi \equiv (\Delta + \Delta_0)\, \tau_R$. In Figs. (1a) and (1b) we show the results of integrating Eqs. (2) with a(0)=0 and $\varphi=\pi/2$ for $|a(t)|^2$ and the total photons absorbed versus time, respectively. The total absorbed photons have some initial regularity, but after roughly $8\tau_R$ begin an average linear growth whose slope is dependent on the initial choice of φ. In previous calculations for similar parameter regimes the observed dynamics were shown to be chaotic by computing the maximal Lyapunov exponent. In Fig. (2a) we show a Fast Fourier Transform (FFT) of the dynamics shown in Fig. (1a) (Note the characteristic broadband spectrum of chaos.). In Fig. (2b) we show another

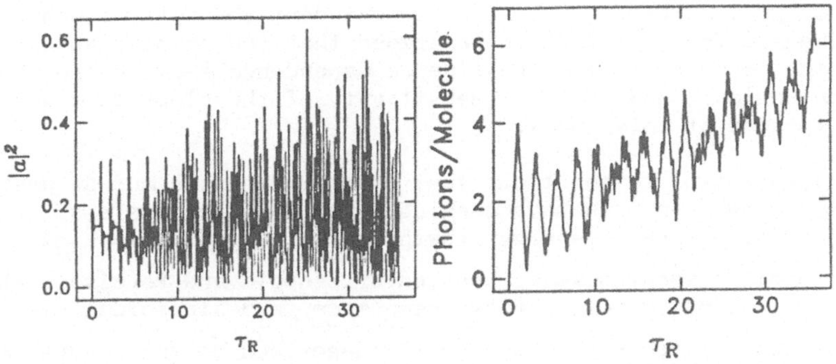

Fig. 1. a) Number of quanta in the pump mode vs. time (τ_R), b) Total quanta in the molecule vs. time (τ_R). Parameters given in the text.

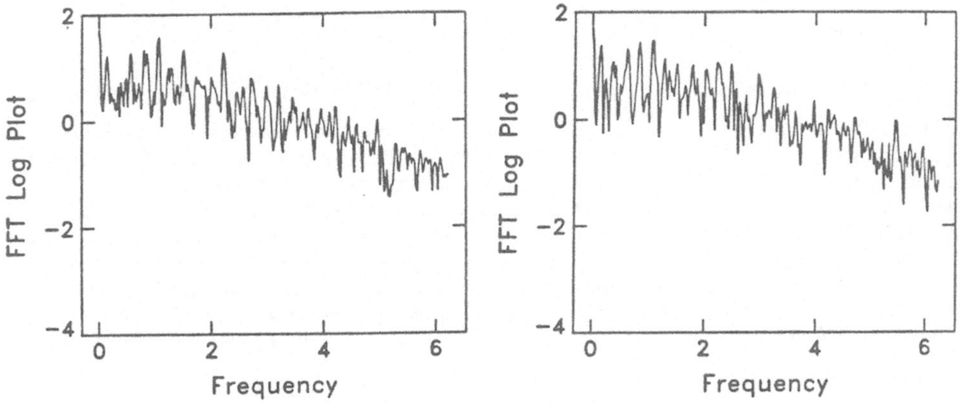

Fig. 2. Logarithmic plot of FFT vs. frequency (cm^{-1}), a) a(0)=0, b) $a(0)=(.05\sqrt{2})e^{i\pi/4}$.

FFT of the same dynamics as shown in Fig. (1a), except the initial condition has been slightly modified, $a(0)=(.05\sqrt{2})e^{i\pi/4}$. The spectra shown in Figs. (2) are very different considering the rather small change in initial condition. This high degree of spectral sensitivity to initial conditions is a nonrigorous but inexpensive measure of chaotic dynamics, as FFT computations are much faster than the corresponding Lyapunov-exponent calculations. Since the laser pulse envelope is square, the linear absorption of photons reflects a fluence-dependent absorption, consistent with experimental trends. In fact we argued that the chaotic absorption of photons is the bond that links the fluence dependence shown by so many different molecules. However, the sensitivity of the slope to φ was an unsettling feature of this model.

It is interesting to point out that chaotic dynamics and an average linear energy growth in time have been observed in other systems. Casati et al. found this same type of energy growth in the periodically kicked pendulum.[6] Similar results were reported by Leopold and Percival, who considered a classical model of a hydrogen atom in a sinusoidal electric field.[7] Further related work was recently described by van Leeuwan et al., who report excellent agreement between their classical computations and their experiments on the microwave ionization of hydrogen.[8] This research coupled with our own work on MPE illustrates the importance of chaos in these driven systems.

In order to make a real comparison with experiment we decided to couple Maxwell's equations to the molecular medium.[9] As interesting as the results were, we again found a strong sensitivity to φ. Thus it became clear that any treatment of MPE which had any hope of comparing directly with experiment must include the molecule's rotational structure.

ROTATIONAL MODEL

Including rotations at lowest order in a MPE model was previously done by Galbraith, et al.[10] In this case the pump mode was harmonic, and no background modes were included in the model. The resulting chaotic dynamics occurred entirely from the (nonlinear) coupling of the rotation of the molecule with the laser. The rotational dynamics were contained in the dipole coupling term which consisted of the laser in the lab frame and the vibrational dipole moment in the molecular frame. In related work Jones and Percival found that even for constant rotational angular momentum the additional time dependence in the dipole coupling term was sufficient to give a fluence dependent absorption, which was not exhibited by their purely vibrational model.[11]

Following the work of Galbraith, et al., we have included rotations in this quasicontinuum model of MPE. The Hamiltonian (1) becomes

$$\hat{H} = \Delta\, \hat{a}^\dagger\hat{a} - \chi\,(\hat{a}^\dagger\hat{a})^2 + \Omega\, P_3\,(\hat{a}+\hat{a}^\dagger) + \sum_m (\Delta+\epsilon_m)\,\hat{b}^\dagger_m\hat{b}_m + \sum_m \beta_m\,(\hat{a}^\dagger\hat{b}+\hat{b}^\dagger\hat{a})$$
$$+ B_0\,(\hat{J}_1^2+\hat{J}_2^2) \qquad (3)$$

where the laser-molecule coupling term has been modified to include the direction cosine matrix of the molecule relative to the laser, oriented along the 3-axis in the lab frame. The last term represents the rotation of the molecule about the 1- and 2-axes. This model is simplified in that the nondegenerate pump mode, oriented along the 3-axis in the molecular frame, is treated as a diatomic molecule, so that it has no rotational angular momentum along that direction. The value of B_0 is chosen as $0.1\ cm^{-1}$ in all cases.

The dynamical equations governing this system, making all the same assumptions as was done for Eqs. (2), are

$$\dot{a}(t) = - i(\Delta-\chi)\, a(t) + 2i\chi\, |a(t)|^2 a(t) - i\Omega\, P_3(t) - (\gamma/2)\, a(t) - \gamma\, s(t) \tag{4a}$$

$$s(t) = e^{-i\varphi}\, [s(t-\tau_R) + a(t-\tau_R)] \tag{4b}$$

$$\dot{J}_1(t) = \Omega\, P_2(t)\, (a(t)+a^\dagger(t)) \tag{4c}$$

$$\dot{J}_2(t) = - \Omega\, P_1(t)\, (a(t)+a^\dagger(t)) \tag{4d}$$

$$\dot{P}_1(t) = - 2B_0\, J_2(t)\, P_3(t) \tag{4e}$$

$$\dot{P}_2(t) = 2B_0\, J_1(t)\, P_3(t) \tag{4f}$$

$$\dot{P}_3(t) = - 2B_0\, (J_1(t)P_2(t) - J_2(t)P_1(t)) \tag{4g}$$

where for ease in comparing results from this model with the earlier vibrational model, we have specified $P_3(0) = 1$, $P_2(0) = P_1(0) = J_2(0) = 0$. Therefore, we recover the previous vibrational model for $J_1(0) = 0$ and have a simple rotational model for $J_1(0) \neq 0$. From Eqs. (4) the absorbed rotational and vibrational quanta in the molecule are

$$J_1^2(t) + J_2^2(t) = - 2\, \frac{\Omega}{B_0} \int_0^t \dot{P}_3(t')\, \mathrm{Re}(a(t'))\, dt' \tag{5a}$$

$$a^\dagger(t)a(t) + \sum_m b_m^\dagger(t)b_m(t) = - 2\Omega \int_0^t P_3(t')\, \mathrm{Im}(a(t'))\, dt' \ . \tag{5b}$$

In Figs. (3a) and (3b) we show the square of the rotational angular momentum and photons absorbed per molecule vs. time, respectively, computed using Eqs. (4) and (5) with $J_1(0) = 6$. All other parameters have been chosen the same as in Figs. (1). Two aspects of these figures should be noted: the regular region shown in Fig. (1b) in the early absorption regime is absent in Fig. (3b), and the rotational absorption to some extent closely parallels the vibrational absorption. We have confirmed that the dynamics

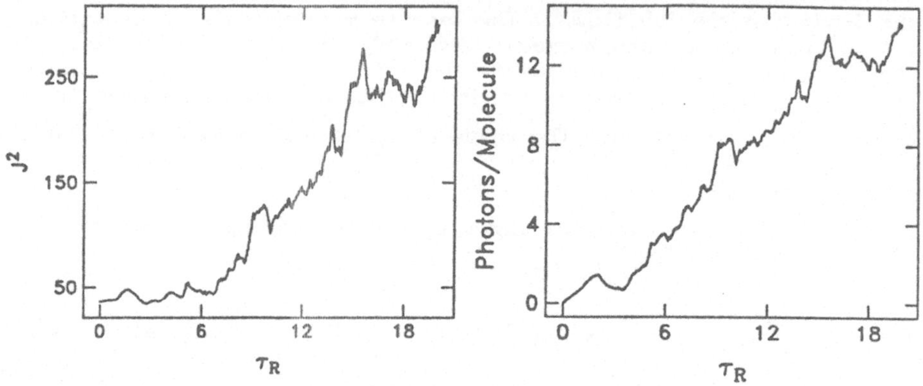

Fig. 3. a) J^2 vs. time (τ_R). b) Total quanta in the molecule vs. time (τ_R), Q-branch excitation (Δ=2.0cm^{-1}).

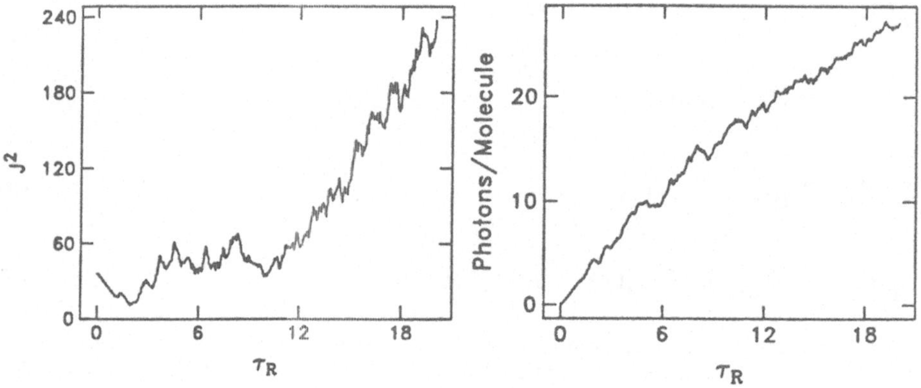

Fig. 4. Same as Fig. (3), but with P-branch excitation (Δ=3.2cm^{-1}).

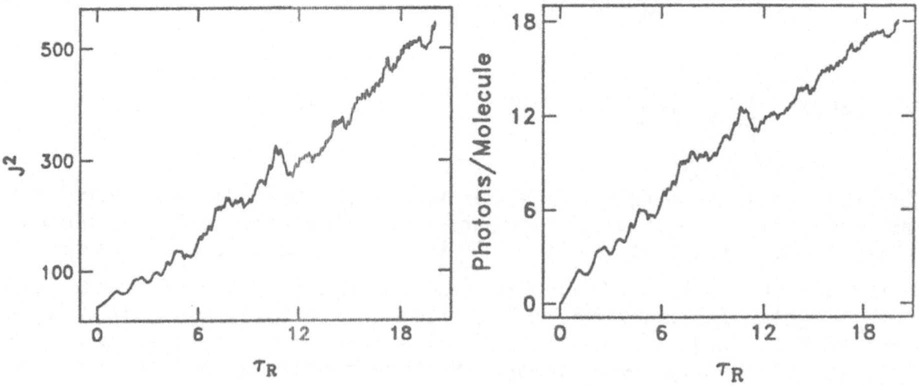

Fig. 5. Same as Fig. (3), but with R-branch excitation (Δ=0.8cm^{-1}).

is chaotic using an FFT as described above. In Figs. (4) and (5) we show the exact same dynamics as in Figs. (3), but for the P- and R-branches, respectively, i.e., $\Delta = 3.2\ cm^{-1}$ and $\Delta = 0.8\ cm^{-1}$. In comparing Figs. (3) through (5) one immediately notices that the overall absorption is sensitive to detuning. In addition, the R-branch rotational absorption curve very closely parallels the corresponding vibrational absorption curve in both shape and number of quantum changes. This is not completely surprising, since in the R-branch $\Delta J=1$, consistent with a one-to-one correspondence between vibrational and rotational absorption of quanta. What is perhaps surprising is that this calculation is completely classical.

In Figs. (3) and (4) one notices that the P-branch ($\Delta J=-1$) curve for J^2 shows an initial decrease, a flat region and an increasing region. This is again entirely consistent with our quantum intuition, since the anharmonicity of the pump mode shifts the absorption feature to the red with increasing vibrational excitation making the rotational dynamics shift from P- to Q- to R-branch absorption. Our quantum intuition also works well in describing the Q-branch results shown in Fig. (3). In all the cases we studied, the correspondence between rotational and vibrational quantum changes was in very good agreement with our expectations from quantum mechanics in both magnitude and direction.

Since the fluence-dependent chaotic absorption for cases where $J_1(0)>0$ showed linear absorption over the entire time interval with no initial regular period, as in Fig. (1), we began a study where the electric field envelope was time-dependent. We chose two profiles to study: one a fairly wide super-Gaussian and the other a moderately narrow Gaussian, both shown in Figs. (6a) and (6b), respectively. The two pulses were chosen to have the same fluence. In Figs. (7) and (8) we show the excitation using the same parameters and format of Fig. (3), but with the excitation due to pulse envelope formats represented by Figs. (6a) and (6b), respectively. A comparison of these figures shows the very strong fluence (rather than intensity) dependence of the absorption. We should mention, however, that the overall absorption is still very sensitive to the exact value of φ.

COMPLETED RESULTS AND CONCLUSIONS

We concluded our calculations by performing a rotationally averaged calculation of the molecular absorption. For simplicity we used a temperature of 11.74 degrees Kelvin so that the peak of the Boltzmann distribution was at J=6 and the largest J needed for the calculations was J=20. The calculations were performed for various electric field envelopes, values of φ, and detunings. <u>Our overall results showed only a very slight dependence on φ and a strong fluence-dependent chaotic absorption for pulse envelopes of compact support</u>. (We noted some anomalies when we used full-windowed square pulse shapes which have an instantaneous turn-on and -off.)

The global implication of these calculations is that the strong (onsets with the turn-on of the electric field) fluence-dependent absorption in IR MPE is due to chaos originating in the interaction of the laser with the rotating molecule's anharmonic pump mode and the intramolecular transfer of energy to the remainder of the molecule. This fluence-dependent absorption had previously been attributed to simply a large density of states in the molecular quasicontinuum, justifying a rate-equation description of the MPE process. <u>It is interesting that this Hamiltonian system does not require ad hoc homogeneous broadening mechanisms in order to obtain incoherent rate dynamics</u>. The observable effect of this MPE chaos in experiments is a

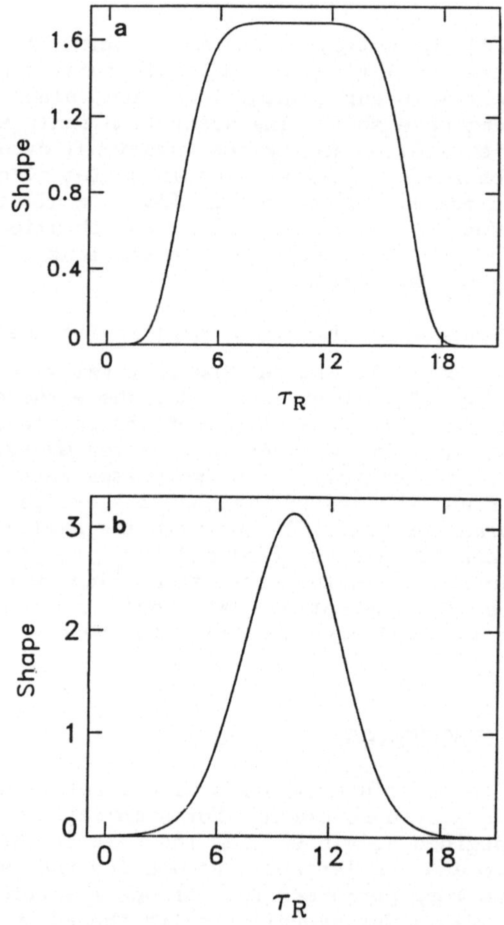

Fig. 6. Electric field pulse shapes vs. time (τ_R), a) Broad high order super-Gaussian, b) narrow Gaussian.

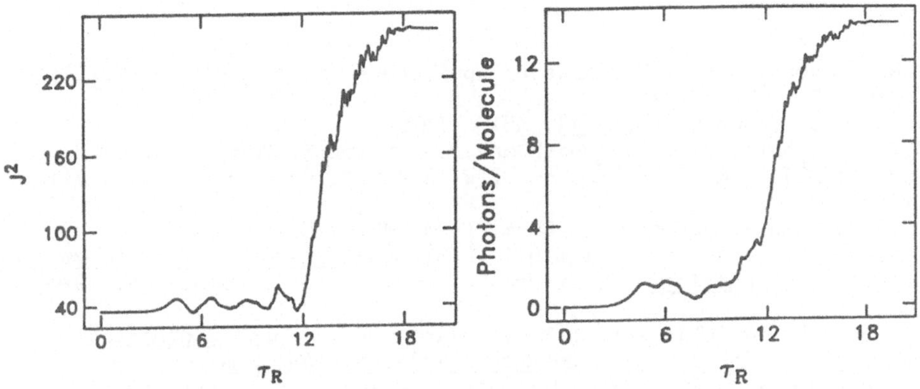

Fig. 7. Same as Fig. (3), but with pulse shape shown in Fig. (6a).

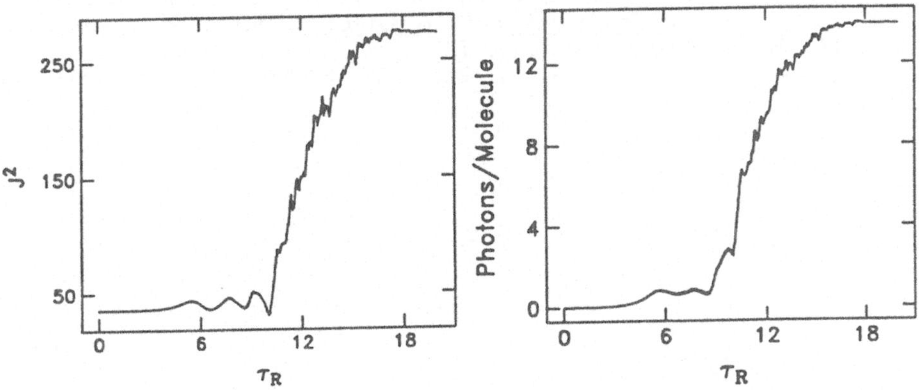

Fig. 8. Same as Fig. (3), but with pulse shape shown in Fig. (6b).

Beer's law absorption of photons, and as many of us already know, this type of absorption was a hallmark of the majority of MPE experiments performed over the last decade. While this is certainly not definitive proof that chaos plays a fundamental role in MPE, it is a completely consistent interpretation of the experimental results.

We would like to acknowledge our previous collaboration with Harold Galbraith.

REFERENCES

1. N.R. Isenor and M.C. Richardson, Appl. Phys. Lett. 18, 224 (1971);Opt. Commun. 3, 360 (1971).
2. O.P. Judd, J. Chem. Phys. 71, 4515 (1979).
3. J.R. Ackerhalt, H.W. Galbraith and P.W. Milonni, Phys. Rev. Lett. 51, 1259 (1983); J.R. Ackerhalt and P.W. Milonni, Phys. Rev. A34, 1211 (1986).
4. M. Bixon and J. Jortner, J. Chem. Phys. 48, 715 (1968).
5. See, for instance, M.J. Lighthill, Introduction to Fourier Analysis and Generalized Functions (Cambridge University Press, Cambridge, 1970), p. 67.
6. G. Casati, B.V. Chirikov, F.M. Izrailev and J. Ford, Stochastic Behavior in Classical and Quantum Hamiltonian Systems, Vol. 93 of Lecture Notes in Physics, edited by G. Casati and J. Ford (Springer, Berlin, 1979).
7. J.G. Leopold and I.C. Percival, Phys. Rev. Lett. 41, 944 (1978).
8. K.A.H. van Leeuwen, G.v. Oppen, S. Renwick, J.B. Bowlin, P.M. Koch, R.V. Jensen, O. Rath, D. Richards and J.G. Leopold, Phys. Rev. Lett. 55, 2231 (1985).
9. J.R. Ackerhalt and P.W. Milonni, paper THB3 presented at the International Laser Science Conference, Seattle, Oct. 21-24 (1986); presented at the conference on Lasers, Molecules and Methods, Los Alamos, July 7-11 (1986).
10. H.W. Galbraith, J.R. Ackerhalt and P.W. Milonni, J. Chem. Phys. 79, 5345 (1983); in Coherence and Quantum Optics V, edited by L. Mandel and E. Wolf (Plenum Press, 1984), p. 721.
11. D.A. Jones and I.C. Percival, J. Phys. B: At. Mol. Phys. 16, 2981 (1983).

ROTATIONAL COHERENCES IN THE IR EXCITATION OF MOLECULAR ROTATIONS AT HIGH INTENSITIES

André D. Bandrauk, Lorraine Claveau

Département de chimie, Faculté des Sciences
Université de Sherbrooke, Sherbrooke, Que, Canada, J1K 2R1

INTRODUCTION

At high intensities, one expects multiphoton transitions to prevail, thus resulting in the coupling of many levels via the radiative perturbation $\vec{\mu}\cdot\vec{E}$, where $\vec{\mu}$ is a transition dipole and E is the electromagnetic field amplitude. In some cases, analytic models of the field-system can be obtained enabling one to visualize the coherences established by the simultaneous excitation of these levels[1,2]. Thus in some of our previous work on direct photodissociation of a molecule in an intense field, new bound states were found to be created by the molecule-field interaction as a result of the field induced superposition of many electronic-vibrational-rotational states called dressed states[2]. Recent interest in these dressed states in atomic[3] and molecular problems stems from the possibility of understanding the nature of quantum chaos in field-matter interactions[4]. Previous work has also indicated a close connection between analytic models of multiphoton rotational excitations and one dimensional exciton models of solid state physics[5,6], in particular the close analogy between barriers to energy absorption in the energy level space and Anderson localization in random one dimensional electronic networks[7-8].

In the present, we reexamine this question by reporting calculations of the time dependent evolution of the energy in a linear molecule and the dressed states of particular molecule-field systems. Analytic models similar to those studied by Zaslavski et al.[9,11] for an understanding of quantum chaos will be shown to apply to various vibrational-rotational excitation schemes. In particular, the classical harmonic vibrator coupled to rotations studied previously by Ackerhalt et al.[12] will be examined from a quantum mechanical aspect. Some of the problems investigated here such as the importance of localization of the dressed states will prove to be useful for a later elaborate discussion of the propagation of short electromagnetic pulses in molecular multi-level systems. As an example, isolated two level systems are known to transmit electromagnetic soliton pulses[13-14]. For molecules such two level configurations are rarely found. One main thrust of the present work is thus an examination of the caracteristics of energy absorption in realistic many level systems such as occurs in molecules, for an eventual application to pulse propagation at high intensities.

Pure Rotations - We consider here resonant transitions in molecules involving rotations only where high intensities and thus extensive rotational coherences can occur. An actual example of this would be optically pumped molecular lasers based on rotational transitions which have been long recognized as sources of efficient and powerful far infrared (FIR) emissions (50 μ - 3 mm)[15].

The rotational energy is quadratic in the rotational quantum number J and as a consequence, the stationary (time independent) dressed states are obtainable for $M_J = 0$ from a Schroedinger equation giving rise to eigenstates which are solutions of Mathieu's equation[5]. For an initial rotational level J_o, one expresses the neighbouring levels as

$$E(m) = B(J_o+m)\ (J_o+m+1) = BJ_o(J_o+1) + B[m(2J_o+1) + m^2] \quad , \tag{1}$$

where $BJ_o(J_o+1) = E(o)$ is the initial rotational energy. The resonance condition is then given by $\hbar\omega = 2BJ_o = E(0) - E(-1)$, i.e., corresponding to the transition $J_o \leftrightarrow J_o-1$. In the molecular field representation, the unperturbed field-molecule eigenstates are expressed as[5]

$$|J,n\rangle = |J_o+m\rangle\ |n\rangle \quad , \tag{2}$$

where $|n\rangle$ are the radiation field eigenstates of energy $n\hbar\omega$. These states are coupled by the radiative interaction for z-polarized light,

$$V_{rad} = \vec{\mu}\cdot\vec{E} = \gamma\mu\cos\Theta \quad , \tag{3}$$

where $\gamma = 1.17 \times 10^{-3}\ (I\ (W/cm^2))^{\frac{1}{2}}\ cm^{-1}/a.u.$, i.e., γ is the conversion factor for calculation V_{rad} in cm^{-1} when the dipole transition moment μ is in atomic units. In the quantum representation, the electromagnetic field is an operator[16],

$$\hat{E} = (\frac{\hbar\omega}{2L})^{\frac{1}{2}}\ [\hat{a}^+ - \hat{a}] \tag{4}$$

where $\hat{a}^+|n\rangle = (n+\frac{1}{2})^{\frac{1}{2}}\ |n+1\rangle$; $\hat{a}|n\rangle = n^{\frac{1}{2}}\ |n-1\rangle$. As a result of equation (4), every rotation-field state $|J_o+m,n\rangle$ is coupled to four states: the quasi-resonant transitions $|J_o+m+1,\ n-1\rangle \leftrightarrow |J_o+m,n\rangle \leftrightarrow |J_o+m-1,n+1\rangle$ (the horizontal couplings in figure 1) and the nonresonant virtual transitions $|J_o,m+1,n+1\rangle \leftrightarrow |J_o+m,n\rangle \leftrightarrow |J_o+m-1,n-1\rangle$ (vertical couplings in figure 2). Figure 2 is drawn on a total energy scale reflecting the degeneracy of $|J_o,n\rangle$ and $|J_o-1,n+1\rangle$ states at resonance as initial condition. Clearly the virtual states are at least at energies $2\hbar\omega$ above or below the resonant states. Their effect on the resonant transitions, which are termed as corrections to the RWA approximation, will depend on the relative magnitudes of the Rabi frequency $\omega_R = \vec{\mu}\cdot\vec{E}/\hbar$ and the virtual state energies, i.e., the parameter

$$\delta = \omega_R/2\omega = \frac{\vec{\mu}\cdot\vec{E}}{2\hbar\omega} \quad , \tag{5}$$

will imply the importance of virtual corrections whenever $\delta \sim 1$ and the accuracy of the RWA approximation for $\delta << 1$.

The case $J_o = 0$, i.e., initial ground state, deserves special mention. For polarized light along the z axis, in view of the selection rules from

$$\langle JM_J|\cos\Theta|J-1,M_J\rangle = [(J+M_J)\ (J-M_J)]^{\frac{1}{2}}\ \frac{1}{2J} \quad , \tag{6}$$

one has that for the particular initial condition $J_o = 0$, the $M_J = 0$ level ladder is the only absorption route available. For this case, the dressed states are exactly Mathieu functions[5]. For other initial conditions $J_o > 0$, one must consider figure 1 for energy M_J with the corresponding correction, equation (6) to the radiative coupling defined in equation (3). Again only the $M_J = 0$ manifold has Mathieu functions as eigenstates in the presence of the field exhibiting the close analogy of this system to one dimension solid state models[5,6]. The $M_J \neq 0$ mainfolds correspond to a ladder (figure 1) with random couplings as a result of the variation of the matrix element (6) as a function of J for fixed M_J (fixed molecular orientation with respect to the field).

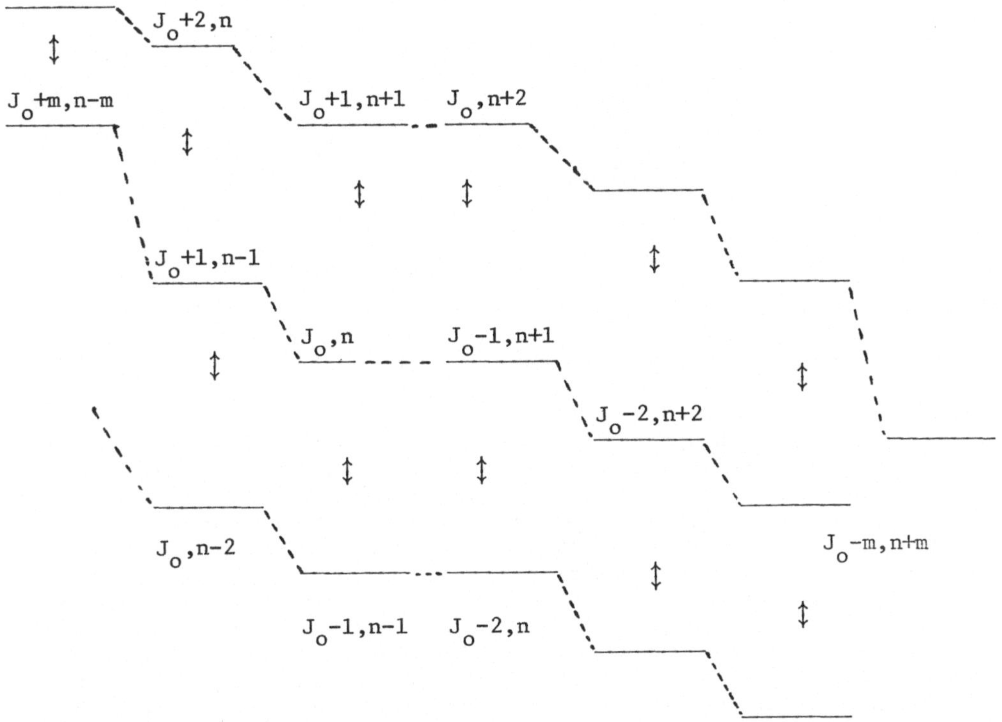

Figure 1. Molecular rotation-photon states $|J+m,n\rangle$ with horizontal (quasiresonant) couplings (----) and vertical (virtual, nonresonant) couplings ($\updownarrow$).

Two Level System with Rotations - This is the case illustrated in figure 2 where an initial vibrational-rotational-photon level $|v,J,n\rangle = |0,J_o,n\rangle$ is coupled resonantly via a P transition ($\Delta J = -1$) of an upper v = 1 level, i.e., $|1,J-1,n-1\rangle$. The R transition ($\Delta J = +1$), is thus slightly off resonance, since the resonance condition is defined now as,

$$\hbar\omega = \hbar\omega_v + 2\,BJ_o \quad , \tag{7}$$

where ω_v = vibrational frequency. Clearly in this case, virtual transitions will be negligible for $\omega_v \sim 1000\ \text{cm}^{-1}$ and $\mu < 1$ a.u. at intensities below I = 10^{12} W/cm^2 (see equations 3 and 5). For this case, the quasiresonant levels illustrated in figure 3 have the same energy spacing as the quasiresonant ladders illustrated in figure 1 for pure rotational transitions. Thus the two vibrational level model including all rotational levels is expected therefore to manifest the same properties as the pure rotational model in the RWA approximation.

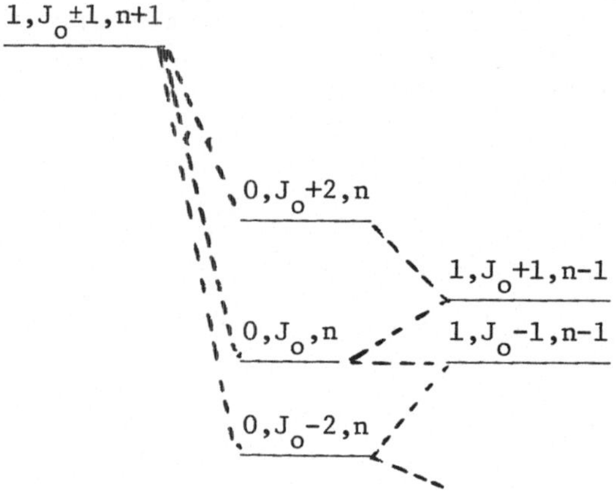

Figure 2. Molecule-field states for two isolated vibrational levels v = 0, 1 including all possible accompanying rotational transitions. The transition $|0,J_o\rangle \leftrightarrow |1,J_o-1\rangle$ is assumed resonant.

<u>Harmonic Oscillator with Rotations</u> - This is the model studied previously classically[12]. For Rabi frequencies ω_R much less than the vibrational spacing, nonresonant virtual transitions can be neglected safely as will be shown for the real cases examined in the next section. The dressed level structure is illustrated in figure 3. The diffusion of the energy up and down the level ladder illustrated in figure 3 at Rabi frequencies less than the rotational spacings should behave as in the two previous models above for $J_o = 0$, since for the case of figure 3 energy can only diffuse up the energy ladder. For larger Rabi frequencies, the many pathways now available will cause the system in figure 3 to deviate considerably from the two previous examples. In particular, one notices the high degeneracies such as the level sequence $|0,J_o,n\rangle$, $|1,J_o-1,n-1\rangle$, $|3,J_o-1,n-3\rangle$, ... $|m,J_o-1,n-m\rangle$, etc.

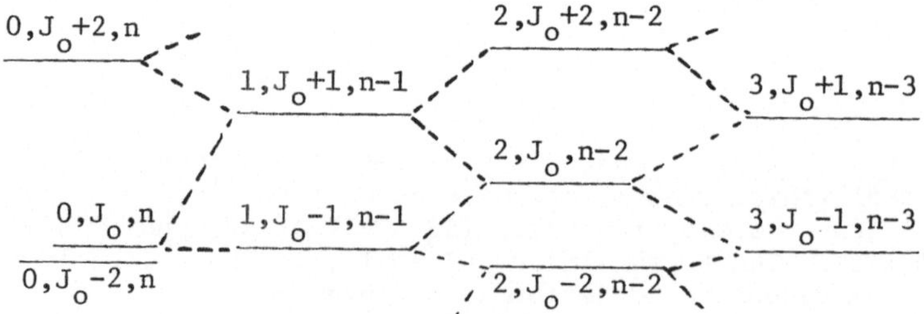

Figure 3. Dressed states in RWA for a harmonic oscillator including all rotations. The transition $|0,J_o\rangle \leftrightarrow |1,J_o-1\rangle$ is assumed resonant.

In our previous work[5], we have shown that the dressed states of particular rotational-vibrational transitions were obtainable as solutions to a Mathieu differential equation. More specifically, the time independent Schroedinger equation for the models in figures 1 and 2 reduces in RWA for the strongest coupling orientation, $M_J = 0$, (molecule // to the z-linearly polarized field) to Mathieu's equation. This corresponds to hindered rotation of the molecule in a cos Θ potential created by the field (equation (3)). Localization or delocalization corresponds to trapping of the dressed states near the initial rotational state or tunnelling to nearby states, creating bands of states[5]. We examine here analytic solutions of the time dependent dressed states to emphasize this point.

Restricting ourselves to the $M_J = 0$ manifold which corresponds to the most strongly coupled molecule orientation in a z-polarized field, the time-dependent Schroedinger equation for the rotational states depicted in figures 1 and 2, $|J\rangle = |J_0+m\rangle = |m\rangle$, becomes[5] ($\hbar = 1$)

$$i\frac{d}{dt} C_m(t) = (E_0 + B(2J_0+1)m + Bm^2)\, C_m(t) + \beta \cos \omega t\, [C_{m+1}(t) + C_{m-1}(t)] \tag{8}$$

where from equation (1) we have $E_0 = BJ_0(J_0+1)$ and we have expanded the total function $\psi(t)$ as

$$\psi(t) = \sum_m C_m(t)\, |J_0+m\rangle = \sum_m C_m(t)\, |m\rangle \quad . \tag{9}$$

In the limit of large J, we can neglect quadratic terms in m in equation (8). Defining new coefficients

$$a_m(t) = C_m(t)\, e^{-iE_0 t} \quad , \tag{10}$$

we obtain the new expression with $\alpha = B(2J_0+1)$,

$$i\,\frac{da_m}{dt} = \alpha m a_m(t) + \beta(t)\,[a_{m+1}(t) + a_{m-1}(t)] \quad . \tag{11}$$

The case $\alpha = 0$ corresponds to a degenerate model, typical of 1D homogeneous systems[17]. In equation (11), we have introduced the general time dependent perturbation $\beta(t)$ since we will show that equation (11) is solvable for any perturbation $\beta(t)$, including pulses[14]. For zero perturbation, i.e., $\beta(t) = 0$, the eigenstates a_m are the rotational eigenstates $e^{im\Theta}$ of the rotation operator $\hat{l}_z = i\partial/\partial\Theta$ with eigenvalue m. We thus look for general solutions

$$a(\Theta,t) = \sum_m a_m(t)\, e^{im\Theta} \quad , \tag{12}$$

so that

$$a_m(t) = \frac{1}{2\pi}\int_0^{2\pi} e^{-im\Theta}\, a(\Theta,t)\,d\Theta \quad , \tag{13}$$

which reduces equation (11) to a single first order partial differential equation

$$[i\,\frac{\partial}{\partial t} + i\alpha\,\frac{\partial}{\partial\Theta}]a = 2\beta(t)a \quad . \tag{14}$$

Using the shift operators

$$\hat{O} = \exp(-\alpha t \partial/\partial\Theta) \quad , \quad \hat{O}^{-1} = \exp(+\alpha t \partial/\partial\Theta) \tag{15}$$

so that

$$\hat{O}\Theta\hat{O}^{-1} = \Theta + \alpha t, \ \hat{O}\frac{\partial}{\partial t}\hat{O}^{-1} = \frac{\partial}{\partial t} - \alpha\frac{\partial}{\partial\Theta} \quad , \tag{16}$$

one readily obtains

$$f(\Theta,t) = \exp(-2i)\ [\int_0^t \beta(t')\cos(\alpha t'+\Theta)dt']\ f(\Theta,o) \tag{17}$$

and

$$a(\Theta,t) = \hat{O}^{-1} f(\Theta,t) = \exp(-2i)\ [\int_0^t \beta(t')\cos\Theta + \alpha(t'-t))\ dt']a(\Theta,o). \tag{18}$$

In the limit β = constant, corresponding to resonant excitation in RWA[5], and setting $\alpha = 0$ for degenerate states $|m\rangle$, one obtains

$$a(\Theta,t) = \exp(-2i\beta t\cos\Theta)\ a(\Theta,0) \tag{19}$$

The case $\alpha = 0$ thus corresponds to a degenerate model with maximum delocalization over the rotational states $|m\rangle$ and conversely to permanent localization at the initial angle Θ, ($|a|^2$ is independent of time). One expects such a case to occur at very high fields, where $\beta \gg \alpha m$, thus justifying the neglect of the diagonal energy αm. For $\alpha \neq 0$, i.e., keeping rotational effects to first order in m, one obtains after use of equations (13) and (18),

$$C_m^\alpha(t) = \exp(im\alpha t/2)\ J_m\ (\frac{4\beta}{\alpha}\sin\frac{\alpha t}{2}) \quad , \tag{20}$$

which reduces for $\alpha = 0$ to

$$C_m^0(t) = J_m(2\beta t) \quad , \tag{21}$$

J_m is the Bessel function of the first kind. The result (21) has previously been obtained in the study of electron migration in periodic one dimensional networks[17], thus once again emphasizing the direct analogy between the rotator in time-periodic fields and one dimensional solid state models.

Summarizing the above, we observe from the asymptotic forms of the Bessel functions[18]

$$\lim_{m\to\infty} J_m(x) = (\frac{ex}{2m})^m\ (2\pi x)^{-\frac{1}{2}} \tag{22}$$

$$\lim_{x\to\infty} J_m(x) = (\frac{2}{\pi x})^{\frac{1}{2}}\cos\left[x - \frac{\pi}{2}(m+\tfrac{1}{2})\right] \tag{23}$$

the different behaviour of the two delocalization regimes. For very short times i.e., $\alpha t \ll 1$, both systems delocalize at the same rate since $|C_m(t)|^2 \sim |J_m(2\beta t)|^2$ in both cases. This is the occupation probability law expected for coherent propagation in one dimensional extended systems[17]. The linear rotation model, applicable to high J_0, equations (11,20) retains the regular rotational periodicity $\tau = 2\pi/\omega_0$ where $\omega_0 = 2B(J_0+1)$. This periodicity is observed in the classical calculations[12]. The degenerate level model, equation (21) does not manifest any regular periodicity. Finally, the linear rotational model exhibits much stronger localization of the dressed eigenstates over the unperturbed rotational states $|m\rangle$. The measure of this delocalization for the linear model is the parameter $2\beta/B(2J_0+1)$. Thus for field strengths such that the Rabi frequency β is much less than the rotational spacing $B(2J_0+1)$, one expects localization to predominate. In the opposite limit, one recovers the extensive delocalization of the degenerate model. These qualitative features will be illustrated in the next section by numerical calculations.

NUMERICAL RESULTS

The time independent states of the full coherent molecule-radiation system are called dressed states[5] and are the eigenstates of the following matrix Hamiltonian for rotations (see figure 1)

$$(E+n\hbar\omega)|J,n\rangle = BJ(J+1)|J,n\rangle + \gamma[(|J+1,n-1\rangle + |J+1,n+1\rangle) + (|J-1,n+1\rangle + |J-1,n-1\rangle)] \quad , \tag{24}$$

where both simultaneous absorptions and emissions of photon to the states $J \pm 1$ are taken into consideration. In RWA, the equations include only resonant terms, i.e.,

$$(E+n\hbar\omega)|J,n\rangle = BJ(J+1)|J,n\rangle + \gamma[|J+1,n-1\rangle + |J-1,n+1\rangle] \quad . \tag{25}$$

For classical fields, (large n), the photon states can be eliminated. The diagonalization of the Hamiltonian matrix equation (25) gives eigenvalues (quasienergies) $E_K = \hbar\omega_K$ and the dressed eigenstates $|\psi_k\rangle = \Sigma\, A_J^K|J\rangle$. In terms of these states, the general time dependent state with initial condition $|\psi(0)\rangle = J_0$ at $t = 0$, is then

$$|\psi(t)\rangle = \sum_{K,J} A_{J_o}^K A_J^K \exp(-i\omega_K t)\, |J\rangle \tag{26}$$

where $|J_o\rangle = |\psi(0)\rangle$, $A_J^K = \langle\psi_K|J\rangle$.

From this one obtains the probability of remaining in the initial state as

$$|\langle J_o|\psi(t)\rangle|^2 = 2 \sum_{KK'} (A_{J_o}^K)^2 (A_{J_o}^{K'})^2 \cos(\omega_K - \omega_{K'})t \tag{27}$$

The population of the initial state will thus have many natural recurrence times $\tau_{KK'} = 2\pi/(\omega_K-\omega_{K'})$ where the $\omega_{K'}$ are the dressed energies. These recurrence times are a manifestation of the coherences induced in the rotational spectrum by the field.

The time dependence of the energy absorption can also be calculated as the average energy of a coherent state $|\psi(t)\rangle$, for a rotational Hamiltonian $\hat{H}_R|J\rangle = BJ(J+1)\;|J\rangle$,

$$\langle E(t)\rangle = \langle\psi(t)|\hat{H}_R|\psi(t)\rangle = 2B \sum_J J(J+1) \sum_{KK'} (A_{J_o}^K A_J^K)(A_{J_o}^{K'} A_J^{K'}) \cos(\omega_K - \omega_{K'})t \quad . \tag{28}$$

Thus the same recurrence times which occur in the population densities will appear in the time dependent energy flow. Of primary importance are therefore the probabilities $|A_{J_o}^K|^2$ and $|A_J^{K'}|^2$ which give the probabilities of finding the initial rotational state $|J_o\rangle$ and any subsequent rotational state $|J\rangle = |J_o+m\rangle$ in the dressed eigenstates ψ_K with eigenenergy ω_K. We have therefore undertaken a numerical diagonalization of equation (24), following the level schemes illustrated in figures 1 and 2, in order to obtain the time dependent energies E(t) for particular values of J_o and M_J, figures 4 and 6, and the corresponding $|A_J^K|^2$, figures 5 and 7.

The molecule we have in mind is the CO_2 molecule for which the rotational constant is $B = 0.39\ cm^{-1}$ and the transition moment between the (1,σ,0) and (0,0,1) vibrational levels is 1.5 x 10^{-2} a.u. (3.4 x 10^{-2} Debye)[19]. Using equations 3 and 6, this gives a Rabi frequency for $M_J = 0$,

$$\omega_R = 8.5 \times 10^{-6} \sqrt{I(W/cm^2)}\ cm^{-1} \quad . \tag{29}$$

Thus at 10^{12} W/cm², $\omega_R = 8.5\ cm^{-1}$ which is about half the rotational spacing $B(2J_0+1) = 16\ cm^{-1}$ at $J_0 = 20$, the lasing level in the CO_2 laser. The results presented in figures 4 to 8 span therefore intensities from 10^{10} W/cm² with $\omega_R = 0.85\ cm^{-1}$ to 10^{13} W/cm² with $\omega_R = 27\ cm^{-1}$. The former corresponds to radiative effects less than the rotational spacing and the letter to radiative effects larger than $B(2J_0+1)$.

Figure 4 illustrates the time dependence of the rotational energy for initial $J_0 = 20$, $M_J = 20$, corresponding to a molecular orientation perpendicular to the field $\varepsilon(J||\varepsilon)$. This is the weakest coupling case possible. The average energy illustrates a single sinusoidal behaviour up to 10^{12} W/cm², which is typical of a two-level system. The localisation of the two resonant states $J_0 = 20$ and $J_0+1 = 21$ is evident for this weak coupling case, thus confirming the two-level behaviour of this system. At 10^{13} W/cm², a second slower frequency becomes superposed, in concordance with the wider spread of J values in the two dressed state ψ_K displaced from the unperturbed energy $BJ_0(J_0+1) = 160\ cm^{-1}$ by $\pm\ \omega_R$. The $M_J = 20$ component clearly remains periodic up to 10^{13} W/cm².

Figure 6 illustrates the strongest coupling case, $M_J = 0$, which corresponds to the molecule parallel to the field $\varepsilon(J\perp\varepsilon)$. In this case, the quasiperiodicity remains up to 10^{12} W/cm². Again at 10^{13} W/cm², the energy behaviour becomes more complex, albeit it is still quasiperiodic since the ω_KS remain a discrete set. The delocalisation of the dressed states with energies near the initial energy $BJ_0(J_0+1)$ is illustrated in figure 7. Thus at 10^{10} W/cm², only the two resonant levels make up the dressed state. Between 10^{12} and 10^{13} W/cm², a transition would seem to occur, resulting in a dressed state at 10^{13} W/cm² delocalized over about 20 rotational states, i.e., $m = \pm\ 10$, equation (8).

A measure of the delocalization of these dressed states can be derived from the following considerations. According to equation (19), the dressed states at $M_J = 0$ are defined by a potential $V = 2\beta \cos\ \Theta$ (which gives rise to the Mathieu equation for the time independent states[5]). Thus the maximum energy spread is expected to be 4β, where $\beta = \omega_R$, the Rabi frequency defined in equation (29). Thus the maximum kinetic energy will be given by $Bm(m+1) = 4\beta$. Using $B = 0.4\ cm^{-1}$ for CO_2, we obtain $m^2 \simeq 10\beta$. At 10^{13} W/cm², $m^2 \simeq 270$, giving $m \simeq 16$. This is close to the observed value of $m \simeq 10$, figure 7d.

The quasiperiodicities observed in figures 4 and 6 correspond to pure coherence effects, i.e., the field creates coherent superposition of states |J> corresponding to the dressed eigenstates, equation (26) for each value of M_J. The total state is an incoherent superposition of these dressed states for the (2J+1) values of M_J. In figure 8 we show the time dependence of the total energy E(t) averaged over all initial sublevels M_{J_0}, i.e.,

$$\langle E(t)\rangle = \sum_{M_{J_0}=0}^{\pm J_0} \langle E_{M_{J_0}}(t)\rangle/(2J_0+1) \tag{30}$$

The quasiperiodicity is now lost due to the fact that one has summed

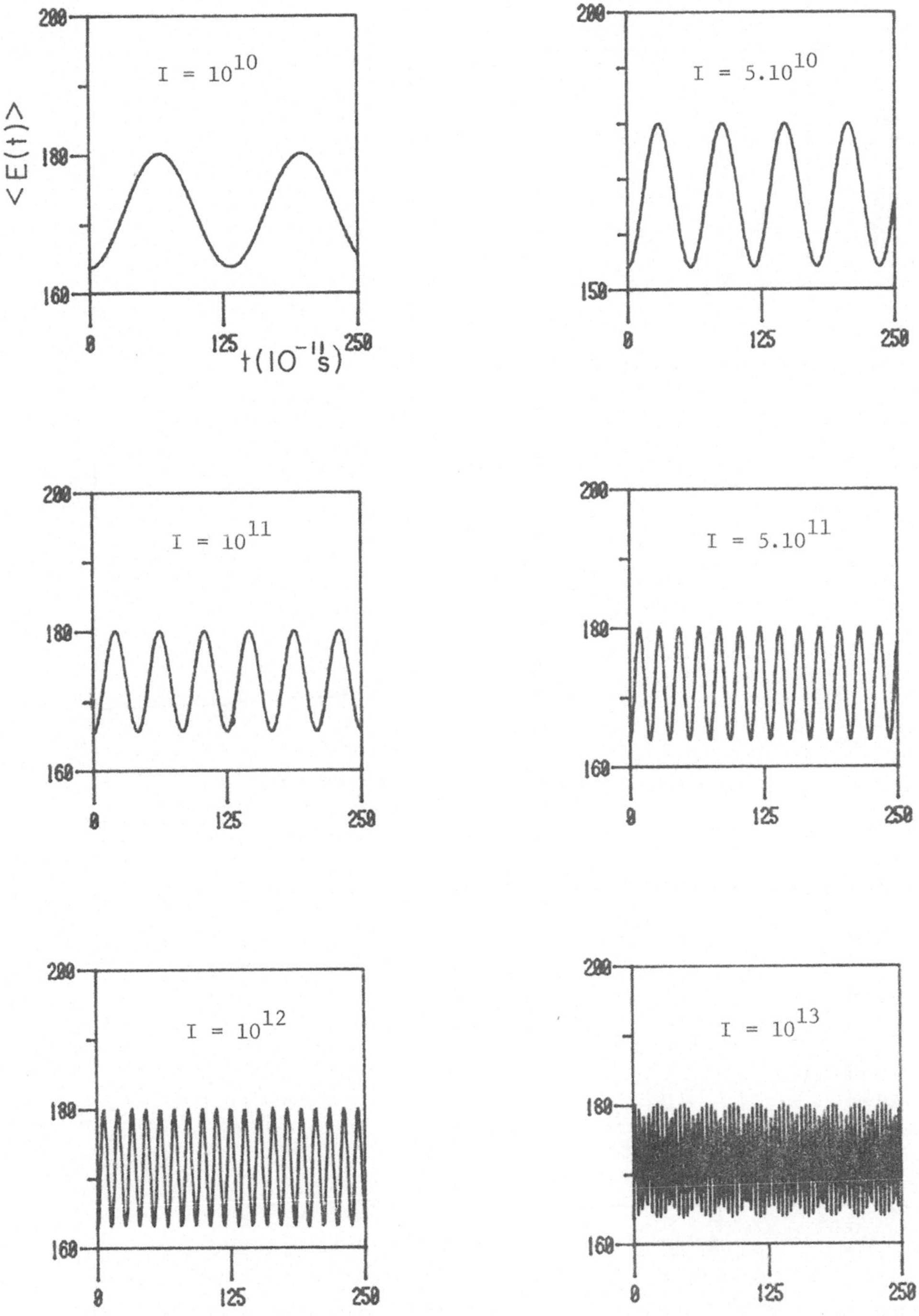

Figure 4. Energy versus time for J_o=20 and M_J=20 with different values of the intensity (in W/cm^2).

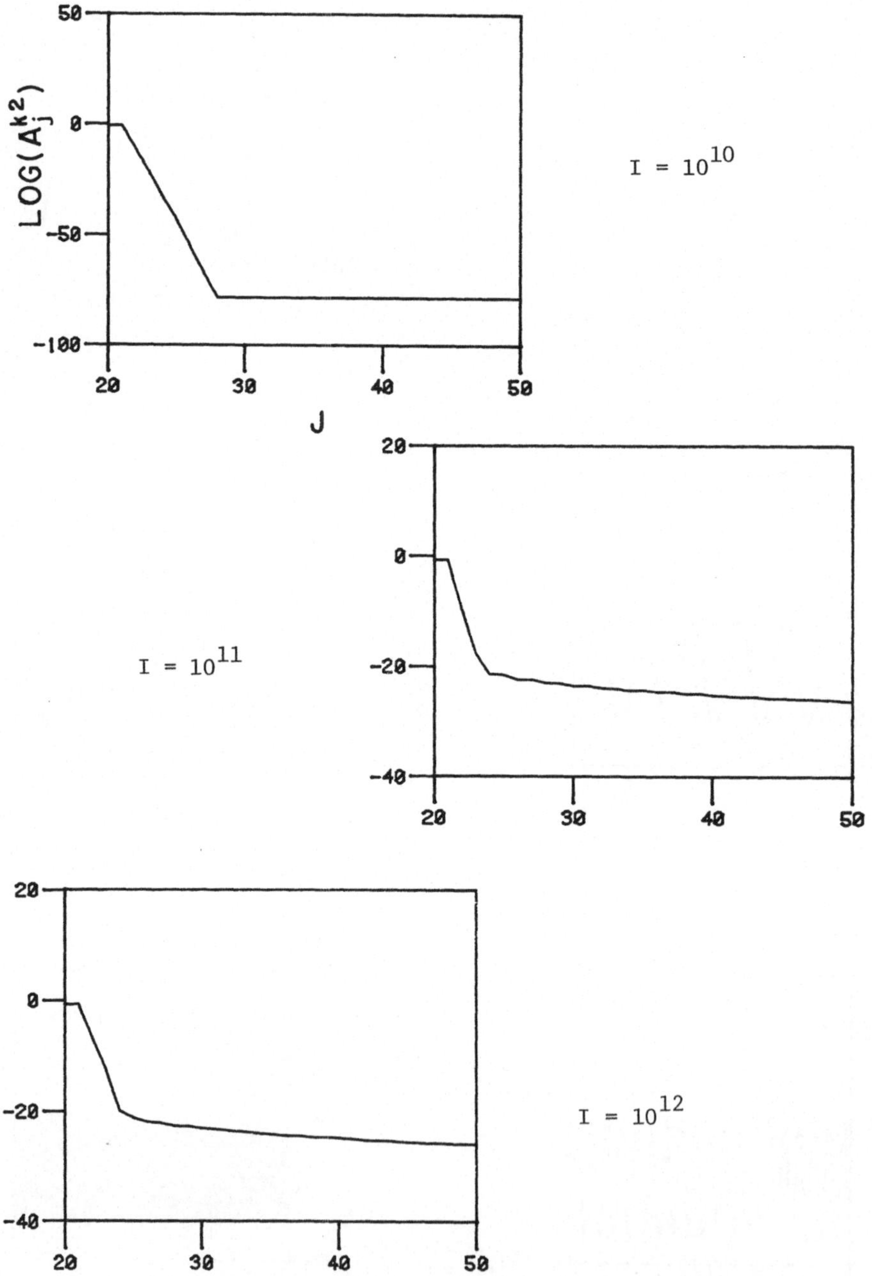

Figure 5. Log($A_j^{K^2}$) versus J for J_o=20 and M_J=20 with different values ofthe intensity (in W/cm^2).

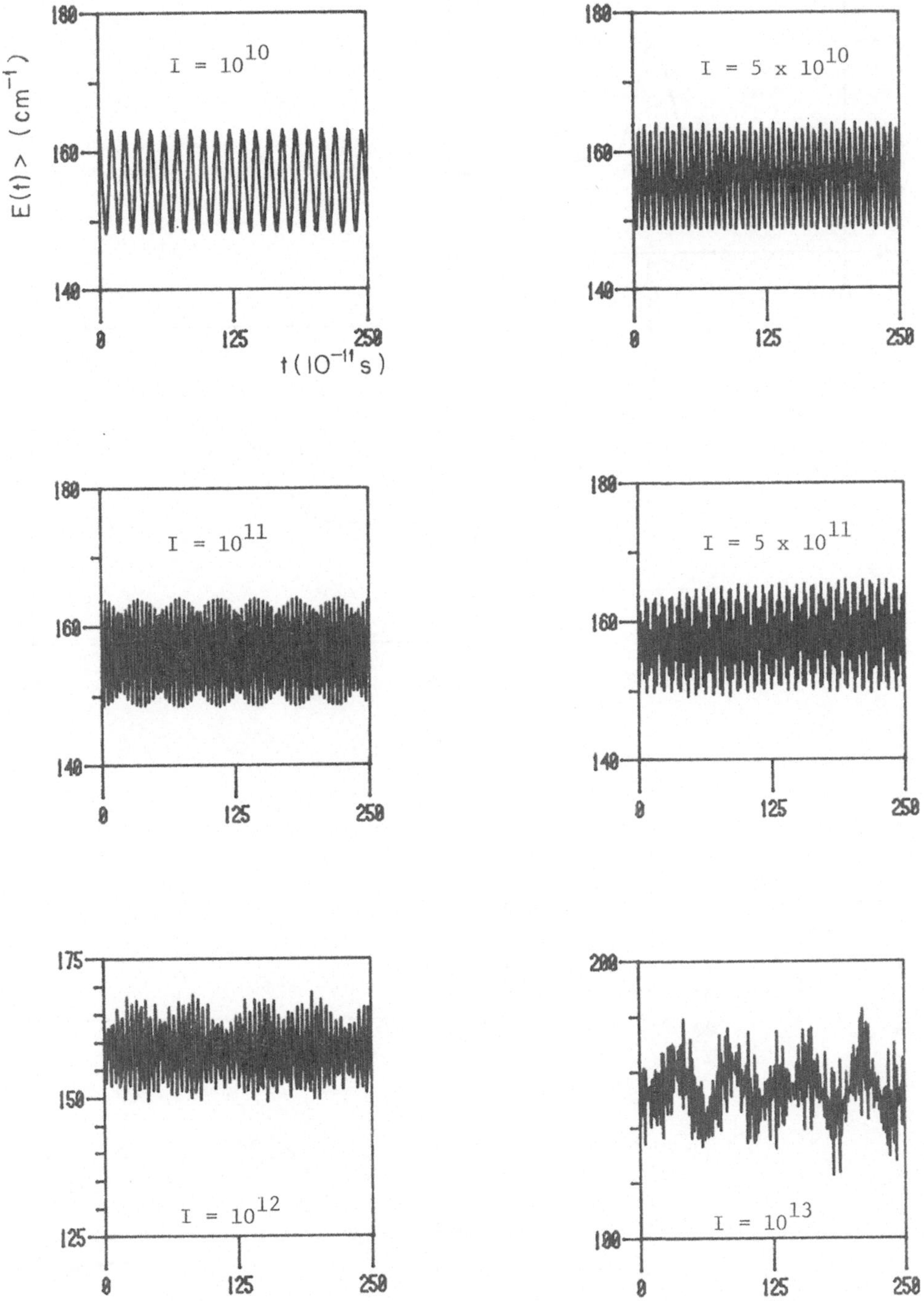

Figure 6. Energy versus time for J_o=20 and M_J= 0 with different values of the intensity (in W/cm^2).

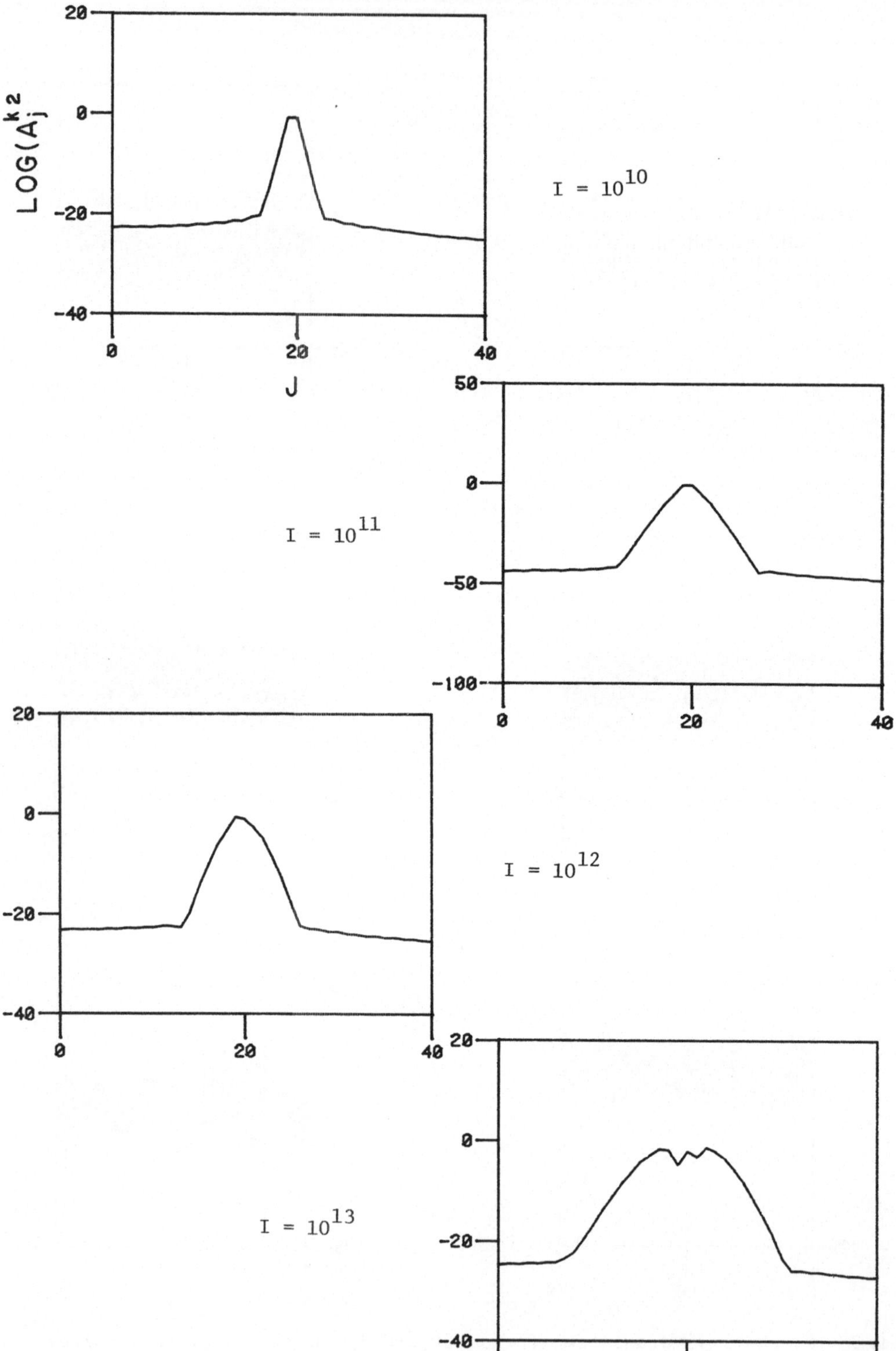

Figure 7. Log($A_j^{K^2}$) versus J for J_o=20 and M_J=0 with different values of the intensity (in W/cm^2).

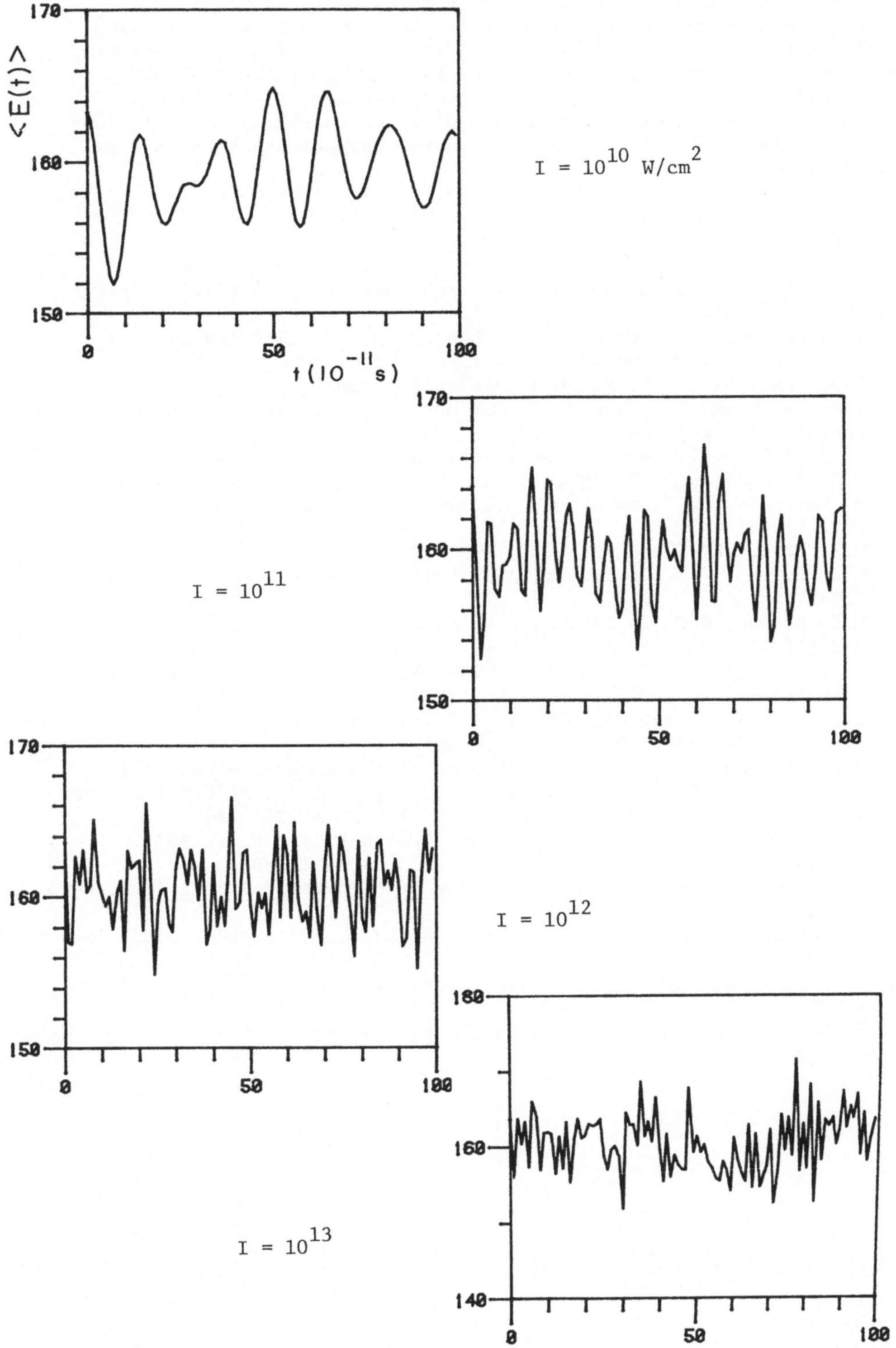

Figure 8. <E(t)> averaged over rotational quantum numbers M_J.

over a large number of disparate Rabi frequencies ω_R. The energy behaviour looks chaotic although it still oscillates around the initial energy. There is no gradual increase of energy with time as is expected of chaotic systems[9-12]. Recently, the nature of quantum chaos has been examined in a two-level system coupled through time-dependent perturbations[20]. It was conjectured that the presence of chaotic Rabi oscillations need not be the signature of chaos in the dynamics[21]. Our results in figures 4, 6 and 8 would seem to substantiate this last observation.

REFERENCES

1. J.H. Eberly, B.W. Shore, Z. Bialynicka-Birula, I. Bialynicki-Birula, Phys. Rev. A16, 2038, 2048,(1977).
2. A.D. Bandrauk, G. Turcotte, J. Chem. Phys. 77, 3867 (1982).
3. R.V. Jensen, S.M. Susskind, see this volume.
4. J.R. Ackerhalt, P.W. Milonni, see this volume.
5. A.D. Bandrauk, L. Claveau, SPIE Proc. High Intensity Laser Processes, 664, 217 (1986).
6. R. Blumel, S. Fishman, U. Smilansky, J. Chem. Phys. 84, 2605 (1986).
7. D.R. Grempel, R.E. Prange, S. Fishman, Phys. Rev. A29, 1634 (1984).
8. M. Feingold, S. Fishman, D.R. Grempel, R.E. Prange, Phys. Rev. B31, 6852 (1985).
9. G.M. Zaslavsky, Phys. Rep. 80, 159 (1981).
10. G.P. Berman, G.M. Zaslawsky, A.R. Kolovsky, Phys. Lett. A87, 152 (1982).
11. G.P. Berman, G.M. Zaslavsky, A.R. Kolovsky, JETP (Sov. Phys) 54, 272 (1981).
12. H.W. Galbraith, J.R. Ackerhalt, P.W. Milonni, J. Chem. Phys. 79, 5345 (1985).
13. G.L. Lamb Jr., Rev. Mod. Phys. 43, 99 (1971).
14. S. Chelkowski, A.D. Bandrauk, see this volume.
15. R.G. Harrison, H.N. Rutt, in Physics of New Laser Sources, edit. N.B. Abraham, F.T. Arecchi, A. Mooradian, A. Sona, NATO ASI Series B, Phys. vol. 132, Plenum Press, N.Y. 1984, pp. 201-216.
16. R. Loudon, Quantum Theory of Light, Oxford Press, U.K. 1983.
17. K. Iguchi, K. Suzuki, J. Chem. Phys. 73, 4026 (1980).
18. M. Abramowitz, I.A. Stegun, Handbook of Mathematical Functions, Dover Publications, N.Y. 1968.
19. C. Cousin, C. Rossetti, C. Meyer, C.R. Acad. Sc. (Paris) 268B, 1640 (1969).
20. Y. Pomeau, B. Darizzi, B. Grammaticos, Phys. Rev. Lett. 56, 681 (1986).
21. R. Badii, P.F. Meier, Phys. Rev. Lett. 58, 1045 (1987).

CHAOTIC DYNAMICS, POLYATOMIC MOLECULES, MULTIPHOTON DISSOCIATION AND ALL THAT

A. Ferretti* and N. K. Rahman+

+Dipartimento di Scienze Chimiche dell'Università di Trieste Piazzale Europa 1, Trieste, Italy
*Dipartimento di Chimica dell'Università di Pisa, Via Risorgimento 35, Pisa, Italy

INTRODUCTION

In this article, we shall be concerned with certain aspects of polyatomic molecules that are relevant to understand both its intrinsic dynamics as well as that involving its coupling with electromagnetic radiation. Since the space is limited, considerations will be brief and the reader will be referred to the original literature for details. As will be clear to the readers, the research results are all obtained within our group.

The final objective of this paper is to try to understand how and why polyatomic molecules dissociate when subject to radiation of frequency different from the natural frequencies of the molecule and the photons having energy at least one order of magnitude less than the dissociation energy of the molecule. This area of investigation has been the object of rather extensive research in the last decade. (We refrain from referring to any paper since the literature is too vast and it is easy for an interested reader to look up appropriate reviews). While the subject has now been explored from many different angles, it appears that there is much to be learned, if fresh approaches to the problem are taken.

The paper is divided into 2 parts which are covered in the next 2 sections. In the first part, we discuss how to confront a dense set of levels of a molecule theoretically as well as from data arising from an extensive set of measurements. The reason for looking at a large set of levels is due to the fact that it is in this region that one expects deviations from the usual Poisson distribution of nearest level spacing. An analysis of this nearest level spacing distribution, the Δ_3 statistics, the correlation coefficients and the Q statistics has been made for a theoretical molecule in which two modes of vibration are coupled. The changes brought in due to the increase of coupling between the two modes are seen rather clearly. In the second part of this section, we take a real molecule, the methanol. Recently, a collaboration between laboratories in Pisa,

Firenze and Bologna has allowed us to obtain the energy levels of the torsional-rotational levels of methanol. We have applied our statistical methods to these and obtained rather interesting qualitative results. It will certainly be more fruitful to be able to obtain data of the kind envisaged in the theoretical model enunciated before. But at this time one has to make do with what is available in the experimental measurements.
The second part of this paper considers the dissociation of polyatomic molecules when subject to infrared radiation especially CO_2 laser radiation with usually short and intense (from few hundred nsec down to even psec, ~ 10 MW/cm^2) pulses. It will be shown how the developments in the field of chaotic dynamics point toward further utilization of the newer techniques to confront the dynamics of such processes. It is really a matter of some satisfaction that one may make some nontrivial qualitative prédiction with the technique of discrete mapping. This technique of mapping is the main content of this second part of this article. It will be argued that mapping can be viewed as an alternative to usual theoretical models that invariably involve solving a large number of coupled differential equations. It will be shown that in particular a study of coupled logistic map with linear coupling is enough to simulate the process of dissociation by two external e.m. fields and provides some key to the utilization of maps to problem of this class.

STATISTICAL ASPECTS OF SPECTRA OF POLYATOMIC MOLECULES

Classical mechanics of models of polyatomic molecules leads invariably to chaotic trajectories. This is simply shown in the following manner. Consider two vibrational degrees of freedom of a polyatomic molecule which are coupled. Let the Hamiltonian be represented by

$$H = H_1(q_1,p_1) + H_2(q_2,p_2) + V(q_1,q_2,p_1,p_2)$$

The equations of motion are the Hamiltonian equations which are

$$\frac{\partial H}{\partial q_1} = -\dot{p}_1 \;;\quad \frac{\partial H}{\partial p_1} = \dot{q}_1 \;;\quad \frac{\partial H}{\partial q_2} = -\dot{p}_2 \;;\quad \frac{\partial H}{\partial p_2} = \dot{q}_2$$

These are 4 equations. q_1, q_2, p_1 and p_2 are not independent since

$$H(q_1,p_1,q_2,p_2) = E$$

is conserved. We are led to 3 differential equations which are in general non-linear and coupled. For given E, the trajectories may be computed and for almost all values of E beyond a critical value E_c, the trajectories are exponentially divergent. These are best seen by making Poincaré sections which give rise to regular continuous curves up to E_c and random points beyond E_c. Thus we know that the classical dynamics of polyatomic molecules for some energy range must be chaotic. But then one may ask: so what?

In fact, a experimentalist does not see a molecule with trajectories as Poincaré sections. He perceives the molecule as a quantum system with partly discrete energy values as well as the presence of the continuum. The question that arises is how a system that is chaotic in principle from classical dynamics manifests itself with in its quantum dynamics.

Among many possible answers to this question, one that we have decided to pursue somewhat in detail is the statistical nature of the energy levels. We start from the knowledge that as the energy increases, the vibrational spectra become dense. Thus it makes sense to look after the statistical regularities of such systems. Indeed, Wigner back in 1967[1] had surmised that one could expect deviations to occur from the usual Poisson distribution due to dynamical effects of level repulsion. Recent times have seen that suggestion taken seriously enough to even link the deviation from the Poisson distribution as a sign of what has been termed quantum chaos. Even though this latter term appears to be privy of a proper definition, the statistical distribution of complex system's eigenvalues appears to be fit for some investigations.

We have investigated this problem both with a purely theoretical construct as well as experimental data. First the theoretical construct[2].

Let us imagine a molecule with two vibrational degrees of freedom and let these two degrees of freedom be coupled in a rather simple manner. The Hamiltonian of the molecule is represented as

$$H = H_1(q_1,p_1) + H_2(q_2,p_2) + \lambda q_1^2 q_2$$

with

$$H_1(q_1,p_1) = \frac{p_1^2}{2m} + V_1(q_1)$$

$$H_2(q_2,p_2) = \frac{p_2^2}{2m} + V_2(q_2)$$

Let $V_2(q_2)$ be a Morse oscillator and $V_1(q_1)$ be an anharmonic oscillator. We have diagonalized such a system and calculated the eigenvalues as a function of λ. To be precise, the Morse oscillator has been fixed to have 33 bound states. The resulting eigenvalues are essentially exact for the system barring the effect of the continuum of the Morse oscillator. Since this latter omission is expected to have some effect on the higher part of the resulting eigenvalues, the statistical analysis has been limited up to those energies where the energies are expected to be accurate.

The results are shown in fig. 1 for the nearest level spacing distribution. It is evident that as λ increases, noticeable deviations from the Poisson distribution occurs. The fact that these deviations are there, is clearly exhibited by the calculation of the Δ_3 statistics of Dyson-Mehta[3] which are shown in fig. 2. The Poisson distribution should correspond to the straight line, especially for low values of L (see definition in ref. 3) which is not quite the case as is clear from the results of the calculation.

It is evident that it would be most interesting to dispose of experimental data which correspond to these theoretical calculations. However, such experimental data suitable for statistical considerations are non-

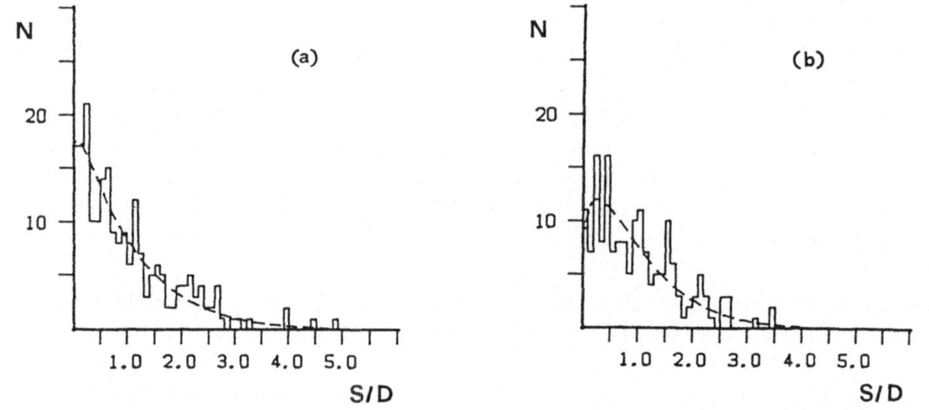

Fig. 1. Level spacing distribution (the spacing S is given in units of the average spacing D) for the eigenvalues of a Morse potential coupled with an anharmonic potential in the energy range $-8<E<-5$ eV and with two different coupling parameters[2] (a) $\lambda=0.3$; (b) $\lambda=0.9$. The dashed line is a fit of the histograms by means of the Brody distribution

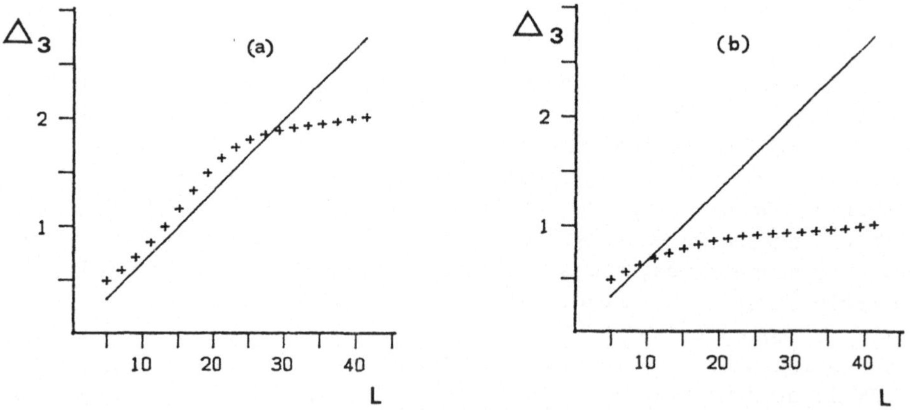

Fig. 2. Δ_3 statistics for the sets of levels considered in fig. 1. (a) $\lambda=0.3$; (b) $\lambda=0.9$. The straight line is the theoretical L/15 behaviour expected for a Poisson distribution[2]

existent. However, the consideration of chaotic dynamics is valid for independent degrees of freedom. We have been recently able to analyze statistically the experimental data of torsional-rotational spectra of the methanol molecule[4]. Here, the deviation from the Poisson distribution would have arisen from the coupling of essentiallytwo distinct types of motion: the rotation and the torsion. Some representative results are shown in fig. 3a and fig. 3b. In a nutshell, deviation from the Poisson distribution appears to exist but not very pronounced. It is, however, most interesting that the Δ_3 statistics reveal a rather striking continuity for each of the groups of energy levels separated by symmetry, which for methanol is only an approximate description.

MULTIPHOTON DISSOCIATION DESCRIBED BY MAPPING

The history and the status of the research in multiphoton dissociation of polyatomic molecules with short and intense pulses of CO_2 lasers need not be recounted here. The pioneering work was done in Canada, even though the implications were note quite realized, and the research later was mostly done in various laboratories scattered all over the world resulting in thousands of papers and even a few dozen review articles. Now, the initial enthusiasm for such processes may have somewhat died down, the interest has been lingering in this area, since now that the economic aspects of isotopic separation have been finally resolved, the fundament aspects of the research in this area is emerging. Note also that the research in the field of photo-dissociation itself has close connection with other areas of research regarding coupling of the bound state with the continuum[5].

We have decided to take a fresh look at this problem from the point of view of chaotic dynamics. The reason is twofold. Firstly, as has been discussed in the previous section, the dynamics of the polyatomic molecules

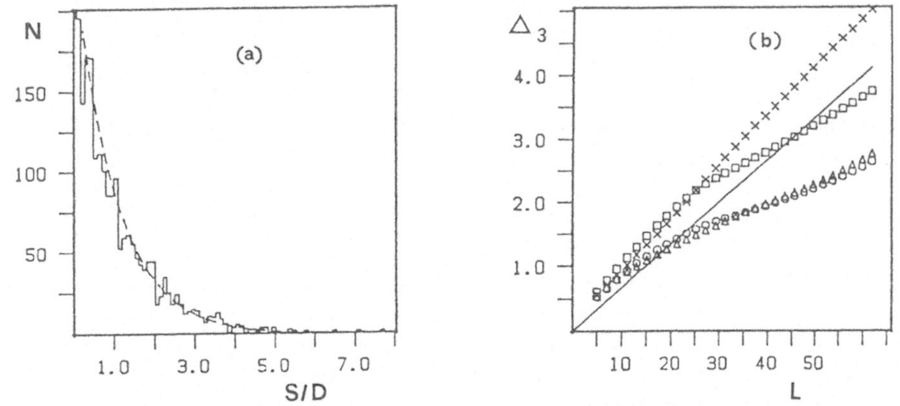

Fig. 3. (a) Level spacing distribution for 2396 torsional-rotational levels of methanol without any distinction of symmetry[4], the dashed line is the Poisson distribution; (b) Δ_3 statistics for torsional-rotational levels of methanol divided by symmetry: □ A , ○ E_1 , △ E_2 , + all the symmetries

is potentially one of chaotic dynamics. The second reason needs a little amplification . Consider first a diatomic molecule with its single vibrational degree of freedom. Couple it an oscillating external electromagnetic field. Carry out the dynamics by for example computing trajectories and Poincaré sections. One sees that the dynamics is chaotic! The reason for this is quite clear. Coupling an external oscillating field to the vibrational motion is tantamount to coupling two vibrating degrees of freedom. Coupling a system with more than one vibrational degree of freedom, as is necessary for simulating polyatomic molecules, the results are even more striking. Thus, we expect chaotic dynamics to be really a basic feature of polyatomic molecules coupled to strong electromagnetic fields such as CO_2 lasers.

In fact, the importance of chaos in these processes is so clear that theoretical models have been constructed that go on to link the process of photodissociation in terms of one of the favourite scenarios of chaotic behaviour, namely the diffusion-like behaviour arising due to the occurrence of overlapping resonances, a criterion called the Chirikov criterion of overlapping resonances.

Finally, it must be mentioned that one of the widely utilized models of multiphoton dissociation is that of a Duffing oscillator, which is a non-linear oscillator forced by an external oscillating field and containing a damping term[6]. Among the generic behaviour of such a system, one notices two of the most widely utilized scenarios of chaos, namely the occurrence of strange attractors and the phenomenon of period doubling. These last two are significant in addition to the fact that if the Duffing oscillator is modified as envisaged recently[7], one expects in certain regimes multiple stable periodic solutions.

With the foreknowledge of the above, it is natural to ask why if chaotic dynamics is intrinsic to the multiphoton dissociation of polyatomic molecules, one of the persuasive and at the same time potent description of chaos has not been attempted for multiphoton dissociation of polyatomic molecules. We are referring to the study of mapping as a simulator of dynamics that is normally otherwise described by coupled differential equations.

While there have been studies with various discrete mapping, the most fruitful of them is the logistic map which is given by

$$X_{n+1} = \lambda X_n(1-X_n) \quad ; \quad 0<\lambda<4$$

It is not the place to describe the rich content of this map, which can be easily looked up in the literature[8]. For our purpose, we note that as λ goes beyond some critical value λ_c, the fixed point of this map goes from oscillating to chaotic regime, the route towards this chaos is by period doubling. It is therefore very suggestive to try to simulate the dissociation of a polyatomic molecule by the logistic map in which λ represents the strength of the external e.m. field's coupling with the molecule and X being one of the dynamic variables which could for example be a particular mode of the molecule. While this is quite correct, there is one particular reason for which this is not very fruitful. The attractiveness of using discrete map such as logistic map is that it allows one to make particular numerical predictions which are in general experimentally measurable. In our case, this is highly problematic, since the dissociation process has been studied with a rather wide range of spatio-temporal pulse shape (some of which are not very well defined), rendering the precise numerical constants arising out of the logistic map not quite useful.

We propose however to circumvent this problem by the following manner. Consider a coupled logistic map

$$X_{n+1} = \lambda_1 X_n (1-X_n) + \beta_1 Y_n$$

$$Y_{n+1} = \lambda_2 Y_n (1-Y_n) + \beta_2 X_n$$

generalizing the meaning of (X,Y) explained before, we now have a system with two dynamical variables (X,Y) which are coupled in the simplest possible manner (linearly) and each coupled to an external field whose strengths are measured by λ_1 and λ_2 respectively. We may then ask what is the consequence of augmenting the description of the system with two variables and additional parameters?

The answer to such a question is rather simple. The additional "mode" allows now to examine what effect it has on the original "mode" X. The coupled logistic map being a much more complicated system than the simple logistic map (now there is a 4-dimensional parameter space to deal with) needs a detailed study. We have completed one such study[9,10]. This has enabled us to make some interesting observations which tend to reinforce our original conjecture. While it is impossible to summarize here all the results of our finding, it will be sufficient here to highlight two of the remarkable results that have been obtained. These are (1) the threshold λ for chaos to occur in the mode X gets remarkably lowered in the presence of the excitation of the other mode Y. (2) the coupled logistic map have solutions that are multiply stable, each of which can lead to chaos by period doubling[11].

We believe that each of these conclusions are of some significance. The first conclusion has experimental consequences. If a polyatomic molecule is subjected to simultaneous irradiation by two different external e.m. fields each of which couple to different modes, then the threshold for dissociation due to only one of the two fields gets considerably lowered. This is certainly a testable preposition and it would appear that experiments done to date are not at least in contradiction with this result. Furthermore, it appears, that some new experiments may be attempted to pin down the effect of two fields compared to only one.

The second conclusion is of a more theoretical interest while not of any lesser importance. The standard models of dissociation by the forced Duffing oscillator if modified to be slightly more realistic give rise to multiple stable solutions. The discrete logistic map itself has no such behaviour. So far multiple stable solutions have been found for the discrete map with cubic terms or in three dimensional systems. The coupled logistic map appears to be the simplest system that possess multiple stable solutions.

We have been able to locate such solutions for some parameter values and have seen that these multiply stable solutions themselves lead to chaos by means of period doubling. The basins of attraction for these solutions are a point of general interest. We have investigated a few of these for certain parameters values and obtained some remarkably rich structures[11] (fig. 4).

CONCLUSIONS

Polyatomic molecules that we have investigated so far both theoretically with models as well as from experimental data appears to us extremely rich in content all of whose implications have yet to be explored. Since polyatomic molecules are rather complicated systems, it is all the more desirable that all the apparatus of modern theoretical and experimental physics be brought into force. The result obtained in our own area certainly have been very encouraging.

ACKNOWLEDGEMENTS

This research is partially funded by Ministry of Public Instruction as well as by I.C.Q.E.M. of C.N.R. at Pisa. The authors are grateful for the collaboration of a number of colleagues, in particular R. Cimiraglia (University of Pisa, Department of Chemistry) and G. Moruzzi (University of Pisa, Department of Physics).

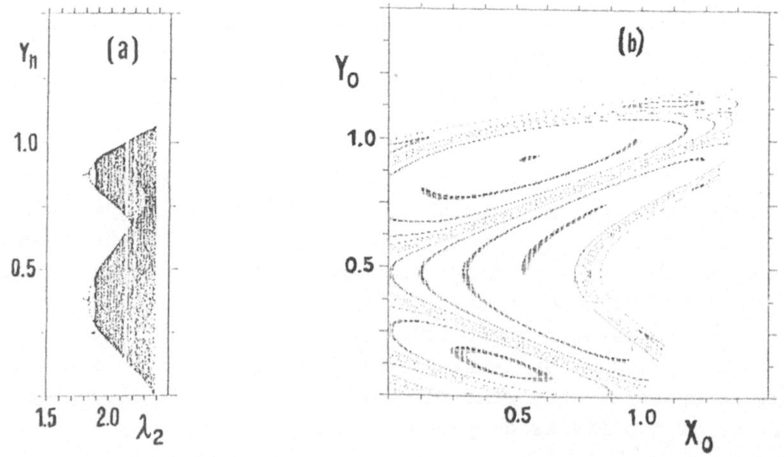

Fig. 4. Coupled logistic map with symmetric coupling ($\beta_1=\beta_2=0.5$) (a) bifurcation sequence of Y_n as a function of λ_2 ($\lambda_1=2$) showing the reduction of the threshold for chaos ($\lambda_c=1.81$) compared to the monodimensional map ($\lambda_c=3.54$); (b) basins of attraction for the two coexisting attractors of period 4 (white areas) and period 6 (black areas) in the space of initial conditions for $\lambda_1=1.1$ and $\lambda_2=2.6148$

REFERENCES

1. E. P. Wigner, Random Matrices in Physics, SIAM Rev., 9:1 (1967).
2. R. Cimiraglia, A.Ferretti, and N. K. Rahman, to be published.
3. F. J. Dyson, and M. L. Mehta, Statistical Theory of the Energy Levels of Complex Systems. IV, J. Math. Phys., 4:701 (1963), also in: "Statistical Theories of Spectra: Fluctuations," C. E. Porter, ed., Academic Press, New York (1965).
4. A. Ferretti, and N. K. Rahman, Statistical Analysis of Spectra of Polyatomic Molecules: Methanol, Phys. Rev. A, 32:3797 (1985).
   A. Ferretti, G. Moruzzi, and N. K. Rahman, Statistical Analysis of the Torsional-Rotational Levels of Methan in the Range 127-1600 cm^{-1}, Phys. Rev. A, to be published.
5. N. K. Rahman, C. Guidotti, and M. Allegrini, "Photons and Continuum States of Atoms and molecules, Springer-Verlag, Berlin (1987).
6. N. Bloembergen, Comments on The Dissociation of Polyatomic Molecules by Intense 10.6 μm Radiation, Optics Commun., 15:416 (1975)
7. F. T. Arecchi, Multiphoton Selective Excitation: On the Role of Chaotic Dynamics with Many Basins of Attraction, in: "Collisions and Half-Collisions with Lasers," N. K. Rahman and C. Guidotti, ed., Harwood Academic Publisher, New York (1984).
   R. Räty, J. von Boehm, and H. M. Isomäki, Chaotic Motion of a Periodically Driven Particle in an Asymmetric Potential Well, Phys. Rev. A, 34:4310 (1986).
8. P. Cvitanovic, " Universality in Chaos," Adam Hilger Ltd, Bristol (1984)
9. A. Ferretti, and N. K. Rahman, A Study of Coupled Logistic Maps and Their Usefulness for Modelling Physico-Chemical Processes, Chem. Phys. Lett., 133:150 (1987).
10. A. Ferretti, and N. K. Rahman, A Study of Coupled Logistic Map and Its Applications in Chemical Physics, Chem. Phys., to be published.
11. A. Ferretti, and N. K. Rahman, Coupled Logistic Maps in Physico-Chemical Processes: Coexisting Attractors and Their Implications, Chem. Phys. Lett., to be published.

THEORETICAL STUDY ON THE POSSIBILITY OF LOCAL MODE(LM) EXCITATION IN POLYATOMICS BY FREQUENCY-MODULATED(FM) PULSE LASERS

Bo-Min Xie and Jia-Qiang Ding

Institute of Mechanics
Chinese Academy of Sciences
Beijing, China

INTRODUTION

Due to its spectacular frequency characteristics, there has been much concern about the possibility of using lasers to realize mode-selective or bond-selective dissociation and chemical reactions[1]. Ordinarily the chemical reactions, not controlled in the molecular level, proceed according to statistical law and often produce unwanted intermediate substances with waste of energy. To control the chemical reactions in the molecular level can provide not only low temperature chemical reactions, but also the possibility of synthesizing some special molecules. The main obstacles for achieving this goal are the anharmonicity in the bond vibration and the fast intramolecular energy transfer among various modes. The exploratory experimental work on t-butyl hydroperoxide[2] while showed some promise of bond-selective chemistry, also demonstrated the poor efficiency of LM excitation by fixed frequency lasers. Various ways of modulation in phase (PM), amplitude(AM) and frequency(FM) of the laser pulse [3-6] have been proposed to circumvent the above mentioned obstacles. The feasibility of PM and AM were investigated theoretically only for systems of few energy levels and technologically they are more difficult than the FM method which we are going to discuss here.

The existence of LM characteristic frequencies in many polyatomic molecules was already pointed out in forties and summarized in Herzberg's book[7]. In recent years, the LM fundamentals and high overtones of C-H, N-H, O-H, Si-H bonds etc. have been repeatedly confirmed with increasing spectroscopic resolution [8-16]. Moreover, the experimental evidence of overtone induced unimolecular fragmentation and isomerization [17-19] showed some peculiarities different from statistical interpretation. Therefore, should bond-selective excitation and bond-selective chemical reactions for an arbitrary molecule be a dream too ambitious, the same attempt for molecules with prominent LM characteristics may well be within our reach. In this regard, this report is hoped to stimulate the interest and initiatives of our experimental colleagues to advance the FM technique of pulse lasers and to further the experimental study on selective chemical reactions.

In concluding this introduction, we would like to cite a recent interesting publication by Tannor, Kosloff and Rice [20] in which they advanced a novel idea of making use of the electronically excited energy surface to control the reaction channel in ground state. We suppose a new arena of selective chemical reactions involving the electronically excited states in molecules, the curve crossing phenomena and selective reactions on solid surface etc. is now making its shape and will bring forth new consequences unforeseen today.

BASIC IDEAS AND RESULTS

The basic idea of exciting the LM of polyatomics is invoked from the classical mechanics that we can use suitable FM external force to push the anharmonic oscillator through the nonlinear resonance[21]. If the response curve of the oscillator is something like that in Fig.1 of a soft nonlinear spring (such as in the case of a Duffing or Morse oscillator) and if an external force of gradually decreasing frequency is applied, starting from some point A with $\omega_e/\omega_0>1$ then the amplitude of the oscillator will follow the zigzagging line about AB in Fig.1 and builds up. For a corresponding quantum anharmonic oscillator, we have shown, both analytically and by numerical computation the same picture holds up to high excitation state. It can be seen from Fig.2 that the FM laser is much superior (curves 22,33) to the fixed frequency excitation (curve 11) in exciting the anharmonic oscillator. Encouraged by this preliminary result, we proceeded to investigate the more complex case of polyatomic molecules. [5,6]

For polyatomic molecules, its rotational frequency at room temperature is usually less than one hundredth of its typical vibrational frequency (let alone the case of low temperature, e.g. in a supersonic stream), while the laser frequency is near the frequency of the LM in order to excite it. Therefore, viewed from the laser, the rotational motion of the molecule is very slow and its effect can be neglected in the first approximation. We also assume that only the LM under consideration has a dipole moment which is coupled directly with the laser, while all other vibrational modes are not coupled effectively with the EM field of the laser due to the frequency mismatch or simply due to the absence of dipole moment. The fast intramolecular energy transfer among various modes in polyatomics is the main obstacle for the excitation of its LM. Hence the problem to be dealt with is essentially the competition between the FM action and this fast energy transfer. Experimentally, this energy transfer is embodied in the broad linewidth of the LM overtones, e.g. for C-H or C-D overtones in benzene, FWHM=50-100cm^{-1} which can not be attributed to rotational effect and is due to the interaction between the LM and other vibrational modes [22-23]. By the reasoning similar to that of Fano [24] in developing the quantum theory of spectral lines, this interaction can be reduced to an equivalent decay rate of the excited LM overtones and it can be further interrelated with the spectral line width of the LM overtones. In this way, the cumbersome problem of specifying the concrete form of the intramolecular interaction can be avoided. We have developed along this line a theory of the LM excitation in polyatomics [6], by which one can judge if it is feasible to excite the LM of a specific polyatomic molecule by FM lasers, and if it does, what FM parameters should we choose. Figs. 3,4 depict the results of excitation of C-H bond in benzene[6], taking the time dependence of the FM to be

$$f(\tau) = \mathrm{Cos}\ 2\pi(1-\beta\tau)\tau$$

where β is an adjustable modulation parameter, $\tau= \Omega_0 t/2\pi$ (Ω_0 the initial frequency of the laser and t the time). It can be seen the value of β is is not very sensitive for the result and this is favorable for designing the experiment. The linewidth of the overtones has a crucial effect on the effective excitation of the LM, but if the laser field strength is sufficiently high, average FWHM linewidth of 50cm^{-1} or even more does not constitute an obstacle (Fig.3 curve 1-3). When the field strength is not

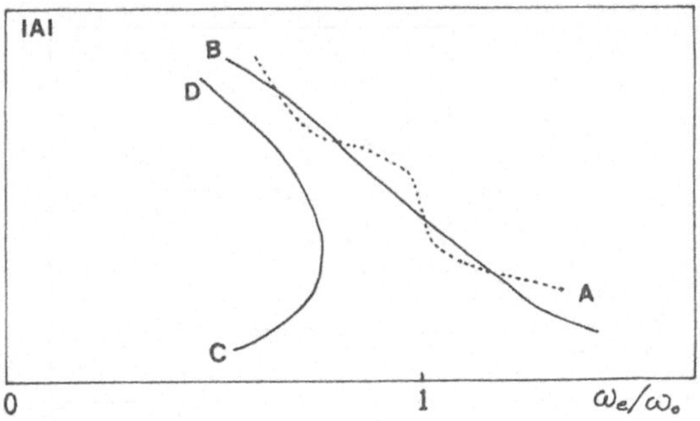

Fig. 1. Response curve of a classical soft nonlinear oscillator
A -- amplitude of oscillation
ω_0-- small amplitude frequency
ω_e-- frequency of the external force

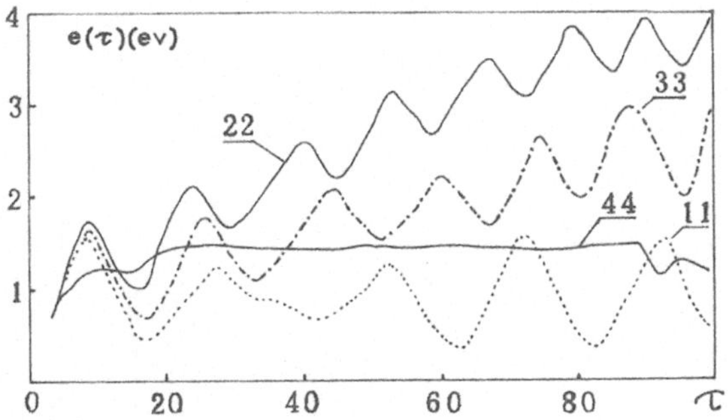

Fig. 2. HF exicited by FM lasers, influence of different modulation speed β . Laser electric field $E(t')=E_0\cos\Omega t'=E_0\cos 2\pi(1-\beta\tau)\tau$. Electric field coupling strength fixed at $\mu E_0=0.3eV/a_0$ The FM is $\Omega=\Omega_0(1-\beta\tau)$,$\tau=\Omega_0 t'/2\pi$. Mechanical frequency ω_H of HF: $\hbar\omega_H=0.5137eV$, $e(\tau)$ the energy absorbed by HF.
11: $\Omega_0=0.9467\omega_H$ $\beta=0$ [WP's result (Ref.25) re-calculated],
22: $\Omega_0=0.9467\omega_H$ $\beta=1.8\times10^{-3}$,
33: $\Omega_0=0.9467\omega_H$ $\beta=0.9\times10^{-3}$,
44: $\Omega_0=0.9467\omega_H$ $\beta=9\times10^{-3}$.

high enough, even the pure LM can not build up very soon, the intramolecular interaction naturally hinder us to achieve the goal of LM excitation (Fig.3 curve 5).

It can be seen from the above that FM laser is a quite promising agent

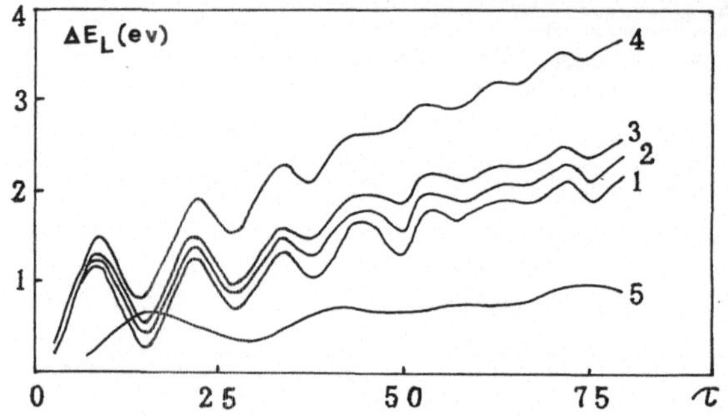

Fig. 3. Effect of overtone line width on the LM excitation of C-H bond in benzene
LM energy absorption $\Delta E_L = E_L(\tau) - E_L(0)$ in eV, $\tau = \Omega_0 t/(2\pi)$, $\Omega_0 = 2988\text{cm}^{-1}$
Field strength $\hat{\mu}E_0/\alpha = 0.26$eV(laser intensity $=10^{14}\text{W/cm}^2$), FM parameter $\beta = 3.2\text{x}10^{-3}$. For curves 1-4, $\gamma = 0.1050$, 0.08, 0.06 and 0(i.e FWHM of overtone lines=50,38,29 and 0 cm^{-1}) respectively. The effect of field strength is shown in curve 5: $\hat{\mu}E_0/\alpha = 0.1$eV, $\beta = 1\text{x}10^{-3}$ and $\gamma = 0$.

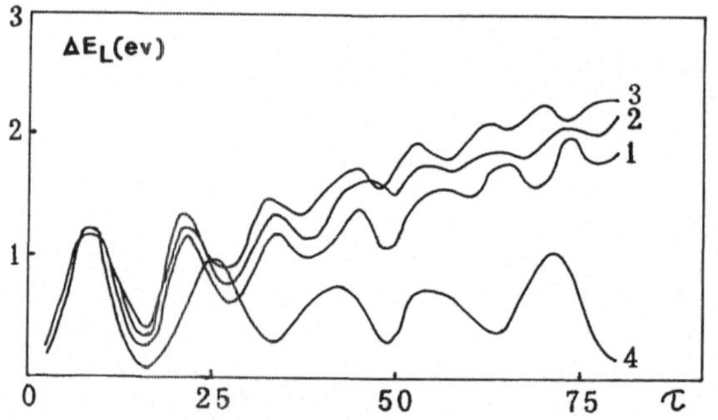

Fig. 4. Effect of FM parameter β on the LM excitation of C-H bond in benzene
Field strength $\hat{\mu}E_0/\alpha = 0.26$eV, $\bar{\gamma} = 0.1051$. For courves 1-3: $\beta = 2.8\text{x}10^{-3}$, $3.2\text{x}10^{-3}$, $3.6\text{x}10^{-3}$, respectively. 4: $\beta = 0$, $\gamma = 0$.

to enhance the LM excitation in polyatomic molecules, so long as the latter possess distinct LM overtones with not excessive linewidths. The technological difficulty is perheps not so tremendous as it first looks, e.g. we can use piezo-electric or electro-optic device to change the optical cavity length during the pulse in a controllable way to get the desired center frequency shift. In order to break a chemical bond effectively and in a desired way, it is necessary to pump it to an energy extent at least equal to the bond energy. Let the duration of the pulse be T, then its frequency bandwidth is $\Delta\nu \approx T^{-1}$. If the necessary frequency change during the pulse is $\Delta\Omega$, the modulation cannot be fruitful unless $\Delta\nu \ll \Delta\Omega$ or $T \gg (\Delta\Omega)^{-1}$. Judged from curve 3 Fig.4, we may excite the C-H bond to nearly dissociation(with dissociation energy=5.4eV) in 200 cycles. During such an interval $\Delta\Omega \approx 200\beta\Omega_0$, so we should have $T \gg (200\beta\Omega_0)^{-1} = 2.5\times10^{-15}$Sec. Obviously, ns even ps pulses suffice for this requirement. But at present, we are still lacking the information of LM overtones of very high quantum number, and here new physical phenomena such as quantum chaos might intervene. More theoretical and especially experimental work await exploration.

This work was supported by the National Science Foundation of the PR of China.

REFERENCES

1. See, for instance, Physics Today 33, No 11 (1980)
2. D.W.Chandler, W.E.Farneth and R.N.Zare J.Chem. Phys. 77 4447 (1982)
3. W.S.Warren and A.H.Zewail J.Chem. Phys. 78 3583 (1983)
4. J.C.Diels and S.Besnainou J.Chem. Phys. 85 6347 (1986) and the references therein
5. Xie Bo-Min, Ding Jia-Qiang and Wei Xie J. Chem. Phys. 84 3819 (1986)
6. Xie Bo-Min and Ding Jia-Qiang J.Chem. Phys. 86 6579 (1987)
7. G.Herzberg, "Molecular Spectra and Molecular Structure" (D.Van Nostrand. 1954) V2, 194-201
8. R.G.Bray and M.J.Berry J.Chem. Phys. 71 4909 (1979)
9. K.V.Reddy, D.F.Heller and M.J.Berry J.Chem.Phys. 76 2814 (1982)
10. J.S.Wong and C.B.Morse J.Chem.Phys. 77 603 (1982)
11. J.W.Perry and A.H.Zewail J.Chem.Phys. 80 5333 (1984)
12. M.S.Burberry and A.C.Albrecht J.Chem.Phys. 70 147 (1979)
13. R.A.Berheim, F.W.Lampe, J.K.O'Keeft and J.R.Qualey Chem. Phys. Lett. 100 45 (1983)
14. H.L.Fang, R.L.Swofford and D.A.C. Compton Chem.Phys. Lett. 108 539 (1984)
15. J.J.Jasinski Chem.Phys.Lett. 109 462 (1984)
16. L.M.Ticich, M.D.Likar and F.F.Crim J.Chem.Phys. 85 2331 (1986)
17. K.V.Reddy and M.J.Berry Chem.Phys.Lett. 66 223 (1979)
18. B.D.Cannon and F.F.Crim J.Chem.Phys. 75 1752 (1981)
19. N.F.Sherer, F.E.Doany, A.H.Zewail and J.W.Perry J.Chem.Phys. 84 1932 (1986)
20. D.J.Tannor, R.Kosloff and S.A.Rice J.Chem.Phys. 85 5805 (1986)
21. A.H.Nayfeh and D.T.Mook "Nonlinear Oscillations" (Miley, New York 1979) Chap.4
22. D.F.Heller and S.Mukamel J.Chem.Phys. 70 463 (1979)
23. L.L.Siebert III, W.P.Reinhardt J.T.Hynes J.Chem.Phys. 81 1115 (1984)
24. U.Fano Phys. Rev. 131 259 (1963)
25. R.B.Walker and R.K.Preston J.Chem.Phys. 67 2017 (1977)

MULTIPHOTON EXCITATION OF BOND MODES

Martin L. Sage

Department of Chemistry
Syracuse University
Syracuse, New York 13244-1200 U.S.A.

INTRODUCTION

In this paper we present a treatment of weakly coupled identical bond modes which allows a detailed computation of spectroscopic intensities based upon the knowledge of properties of the isolated bond modes and their interactions. Our approach is useful in that we can explicitly keep track of which terms in the bond mode potential, in the coupling, and in the dipole expression will give rise to a given multiphoton transition. In our discussion we use the problem of two coupled Morse oscillators but the method works equally well for more realistic potentials. The hamiltonian for our model problem is[1]

$$H = (p_1^2 + p_2^2 + 2\,\alpha\, p_1p_2)/2 + (f_1^2 + f_2^2 + 2\,\beta\, f_1f_2)/2\,\lambda^2 \qquad (1)$$

where $f_j = 1 - \exp[-\lambda\,(q_j - q_0)]$. The width of the Morse potential is proportional to $1/\lambda$. In these units the energy levels of the individual local modes are $(n_j + ½) - \lambda^2(n_j + ½)^2/2$. After outlining our method we will illustrate using two photon excitation of the $n_1 + n_2 = 4$ states of the system.

The obvious way to proceed is to use a Morse oscillator basis set.[1,2] However we find it instructive to expand the Morse potential about its equilibrium position, use a harmonic basis, and approximately diagonalize this hamiltonian by means of a series of van Vleck transformations[3,4] using computer algebra. The same tranformations applied to the dipole operator yield the transition dipole in terms of the new basis.

The advantage of this procedure over our previous approach is that our new method leads to an effective hamiltonian for the coupled oscillators and can be applied to a wide variety of vibrational potentials. The effective hamiltonian can then be used in quantum or semiclassical calculations. Furthermore the method can be extended to a more complete hamiltonian in which bond modes are coupled to the remaining degrees of freedom in the molecule and thus used to treat intramolecular vibrational relaxation of bond modes.[5]

VAN VLECK TRANSFORMATION OF THE HAMILTONIAN

To begin with we shall ignore the coupling of the two oscillators. The hamiltonian for each Morse oscillator is

$$H_j = \frac{1}{2}(p_j^2 + x_j^2) - \frac{\lambda}{2} x_j^3 + \frac{7\lambda^2}{24} x_j^4 - \frac{\lambda^3}{8} x_j^5 + \ldots \qquad (2)$$

where $x_j = q_j - q_0$. We look for a unitary transformation of this hamiltonian

$$H'_j = U_{1j}^+ H_j U_{1j} = e^{-i\lambda S_{1j}} H_j e^{i\lambda S_{1j}} \qquad (3)$$

in which the terms first order in λ vanish. Eq(3) can easily be expanded as a power series in λ.[6]

$$H'_j = \sum_{n=0}^{\infty} \frac{\lambda^n i^n}{n!} [H_j, S_{1j}]_n \qquad (4)$$

where

$$[H_j, S_{1j}]_0 \equiv H_j$$

and

$$[H_j, S_{1j}]_n \equiv [[H_j, S_{1j}]_{n-1}, S_{1j}] \qquad (n \geq 1).$$

After this first transformation energy levels are correct to second order in λ.* The procedure can be repeated to remove off diagonal λ^2 terms and give energy levels correct to fourth order. A third transformation yields energy levels to sixth order, etc. We find it most convenient to switch annihilation, creation, and number operators. In terms of these operators the bond mode hamiltonian is

$$H_j = N_j + \frac{1}{2} - \frac{\lambda}{4\sqrt{2}}(a_j + a_j^+)^3 + \frac{7\lambda^2}{96}(a_j + a_j^+)^4 - \ldots \qquad (5)$$

The most convenient way to evaluate the multiple commutators needed to transform the hamiltonian is to use a computer algebra system. We have used

*Of course, for the one-dimensional Morse oscillator these are the exact energy levels for the bound states.

Table 1. Generators of van Vleck Transformations

$$S_{j1} = \frac{i}{12\sqrt{2}} \left(a_j^{\,3} + 9\, a_j N_j - 9\, N_j a_j^{\,+} - a_j^{\,+3}\right)$$

$$S_{j2} = \frac{i}{48} \left(- 2\, a_j^{\,4} + a_j^{\,2}(2\, N_j - 1) - (2\, N_j - 1)a_j^{\,+2} + 2\, a_j^{\,+4}\right)$$

$$S_{j3} = \frac{i}{480\sqrt{2}} \Big(- 7\, a_j^{\,5} - 5\, a_j^{\,3}(N_j - 1) - 5\, a_j(358\, N_j^{\,2} + 51)$$
$$+ 5\,(358\, N_j^{\,2} + 51)a_j^{\,+} + 5\,(N_j - 1)a_j^{\,+3} + 7\, a_j^{\,+5}\Big)$$

$$T_1 = \frac{i(\alpha - \beta)}{4} \left(a_1 a_2 - a_1^{\,+} a_2^{\,+}\right)$$

REDUCE and discussed the details elsewhere.[4] The generators of the first three van Vleck transformations are given in Table 1. The diagonal part of H_j is the exact energy for bound states of the Morse oscillator. The sum of the many diagonal λ^4 and λ^6 terms vanish.

We now look at the interaction terms. Before the unitary transformations, $U_1 = U_{11}U_{12}$, $U_2 = U_{21}U_{22}$, ...

$$H_{int} = \alpha\, p_1 p_2 + \beta\, x_1 x_2 [1 - \frac{\lambda}{2}(x_1 + x_2) + \frac{\lambda^2}{12}(2\, x_1^{\,2} + 3\, x_1 x_2 + 2\, x_2^{\,2}) + \ldots] \tag{6}$$

The terms independent of λ can be written as

$$\frac{\alpha + \beta}{2}(a_1^{\,+} a_2 + a_1 a_2^{\,+}) - \frac{\alpha - \beta}{2}(a_1^{\,+} a_2^{\,+} + a_1 a_2)$$

in which the first group does not change the quantum number of $N = N_1 + N_2$.

After the transformations H_{int} contains many additional terms. To find the contributions to the block diagonal hamiltonian which are second order in α and β and independent of λ and those linear in α and β and up to second order in λ, we must carry out an additional van Vleck transformation. Its generator, T_1, is given in Table 1. The resulting effective hamiltonian is

$$\begin{aligned} H_{eff} = {} & \left(1 - \frac{(\alpha - \beta)^2}{8}\right)(N + 1) - \frac{\lambda^2}{2}\left(1 - \frac{9\,\beta}{8}\right)(N + 1)^2 - \frac{\lambda^2}{2}\left(1 + \frac{9\,\beta}{8}\right)(\Delta N)^2 \\ & + \left(\frac{\alpha + \beta}{2} - \frac{\lambda^2(11\,\alpha - 3\,\beta)}{32}(N + 1)\right)(a_1^{\,+} a_2 + a_1 a_2^{\,+}) \\ & + \frac{\lambda^2(4\,\alpha + \beta)}{16}(a_1^{\,+2} a_2^{\,2} + a_1^{\,2} a_2^{\,+2}) \end{aligned} \tag{7}$$

Table 2. Dipole Matrix Elements Contributing to Two Photon Transitions from the Ground State to 4 a Manifold

I	J	< I | μ | J >
0 0:s	2 0:s	$-\frac{\lambda}{2}\mu_s(1,0) + (1-\lambda^2)\mu_s(2,0) - \frac{8+\lambda^2}{8}\mu_s(1,1)$
0 0:s	2 0:a	$-\frac{\lambda}{2}\mu_a(1,0) + (1-\lambda^2)\mu_a(2,0)$
0 0:s	1 1:s	$\frac{2+\lambda^2}{2}\mu_s(2,1) + (\alpha-\beta)\mu_s(2,0) + \frac{2+\lambda^2}{4}\mu_s(1,1)$
2 0:s	4 0:a	$-\frac{\sqrt{3}\,\lambda}{2}\mu_a(1,0) + \frac{6-14\lambda^2}{\sqrt{3}}\mu_a(2,0)$
2 0:s	3 1:a	0
2 0:a	4 0:a	$-\frac{\sqrt{3}\,\lambda}{2}\mu_s(1,0) + \frac{6-14\lambda^2}{\sqrt{3}}\mu_s(2,0) - \frac{3\sqrt{6}\,\lambda^2}{16}\mu_s(1,1)$
2 0:a	3 1:a	$\sqrt{3}\,(\alpha-\beta)\mu_s(2,0) + \frac{3+2\lambda^2}{2\sqrt{3}}\mu_s(1,1)$
1 1:s	4 0:a	0
1 1:s	3 1:a	$-\frac{\sqrt{3}\,\lambda}{2}\mu_a(1,0) + \frac{3-5\lambda^2}{\sqrt{3}}\mu_a(2,0)$

where we have written $\Delta N = N_1 - N_2$. In general, eigenfunctions of H_{eff} involve coupled bond modes.

DIPOLE OPERATOR

Instead of using the exponential dipole form which we have applied to one photon local mode overtone spectra[7,8] we shall expand the dipole operator in a power series in x_1 and x_2. The dipole has two components, μ_s and μ_a,

$$\mu_\chi = \sum_{k=0}^{\infty} \sum_{j=0}^{k} \mu_\chi(k-j,j)\, x_1^{k-j} x_2^{j} \tag{8}$$

where χ is s or a, $\mu_s(k-j,j) = \mu_s(j,k-j)$, and $\mu_a(k-j,j) = -\mu_a(j,k-j)$. If the bond dipole approximation can be used, all terms with two non-zero indices will vanish. Furthermore the ratio $\mu_s(k,0)/\mu_a(k,0)$ will be independent of k and depend only on geometry. In this case all dipole matrix elements between basis functions in which both quantum numbers change arise from the coupling terms in the original hamiltonian, eq(6).

Using REDUCE the dipole operator can readily be written in the transformed basis that led to the effective hamiltonian using REDUCE. Table 2 lists transition dipole matrix elements which contribute to two photon transitions from the ground state to n = 4 antisymmetric states and involve nearly resonant intermediate levels. Basis functions are written $n_> \, n_<{:}\chi$

where $n_>$ is the larger of n_1 and n_2, $n_<$ the smaller, and χ is s or a for symmetric or antisymmetric functions. Eigenfuntions of H_{eff} will involve coupling of functions with the same symmetry and quantum number n. Only terms through second order in the expansion of the dipole operators have been kept. Expressions independent of α and β are given to order λ^2 while those linear in α and β to order λ^0.

High order van Vleck transformations or high order terms in the dipole expansion are needed to induce transitions between states with large differences in n. Actual transition dipole matrix elements between states in which both quantum numbers change involve a complicated sum of terms which may largely cancel for some states and reinforce for others.

CONCLUSIONS

In this paper we have presented a method of generating an effective hamiltonian and finding the energy levels and dipole matrix elements for two coupled anharmonic oscillators. The specific system dealt with was two weakly coupled Morse oscillators, however the algebraic procedure presented can be used for other anharmonic potentials, for more strongly interacting oscillators, and for a larger number of degrees of freedom. In addition it can readily be generalized to include rotational degrees of freedom and Coriolis coupling.[9]

REFERENCES

1. M. L. Sage and J. A. Williams III, Energetics, wave functions, and spectroscopy of coupled anharmonic oscillators, J. Chem. Phys. 78:1348 (1983).
2. M. L. Sage, Morse oscillator transition probabilities for molecular bond modes, Chem. Phys. 35:375 (1978).
3. E. C. Kemble, "The Fundamental Principles of Quantum Mechanics", McGraw Hill, New York (1937).
4. M. L. Sage, An algebraic treatment of quantum vibrations using REDUCE, J. Symb. Comp., in press.
5. E. L. Sibert, W. P. Reinhardt, and J. T. Hynes, Intramolecular vibrational relaxation and spectra of CH and CD overtones in benzene and perdeuterobenzene, J. Chem. Phys. 81:1115 (1984).
6. E. Merzbacher, "Quantum Mechanics", John Wiley, New York (1961).
7. I. Schek, J. Jortner, and M. L. Sage, Intensities of high-energy molecular C-H vibrational overtones, Chem. Phys. Lett. 64:209 (1979).
8. M. L. Sage, High overtone C-H and O-H transitions in gaseous methanol, J. Chem. Phys. 80:2872 (1984).
9. M. L. Sage and M. S. Child, to be published.

TEMPORAL AND STEADY STATE TRANSITION PROBABILITIES FOR MOLECULES INTERACTING WITH OSCILLATING AND STATIC ELECTRIC FIELDS AND SURFACES

William J. Meath and R. A. Thuraisingham

Department of Chemistry, University of Western Ontario
London, Ontario, Canada, N6A 5B7

1. INTRODUCTION

The purpose of this paper is to illustrate the importance of, and the effects caused by, (1) averaging over molecular orientations and (2) the presence of permanent dipole moments in calculating and understanding single- and multi-photon steady state and temporally resolved populations of molecular states. Examples involving continuous wave lasers, static electric fields, surfaces, and fixed molecule-field, free, or hindered rotationally averaged spectra, are discussed in Sec. 2. Definitions and expressions needed to help interpret the results are discussed in this section and some of the implications of the examples are summarized in Sec. 3.

Two-level models, with the molecule interacting with a sinusoidal plane-polarized electromagnetic field (EMF), are used for illustrative purposes; $\hat{e}$, ω and E will represent the unit vector of polarization, the circular frequency, and the strength of the EMF. For fixed molecule-field configurations, the (EMF phase averaged) temporal population of the excited state 2, $P_2(t)$, and its long time or steady state average, $\bar{P}_2$ will be used to monitor the system; $\bar{P}_2$, as a function of ω, is the absorption spectrum or resonance profile of the molecule. For molecules that can assume various orientations with respect to the probing EMF, the orientational or rotational averages, $P_2^{rot}(t)$ and $\bar{P}_2^{rot}$ respectively, of these fixed configuration populations are the observable quantities. Free rotational averages are appropriate for "gas phase" molecules in the absence of static fields whereas Boltzmann averaged results, which are a function of temperature T, are suitable for molecules undergoing hindered rotations due to their interaction with static fields, including surfaces. $\bar{P}_2^{\tau}$ and $\bar{P}_2^{rot,\tau}$ are the averages of $P_2(t)$ and $P_2^{rot}(t)$ up to time τ; $\bar{P}_2^{\infty} = \bar{P}_2$ and $\bar{P}_2^{rot,\infty} = \bar{P}_2^{rot}$. The computational techniques, which include counter-rotating EMF effects, have been discussed in detail for steady state populations previously (refs. 1-3 and references therein). All calculations correspond to $P_2(0) = 0$ and atomic units are used throughout unless indicated otherwise.

Rotating wave approximations, including permanent dipole effects, are available[4] for $P_2(t)$ and $\bar{P}_2$ for N=1,2,3,... photon transitions:

$$P_2^N(t) = 2\bar{P}_2^N \sin^2\{\tfrac{1}{2}[(\Delta E - N\omega)^2 + |C(N)|^2]^{\frac{1}{2}}t\} \quad (1)$$

$$\bar{P}_2^N = \frac{|C(N)|^2}{2[(\Delta E-N\omega)^2 + |C(N)|^2]} \quad , \quad \omega^N_{res} = \Delta E/N \tag{2}$$

$$C(N) = \underline{\mu}_{12}\cdot\hat{e}E\ 2N\ (\underline{d}\cdot\hat{e}E/\omega)^{-1}\ J_N(\underline{d}\cdot\hat{e}E/\omega) \quad , \quad \underline{d} = \underline{\mu}_{22} - \underline{\mu}_{11} \tag{3}$$

$$p_R(N) = 2\pi/[(\Delta E-N\omega)^2 + |C(N)|^2]^{\frac{1}{2}} \tag{4}$$

$$(FWHM)_R^N = \frac{2}{N}\,|C(N)|_{\omega=\Delta E/N} \text{ (for reasonably narrow resonances)} \tag{5}$$

Here C(N) is the coupling between the molecule and the EMF, $p_R(N)$ the period of $P_2^N(t)$, and $(FWHM)_R^N$ the full width at half maximum of $\bar{P}_2^N$ versus ω, all in the RWA, and $\underline{\mu}_{ij}$ are the transition (i≠j) or permanent (i=j) dipoles of the molecule. The RWA neglects counterrotating effects and therefore is reasonable only for relatively weak coupling strengths $\beta(N) = |C(N)|/\Delta E < .1$. It predicts a resonance frequency of $\omega^N_{res} = \Delta E/N$ and not the shifts from this weak field limit. The coupling given by Eq. (3) is not valid for N > 2 if d→0 and must be augmented by "atomic" results[5] under these conditions. Analogous RWA results are available[2] for problems (e.g. Sec. 2(b)) involving static electric fields ($\hat{e}_s E_s$). The RWA is useful in interpreting the examples that follow, yielding reasonable predictions for those of Secs. 2a,b,c and qualitative predictions for the strong β example of Sec. 2d.

2. EXAMPLES

A more detailed discussion for most of the steady state results can be found in cited references[1-3]. $\bar{P}_2$ a $\bar{P}_2^{rot}$, as a function of ω, can be read off the steady state spectra. Similarly $\bar{P}_2^{\tau}$ and $\bar{P}_2^{rot,\tau}$ can be estimated, as a function of τ, from the relevant temporal plots at fixed frequency. In what follows $\underline{R}$ represents the symmetry axis of the molecule.

2.a Free Rotational or Orientational Averaging

The free orientational averaged one- and two-photon spectra are shown in Fig. 1a for a 2-level molecule with $\mu_{12} = 5$, d = 2.5, ΔE = .1, $\underline{\mu}_{12} \parallel \underline{R} \parallel \underline{d}$, and for $E = 2 \times 10^{-3}$; the resonance frequencies are 0.1000 and 0.0503 respectively. For this example $C(1) \simeq \underline{\mu}_{12}\cdot\hat{e}E = \mu_{12}\,E\,X$ and $|C(2)| << |C(1)|$ since the relevant values of $\underline{d}\cdot E/\omega$ are small, see Eq. (3). The fixed configuration spectrum for the maximum EMF-molecule coupling, X = 1, is included in this figure; the resonances are at 0.1003 and 0.0507. Some of the effects of rotational averaging, relative to the X = 1 spectrum, are the sharpening and narrowing of the resonance profiles, the marked reduction in the height of the two-photon peak, and the reduction in the Block-Siegert shifts.

Various fixed configuration $P_2(t)$ are illustrated for $t \leq 1.5 \times 10^{-12}$ sec., namely for ω = 0.1000 with X = 1,.05 and ω = 0.0503 with X = 1,.7, in Figs. 1b,c and 1e,f respectively. Inspection of these results leads to the following comments: (1) When $\omega = \omega_{res}$, $P_2(t)$ attains the maximum value of unity after each period; (2) At ω = .0503, fixed configurations with X ≃ .7, and not X = 1, make major contributions to $\bar{P}_2^{rot}$ of fig. 1a (for X = 1, ω_{res} = .0507 for N = 2); (3) The period of $P_2(t)$, for a given ω, decreases with increasing molecule-EMF coupling, see Eq. (4); (4) The period for N = 2 is much greater than for N = 1 for similar molecule-EMF configurations; |C(2)|<<|C(1)|; (5) In general the shorter the period, the more rapidly steady state behaviour is attained; (6) For N = 2, the fixed configurations that make the major contribution to $\bar{P}_2^{rot}$ at resonance do not correspond to the $P_2(t)$ with the shortest period even though they have the highest maxima (~1); see also point (2); (7) The effects discussed in points (2) and (6), and the marked reduction of the height of the two-photon peak in Fig. 1a upon orientational averaging, are caused by the interplay between the narrow resonance profiles and the relatively large Block-Siegert shifts associated with two versus one photon transitions[1]. The RWA does not support Block-Siegert shifts.

The rotationally averaged populations of state 2, for ω = .1000 and ω = .0503, are shown in Figs. 1d,g. As expected the time to reach steady state behaviour is much shorter for the one photon case. For small times $\bar{P}_2^{rot}$ is dominated by configurations with the strongest EMF-molecule couplings corresponding to shorter periods for $P_2(t)$. As time increases more weakly coupled configurations contribute to the average; examples can be obtained by comparing Figs. 1b,c with 1d and Figs. 1e,f with 1g.

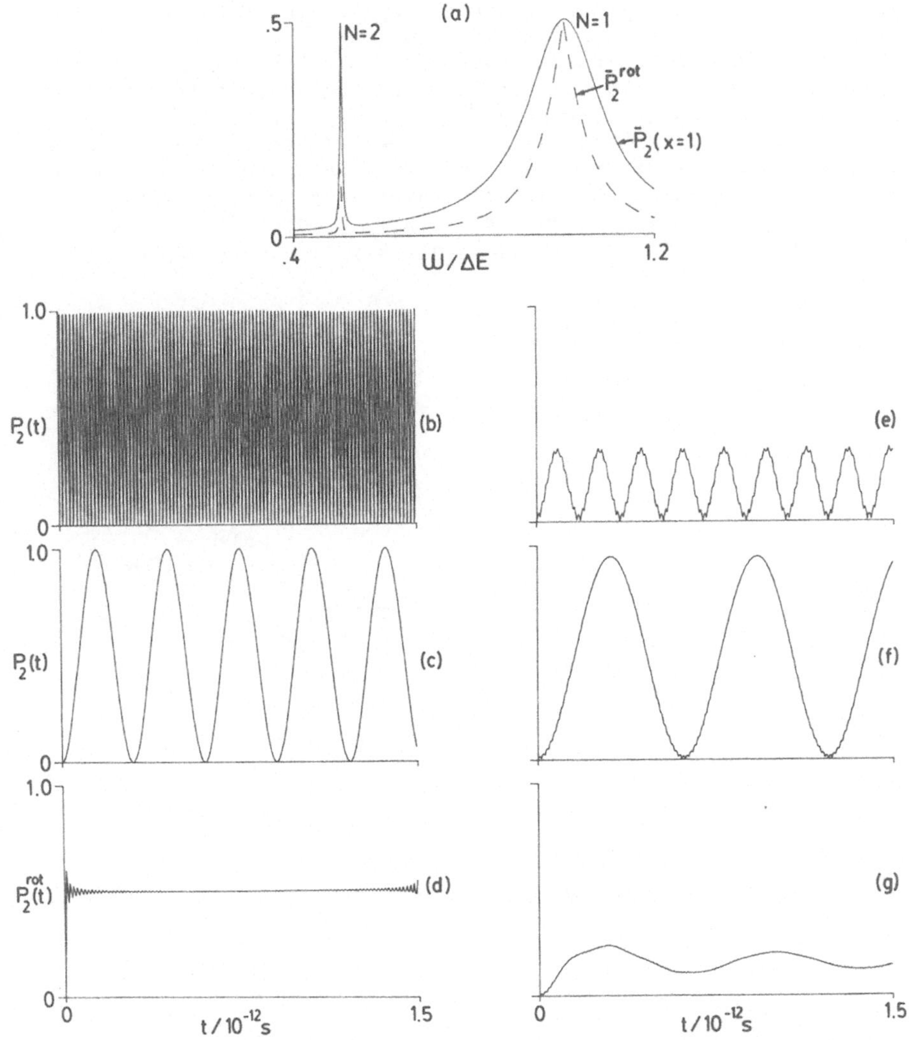

Fig.1. (a) The absorption spectra for free rotation, and the fixed molecule-EMF configuration X=1, for the molecule-EMF interaction of Sec. 2a. Also shown are the temporal populations of the excited state for ω = .1000 and (b) X = 1, (c) X = .05, (d) free rotation, and for ω = .0503 and (e) X = 1, (f) X = .7, and (g) free rotation. $X = \underline{\mu}_{12}\cdot\hat{e}/\mu_{12}$.

2b. Hindered Rotations Caused by a Static Electric Field

Consider a two-level molecule with $\Delta E = .158$, $\mu_{12} = .31$, $d = 1.18$, interacting with an EMF and a static electric field ($\underline{E} = \underline{E}_s = 10^{-3}$, $\hat{e} \parallel \hat{e}_s$). If $\underline{\mu}_{12} \perp \underline{d} \parallel \underline{R}$ the molecular parameters correspond to the $S_0 \rightarrow S_2$ electronic transition in nitrobenzene. The (hindered) rotationally averaged one-photon spectra for $\underline{\mu}_{12} \parallel \underline{d}$ and $\underline{\mu}_{12} \perp \underline{d}$, as a function of T, are shown in Figs. 2a and b respectively. The spectral features, as a function of ω and T, can be understood[2] by working in the representation that diagonalizes the static part of the Hamiltonian and are sensitive to the magnitude and relative orientations of $\underline{\mu}_{12}$ and $\underline{d}$; this carries over into the temporal behaviour of the system as well.

The corresponding rotationally averaged temporal population of excited state 2 is given in Figs. 2c,d for ω = .1570, the resonance frequency for the low temperature absorption spectra of Figs. 2a,b. For $\underline{\mu}_{12} \parallel \underline{d}$ the heights of the maxima in $P_2^{rot}(t)$ decrease continuously as T increases from 20K and these results show a fixed configuration type of temporal behaviour, especially for low T, where the average is dominated by the most attractive molecule-static field configuration ($\underline{d} \parallel \hat{e}_s$) with $\omega_{res} = .1570$. On the other hand for $\underline{\mu}_{12} \perp \underline{d}$ the heights of the maxima in $P_2^{rot}(t)$ at first increase and then decrease as T increases from 20 K and even for low T these results show definite averaging characteristics; there is more orientational freedom for the $\underline{\mu}_{12} \perp \underline{d}$ molecule. Other interesting temporal results can be qualitatively deduced from the steady state spectra of Figs. 2a,b.

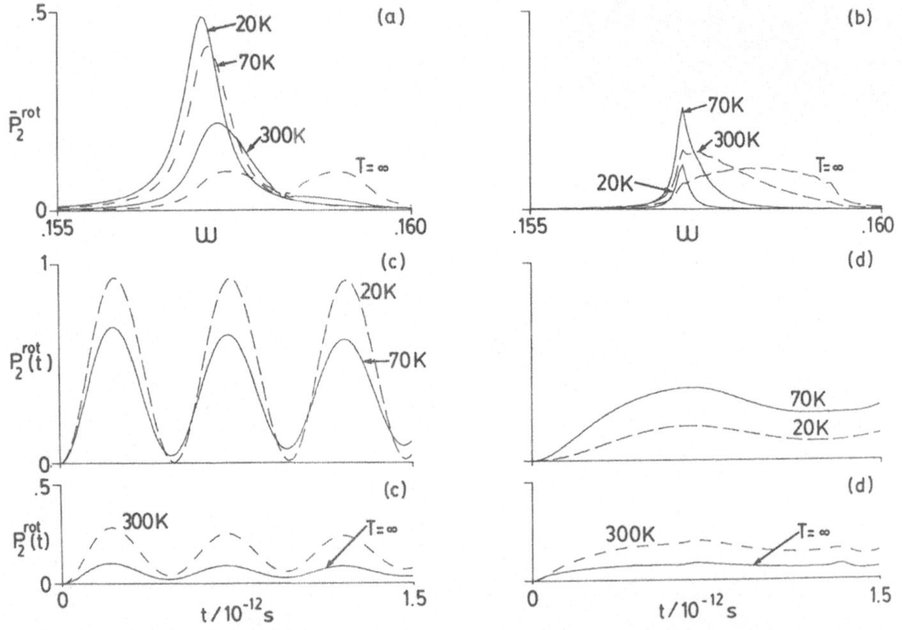

Fig. 2. The hindered rotationally averaged absorption spectrum, as a function of T, for (a) $\underline{\mu}_{12} \parallel \underline{d}$ and (b) $\underline{\mu}_{12} \perp \underline{d}$, for the molecule-EMF-static electric field interaction of Sec. 2b. Also shown, in (c) and (d) respectively, are the corresponding orientationally averaged temporal populations of the excited state for ω = .1570. T = ∞ indicates a free rotational average.

2c. Hindered Rotations Caused by Molecule-Surface Interactions

Consider a two-level model for N_2 adsorbed, in its ground state 1 ($X\ ^1\Sigma_g^+$), over the centre of a carbon hexagon on a graphite surface. The idea is to probe the adsorbed molecule by monitoring the electronic $X\ ^1\Sigma_g^+ \rightarrow C'\ ^1\Sigma_u^+$ (state 2) transition where $\Delta E = .474$, $\mu_{12} = .5$, $d = 0$, $\underline{\mu}_{12} \parallel \underline{R}$. An N_2-graphite potential energy[6] is used to evaluate the Boltzmann factors needed in our calculations of $P_2^{rot}(t)$ and $\bar{P}_2^{rot}$ which include the modification of the EMF by the substrate[7] and an average with respect to the height of the molecule above the surface.

The rotationally averaged single-photon absorption spectra, for ε = 5° and 90° where ε is the angle between $\hat{e}$ and the surface normal (p-polarization), are shown in Fig. 3a for $E = 1.8 \times 10^{-3}$ and T = 5K and indicate that the molecule is adsorbed with $\underline{R} \parallel$ surface[3,6]. The FWHM for ε = 90° is much larger than that for ε = 5°. The corresponding $P_2^{rot}(t)$ are shown in Figs. 3b,c for the resonance frequency, $\omega = .474$, of the absorption spectra (steady state behaviour begins to be established at t $\approx 10^{-11}$ sec.). The approach to the steady state is much more rapid, that is the period of $P_2^{rot}(t)$ is much smaller, for ε = 90° than for ε = 5° indicating the molecule-EMF coupling is much stronger for ε = 90° where $\hat{e} \parallel$ surface. Hence N_2 is adsorbed with its bond axis $\underline{R} \parallel$ surface since $\underline{\mu}_{12} \parallel \underline{R}$ and $p_R(1)\ \alpha\ (\underline{\mu}_{12} \cdot \underline{E})^{-1}$. The temporal behaviour of $P_2^{rot}(t)$ is dominated by those configurations with shortest periods (maximum EMF-molecule couplings) for small times. The periods, $\sim 2.3 \times 10^{-12}$ and 6.1×10^{-13} sec for ε = 5° and 90° respectively, estimated in the RWA approximation assuming $\underline{\mu}_{12} \parallel$ surface, are in qualitative agreement with the very early time behaviour of $P_2^{rot}(t)$ in Figs. 3b,c. More weakly coupled configurations contribute to the rotational average as time increases; relative to the static field example of Sec. 2b there are many probable orientations of $\underline{\mu}_{12}$ relative to $\hat{e}$ since, due to the nature of the molecule-surface forces, the molecule is free to rotate with $\underline{R} \parallel$ surface.

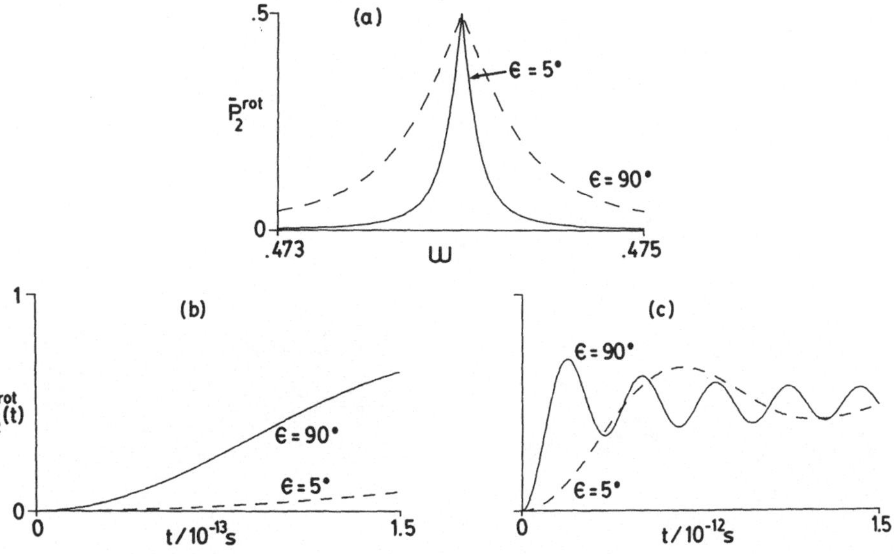

Fig. 3. (a) Hindered rotationally averaged absorption spectra, for ε = 5°,90°, for the N_2-graphite-EMF interaction of Sec. 2c. Also included, in (b) and (c), are the corresponding temporal populations of the excited state for $\omega = .474$.

2d. Effects of Permanent Dipoles ($d \neq 0$)

The multi-photon resonance profile, P_2 versus ω, for the two-level model characterized by $\Delta E = .1$, $d = 6.5$, $\mu_{12} = 3$, $E = 5 \times 10^{-2}$, and for a fixed molecule-EMF configuration specified by $\underline{\mu}_{12} \parallel \underline{d} \parallel \underline{R} \parallel \hat{e}$, is shown in Fig. 4a; the molecular parameters are representative of substituted aromatic molecules exhibiting intense one-photon transitions and, in the figure, the corresponding "atomic" spectrum ($d = 0$) is given for comparative purposes. These results show large positive (Block-Siegert) shifts in ω_{res} from the RWA results of Eq. (2) for $d = 0$ while for $d \neq 0$ the shifts are to the low frequency side of the RWA predictions[2,4]. Other features of the spectra, such as the appearance of even as well as odd photon transitions for $d \neq 0$, the marked narrowing of the resonances and the introduction of oscillatory resonance fringes into the spectra for $d \neq 0$ relative to the atomic results, and the filling in of the fringes by rotationally averaging[1], can be explained/predicted[1,2,4], only qualitatively for this strong molecule-EMF coupled example, via the RWA results of Sec. 1. They arise because of the coupling between $\underline{d}$ and $\hat{e}E$ arising in the Bessel function J_N in Eq. (3) - these functions have nodes and oscillate with decreasing amplitude as their argument increases. The molecule-EMF coupling is frequency dependent and can be small depending on the value of dE/ω.

Only one example of the temporal populations of state 2 associated with Fig. 4a will be discussed. $P_2(t)$ for the resonance frequency ($\omega = .1540$) associated with the one-photon peak for the "atom" ($d = 0$) is shown in Figs. 4b and 4c for $d = 6.5$ and $d = 0$, respectively. The period is much greater for $d \neq 0$ since the molecule-EMF coupling is much smaller for $d = 6.5$ than for $d = 0$ in agreement with the predictions of the RWA. The maxima for $P_2(t)$ for $d = 0$ does not reach unity, even though $\omega = \omega_{res}$, but the steady state value of 0.5 is still attained since the molecule spends more time in the excited state on the average (analogous behaviour is also observed on resonance for $d = 6.5$).

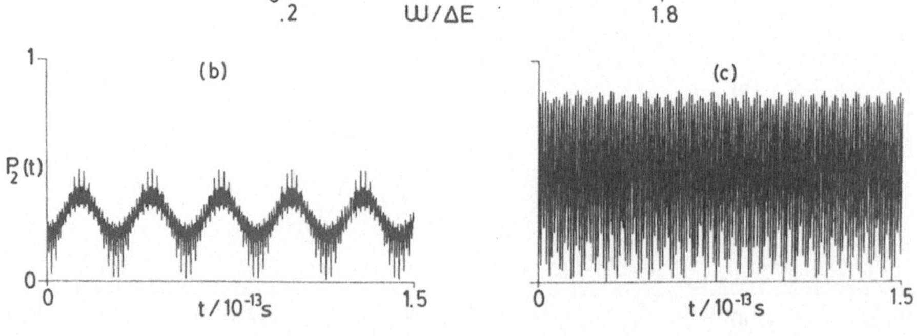

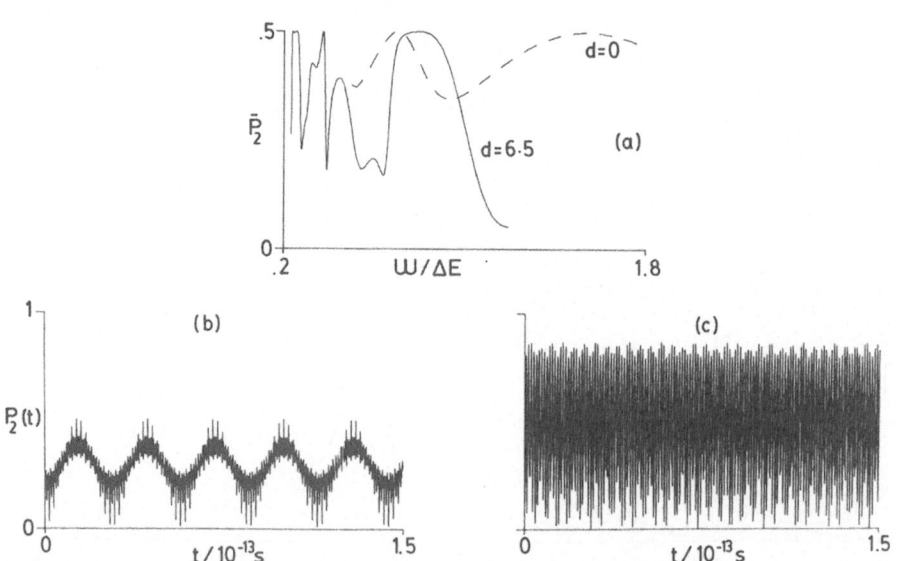

Fig. 4. (a) Fixed configuration absorption spectrum for $\hat{e} \parallel \underline{\mu}_{12} \parallel \underline{d}$, for $d = 6.5$ and $d = 0$, for the molecule-EMF interaction of Sec. 2d. Also shown, in (b) and (c), are the corresponding temporal populations of the excited state for $\omega = .1540$.

3. CONCLUSIONS

An examination of the steady state absorption spectrum gives only a global or average idea of the dynamics of state populations as a function of time. Augmenting this by temporal results for the populations gives a much better understanding of the dynamics and important insights of the short time behaviour of the system. The importance of investigating the temporal development of molecular states has of course been discussed previously by many authors and is a subject of much interest in reaction dynamics.

The model calculations of Sec. 2 have many interesting features, some of which have been discussed. In addition they, for example, indicate:
(1) The short time behaviour of the system provides information regarding the fixed configurations with the strongest EMF-molecule couplings. This is particularly so for rotationally averaged spectra dominated by a few fixed configurations but also applies for less hindered cases and even to free rotations (in a more limited way).
(2) In the presence of a static electric field both the absorption spectra and the time-dependence of the populations of molecular states, as a function of ω and T, are sensitive functions of the magnitude and relative orientations of the relevant transition and permanent dipoles. Such spectra should yield useful information about molecular structure and properties.
(3) The temporal behaviour of the states of molecular adsorbates, induced by interactions with lasers, can provide useful information that augments that provided by absorption spectra. For example the short time behaviour, $t \lesssim 10^{-13}$ sec. re Figs. 3b,c, indicates the orientation of the adsorbed molecule on the surface; the critical limits on the time can be varied by changing, for example, the strength of the EMF. In some cases the observation of a steady state spectrum may not be possible since the lifetime of adsorbate excited states can be markedly reduced by the presence of the substrate[8]; qualitatively the population of the excited state is given by $P_2^{rot}(t)\ \exp(-t/\tau_\ell)$ for finite lifetime τ_ℓ. Making temporal observations for $t \lessgtr \tau_\ell$ can avoid this difficulty and lead to estimates of the lifetimes.
(4) For molecules with permanent dipole moments the molecule-EMF coupling can be much less than expected from atomic theory. The experimental implications of this are significant; for example EMFs with large field strengths can actually interact with molecules weakly. The effects of $d \neq 0$ should be particularly interesting for giant dipole molecules[9].
(5) Approaches for altering the time evolution of molecular states, by using static electric fields, surfaces, and the connection between the structure of a molecule and/or its transition and permanent dipoles.

While the results discussed here illustrate interesting effects, it almost goes without saying that often it is not possible to isolate two levels. This can sometimes be achieved by the choice of field strengths, but in general other dipole-connected states must often be considered. Nevertheless two-level results can furnish a starting point for investigating and understanding many types of "many level" systems.

Acknowledgement: This work was supported by the Natural Science and Engineering Research Council of Canada.

REFERENCES

1. R.A. Thuraisingham and W. J. Meath, Phase and rotational averaged transition probabilities for molecules in a sinusoidal field using the Floquet formalism, Molec. Phys. 56:193 (1985).

2. M. A. Kmetic, R. A. Thuraisingham and W. J. Meath, Two-level rotating-wave approximations for molecules in a static electric field with an application to rotationally averaged spectra, Phys. Rev. A 33:1688 (1986).
3. R. A. Thuraisingham and W. J. Meath, Orientationally averaged spectra: an application to molecules adsorbed on a surface, Surf. Sci., submitted.
4. M. A. Kmetic and W. J. Meath, Permanent dipole moments and multi-photon resonances, Phys. Lett. 108A:340 (1985); G. F. Thomas, Effects of permanent dipole moments on the interaction of a two-level system with a laser and a static field, Phys. Rev. A 33:1033 (1986).
5. J. H. Shirley, Solution of the Schrödinger equation with a Hamiltonian periodic in time, Phys. Rev. 138:B979 (1965).
6. J. Talbot, D. J. Tildesley and W. A. Steele, A molecular dynamics simulation of nitrogen adsorbed on graphite, Molec. Phys. 51:1331 (1984).
7. S. A. Frances and A. H. Ellison, Infrared spectra of monolayers on metal mirrors, J. Opt. Soc. Am. 49:31 (1959).
8. Ph. Avouris and J. E. Demuth, Electronically excited states of adsorbates on metal surfaces, in "Surface Studies With Lasers", F. R. Aussenegg, A. Leitner and M. E. Lippitsch, eds., Springer Verlag (1983).
9. M. Terauchi and T. Kobayashi, Lasing and gain properties of a giant dipole molecule, Chem. Phys. Lett. 137:319 (1987).

SOME THEORETICAL ASPECTS OF INFRARED MOLECULAR LASERS PUMPED BY ELECTRONIC-TO-VIBRATIONAL ENERGY TRANSFER

J.Campos-Martínez, O. Roncero, S. Miret-Artés, P.Villarreal and G. Delgado-Barrio

Instituto de Estructura de la Materia,Serrano, 123
28006-Madrid, Spain

ABSTRACT

Pulsed infrared lasers in which the pumping mechanism is electronic-to-vibrational energy transfer are well known today in many practical situations. However, this selective process is still submitted to many theoretical objections in order to reach a close confrontation with the experimental data. This work can be seen as a contribution more to this respect. The Na-N_2 system is one of the main examples chosen in this field. Its study has been carried out in the C_{2v} configuration within the close-coupling framework, describing the N_2 molecule as an anharmonic oscillator.

INTRODUCTION

Intermolecular energy transfer is one of the elementary processes more carefully studied in the nonadiabatic molecula collisions. Within the low-energy regime, where the kinetic energy associated with nuclear motion is less than $\sim$20 eV /1/, a great amount of phenomena of this type is found in planetary atmospheres, interstellar space, magnetohydrodynamics systems, chemical reaction dynamic and gas lasers. Special attention is receiving the quenching of resonance radiation where electronic to vibrational and rotational (E-V-R) energy transfer occurs in the collision of electronically excited alkali atoms with different diatomic or poliatomic partners. The most sophisticated experimental techniques, at present, are being applied /2a/ in order to improve our understanding of this process. Unfortunately, this tendency is not closely followed at theoretical level /2b/ or at least with the same intensity. A lack of availability of accurate potential energy surfaces and reasonable interelectronic couplings prevents more refined dynamical calculations.

It is well known that E-V-R energy transfer leads to a population inversion among final states. This process is selective and pumps the molecular partner into specific energy states. Infrared lasers have been built in the last years /3/

based on the following reactions:

$$X_2 \xrightarrow{h\nu} X + X^* \qquad \text{(flash photolysis)}$$

$$X^* + M \longrightarrow X + M^{\dagger} + \Delta E$$

where X are halogen atoms, M represents a polyatomic, the symbols * and † denote electronical and vibrational excitation, respectively, and Δ E is the energy defect. Triatomic or larger polyatomic molecules show a more resonant behavior than in the diatomic cases. The choice of the X atoms, halogen or not, affects to the values of the quenching cross sections observed. The corresponding energy transfer mechanisms are expected to be very different /2b/.

In this work, we have proposed to investigate some theoretical aspects about the quenching alkali resonance radiation in E-V energy transfer collisions with diatomic molecules. The existence of chemical lasers or even experimental works is well known, but not many theoretical calculations are to be present in the literature and, therefore, to get new information about the intimate nature of this process, we have chose the Na-N_2 system. It is one of the prototype molecular systems in this context for which reasonable three-dimensional potential energy surfaces are available/4/. As far as we know, potential energy surfaces together with the relevant interelectronic couplings have only been reported for a fixed N_2 bond length /5/, neglecting then the diatomic vibration. Following the ideas of some authors /4,6-9,12 /, we have proposed elsewhere /10/ an interelectronic coupling based on a curve-crossing model in order to stress the role of the N_2-vibration in the quenching process. This model yielded a moderated inversion of the diatomic vibrational population after the collision when the molecular subunit was described as an harmonic oscillator /10/ at the perpendicular configuration. On the contrary, in this paper, the diatom is taken as an anharmonic oscillator and a stronger inversion population and greater efficiency is obtained as compared with the harmonic case.

COUPLED EQUATIONS AND RESULTS

The dynamic of the quenching

$$\text{Na}\,(3\,^2P) + N_2(^1\Sigma_g^+, v=0) \longrightarrow \text{Na}\,(3\,^2S) + N_2(^1\Sigma_g^+, v'=n) \quad (1)$$

at low-energy regime may be described fairly well on the C_{2v} symmetry, which is energetically the most favourable /2b/ for this reaction, the rotational degree of freedom being ignored.

It has been estimated /2b/ that a small amount of ~ 9% of the energy transferred into internal motion of N_2 is going to the rotational degree of freedom.

In our calculations, we have taken the potential energy surfaces reported inf ref. 4 which are expressed in terms of the internal coordinates of the molecular system: an angle and two distances, one describing the relative motion of Na with respect to the center of mass of N_2 and the other one the diatomic vibrational motion. Unlike Ref.4, we have assumed the N_2 molecule as an anharmonic oscillator. The corresponding parameters were taken from Herzberg /11/.

When the total wave function is expanded in terms of a basis set of electronic wave functions, which depend parametrically on the nuclear positions, and the two-dimensional nuclear wave functions are also expanded in a basis of anharmonic oscillators, we get, after substitution into the corresponding Schrödinger equation, the usual set of coupled equations for the relative motion /1/

$$\left[\frac{\hbar^2}{2\mu}\frac{d^2}{dR^2} + E - \varepsilon_v - U_{\alpha v,\alpha v}(R)\right]\phi_{\alpha v}(R) = \sum_{\substack{\alpha' \neq \alpha \\ v' \neq v}} U_{\alpha v,\alpha' v'}(R)\,\phi_{\alpha' v'}(R) \qquad (2)$$

In this expression, μ is the reduced mass for the Na-N_2 system, ε_v are the eigenvalues for the anharmonic oscillator, E is the total energy, $U_{\alpha v,\alpha v}(R)$ is an effective potential energy curve at each α-electronic state and $U_{\alpha v,\alpha' v'}(R)$ represent the interelectronic and internuclear couplings. The interelectronic coupling is modelled by a gaussian function and only two electronic states (the ground and the first excited state) are taken into account /4/.

Concerning with the electronic coupling, it is well accepted, in the light of many studies carried out in the last ten years, that the energy transfer is determined by the location of the crossing between the electronic surfaces. This idea was recognized firstly by Bjerre and Nikitin /6/ and after by Bauer, Fisher and Gilmore /7/. In a second step, Mc Guire and Bellum /8/ treated this problem by close-coupling calculations, modelling also this coupling. In the same spirit, two recent works /9,10/ have accomplished some calculations with the same goal. We have reported a curve-crossing model elsewhere, using a gaussian-type function. Doing that, our aim was to collect the vibrational arguments given about the quenching mechanism. As previously, the off-diagonal elements of the equation (2) are expressed by

$$U_{\alpha v,\alpha' v'}(R) = B \langle \chi_v(r) | e^{-\beta[R - R_c(r)]^2} | \chi_{v'}(r) \rangle \qquad (3)$$

where χ_v are now, eigenfunctions of the N_2 anharmonic oscillator and B, β and $R_c(r)$ the three parameters of the gaussian function. $R_c(r)$ represents the crossing curve of the two electronic surfaces and this, at first approximation, can be fitted to a straight line around the equilibrium position of the diatomic oscillator. In this way, the important role of the vibration, that was already pointed out by different authors /4,12/, is explicitely included in our calculations.

In table I we present partial quenching probabilities averaged by a Boltzmann distribution of the temperature. This calculation was performed using the value B =0.001 and β = 8.(atomic units), that were chosen after some tests to reproduce the main features of the experiment. Convergence was achieved with thirteen channels in the upper state and nine the lower one.

TABLE I. Quenching probabilities $P_{0 \to n}$, where v=0 and v'=n stand for the entry and exit vibrational channels, at different temperatures.

T(K)	$P_{0\to 0}$	$P_{0\to 1}$	$P_{0\to 2}$	$P_{0\to 3}$	$P_{0\to 4}$	$\sum_n P_{0\to n}$
200	3.9(-3)	2.6(-2)	3.6(-2)	3.0(-2)	1.8(-3)	9.67(-2)
400	4.3(-3)	2.3(-2)	4.0(-2)	3.1(-2)	4.4(-4)	9.86(-2)
600	4.5(-3)	2.0(-2)	4.2(-2)	3.0(-2)	7.4(-4)	9.84(-2)
800	4.6(-3)	1.8(-2)	4.3(-2)	2.9(-2)	1.0(-3)	9.75(-2)
1000	4.7(-3)	1.7(-2)	4.4(-2)	2.9(-2)	1.2(-3)	9.64(-2)
1200	4.8(-3)	1.6(-2)	4.4(-2)	2.9(-2)	1.4(-3)	9.54(-2)

Only one channel in the initial state and eight in the final one are open in the range of energies studied. Te inversion of the final vibrational population appearing in this table is mainly due to the excited closed channels. The existence of quasibound levels could be responsible of this highly resonant process. In fact, some preliminary calculation about the lifetimes of the levels supported by the v=1 vibrational channel of the excited electronic state gave values comprises in the range of 8-14 ps. The approximation applied was within the "Golden-Rule" framework and the exit channels with a major population were v'=3,4. This fact could explain the placing of peaks of our distributions. Moreover, the behavior of these probabilities with the temperature for the harmonic case /10/ is exactly the opposite to the Table I, where the inversion of the vibrational population increases lightly and the total inelastic quenching probabilities decrease very smoothly in the range 200-1200 K. Once more, the location of the crossing-points is very important in order to understand the mechanism of this process. The diatomic anharmonicity shifts the potential energy surfaces and changes the interelectronic coupling yielding distributions peaked at higher levels and a stability of them in a wider range of temperature.

The experimental results of Silver et al. /13/ and Reiland et al. /14/ are taken to be compared with our results at T=1200 K. We obtain a bit narrower distribution of final internal energies for N_2 and the peak shifted to smaller vibrational energies (one /13/ or two /14/ quanta) with respect to these experimental works. Also, as regards the efficiency of the quenching process, we calculate that ~26% of the available total energy is transferred to N_2 faced to the reported values there, ~30-40% /13/ or ~50% /14/. Due to the minor incidence of the rotational degrees of freedom, as was already mentioned, these discrepances may be attributed mainly to the potential energy surfaces used here. In fact, the peaking of the final vibrational diatomic distribution, in this model, is strongly dependent on the crossing of both surfaces meanwhile the corresponding spreading and the percentage of energy transferred to the internal motions of N_2 can be monitored by varying the values of the β and B parameters, respectively.

In conclusion, we have presented here some theoretical results on infrared molecular lasers pumped by E-V energy transfer. We have taken an hipotetic laser system (Na-N_2) that has been studied from a methodological point of view. Many other quencher molecules may be even much more favourable in this respect. For example, triatomic (CO_2, N_2O, etc.) or organic molecules with double or triple bonds are known to be extremely good quenchers of alkali resonance radiation and the observed population inversion among vibrational levels of the molecules is very important for this goal. Despite of a detailed study of a simple system, like Na-N_2, already presents many difficulties, it is an useful starting point to treat more complicated molecular laser systems based on this mechanism of pumping.

REFERENCES

/1/ J.C. Tully, in "Dynamics of Molecular Collision", edited by W.H. Miller (Plenum, New York, 1976), part B, pag.217.

/2/ a) I.V. Hertel, in "The Excited State in Chemical Physics", Adv. Chem. Phys., XLV (Part 2), edited by J.W. Mc Gowan (Wiley, New York, 1981) pag. 341.
b) I.V. Hertel, in "The Dynamics of the Excited State", edited by K.P. Lawley (Wiley, New York, 1982) pag. 475.

/3/ A.B. Petersen, C. Witting and S.R. Leone, Appl.Phys.Lett. 27 (1975) 305; J. App. Phys., 47 (1976) 1051.

/4/ P. Habitz, Chem. Phys., 54 (1980) 131; P. Archirel and P. Habitz, Chem. Phys., 78 (1983) 213.

/5/ M. Persico, in "Spectral Line Shapes", Vol. III, De Grunter, Berlin, 1985.

/6/ A. Bjerre and E.E. Nikitin, Chem. Phys.Lett., 6 (1967) 438.

/7/ E. Bauer, E.R. Fisher and F.R. Gilmore, J. Chem. Phys., 51 (1969) 4173.

/8/ P. Mc. Guire and J.C. Bellum, J. Chem. Phys., 71 (1979) 1975.

/9/ D. Poppe, D. Papierowska-Kaminski and V. Bonacic-Koutecky, J.Chem.Phys., 86 (1987) 822.

/10/ J.Campos-Martínez, S. Miret-Artés, P. Villareal and G. Delgado-Barrio, Chem. Phys. (submitted).

/11/ G. Herzberg, "Spectra of Diatomic Molecules", Van Nostrand Reinhold Company, New York, 1979.

/12/ P. Botchwina, W. Meyer, I.V. Hertel and W. Reiland, J.Chem. Phys., 75 (1981) 5438.

/13/ J.A. Silver, N.C. Blais and G.H. Kwei, J. Chem. Phys., 71 (1979) 3412.

/14/ W. Reiland, C.P. Shulz, M.V. Tittes and I.V. Hertel,Chem. Phys., Lett., 91 (1982) 329.

SOME INTENSE FIELD LASER-MOLECULE THEORETICAL TECHNIQUES

Joseph O. Hirschfelder

Departments of Chemistry at the Univerities of California, Santa Barbara, CA 93106, (October-April) and Wisconsin, Madison, WI 53706 , (April-October)

I. SOME THOUGHTS ABOUT THE NATO SYMPOSIUM

In 1966 I went to a symposium which was arranged by Peter Debye. The purpose of the meeting was to determine on which points the experts agreed, on which ones they diagreed, and what needed to be done to establish agreement. At our recent NATO symposium, we learned the points of agreement and disagreement. Now, each of us is trying to decide what we can do to establish agreement on the mechanism of short laser pulse interactions.

A. Effects of Ultra-Intense Short Pulses

When atoms are irradiated by intense monochromatic pulses, ATI, Multiply Charged Positive Ion Formation, and Harmonic Photon Generation occur. Low order perturbation theory explains these effects for picosecond 10^{15} W/cm^2 pulses. However, there is no agreement about the mechanism of shorter more intense pulses which we call ultra-intense since perturbations cannot explain their effects! At the present time there is need for rigorous computations of pulse effects to compare with the experimental data. Fortunately, a number of excellent theoretical research programs are under way.

1). Above Threshold Ionization (A.T.I.)

The atom absorbs more than the least number of photons needed to reach the ionization threshold. The photo-electrons produced by ionization gain kinetic energy by absorbing additional photons. For example, Rhodes et al[1] (using a 193 nm 10^{15} W/cm^2 picosec pulse) observed the following numbers of photons absorbed per ion formed:

He 4 to 6; Ne 4 and 5; A 3 to 6; Kr 3 to 5; and Xe 2 or 3.

Many explanations of the A.T.I. have been proposed. Both Eberly[2] and Dulcic[3] have concluded that the bound-continuum transitions determine the overall ionization rate while the continuum-continuum transitions redistribute the continuum states into the ATI peaks! Everyone agrees that a very large atomic dipole moment is produced by the oscillating field coherently driving the loosely bound outer atomic shell. However there is an argument as to whether the pondermotive or the quiver energy is more important in producing the A.T.I.

2). Multiply charged positive ions are formed in surprisingly large numbers. Rhodes et al[1,4] (using 193 nm 10^{15} Watts/cm^2 picosecond pulses) observed the following multiply charged positive ions: He^{++}, Ne^{++}, Ar^{6+}, Xe^{8+}, Hg^{4+}, and U^{10+}.

Rhodes[1] proposed that the electrons in the inner shells are excited while the loosely bound outer shell is coherently driven by the laser field. Two or more of the outer shell electrons become coupled and move as a unit. Then the excitation energy of the inner shells is transferred to the outer shell electrons which are thereby simultaneously ejected from the atom by a Reverse Coster-Kronig type Auger effect[4,34]. However, the experimental evidence for this mechanism is not conclusive at the present time.

Surprisingly little is known about electron couplings except that two electron couplings occur --- for example, the Cooper pairs in superconductors and the spin-orbit interactions of almost degenerate moilecular excited states[5]. I have been told that three electron couplings would not be strong and more than three electron correlations are unlikely.

3). Harmonic photon generation is produced by the scattering mechanism, $N\gamma(\omega) + X \rightarrow \gamma(N\omega) + X$, arising from the nonlinear nature of the atomic susceptibility at the high field strength. Here X can either be an atom or an ion. In the electric dipole approximation, the harmonic order N is restricted to odd integers. Rhodes et al[6] (using 248 nm 10^{16} Watts/cm^2 picosecond pulses) observed the following maximum harmonics: He 13 ; Ne 17; Ar, Kr 7 ; and Xe 9.

B. The Theory of the Intense Pulse Effects

1). The lowest order of perturbation theory explains the effects of ultraviolet pulses which are longer than 0.1 picosecond and have maximum intensity less than 10^{15} W/cm^2. Indeed, Lambropoulos and Tang[7] have developed an excellent very general first order perturbation procedure which agrees with the experimental data for the number of multi-charged positive ions which are formed.Their results show that the sequential ionization (stripping of one electron at a time) is by far the most dominant mechanism. Furthermore, most of the electrn stripping occurs at the beginning of the pulse before the field has reached its peak intensity. For pulses shorter than 0.1 picosecond or for pulse intensities greater than 10^{15} Watts/cm^2, their formalism allows for two or more electrons to be ejected simultaneously.

2). Effects of a very intense monochromatic field on a hydrogen atom have been studied by Shih-I Chu and Cooper[8]. They used an extension of the Chu and Reinhardt L_2 non-Hermitian Floquet formalism to determine the multiphoton ionization and A.T.I. of a ground state hydrogen atom in a very intense (10^{13} < I > 10^{15} W/cm^2) field with frequency 0.5 a.u.= 2×10^{16}sec^{-1}. To the extent that enough Floquet blocks are included, this procedure gives exact non-perturbative results and includes all relevant processes involving various photon numbers. They found that in this intensity regime, the two and three photon ionization rates do not obey the power laws predicted by lowest-order perturbation theory. Kulander[9] made a direct integration of the hydrogen atom time dependent Schrodinger equation and obtained good agreement with Chu and Cooper's results.

3). Effects of very intense pulses on He atoms have been studied by Kulander[10]. He numerically integrated time dependent Hartree Fock (TDHF) equations (for intensities up to 10^{15} W/cm^2) to determine the effects of frequency, intensity, pulse length, and pulse shape on the multiphoton ionization of He atoms in very intense laser fields. The principal disadvantage of this model is that the instantaneous correlations of the electrons are smeared out or averaged over. This mean field approximation makes the calculations manageable, but it reduces the accuracy of the results. A good feature of the TDHF is that it gives an indication of the extent to which multiple ionization is a sequential process. Kulander found that the preionization dynamics (PD) [which Lambropoulos says is most important] varies dramatically with the wavelengths. For very short wavelengths, he had evidence that two-electron collective modes are excited in the PD and produce A.T.I. Clearly this work is the beginninmg of a very interesting series of papers on multiphoton ionizations.

4). A very thorough theoretical problem has been undertaken recently by Abraham Szöke[11] who is developing a rigorous but manageable theoretical treatment of multi-photon excitations of atoms: First, he decided to use a classical description of the field and make suitable corrections for the scattering and attenuation of the incident radiation by the atom and for the field momentum transferred to the electrons. Next he decided to treat the motions of the bound and the free electrons separately in a non-perturbative manner, while the ionization is treated as a perturbation.
By restricting himself to laser pulses which have a narrow spectral width such that $\epsilon = \overline{\Delta\omega}/\overline{\omega} << 1$ where $\overline{\Delta\omega}$ is the average spectral width and $\overline{\omega}$ is the average frequency, Szöke is able to make use of the *METHOD OF MULTIPLE TIME SCALES*[12] in which there is a "fast time", $\tau_0 = \int_0^t \omega(t')\,dt'$ = "optical phase" and a "slow time", $\tau_1 = \epsilon\,\tau_0$. The Schrodinger equation can then be expressed in terms of these two times and solved by successive expansions in powers of τ_0 and τ_1. In carrying this out formally, Szöke makes use of Floquet theory, strong-field scattering theory, and many-body diagrammatic perturbation theory. In reading his preprint, I was fascinated by his ingenuity and I am sure that this will become a very useful technique.

C. Nuclear Decay Processes

There is a great resemblence between ultra-intense laser effects and radioactive decay processes!

1). Nuclear Isomeric Transitions with Internal Conversion of a γ-ray. The γ-ray removes the electrons from the K-shell and the subsequent vacancy cascades (Auger Effect) leaves the molecule with a very high positive charge[13]. In this reference, Cooper and Burhop explain their concept of the mechanism and estimate the time required for an excited electron to transfer from one atomic shell to another. For example, when $^{80m}_{35}Br$(4.4 hr) decays to $^{80}_{35}Br$(18m), the bromine atom from CH_3Br has up to 13+ charges (with an average of 6+)[14]. From the widths of x-ray fluorescence lines and absorption edges, the time required for all of the Auger electrons to be ejected was estimated to be between 10^{-15} and 10^{-16} sec. Since this is much shorter than the vibrational period (10^{-14}sec) of the methyl bromide, the molecule holds together until after the positive ions have been formed. Then there must be an internal transfer of electrons from the CH_3 to the Br prior to fragmentation that occurs as the result of Coulombic repulsion of the redistributed charges.

2). Photionization of K- or L-Shell Electrons by X-Rays. Such experiments[16] agree with radioactive decay process results, but greater control over the mechanism of the multicharged positive ion formation is obtained by using the x-rays. The wave length is the principal difference between our present ultraviolet laser pulses and these x-rays.

3). Comparison of this Radiochemistry and Ultra-Intense Laser Pulse Effects. In the limit as pulses get shorter (so that the preionization period becomes less important) and as our present ultraviolet wavelengths climb into the x-ray range, the radiochemical and laser effects will be the same.. Furthermore, the mechanism of the multicharged ion formation may be similar to Rhodes present explanation[4]!

Whereas, the radiochemists were primarily interested in fragmentation products and chemical changes in molecules, most of present ultra-intense laser pulse research has been applied to noble gas atoms. In the future, some of the multicharged molecular fragments produced by laser pulses will be used as catalysts; other fragments will make magnetic hydrodynamics (to convert combustion products directly into electrical energy) practical; etc.

II. NEXT, LET'S DISCUSS SOME THEORETICAL TOPICS

A. The Spin-Generalized Classical Poisson Bracket

There is considerable interest in semi-classical techniques at the present time. For example, Shih-I Chu[17] has shown how to use the Gerber-Ratner[18] semi-classical time dependent self consistent field technique to treat infrared laser multiphoton excitation problems --- Furthermore this technique is numerically efficient! This brings up the question of how you could use it to solve a problem involving spin?

The answer is that Tom Yang and I were able to generalize the classical Poisson bracket to include spin[19] on the basis of our analysis of Breit-Pauli Hamiltonian molecular dynamics[20].

We assumed that a classical electron is a finite size spinning sphere whose angular momentum (in its rest frame) is its spin ---If an electron were not a sphere it would be necessary to know its angular velocity as well as its spin in order to describe its dynamics . [In 1962, Dirac[21] used a similar classical model of an electron together with the Bohr-Sommerfeld quantization conditions. He found that the rest mass of the first excited state was similar to a muon's].

If both $f(t,\underline{p},\underline{r},\underline{s})$ and $g(t,\underline{p},\underline{r},\underline{s})$ are any general functions of time as well as momenta, coordinates, and spins of each of the particles [note that $\underline{p}$, $\underline{r}$, and $\underline{s}$ are vectors], then <u>the generalized Poisson bracket is</u>

$$\{\underline{f},g\} = \sum_j \left[-\frac{\partial f}{\partial \underline{p}_j}\cdot\frac{\partial g}{\partial \underline{r}_j} + \frac{\partial f}{\partial \underline{r}_j}\cdot\frac{\partial g}{\partial \underline{p}_j} - \frac{\partial f}{\partial \underline{s}_j}\cdot\left[\underline{s}_j \times \frac{\partial g}{\partial \underline{s}_j} \right]\right]$$

This generalized Poisson Bracket should apply to any system having a Hamiltonian such that $d\underline{s}_j/dt = -\underline{s}_j \times \frac{\partial H}{\partial \underline{s}_j}$.

A desireable feature of this Poisson bracket is that

$$\frac{df}{dt} = \frac{\partial f}{\partial t} + \{f,H\},$$

so that we obtained the three <u>*Generalized Canonical Equations*</u>:

$d\underline{p}_j/dt = \{\underline{p}_j,H\}$, $d\underline{r}_j/dt = \{\underline{r}_j,H\}$, and $d\underline{s}_j/dt = \{\underline{s}_j,H\}$.

Previously, Sudarshan and Mukunda[22] had derived the same spin generalized Poisson bracket by using purely group theoretic arguments; whereas our derivation is based upon a dynamical approach. The advantage of the group theoretic approach is its simplicity and elegance; whereas the dynamical approach seems to be required for systems in external electromagnetic fields.

<u>B). Nonlinear Integrable Equations for Nonlinear Optics</u>

Nonlinear optics is one of the most important applications of short very intense laser pulses. Some examples are: *Kerr Effect, Josephson Juncture, Self-Induced-Transparency, 3 and 4 Wave Mixing, Parametric Oscillations, etc.* [See Ref.(23), Chapt.4]. Most nonlinear optical effects result from a <u>coherent response of a resonant medium to an optical laser pulse</u>[24]. The nonlinearity is caused by the nonlinear corrections to the polarization of the dielectric medium. The nonlinearity of these effects increases sharply as the pulse intensity becomes larger and the pulse length becomes shorter. The pulse needs to be short in order that the high frequency polarization induced in the medium can retain a definite phase relationship with the molecules[24].

Up to the present time, the electric dipole approximation for the atom-field interaction has sufficed. However, as laser intensities become more intense, relativistic effects will become important and it will become necessary to go beyond the electric dipole approximation and use either the electric-quadrupole + magnetic-dipole approximation or the semi-relativistic Breit-Pauli Hamiltonian.

In order to determine nonlinear optical effects, it is necessary to simultaneously solve both the optical Bloch equation (with relaxation terms) and Maxwell equations. The resulting (somewhat idealized) nonlinear system of equations are closely related to the Korteweg-de Vries (KDV) equation[23,25,26]

$$K(u) = \frac{\partial u}{\partial t} + 6\,u\,\frac{\partial u}{\partial x} + \frac{\partial^3 u}{\partial x^3} = 0. \qquad (1)$$

which determines the evolution of *soliton waves*.

The recent interest in the KdV started with the discovery of the *inverse scattering transformations* (IST) which is a very powerful tool for obtaining nonperturbative solutions of many types of nonlinear integrable differential equations. Also, the IST makes it possible to investigate the remarkable properties which these equations possess[26]: such as soliton solutions, infinite sets of integrals of motion, infinite symmetry groups, Bäckland transformations, Hamiltonian structures, etc. Thus, the IST can provide a great deal of information about the nature of the solutions of the nonlinear partial differential equations!

The IST was motivated[23] by the transformation, $v = \frac{\partial \psi}{\partial x} / \psi$ which converts the nonlinear Ricotti equation, $\frac{\partial v}{\partial x} + v^2 + u = 0$, into the linear equation, $\frac{\partial^2 \psi}{\partial x^2} + u\,\psi = 0$.

In 1968, Miura, Gardner and Kruskal generalized this procedure by starting with the Schrodinger-like equation,

$$\frac{\partial^2 \psi}{\partial x^2} + (\lambda + u)\psi = 0 \qquad (2)$$

and postulating the associated time evoluton equation,

$$\frac{\partial \psi}{\partial t} = A(x,t)\psi + B(x,t)\,\frac{\partial \psi}{\partial x} \qquad (3)$$

where A and B can be any general scalar functions independent of $\psi(x,t)$.

For example, if $A(x,t) = \partial u/\partial x$; $B(x,t) = 4\lambda - 2u$; and $\partial\lambda/\partial t = 0$. Then since $\frac{\partial^3 \psi}{\partial x \partial t^2} = \frac{\partial^3 \psi}{\partial t \partial x^2}$, it follows that the nonlinear KvD equation, Eq(1) , is satisfied.

Given a nonlinear differential equation, the IST is a procedure for determining the functions A(x,t) and B(x,t) in the linear time evolution equation. Because the choice of A(x,t) and B(x,t) is not unique, the differential equations have an infinite number of integrals of motion.

The best example of the use of the IST to solve nonlinear optical problems is the exact solution of the McCall-Hahn *self-induced transparency* effect which Ablowitz, Kaup, and Newell[23,,24,27] obtained for their atomic two-state model. Their results agreed with the experimental data much better than previous perturbative treatments. Furthermore, the analytical form of their solution helped to explain the <u>significance</u> of the experimental results. From one of the IST conservation laws, they determined the absolute value of the time delay in the pulse transmission --- I found this very interesting since I showed that <u>Wigner's Time Delay</u> is the same as the quantum mechanical virial theorem for scattering processes[28].

Some of the other nonlinear optical effects which have been successfully treated by the IST formalism[23] are: linearly dispersive systems such as 3 and 4 wave mixing, parametric oscillations, the "orientational" Kerr effect, and "frequency-up" conversions which are closely related to the so-called *"Nonlinear Schrodinger Equation"*,

$$i\hbar \frac{\partial \psi}{\partial t} = \frac{\partial^2 \psi}{\partial x^2} + 2\,\sigma\,|\psi|^2\,\psi \qquad \text{where } \sigma = \pm 1.$$

Furthermore, the Josephson juncture and self-induced transparency are related to the *Sine-Gordon Equation.*

$$\frac{\partial^2 \psi}{\partial x^2} - \frac{\partial^2 \psi}{\partial t^2} = \sin(\psi).$$

Both the "nonlinear Schrodinger" and the Sine-Gordon equation can be transformed into the the Korteweg-de Vries equation.

C. Gisin's Nonlinear Schrodinger-Like Equations

Gisin[29-32] has used the *master equation formalism* to develop a large class of nonlinear dissipative Schrodinger-like equations which involve both friction and fluctuations. Furthermore, he determined the formal solutions of these equations! It remains to be seen, how successfully the Gisin formalism can compete with the conventional super-matrix technique for treating relaxation problems. I am wondering what (if any) relationship exists between the soliton and the Gisin types of nonlinear integrable differential equations --- they both can be used to solve evolution equations for systems which interact with a medium or bath!

D. Relaxation and Dissipative Problems where H is Constant

Gisin[29] considers a system, s, interacting with a bath, b. The (constant) Hamiltonian for s + b is

$$H = H_s + H_b + \lambda H_{sb} \qquad (1).$$

After making the usual Markovian approximation, the Schrodinger equation for the evolution of the unnormalized system wave function is

$$i\hbar\, d\varphi/dt = \{ H_s + \lambda P H_{sb} P - i\lambda^2 (B + i A) \}\, \varphi(t) \qquad (2)$$

where the operator P projects the whole onto the system space and A and B are two formally self-adjoint operators defined by

$$B + i A = \int_0^{\infty} P\, e^{i(H_s+H_b)\tau}\, H_{sb} [I-P]\, e^{-i(H_s+H_b)s}\, [I-P] H_{sb} P\, d\tau \qquad (3)$$

Although the functions $\varphi(t)$ can decay to zero, the coherent state wave functions, $\psi(t) = \varphi(t) / |\varphi(t)|$, are normalized to unity. The nonlinear Schrodinger equation for $\psi(t)$ is

$$i\hbar\, d\psi/dt = H_o\, \psi(t) + i\lambda^2 [\langle \psi(t) | B | \psi(t) \rangle - B]\, \psi(t) \qquad (4)$$

where $H_o = H_s + \lambda P H_{sb} P + \lambda^2 A.$

Furthermore, if both H_o and B are bounded and B is positive, Eq.(4) has the unique formal solution,

$$|\psi(t)\rangle = \frac{\exp[- i H_o t - \lambda^2 B t]\ |\psi(0)\rangle}{\{\langle\psi(0)| \exp[iH_o t - \lambda^2 Bt] \cdot \exp[-iH_o t - \lambda^2 Bt] |\psi(0)\rangle\}^{1/2}} \qquad (5)$$

If H_o and B commute, then

$$\frac{d}{dt} < \psi(t) \mid B \mid \psi(t) > \; = - 2\lambda^2 < \psi(t) \mid B^2 \mid \psi(t) > + 2\lambda^2 [< \psi(t) \mid B \mid \psi(t) >]^2 \leqslant 0 \quad (6)$$

As t approaches infinity, $\psi(t)$ approaches the normalized projector of $\psi(0)$ on the eigenspace of B and its eigenvalue becomes the lowest non-vanishing one of the projection.

If, on the otherhand, H_o and B do not commute but λ is sufficiently small, then $\psi(\infty)$ is close to the H_o eigenvector for which $< \psi(\infty) \mid B \mid \psi(\infty) >$ is minimal.

Gisin[29] has applied this formalism successfully to a number of spin systems in a magnetic field. The quantum mechanical analog of the damped simple harmonic oscillator is his simplest example.

2). The Damped Harmonic Oscillator[30]

Allen and Eberly[33] and others have used the classical damped simple harmonic oscillator as a prototype of a two state atomic system interacting with both a resonant laser field and a relaxation bath. Since the correspondence between classical optics and modern quantum mechanics is very close, it is surprising that it was not until 1981 that anyone succeeded in deriving a quantum mechanical analog of the equation for the damped simple harmonic oscillator --- The difficulty was in knowing how to represent quantum mechanical friction! However, N.Gisin[30] knew this from his master equation formulation and Eq.(4).

$$i\hbar\, \partial\psi/\partial t \;=\; H_o\psi + ik[< \psi \mid H_o\psi > - H_o]\psi \;, \qquad (7a)$$

or the equivalent nonlinear Louisville equation,

$$i\hbar\, d\varrho/dt = - [\varrho, H_o] + i(k/\hbar)\, [\, [\varrho, H_o],\, \varrho] \qquad (7b)$$

where H_o is the quantum mechanical simple harmonic oscillator Hamiltonian, $H_o = (2m)^{-1}p^2 + (m\omega^2/2)\, q^2$. (8)

which has the eigenfunctions $\mid \varphi_n >$ and eigenvalues $\epsilon_n = [n + (1/2)]\hbar\omega$.

If Eq.(7) is multiplied by $< \psi \mid H_o \mid$, it is obvious that

$$d[< \psi \mid H_o \mid \psi >]/dt \;=\; 2k[< \psi \mid H_o \mid \psi >^2 - < \psi \mid H_o^2 \mid \psi >].$$

From this, it is easy to show that the formal coherent solutions to Eq.(7) are given by

$$\mid \psi(t) > \;=\; \frac{\exp[-(i + k)\, H_o t] \mid \psi(0) >}{< \psi(0) \mid \exp[-\, 2kHt\,]\; \psi(0) >^{1/2}} \qquad (8)$$

The eigenfunctions $\mid \psi_\alpha(t) >$ can then be expanded as a linear combination of the simple harmonic oscillator functions $\mid \varphi_n >$ so that $\mid \psi_\alpha(t) > \;=\; \sum_{\alpha,n} C_{\alpha,n}(t) \mid \varphi_n >$ where

$$C_{\alpha,n}(t) \;=\; C_{\alpha,n}(0)\, \exp[-\, (i+k)(2n+1)\omega t/2\,]\; N_\alpha(t)^{-1/2} \qquad (9)$$

and $N_\alpha(t) = \sum_{s=0} \mid C_{\alpha,s}(0) \mid^2 \exp[-\, k(2s+1)\omega t\,]$.

Gisin used a Weyl transformation[34] of Eq.(7) into semi-classical phase space to prove that Eq.(7) agrees (in the limit as $\hbar$ approaches zero) with the classical damped harmonic oscillator equation

$$d^2q/dt^2 + 2k\omega\, dq/dt + (1+k^2)\omega^2 q = 0. \qquad (10)$$

The most significant difference between the behavior of this quantum mechanical model and that of a classical damped oscillator is that the coherent states of the quantum system evolve in semi-stable limit cycles to the lowest eigenstate of the simple harmonic oscillator. It is surprising that the coherent states remain coherent during their evolution.

3). Periodic Dissipative Systems

I believe that periodic dissipative system problems can be solved by first transforming the system into the usual Shirley[35]-Sambe[36] form of the Floquet representation in which the Hamiltonian is time-independent and expressed in terms of infinite dimensional pseudostate space. Then the periodic state problem can be solved by the same master equation formulation which Gisin[29] developed for constant Hamiltonian problems.

Instead of which, Gisin[31] abandoned his master equation formulation and treated the periodic problems as generalized damped harmonic oscillators.

REFERENCES

1). U.Johann, T.S.Luk, H.Egger, and C.K.Rhodes, Phys.Rev.A34, 1084 (1986).

2). Z.Deng and J.H.Eberly, J.Opt.Soc.Am.B2, 486 (1985); J.Phys.B18, L287 (1985).

3).A.Dulcic, Phys.Rev.A35, 1673 (1987).

4). T.S.Luk, U.Johann, H.Egger, H.Pummer, and C.K.Rhodes, Collision-Free Multiple Photon Ionization of Atoms and Molecules at 193nm, Phys.Rev. A32, 214 (1985).

5). David R. Herrick, New Symmetry Properties Of Atoms and Molecules, Adv. in Chem.Phys.52, 1 (1983).

6). A.McPherson, G.Gibson, H.Jara, U.Johann, T.S.Luk, I.A.McIntyre, K.Boyer, and C.K.Rhodes, Studies of Multiphoton Production of Vacuum Ultraviolet Radiation in Rare Gases, Submitted to J.Opt.Soc.Am. B.

7). P.Lambropoulos and X.Tang, Multiple Excitation and Ionization of Atoms by Strong Lasers, J.Opt.Soc.Am.B4, 821 (1987) and Phys.Rev.Lett. 58,108 (1987). See also, P.Lambropoulos, Generalized Atomic K-Photon Ionization Cross-Sections, Adv. Atom and Molec. Phys. 12, 87 (1976).

8). Shih-I Chu and J.Cooper, Phys.Rev.A32, 2769 (1985).

9). K.C.Kulander, Ionization of H Atoms by Strong Lasers Phys.Rev.A35, 445 (1987).

10). K.C.Kulander, Time Dependent Hartree Fock Theory of Multiphoton Ionization: Helium, Phys.Rev. A00, 0000 (1987).

11). A.Szöke, Theory of Atoms in Strong, Pulsed Electromagnetic Fields: General Considerations, Phys.Rev.A00, 0000 (1987).

12). C.M.Bender and S.A.Orszag, Adv. Math. Methods for Scientists and Engineers, (McGraw-Hill, New York, 1978).

13). E.P.Cooper, Separation of Nuclear Isomers, Phys.Rev.61, (1942). See also E.H.S.Burhop, The Auger Effect,(Cambridge, London, 1952).

14). S.Wexler and G.R.Anderson, Dissociation of Methyl Bromide by Isomeric Transition of 4.4-hr Br^{80m} ,J.Chem.Phys.33, 850 (1960).

15). T.A.Carlson and R.M.White, Fragment Ions from CH_3Te^{125} Followed by Decay of CH_3I^{125}, J.Chem.Phys.38, 2930 (1963).

16). T.A.Carlson and R.M.White, Abundances and Recoil Energy Spectra of Fragment Ions Produced by X-Ray Interaction with CH_3I, J.C.P.44, 4510 (1966).

17). J.Needels and Shih-I Chu, Time-Dependent Self-Consistent Field Approach to Infrared Laser Multiphoton Excitation, Chem.Phys.Lett.00, 000 (1987).

18). R.B.Gerber, V.Buch, and M.A.Ratner, Time-Dependent Self-Consistent Field for Intramolecular Energy Transfer: I. Application to Dissociation of Van der Waals Molecules, J.Chem.Phys.77, 3022 (1982); M.A.Ratner and R.B.Gerber, Excited States of Polyatomic Molecules: Semiclassical SCF Approach, J.Phys.Chem.90, 20 (1986).

19). K.-H.Yang and J.O.Hirschfelder, Generalizations of Classical Poisson Brackets to Include Spins, Phys.Rev.A22, 1814 (1980).

20). K.-H.Yang and J.O.Hirschfelder, Interaction of Molecules in Electromagnetic Fields: I. Classical Particles and Fields, J.Chem.Phys.72, 5863 (1980).

21). P.A.M. Dirac, Proc.Roy.Soc.(London) A264, 57 (1962). See also, R.G.Nadig, Z.Kunszt, P.Hasenfratz, and J.Kuti, Ann.Phys.116, 380 (1978).

22). E.C.G.Sudarshan and N.Mukunda, Classical Dynamics: A Modern Perspective (Wiley, New York, 1974). See Chaps.17 and 18, especially pp.316-320 and 365-369.

23). M.J.Ablowitz and H.Segur, Solitons and the Inverse Scattering Transformation, (SIAM, Philadelphia, 1981).

24). G.L.Lamb.Jr, Description of Ultrashort Optical Pulse Propagation in a Resonant Medium, Rev.Mod.Phys.43, 99 (1971).

25). P.A.Clarkson and C.M.Cosgrove, PainLeve Analysis of the Nonlinear Schrodinger-Like Equations, J.Phys.A20, 2003 (1987).

26). B.G.Konopelchenko, Nonlinear Integrable Equations, (Lecture Notes # 270, Springer Verlag, Berlin, 1987).

27). D.J.Kaup, Coherent Pulse Propagatotion: A Comparison of the Complete Solution with the McCall-Hahn Theory and Others, Phys.Rev.A16, 704 (1977).

28). J.O.Hirschfelder, Similarity of Wigner's Delay Time to the Virial Theorem for Scattering by a Central Field, Phys.Rev.A19, 2463 (1979)

29). N.Gisin, Derivation of a Class of Nonlinear Dissipative Schrodinger-Like Equations, Physica 111A, 364 (1982).

30). N.Gisin, Simple Nonlinear Dissipative Quantum Evolution Equation. J.Phys.A14, 2259 (1981).

31). N.Gisin, Dissipative Quantum Dynamics for Systems Periodic in Time. Found.Phys.13, 643 (1983).

32). N.Gisin, Generalisation of Wigner's Theorem for Dissipative Systems, J.Phys.A19, 205 (1986).

33). L.Allen and J.H.Eberly, Optical Resonance and Two-Level Atoms, (John Wiley, New York, 1975).

34). S. de Groot, La Transformation de Wetl at la Fonction de Wigner, (Universite de Montreal, Montreal, 1975). See p.41

35). J.H.Shirley, Phys.Rev.138B, 979 (1965) and (unpublished) Ph.D. Thesis, Cal.Inst.Tech.(1963).

36). H.Sambe', Phys.Rev. A7, 2203 (1973)

AUTHOR INDEX

SUBJECT INDEX